NELSON

FUNCTIONS and APPLICATIONS 11

Online Student E-version PDF Files

You are just a few steps away from accessing your myNelson digital resource. Go to **www.mynelson.com**

New myNelson Users

- Press "Click here to register."
- Enter the Access Code for this resource.
- Follow the onscreen instructions to enter your account information, including your email address, password, province, and school information.
- Log in using your email address and password.

Existing myNelson Users

- Log in using your email address and password.
- Enter the Access Code for this resource in the "Add new resource" field. Click "Submit."

0001993

Your Access Code is   nelson HMPK 7428

IMPORTANT: Please retain this Access Code for future use.

You can now access this myNelson resource using your email address and password.

Customer Support

If you require assistance at any time while using the myNelson website, please email **nelson.schoolebooks@nelsoneducation.ca**, or call 1-800-268-2222.

0-17-667821-2

myNelson
dynamic digital learning

FUNCTIONS and APPLICATIONS 11

Series Author and Senior Consultant
Marian Small

Lead Author
Chris Kirkpatrick

Authors
Andrew Dmytriw • Beverly Farahani • Angelo Lillo

Kay Minter • David Pilmer • Noel Walker

NELSON EDUCATION

NELSON EDUCATION

Functions and Applications 11

Series Author and Senior Consultant
Marian Small

Lead Author
Chris Kirkpatrick

Authors
Andrew Dmytriw, Beverly Farahani, Angelo Lillo, Kay Minter, David Pilmer, Noel Walker

Contributing Authors
Kathleen Kacuiba, Ralph Montesanto, Krista Zupan

Senior Consultant
David Zimmer

Math Consultant
Kaye Appleby

General Manager, Mathematics, Science, and Technology
Lenore Brooks

Publisher, Mathematics
Colin Garnham

Associate Publisher, Mathematics
Sandra McTavish

Managing Editor, Mathematics
David Spiegel

Product Manager
Linda Krepinsky

Project Manager
Sheila Stephenson

Developmental Editors
Nancy Andraos, Colin Bisset, Ingrid D'Silva, Tom Gamblin, Anna-Maria Garnham, Betty Robinson

Contributing Editors
Amanda Allan, Alasdair Graham, First Folio Resource Group, Inc., Tom Shields, Robert Templeton, First Folio Resource Group, Inc., Caroline Winter

Editorial Assistant
Caroline Winter

Executive Director, Content and Media Production
Renate McCloy

Director, Content and Media Production
Linh Vu

Senior Content Production Editor
Debbie Davies-Wright

Production Manager
Cathy Deak

Senior Production Coordinator
Sharon Latta Paterson

Design Director
Ken Phipps

Interior Design
Peter Papayanakis

Cover Design
Eugene Lo

Cover Image
Pierre Berton Resource Library, Vaughan, Tom Arban Photography Inc.

Production Services
Pre-Press Company Inc.

Director, Asset Management Services
Vicki Gould

Photo/Permissions Researcher
Lynn McLeod

Photo Shoot Coordinator
Lynn McLeod

Set-up Photos
Dave Starrett

Printer
Transcontinental Printing Ltd.

Reviewers and Advisory Panel

Table of Contents

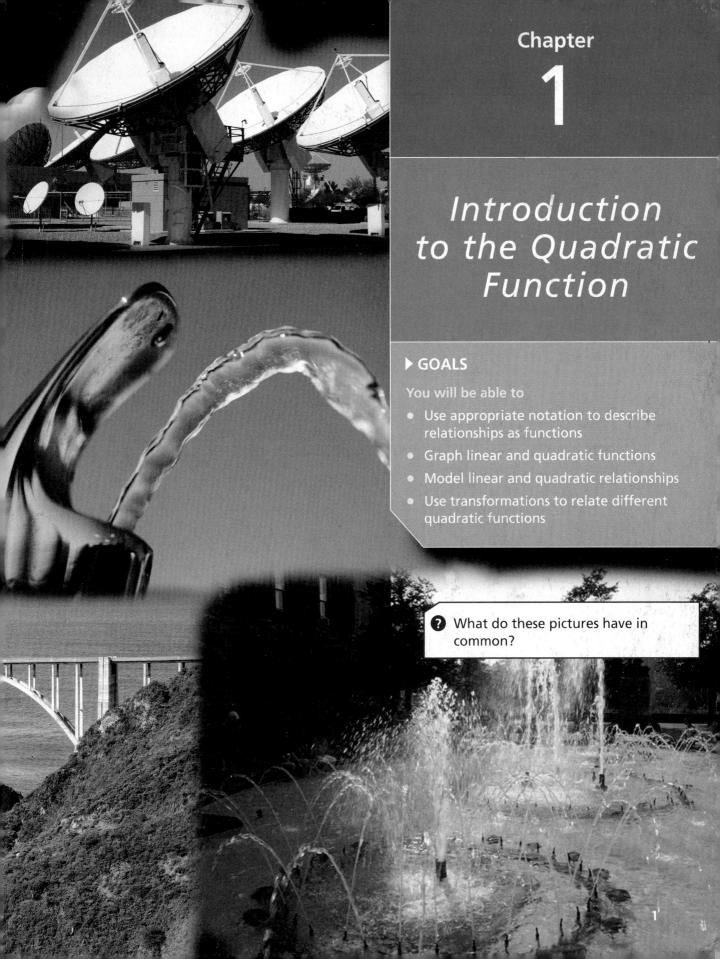

Chapter 1

Introduction to the Quadratic Function

▶ GOALS

You will be able to

- Use appropriate notation to describe relationships as functions
- Graph linear and quadratic functions
- Model linear and quadratic relationships
- Use transformations to relate different quadratic functions

❓ What do these pictures have in common?

WORDS *You Need to Know*

1. Match the term with the picture or example that best illustrates its definition.

a) linear relation

b) quadratic relation

c) vertex of a parabola

d) axis of symmetry of a parabola

e) line of best fit

f) intercepts

i)

iii)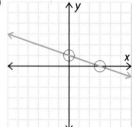

v)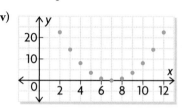

ii)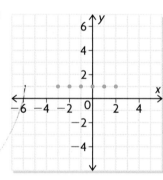

iv)

vi)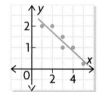

SKILLS AND CONCEPTS *You Need*

Study | *Aid*

For help, see Essential Skills Appendix, A-5.

Evaluating Algebraic Expressions

An expression may be evaluated for different values of the variables. Substitute the given numerical value of the letter in brackets and evaluate.

EXAMPLE

Evaluate $3x^2 - 2x + 1$, when $x = -3$.
Substitute $x = -3$ (in brackets) into the above algebraic expression.

$3(-3)^2 - 2(-3) + 1$
$= 3(9) + 6 + 1$
$= 27 + 7$
$= 34$

2. Evaluate each algebraic expression if $a = 0$, $b = 1$, $c = -1$, and $d = 2$.

a) $b + 3c$

b) $3b + 2c - d$

c) $2a^2 + b^2 - d^2$

d) $(a + 3b)(2c - d)$

Creating a Table of Values and Sketching Graphs of Quadratic Relations

Study | *Aid*

For help, see Essential Skills Appendix, A-8.

When creating a table of values, select various x-values and substitute them into the quadratic relation and solve for y. Plot the ordered pairs to determine the graph.

EXAMPLE

Create a table of values and sketch the graph of $y = 3x^2$.
Select x-values ranging from -2 to 2 and solve for each value of y.

x	$y = 3x^2$	Ordered Pair
-2	$y = 3(-2)^2 = 12$	$(-2, 12)$
-1	$y = 3(-1)^2 = 3$	$(-1, 3)$
0	$y = 3(0)^2 = 0$	$(0, 0)$
1	$y = 3(1)^2 = 3$	$(1, 3)$
2	$y = 3(2)^2 = 12$	$(2, 12)$

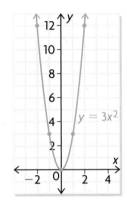

3. Create a table of values and sketch the graph of each quadratic relation.
a) $y = 2x^2$
b) $y = -4x^2 - 3$
c) $y = 0.5x^2 + 5$
d) $y = -5x^2 + 5$

PRACTICE

Study | *Aid*

• For help, see Essential Skills Appendix.

Question	Appendix
4, 5, 6	A-11
7	A-6
10	A-7
11	A-13

4. Solve each equation for y. Then evaluate y for the given value of x.
a) $y - 5 = -3x$; $x = 2$
b) $3x + y = 3$; $x = 2$

5. a) Does the point $(2, -1)$ lie on the line $y = -3x + 5$?
b) Does the point $(-4, 10)$ lie on the parabola
$y = -2x^2 - 5x + 22$?

6. a) Is $(2, -1)$ a solution of $2x - y = 5$? Explain.
b) Is $(-1, 29)$ a solution of $y = -2x^2 - 5x + 22$? Explain.

7. For each linear relation, determine the x-intercept, the y-intercept, and the slope.
a) $2x + 3y = 12$
b) $-x + 4y = 8$

8. For each quadratic relation, determine the y-intercept and the axis of symmetry.

   a) $y = 2x^2 + 2$ b) $y = -x^2 - 4$

9. Match the equation with its graph.

   a) $y = 4x - 8$ B   c) $y = x^2 - 3$ D

   b) $y = -4x - 4$ C   d) $y = -x^2 + 4$ A

A.
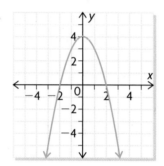

C.
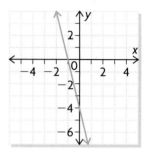

B.
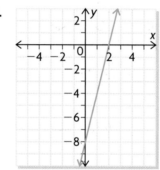

D.
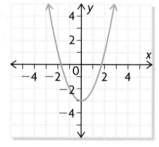

10. Use the suggested method to sketch the graph of each linear relationship.

   a) $y = -4x + 3$ (slope, y-intercept method)

   b) $2x - y = 4$ (x-, y-intercept method)

   c) $3x = 6y - 3$ (table of values)

11. Create a scatter plot, and then use it to find the equation that defines the relation shown in the table.

x	−3	−2	−1	0	1	2	3
y	11	9	7	5	3	1	−1

12. Complete the chart by writing down what you know about the term Quadratic Relation.

Definition:	Characteristics:
	Quadratic Relation
Examples:	Non-examples:

APPLYING *What You Know*

Selling Tickets

YOU WILL NEED
• graph paper

The student council is raising money by holding a raffle for an MP3 player. To determine the price of a raffle ticket, the council surveys students and teachers to find out how many tickets would be bought at different prices. The council found that
• if they charge $0.50, they will be able to sell 200 tickets; and
• if they raise the price to $1.00, they will sell only 50 tickets.

? **What ticket price will raise the most money?**

A. Create a scatter plot that shows the relationship between the price per ticket (p) and the number of tickets sold (n) by graphing the two pieces of information surveyed.

B. Assume that the relationship is linear. Draw a line joining the two given points. Calculate the slope of the line. Use the slope and one of the points to write an equation for the relationship.

C. Use the equation from part B to complete the table of values for other ticket prices.

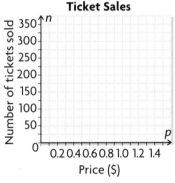

Ticket Sales

Price ($)	0.1	0.2	0.3	0.4	0.50	0.60	0.70	0.80	0.90	1.00
Number of Tickets Sold					200					50

D. Determine the total amount of money raised for each ticket price. (amount raised = price × number sold)

Amount Raised ($)										

E. Use the table to estimate the price per raffle ticket that will raise the most money.

F. Sketch a new graph that shows the relationship between the price per ticket (p) and the amount of money raised (A).

G. How could you use the graph to determine the best price to charge per ticket?

The Characteristics of a Function

mapping diagram

a drawing with arrows to show the relationship between each value of *x* and the corresponding values of *y*

Tech | **Support**

For help using the graphing calculator to create a scatter plot, see Technical Appendix, B-10.

GOAL

Identify the difference between a relation and a function.

INVESTIGATE the Math

Nathan examines two temperature **relations** for the month of March. In relation *A*, the **dependent variable** *y* takes on the same or opposite value of the **independent variable** *x*. In relation *B*, the dependent variable *y* takes on the same value of the independent variable *x*. Each relation is represented by a table of values and a **mapping diagram**.

Relation *A*

Number of Days in the Month, *x*	0	2	2	4	4
Minimum Daily Temperature (°C), *y*	0	2	−2	−4	4

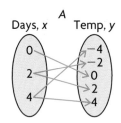

Relation *B*

Number of Days in the Month, *x*	0	2	4	6	7
Maximum Daily Temperature (°C), *y*	0	2	4	6	7

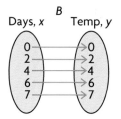

? Which relation is also a function?

A. The table of values of relation A represents the set of ordered pairs $(x, y) = $ (days, minimum temp.). Create a scatter plot of the ordered pairs of A on a coordinate grid.

B. The table of values of relation B represents the set of ordered pairs $(x, y) = $ (days, maximum temp.) Create a scatter plot of the ordered pairs of B on a different coordinate grid.

C. The **domain** of relation A can be expressed using **set notation**, $D = \{0, 2, 4\}$. What do these numbers represent? Express the domain of relation B in set notation.

D. The **range** of relation A is $R = \{-4, -2, 0, 2, 4\}$. What do these numbers represent? Express the range of relation B in set notation.

E. Examine the mapping diagram for relation B. Each x-value in the domain (number of days) is mapped to exactly one y-value (maximum temp.) in the range by an arrow. Examine the mapping diagram for relation A. How are the mappings in A different from the mappings in B?

F. Explain how following the arrows of a mapping diagram helps you see which relation is a **function**.

G. Draw the vertical line $x = 2$ on your scatter plots of relation A and relation B. What do you notice? Explain how drawing a vertical line helps you see which relation is a function.

domain
the set of all values for which the independent variable is defined

set notation
a way of writing a set of items or numbers within curly brackets, { }

range
the set of all values of the dependent variable. All such values are determined from the values in the domain

function
a relation in which there is only one value of the dependent variable for each value of the independent variable (i.e., for every x-value, there is only one y-value)

Reflecting

H. Relations can be expressed by equations. Write the equations for relations A and B that define the relationship between the number of days in the month and the maximum or minimum temperature.

I. A relation can be represented as a table of values, a mapping diagram, a graph, and an equation. Which methods make it easier to help you decide whether a relation is a function? Explain.

J. Write a definition for the term *function* in terms of the domain and range.

APPLY the Math

EXAMPLE **1**	Using reasoning to decide whether a relation is a function

For each of the following relations, determine
- the domain and the range
- whether or not it is a function

a) G: (x, y) = (number of golfers, score below or above par)
$$= \{(0, -2), (0, -1), (0, 0) \ (0, 5)\}$$

b)

x	y
−1	−3
0	1
1	5
2	9

c)

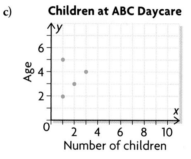

Children at ABC Daycare

Anisha's Solution

a)

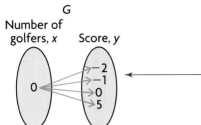

I drew a mapping diagram to represent the ordered pairs. The x-coordinate, 0, maps onto four different y-coordinates.

Domain = $\{0\}$
Range = $\{-2, -1, 0, 5\}$

This relation is not a function.

b)

x	y
−1	−3
0	1
1	5
2	9

Each x-value in the table corresponds to one unique y-value.

Domain = $\{-1, 0, 1, 2\}$
Range = $\{-3, 1, 5, 9\}$

This relation is a function.

c)

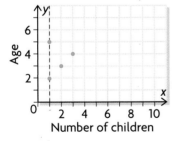

Children at ABC Daycare

The vertical line crosses two points, (1, 2) and (1, 5), when x equals 1. The x-coordinate, 1, corresponds to two different y-coordinates.

Domain = $\{1, 2, 3\}$
Range = $\{2, 3, 4, 5\}$

This relation is not a function.

For any relation that is a function, if you know the value of the independent variable, you can predict the value of the dependent variable.

<table>
<tr><td>**EXAMPLE 2**</td><td>**Connecting mapping diagrams with functions**</td></tr>
</table>

For each relation, determine
• the domain and the range
• whether or not it is a function

a)

Time, *x* (h)   Number of e-mails, *y*

```
2 ────────→ 4
3 ────────→ 6
4 ────────→ 8
5 ────────→ 10
```

c)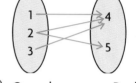

Age of cat, *x*   Number of kittens, *y*

b) Model type, *x*   Manufacturer, *y*

```
Jetta
Beetle ──→ Volkswagen
Passat
```

d)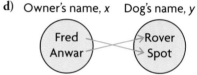

Owner's name, *x*   Dog's name, *y*

Kinson's Solution

a) The domain is {2, 3, 4, 5}. ◄──────

The range is {4, 6, 8, 10}.

This relation is a function.

> Each *x*-value corresponds to only one *y*-value. If I know the quantity of time, I can predict the number of e-mails received.

b) The domain is {Jetta, Beetle, ◄──────
Passat}.

The range is {Volkswagen}.

This relation is a function.

> Each model of car in the domain corresponds with only one manufacturer in the range. If I know the model of car, I can predict the manufacturer.

c) The domain is {1, 2, 3}.

The range is {4, 5}.

This relation is not a function. ◄──────

> The 2 in the domain corresponds to two different values in the range. I can't predict the number of kittens born from knowing the age of the mother.

d) The domain is {Fred, Anwar}.

The range is {Rover, Spot}.

This relation is a function. ◄──────

> Each name in the domain corresponds to one name in the range. If I know the owner's name, I can predict the dog's name.

If you are given the graph of a relation, you can also identify if it is a function by comparing the values of the independent variable with the values of the dependent variable.

EXAMPLE **3**
Connecting functions with graphs

For each of the following relations, determine the domain and the range, using **real numbers**. State whether or not the relation is a function.

a)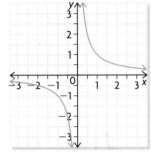

c)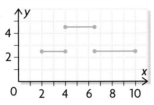

b)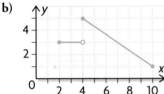

d)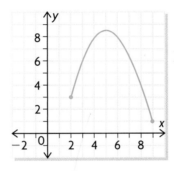

real numbers

the set of real numbers is the set of all decimals—positive, negative, and 0, terminating and nonterminating. This statement is expressed mathematically with the set notation $\{x \in \mathbf{R}\}$

Communication | *Tip*

$\{x \in \mathbf{R} \mid x \neq 0\}$ is the mathematical notation for saying "the set of x such that x can be any real number except 0."

Dimitri's Solution

a) The domain is $D = \{x \in \mathbf{R} \mid x \neq 0\}$.

The range is $R = \{y \in \mathbf{R} \mid y \neq 0\}$.

This relation is a function. ◄——— This relation passes the **vertical-line test**. This means that each x-coordinate corresponds to exactly one y-coordinate.

b) The domain is
$D = \{x \in \mathbf{R} \mid 2 \leq x \leq 10\}$.

The range is $R = \{y \in \mathbf{R} \mid 1 \leq y \leq 5\}$.

This relation is a function. ◄——— This relation passes the vertical-line test. The open dot at the point (4, 3) means that $y \neq 3$ when $x = 4$, and the closed dot at the point (4, 5) means that $y = 5$ when $x = 4$. So when $x = 4$, there is only one y-value.

c) The domain is
$D = \{x \in \mathbf{R} \mid 2 \leq x \leq 10\}$.

The range is
$R = \{y \in \mathbf{R} \mid y = 2.5, 5.5\}$.

This relation is a not function. ◄——— This relation fails the vertical-line test. At $x = 4$, $y = 2.5$ and 5.5; and at $x = 6.5$, $y = 2.5$ and 5.5.

vertical-line test

a test to determine whether the graph of a relation is a function. The relation is not a function if at least one vertical line drawn through the graph of the relation passes through two or more points

d) The domain is
$$D = \{x \in \mathbf{R} \mid 2 \le x \le 9\}.$$

The range is
$$R = \{y \in \mathbf{R} \mid 1 \le y \le 8.5\}.$$

This relation is a function. ◄──── The relation passes the vertical-line test; for each *x*-value, there is exactly one *y*-value.

In Summary

Key Ideas

- A function is a relation where each value of the independent variable corresponds with only one value of the dependent variable. The dependent variable is then said to be a function of the independent variable.
- For any function, knowing the value of the independent variable enables you to predict the value of the dependent variable.
- Functions can be represented in words, by a table of values, by a set of ordered pairs, in set notation, by a mapping diagram, by a graph, or by an equation.

Need to Know

- The set of all values of the independent variable is called the domain. On a Cartesian graph, the domain is the set of all possible values of the independent variable *x*. *Example*: The relation $y = \sqrt{x}$ exists only for positive values of *x*. The domain of this relation is $x = 0$ and all positive values of *x*.

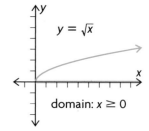

- The set of all values of the dependent variable is called the range. On a Cartesian graph, the range is the set of all possible values of the dependent variable *y*. *Example*: $y = x^2$. The range is $y = 0$ and all positive values of *y*.
- A function can also be defined as a relation in which each element of the domain corresponds to only one element of the range.

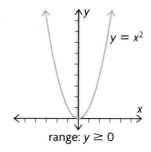

- The vertical-line test can be used to check whether the graph of a relation represents a function. If two or more points lie on the same vertical line, then the relation is not a function.
- It is often necessary to define the domain and range using set notation. For example, the set of numbers ..., $-2, -1, 0, 1, 2, \ldots$ is the set of integers and can be written in set notation as $\{x \mid x \in \mathbf{I}\}$, where the symbol "|" means "such that" and the symbol "∈" means "belongs to" or "is a member of." So set notation for the integers would be read as "The set of all *x*-values such that *x* belongs to the integers."

CHECK *Your Understanding*

1. For each relation, state
   i) the domain and the range
   ii) whether or not it is a function, and justify your answer

   a) $g = \{(1, 2), (3, 1), (4, 2), (7, 2)\}$
   b) $h = \{(1, 2), (1, 3), (4, 5), (6, 1)\}$
   c) $f = \{(1, 0), (0, 1), (2, 3), (3, 2)\}$
   d) $m = \{(1, 2), (1, 5), (1, 9), (1, 10)\}$

2. For each relation, state
   i) the domain and the range
   ii) whether or not it is a function, and justify your answer

   a) 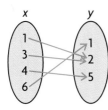   c)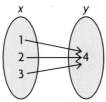

   b) 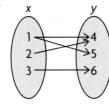   d)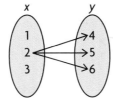

3. For each relation, state
   i) the domain and the range
   ii) whether or not it is a function, and justify your answer

   a) 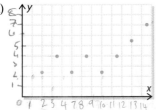   c)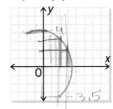

   b) 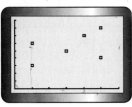   d)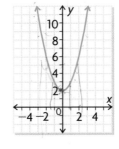

PRACTISING

4. Is the relation a function? Why or why not?

a) $h\colon (x, y) = $ (age of child, number of siblings)
$$= \{(1, 4), (2, 4), (3, 3), (4, 2)\}$$

b) $k\colon (x, y) = $ (number of households, number of TV's)
$$= \{(1, 5), (1, 4), (2, 3), (3, 2)\}$$

c)

x	y
0	−2
2	0
5	−2

d)

Number of Drivers, x	Number of Speeding Tickets, y
1	6
1	4
2	3
7	8

5. Is each relation a function? If not, which ordered pair should be removed to make the relation a function?

a)

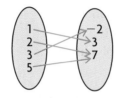

b)

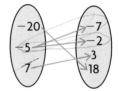

6. Which scatter plot represents a function? Explain. For each graph, state the domain and range.

a)

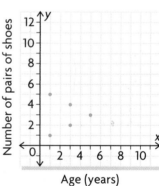

Width (cm)

c)

Age (years)

b)

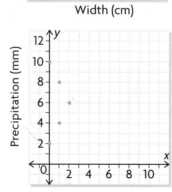

Number of days

d)

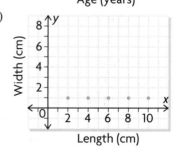

Length (cm)

7. Which relations are functions? Explain. For each graph, state the domain and range.

a)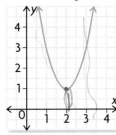

c)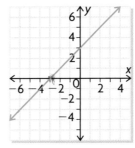

b)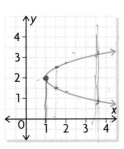

d)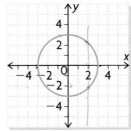

8. a) Draw a line that is both a relation and a function.
b) Draw a line that is a relation but is not a function.
c) Which type of line cannot be the graph of a linear function? Explain.

9. The mapping diagram at the right shows the relation between students and their marks on a math quiz.
a) Write the relation as a set of ordered pairs.
b) State the domain and range.
c) Is the relation a function? Explain.

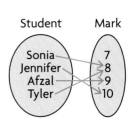

10. The table gives typical resting pulse rates for six different mammals.

Mammal, x	Cat	Elephant	Human	Mouse	Rabbit	Rat
Pulse Rate, y (beats/minute)	176	27	67	667	260	315

a) Is the resting pulse rate a function of the type of mammal? Explain.
b) If more mammals and their resting pulse rates were included, would the extended table of values be a function? Explain.

11. Which variable would be associated with the domain for the following pairs of related quantities? Which variable would be associated with the range? Explain.
a) heating bill, outdoor temperature
b) report card mark, time spent doing homework
c) person, date of birth
d) number of slices of pizza, number of cuts

12. Bill called a garage to ask for a price quote on the size and type of tire he needed.
   a) Explain why this scenario represents a function.
   b) If Bill had given the clerk the tire price, would the clerk be able to tell Bill the tire size and type? Would this scenario represent a function?

13. T Dates and outdoor temperatures are related. The hottest temperature recorded in Canada was 45 °C at Midvale and Yellow Grass, Saskatchewan, on July 5, 1937. The coldest temperature recorded in Canada was − 63 °C, in Snag, in the Yukon Territories, on February 3, 1947.
   a) What is the independent variable in this relation? What is the dependent variable?
   b) What is the domain?
   c) What is the range?
   d) Is one variable a function of the other? Explain.

14. C Summarize your understanding of functions in a chart similar to the one shown. Use at least three different types of representation for the examples and the non-examples.

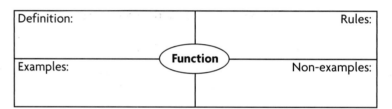

Extending

15. A rock rolls off a cliff 66 m high. Which set of values best represents the range in the relationship between the time elapsed, in seconds, and the resultant height, in metres? Explain.
   a) 100 to 200
   c) − 100 to 100
   b) − 100 to 0
   d) 0 to 66

16. If a company's profit or loss depends on the number of items sold, which is the most reasonable set of values for the domain in this relation? Explain.
   a) negative integers
   b) positive integers
   c) integers between − 1500 and 1500

1.2 Comparing Rates of Change in Linear and Quadratic Functions

YOU WILL NEED
- graph paper

GOAL

Identify and compare the characteristics of linear and quadratic functions.

LEARN ABOUT the Math

Stopping distance is the distance a car travels from the time the driver decides to apply the brakes to the time the car stops. Stopping distance is split into two types of distances.

1. **Reaction Distance**: the distance the car travels from the time you first think "stop" to the time your foot hits the brake

2. **Braking Distance**: the distance the car travels from the time you first apply the brakes to the time the car stops

   Stopping Distance = Reaction Distance + Braking Distance

Reaction distance + Braking distance = Stopping distance

The table shows some average values of these three distances, in metres, for different speeds.

Speed (km/h)	Speed (m/s)	Reaction Distance (m)	Braking Distance (m)	Stopping Distance (m)
0	0.00	0.00	0.00	0.00
20	5.56	8.33	1.77	10.11
40	11.11	16.67	7.10	23.76
60	16.67	25.00	15.96	40.96
80	22.22	33.33	28.38	61.71
100	27.78	41.67	44.35	86.01

? How does the relationship between speed and reaction distance compare with the relationship between speed and braking distance?

EXAMPLE 1	Representing and comparing linear and quadratic functions

Compare the relationship between speed and reaction distance with the relationship between speed and braking distance.

Kayla's Solution

Part 1: Comparing speed and reaction distance

Speed (m/s)	Reaction Distance (m)	First Differences
0.00	0.00	
		$8.33 - 0 = 8.33$
5.56	8.33	
		$16.67 - 8.33 = 8.34$
11.11	16.67	
		$25.00 - 16.67 = 8.33$
16.67	25.00	
		$33.33 - 25.00 = 8.33$
22.22	33.33	
		$41.67 - 33.33 = 8.34$
27.78	41.67	

I calculated the first differences of the dependent variable, reaction distance.

This relationship is linear. ← The first differences appear to be close to a constant, about 8.33.

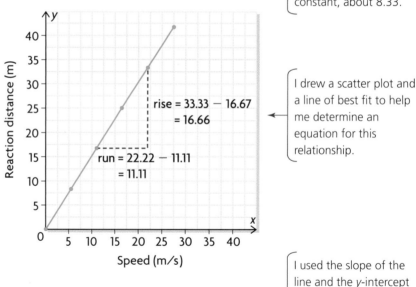

I drew a scatter plot and a line of best fit to help me determine an equation for this relationship.

rise = $33.33 - 16.67$ = 16.66

run = $22.22 - 11.11$ = 11.11

degree

the degree of a polynomial with a single variable, say, x, is the value of the highest exponent of the variable. For example, for the polynomial $5x^3 - 4x^2 + 7x - 8$, the highest power or exponent is 3; the degree of the polynomial is 3

slope $= \dfrac{16.66}{11.11} \doteq 1.5$ and y-intercept $= 0$.

I used the slope of the line and the y-intercept to determine that the equation of the line of best fit is $y = 1.5x$.

So, $y = 1.5x$ is the equation that relates speed to the reaction distance.

The **degree** of this relation is 1.

$$y = f(x) = 1.5x$$

Since there is a unique reaction distance for each value of speed, the reaction distance is a function of speed. I can express this relationship in **function notation**.

function notation

$f(x)$ is called function notation and represents the value of the dependent variable for a given value of the independent variable, x

Part 2: Comparing speed and braking distance

Speed (m/s)	Braking Distance (m)	First Differences	Second Differences
0.00	0.00		
		1.77	
5.56	1.77		3.56
		5.33	
11.11	7.10		3.53
		8.86	
16.67	15.96		3.56
		12.42	
22.22	28.38		3.55
		15.97	
27.78	44.35		

I calculated the first differences. They are not constant, so the relationship is nonlinear.

I calculated the second differences. They are almost the same value, so the relationship is quadratic.

Communication | **Tip**

$f(x)$ is read as "f of x" or "f at x." The symbols $f(x)$, $g(x)$, and $h(x)$ are often used to name functions, but other letters may be used. When working on a problem, it is sometimes easier to choose letters that match the quantities in the problem. For example, use $v(t)$ to write velocity as a function of time.

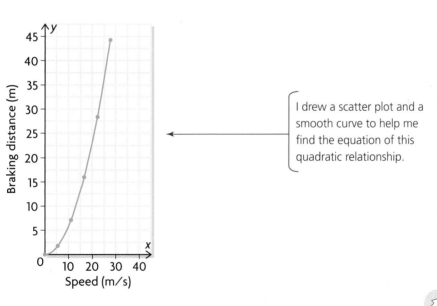

I drew a scatter plot and a smooth curve to help me find the equation of this quadratic relationship.

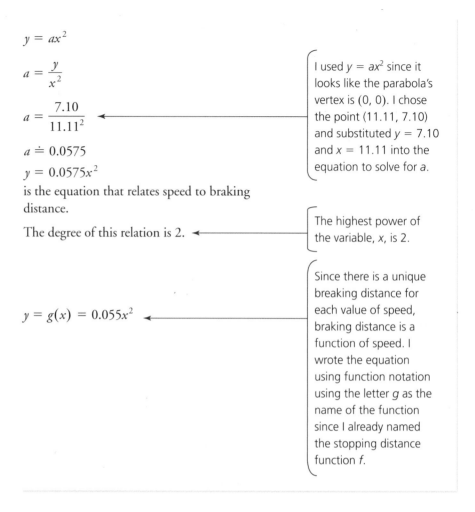

$$y = ax^2$$

$$a = \frac{y}{x^2}$$

$$a = \frac{7.10}{11.11^2}$$

$$a \doteq 0.0575$$

$$y = 0.0575x^2$$

is the equation that relates speed to braking distance.

The degree of this relation is 2. ◄—

$$y = g(x) = 0.055x^2 ◄—$$

I used $y = ax^2$ since it looks like the parabola's vertex is (0, 0). I chose the point (11.11, 7.10) and substituted $y = 7.10$ and $x = 11.11$ into the equation to solve for a.

The highest power of the variable, x, is 2.

Since there is a unique breaking distance for each value of speed, braking distance is a function of speed. I wrote the equation using function notation using the letter g as the name of the function since I already named the stopping distance function f.

Reflecting

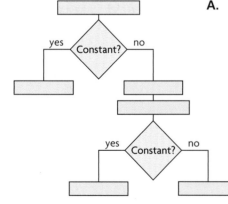

A. The chart at the left summarizes the sequence of steps used to decide whether data for a function are linear, quadratic, or neither. The diamonds in the chart represent decisions that have yes-or-no answers. The rectangles represent steps in the sequence.

   i) Copy the chart. Then complete it by matching each empty rectangle with one of the six phrases.

   1. calculate second differences

   2. linear

   3. neither quadratic nor linear

   4. nonlinear

   5. calculate first differences

   6. quadratic

ii) Use the steps in the chart to determine whether stopping distance is a linear or quadratic relation. The data for stopping distance are provided in the table on page 17.

B. What is the relationship between the differences in a table of values and the degree of a function?

C. How can you tell whether a function is linear or quadratic if you are given
- a table of values?
- a graph?
- an equation?

APPLY the Math

EXAMPLE **2**	Representing volume as a function of time

Water is poured into a tank at a constant rate. The volume of water in the tank is measured every minute until the tank is full. The measurements are recorded in the table.

Time (min)	0	1	2	3	4	5	6	7
Volume (L)	0.0	1.3	2.6	3.9	5.2	6.5	7.8	9.1

a) Use difference tables to determine whether the volume of water poured into the tank, $V(t)$, is a linear or quadratic function of time. Explain.

b) State the domain and range using set notation.

Brian's Solution

a)

Time (min)	0	1	2	3	4	5	6	7
Volume (L)	0.0	1.3	2.6	3.9	5.2	6.5	7.8	9.1
First Differences		1.3	1.3	1.3	1.3	1.3	1.3	1.3

Every minute, an additional 1.3 L of water is added to the tank. Since the first differences are constant, $V(t)$ is a linear function.

b) Domain $= \{t \in \mathbf{R} \mid 0 \leq t \leq 7\}$

It takes 7 min to fill the tank, and time can't be negative.

Range $= \{V(t) \in \mathbf{R} \mid 0 \leq V(t) \leq 9.1\}$

The tank holds a minimum of 0 L when empty and a maximum of 9.1 L of water when full.

EXAMPLE **3**

Representing distance travelled as a function of time

A migrating butterfly travels about 128 km each day. The distance it travels, $g(t)$, in kilometres, is a function of time, t, in days.

a) Write an equation using function notation to represent the distance a butterfly travels in t days.

b) State the degree of this function and whether it is linear or quadratic.

c) Use the function to calculate the distance the butterfly travels in 20 days.

d) What are the domain and range of this function, assuming that a butterfly lives 40 days, on average? Express your answers in set notation.

Rashmika's Solution

a) $g(t) = 128t$ ◄───────────── In this equation, t is the number of days and $g(t)$ is the total distance travelled for the given value of t.

b) The degree is 1. ◄───────────── The highest exponent of t is 1. This means the function $g(t)$ is linear.

The function is linear.

c) $g(t) = 128t$

$g(20) = 128(20)$ ◄───────────── Substitute $t = 20$ into the function $g(t)$.

$= 2560$

The butterfly travels 2560 km in 20 days.

d) Domain $= \{t \in \mathbf{R} \mid 0 \le t \le 40\}$ ◄───────────── The maximum number of days is 40.

Range $= \{g(t) \in \mathbf{R} \mid 0 \le g(t) \le 5120\}$ ◄───────────── The maximum distance is $g(40) = 5120$ km.

In this case, neither distance nor time can be negative.

If you know the equation of a function, its degree indicates whether it is linear or quadratic.

EXAMPLE 4 | Connecting degree to type of function

State the degree of each function, and identify which functions are linear and which are quadratic.

a) $k(x) = 3x(x + 1)$ **b)** $m(x) = (x + 2)^2 - x^2$

Akiko's Solution

a) $k(x) = 3x(x + 1)$ ⟵ | I used the distributive property to expand.

$\quad\quad = 3x^2 + 3x.$

The degree is 2. ⟵ | The highest exponent of x is 2.

The function is quadratic.

b) $m(x) = (x + 2)^2 - x^2$ | I used the distributive property to expand.

$\quad\quad = (x + 2)(x + 2) - x^2$ ⟵ | I simplified by collecting like terms.

$\quad\quad = x(x + 2) + 2(x + 2) - x^2$

$\quad\quad = x^2 + 2x + 2x + 4 - x^2$

$\quad\quad = x^2 + 4x + 4 - x^2$

$\quad\quad = 4x + 4$

The degree is 1. ⟵ | The highest exponent of x is 1.

The function is linear.

Study | *Aid*

- For help, see Essential Skills Appendix, A-9.

In Summary

Key Ideas

- Linear functions have constant first differences, a degree of 1, and graphs that are lines.
- Quadratic functions have constant second differences, a degree of 2, and graphs that are parabolas.

Need to Know

- $f(x)$ is called function notation and is used to represent the value of the dependent variable for a given value of the independent variable, x.
- The degree of a function is the highest power of the independent variable.

CHECK Your Understanding

1. A ball is dropped from a distance 10 m above the ground. The height of the ball from the ground is measured every tenth of a second, resulting in the following data:

Time (s)	0.0	0.1	0.2	0.3	0.4	0.5	0.6	0.7	0.8
Height (m)	10.00	9.84	9.36	8.56	7.44	6.00	4.24	2.16	0.00

   a) Use difference tables to determine whether distance, $d(t)$, is a linear or quadratic function of time. Explain.
   b) What are the domain and range of this function? Express your answer in set notation.

2. A math textbook costs $60.00. The number of students who need the book is represented by x. The total cost of purchasing books for a group of students can be represented by the function $f(x)$.
   a) Write an equation in function notation to represent the cost of purchasing textbooks for x students.
   b) State the degree of this function and whether it is linear or quadratic.
   c) Use your equation to calculate the cost of purchasing books for a class of 30 students.
   d) What are the domain and range of this function, assuming that books can be purchased for two classes of students? Assume that the maximum number of students in a class is 30. Express your answers in set notation.

3. State the degree of each function, and identify which are linear and which are quadratic.
   a) $f(x) = -7 + 2x$
   b) $g(x) = 3x^2 + 5$
   c) $g(x) = (x - 4)(x - 3)$
   d) $3x - 4y = 12$

PRACTISING

4. Use difference tables to determine whether the data represent a linear or quadratic relationship.

Time (s)	0	1	2	3	4	5
Height (m)	0	15	20	20	15	0

Time (h)	Bacteria Count
0	12
1	23
2	50
3	100

5. The population of a bacteria colony is measured every hour and results in the data shown in the table at the left. Use difference tables to determine whether the number of bacteria, $n(t)$, is a linear or quadratic function of time. Explain.

6. State the degree of each function and whether it is linear or quadratic.
   a) $f(x) = -4x(x-1) - x$
   c) $g(x) = 3x^2 + 35$
   b) $m(x) = -x^2 + (x+3)^2$
   d) $g(x) = 3(x-5)$

7. A function has the following domain and range:

$$\text{Domain} = \{t \in \mathbf{R} \mid -3 \le t \le 5\}$$
$$\text{Range} = \{g(t) \in \mathbf{R} \mid 0 \le g(t) \le 10\}$$

   a) Draw a sketch of this function if it is linear.
   b) Draw a sketch of this function if it is quadratic.

8. A golf ball is hit and its height is given by the equation
   $h = 29.4t - 4.9t^2$, where t is the time elapsed, in seconds, and h is the height, in metres.

   a) Write an equation in function notation to represent the height of the ball as a function of time.
   b) State the degree of this function and whether the function is linear or quadratic.
   c) Use difference tables to confirm your answer in part (b).
   d) Graph the function where $\{t \in \mathbf{R} \mid t \ge 0\}$.
   e) At what time(s) is the ball at its greatest height? Express the height of the ball at this time in function notation.
   f) At what time(s) is the ball on the ground? Express the height of the ball at this time in function notation.

9. List the methods for determining whether a function is linear or quadratic. Use examples to explain the advantages and disadvantages of each method.

Extending

10. A baseball club pays a vendor $50 per game for selling bags of peanuts for $2.50 each. The club also pays the vendor a commission of $0.05 per bag.
    a) Determine a function that describes the income the vendor makes for each baseball game. Define the variables in your function.
    b) Determine a function that describes the revenue the vendor generates each game for the baseball club. Define the variables in your function.
    c) Determine the number of bags of peanuts the vendor must sell before the baseball club makes a profit from his efforts.

Using Straight Lines to Draw a Parabola

How can you use a series of lines to create a parabola?

1. Draw lines AB and FD so that line AB is perpendicular to FD and $AF = FB$. Mark O on line DF so that $DO = OF$.

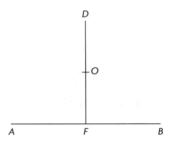

2. Draw lines DA and DB.

3. Divide DA into 8 equal parts. Divide DB into 8 equal parts.

4. Connect the new points as shown in the diagram to create additional lines. What do you notice?

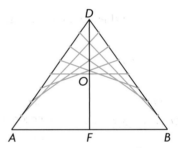

5. Divide lines DA and DB into smaller equal parts. Connect the new points as in step 4. What do you notice?

Here are two similar sketches. The one on the left uses few lines, and the one on the right uses many. As the number of lines used increases, the shape of the parabola becomes more noticeable.

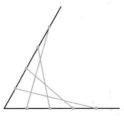 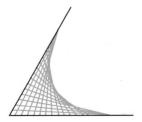

Why do you think this series of lines creates a parabola?

1.3 Working with Function Notation

GOAL

Interpret and evaluate relationships expressed in function notation.

YOU WILL NEED

• graph paper

LEARN ABOUT the Math

A rocket is shot into the air from the top of a building. Its height, in metres, after t seconds is modelled by the function $h(t) = -4.9t^2 + 19.6t + 34.3$.

? What is the meaning of $h(t)$ and how does it change for different values of t?

EXAMPLE 1   Connecting the rocket's height to function notation

Use function notation and a table of values to show the height of the rocket above the ground during its flight.

Leila's Solution: Using a Table of Values

$h(t) = -4.9t^2 + 19.6t + 34.3$

$h(0) = -4.9(0)^2 + 19.6(0) + 34.3$ ←   I substituted 0 for t in the function. I started at $t = 0$ because time can never be negative.

$\quad = 0 + 0 + 34.3$

$\quad = 34.3$

$h(0) = 34.3$ m   ← $h(0) = 34.3$ means that when the rocket was first shot, at 0 s, it was 34.3 m above the ground. This is the height of the building.

$h(3) = -4.9(3)^2 + 19.6(3) + 34.3$ ←   I evaluated the height of the rocket at $t = 3$.

$\quad = -4.9(9) + 58.8 + 34.3$

$\quad = -44.1 + 58.8 + 34.3$

$\quad = 49$

$h(3) = 49$ m   ← $h(3) = 49$ means that after 3 s, the rocket is 49 m above the ground.

t	0	1	2	3	4	5	6
h(t)	34.3	49.0	53.9	49.0	34.3	9.8	−24.5

I found the height of the rocket at different times and created a table of values. $h(6)$ gave a negative result. Since the height is always positive, the rocket hit the ground between 5 s and 6 s.

t	5.1	5.2	5.3	5.4
h(t)	6.81	3.72	0.54	−2.744

I evaluated the function for different times between 5 s and 6 s. The rocket hit the ground between 5.3 s and 5.4 s after it was shot into the air.

Tina's Solution: Using a Graph of the Height Function

Tech | *Support*

For help graphing and tracing along functions, see Technical Appendix, B-2.

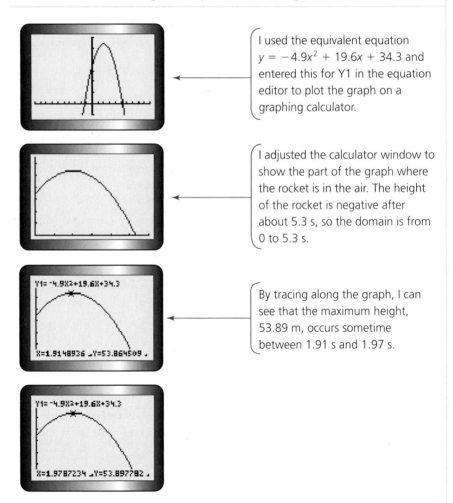

I used the equivalent equation $y = -4.9x^2 + 19.6x + 34.3$ and entered this for Y1 in the equation editor to plot the graph on a graphing calculator.

I adjusted the calculator window to show the part of the graph where the rocket is in the air. The height of the rocket is negative after about 5.3 s, so the domain is from 0 to 5.3 s.

By tracing along the graph, I can see that the maximum height, 53.89 m, occurs sometime between 1.91 s and 1.97 s.

Reflecting

A. How are the two methods similar? How are they different?

B. How did each student determine the domain? What is the domain of the function $h(t)$?

C. How did each student determine the valid values for the range? What is the range of the function $h(t)$?

D. Tina determined that the maximum height of the rocket would occur between 1.91 s and 1.97 s. Explain how Tina could find a more exact answer.

APPLY the Math

EXAMPLE 2   Using a substitution strategy to evaluate a function

Given $f(x) = 2x^2 + 3x - 1$, evaluate

a) $f(3)$ **b)** $f\left(\dfrac{1}{2}\right)$ **c)** $f(5 - 3)$ **d)** $f(5) - f(4)$

Richard's Solution

a) $f(x) = 2x^2 + 3x - 1$

$f(3) = 2(3)^2 + 3(3) - 1$

$\quad = 2(9) + 9 - 1$

$f(3) = 26$

> $f(3)$ means "find the value of the dependent variable or output when the independent variable or input is $x = 3$." I substituted 3 for x and then evaluated the expression.

b) $f(x) = 2x^2 + 3x - 1$

$f\left(\dfrac{1}{2}\right) = 2\left(\dfrac{1}{2}\right)^2 + 3\left(\dfrac{1}{2}\right) - 1$

$\quad = 2\left(\dfrac{1}{4}\right) + \dfrac{3}{2} - 1$

$\quad = \dfrac{2}{4} + \dfrac{6}{4} - \dfrac{4}{4}$

$f\left(\dfrac{1}{2}\right) = 1$

> I substituted $\dfrac{1}{2}$ for x and then evaluated the expression using a common denominator of 4.

c) $f(5 - 3) = 2(5 - 3)^2 + 3(5 - 3) - 1$

$\quad = 2(2)^2 + 3(2) - 1$

$\quad = 2(4) + 6 - 1$

$f(5 - 3) = 13$

> $f(5 - 3)$ is the same as $f(2)$, so I evaluated $f(5 - 3)$ by evaluating $f(2)$.

d) $f(5) - f(4)$

$= [2(5)^2 + 3(5) - 1] - [2(4)^2 + 3(4) - 1]$

$= [2(25) + 15 - 1] - [2(16) + 12 - 1]$

$= 64 - 43$

$f(5) - f(4) = 21$

> $f(5) - f(4)$ is the difference in the value of the function evaluated at $x = 5$ and at $x = 4$.
>
> So, I subtracted the value for $f(4)$ from the value for $f(5)$.

EXAMPLE 3 | **Representing and comparing the value of a function**

Sales of Sunglasses

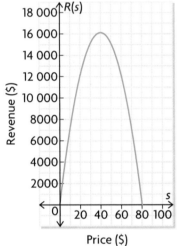

Price ($)

The relationship between the selling price of a new brand of sunglasses and revenue, $R(s)$, is represented by the function $R(s) = -10s^2 + 800s + 120$ and its graph at the left.

a) Determine the revenue when the selling price is $5.

b) Explain what $R(20) = 12\ 120$ means.

c) If $R(s) = 16\ 120$, determine the selling price, s.

Kedar's Solution

a) $R(s) = -10s^2 + 800s + 120$

$R(5) = -10(5)^2 + 800(5) + 120$

$= -10(25) + 4000 + 120$

$= 3870$

> $R(5)$ is the revenue generated at a selling price of $5. I substituted 5 for s into the equation and evaluated the expression.

When the selling price is $5, the revenue is $3870.

b) $R(20)$ represents the revenue generated from a selling price of $20. In this case $R(20)$ is $12\ 120. This corresponds to the point $(20, 12\ 120)$ on the graph.

> When the value of the independent variable, s, is 20, the dependent variable, $R(s)$, has a value of 12 120.

Sales of Sunglasses

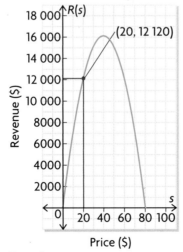

(20, 12 120)

Price ($)

c)

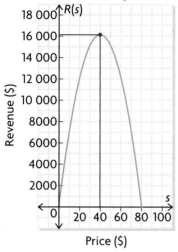

Sales of Sunglasses

I used the graph to estimate the selling price that corresponds to a revenue of $16 120. I drew a horizontal line from 16 120 on the revenue axis until it touched the curve. From this point I drew a vertical line down to the selling price axis.

For a revenue of $16 120 to occur, the selling price must be $40.

EXAMPLE 4 | Representing new functions

If $g(x) = 2x^2 - 3x + 5$, determine
a) $g(m)$ **b)** $g(3x)$

Sinead's Solution

a) $g(x) = 2x^2 - 3x + 5$

$g(m) = 2m^2 - 3m + 5$

In this case, x is replaced by the variable m, not by a number.

b) $g(3x) = 2(3x)^2 - 3(3x) + 5$

$= 2(9x^2) - 9x + 5$

$g(3x) = 18x^2 - 9x + 5$

I substituted $3x$ for x in the equation for $g(x)$ and then simplified.

In Summary

Key Idea

- $f(x)$ is called function notation and is used to represent the value of the dependent variable for a given value of the independent variable, x. For this reason, y and $f(x)$ are interchangeable in the equation or graph of a function, so $y = f(x)$.

Need to Know

- When a function is defined by an equation, it is convenient to name the function to distinguish it from other equations of other functions.

  For example, the set of ordered pairs (x, y) that satisfies the equation $y = 3x + 2$ forms a function. By naming the function f, we can use function notation.

(x, y) Notation	Function Notation
(x, y) is a solution of $y = 3x + 2$.	$(x, f(x))$ is a solution of $f(x) = 3x + 2$.

- $f(a)$ represents the value or output of the function when the input is $x = a$. The output depends on the equation of the function. To evaluate $f(a)$, substitute a for x in the equation for $f(x)$.
- $f(a)$ is the y-coordinate of the point on the graph of f with x-coordinate a. For example, if $f(x)$ takes the value 3 at $x = 2$, then $f(2) = 3$ and the point $(2, 3)$ lies on the graph of f.

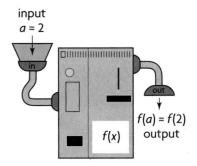

CHECK Your Understanding

1. Explain the meaning of $f(3) = \frac{1}{2}$.
2. Evaluate $f(3)$ for each of the following.
   a) $\{(1, 2), (2, 0), (3, 1), (4, 2)\}$ c)

   b)

x	1	2	3	4
y	2	3	4	5

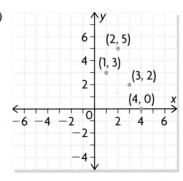

3. Determine $f(-2), f(0), f(2)$, and $f(2x)$ for each function.

   a) $f(x) = -3x^2 + 5$ **b)** $f(x) = 4x^2 - 2x + 1$

4. A stone is thrown from a bridge into a river. The height of the stone above the river at any time after it is released is modelled by the function $h(t) = 72 - 4.9t^2$. The height of the stone, $h(t)$, is measured in centimetres and time, t, is measured in seconds.

   a) Evaluate $h(0)$. What does it represent?

   b) Evaluate $h(2.5)$. What does it represent?

   c) If $h(3) = 27.9$, explain what you know about the stone's position.

PRACTISING

5. Evaluate $f(4)$ for each of the following.

   a) $f = \{(1, 5), (3, 2), (4, 1), (6, 2)\}$ **d)**

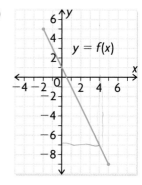

   b)

x	2	4	6	8
f(x)	4	8	12	16

   c) $f(x) = 3x^2 - 2x + 1$

6. The graph of $y = f(x)$ is shown at the right.

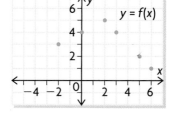

   a) State the domain and range of f.

   b) Evaluate.

      i) $f(3)$ **ii)** $f(5)$ **iii)** $f(5 - 3)$ **iv)** $f(5) - f(3)$

   c) In part (b), why is the value in (iv) not the same as that in (iii)?

   d) $f(2) = 5$. What is the corresponding ordered pair? What does 2 represent? What does $f(2)$ represent?

7. If the point $(2, 6)$ is on the graph of $y = f(x)$, what is the value of $f(2)$? Explain.

8. If $f(-2) = 6$, what point must be on the graph of f? Explain.

9. Evaluate each function for the given x-values.

   a) $f(x) = 9x + 1; x = 0, x = 2$

   b) $f(x) = -2x - 3; x = -1, x = 3$

   c) $f(x) = 2x^2 + 5; x = 2, x = 3$

   d) $f(x) = 3x^2 - 4; x = 0, x = 4$

10. Consider the function $g(t) = 3t + 5$.
   a) Determine
   - **i)** $g(0)$
   - **ii)** $g(1)$
   - **iii)** $g(2)$
   - **iv)** $g(3)$
   - **v)** $g(1) - g(0)$
   - **vi)** $g(2) - g(1)$

   b) In part (a), what are the answers to (v) and (vi) commonly called?

11. Consider the function $f(x) = s^2 - 6s + 9$.
   a) Determine each value.
   - **i)** $f(0)$
   - **ii)** $f(1)$
   - **iii)** $f(2)$
   - **iv)** $f(3)$
   - **v)** $[f(2) - f(1)] - [f(1) - f(0)]$
   - **vi)** $[f(3) - f(2)] - [f(2) - f(1)]$

   b) In part (a), what are the answers to (v) and (vi) commonly called?

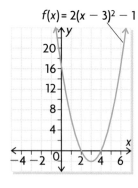

$f(x) = 2(x - 3)^2 - 1$

12. The graph shows $f(x) = 2(x - 3)^2 - 1$.
   a) Evaluate $f(0)$.
   b) What does $f(0)$ represent on the graph of f?
   c) If $f(x) = 6$, determine possible values of x.
   d) Does $f(3) = 4$ for this function? Explain.

13. The sum of two whole numbers is 10. Their product can be modelled by the function $P(x) = x(10 - x)$.
   a) What does x represent? What does $(10 - x)$ represent? What does $P(x)$ represent?
   b) What is the domain of this function?
   c) Evaluate the function for all valid values of the domain. Show the results in a table of values.

x	0										
P(x)	0										

   d) What two numbers do you estimate give the largest product? What is the largest product? Explain.

14. A farming cooperative has recorded information about the relationship between tonnes of carrots produced and the amount of fertilizer used. The function $f(x) = -0.53x^2 + 1.38x + 0.14$ models the effect of different amounts of fertilizer, x, in hundreds of kilograms per hectare (kg/ha), on the yield of carrots, in tonnes.
   a) Evaluate $f(x)$ for the given values of x and complete the table.

Fertilizer, x (kg/ha)	0.00	0.25	0.50	0.75	1.00	1.25	1.50	1.75	2.00
Yield, y(x) (tonnes)									

   b) According to the table, how much fertilizer should the farmers use to produce the most tonnes of carrots?
   c) Check your answer with a graphing calculator or by evaluating $f(x)$ for values between 1.25 and 1.50. Why does the answer change? Explain.

15. Sam read the following paragraph in his textbook. He does not
C understand the explanation.

> "$f(x)$ represents an expression for determining the value of the function f for any value of x. $f(3)$ represents the value of the function (the output) when x (the input) is 3. If $f(x)$ is graphed, then $f(3)$ is the y-coordinate of a point on the graph of f, and the x-coordinate of that point is 3."

Create an example (equation and graph) using a quadratic function to help him understand what he has read.

Extending

16. A glider is launched from a tower on a hilltop. The height, in metres, is negative whenever the glider is below the height of the hilltop. The equation representing the flight is $h(t) = \frac{1}{4}(t-3)(t-12)$, where time, t, is measured in seconds.

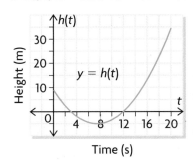

a) What does $h(0)$ represent?
b) What does $h(3)$ represent?
c) When is the glider at its lowest point? What is the vertical distance between the top of the tower and the glider at this time?

17. Babe Ruth, a baseball player, hits a "major league pop-up" so that the height of the ball, in metres, is modelled by the function $h(t) = 1 + 30t - 5t^2$, where t is time in seconds.

a) Evaluate the function for each of the given times to complete the table of values below.
b) Graph the function.
c) When does the ball reach its maximum height?
d) What is the ball's maximum height?
e) How long does it take for the ball to hit the ground?

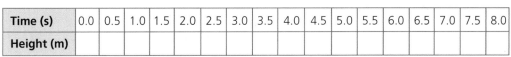

Time (s)	0.0	0.5	1.0	1.5	2.0	2.5	3.0	3.5	4.0	4.5	5.0	5.5	6.0	6.5	7.0	7.5	8.0
Height (m)																	

FREQUENTLY ASKED Questions

Study | **Aid**

- See Lesson 1.1, Examples 1, 2, and 3.
- Try Mid-Chapter Review Questions 1, 2, and 3.

Q: **What is the difference between a relation and a function?**

A: A function is a special kind of relation in which there is only one value of the dependent variable for each value of the independent variable. This means that for each input number, the equation generates only one output value. As a result, every number in the domain of the function corresponds to only one number in the range.

Q: **What are some ways to represent a function?**

A: You may describe a function as
- a set of ordered pairs: $\{(0, 1), (3, 4), (2, -5)\}$
- a table of values:

x	1	2	3
y	5	7	9

- a description in words: the height of a ball is a function of time
- an equation: $y = 2x + 1$
- function notation: $f(x) = 2x + 1$
- a mapping diagram: • a graph or a scatter plot:

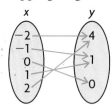 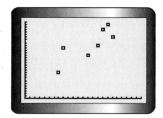

Q: **What do first and second differences indicate about a function?**

Study | **Aid**

- See Lesson 1.2, Examples 1, 2, and 3.
- Try Mid-Chapter Review Questions 4 and 5.

A: For linear functions, the first differences are constant, indicating a function of degree 1.

For quadratic functions, the first differences are not constant but the second differences are, indicating a function of degree 2.

Q: **What is function notation and what does it mean?**

Study | **Aid**

- See Lesson 1.3, Examples 1, 2, 3, and 4.
- Try Mid-Chapter Review Questions 6 and 7.

A:
- $f(x)$ is called function notation and is read "f at x" or "f of x."
- $f(x)$ represents the value of the dependent variable for a given value of the independent variable, x. This means y and $f(x)$ are interchangeable, so $y = f(x)$.
- $f(3)$ represents the value of the function when x is 3. So $(3, f(3))$ is the ordered pair for the point on the graph of f.

PRACTICE Questions

Lesson 1.1

1. Determine whether the following relations are functions. Explain.

a) $(1, 3), (2, 3), (3, 2), (1, 4), (4, 1)$

b)

c)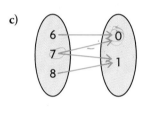

d)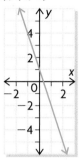

e)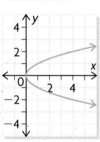

2. Relations f and g are defined by

$f = \{(1, 2), (2, 3), (3, 4)\}$ and
$g = \{(1, 2), (2, 1), (2, 3), (3, 0), (3, 4)\}$.

a) State the domain and range of each relation.

b) Is f a function? Is g? Explain.

3. Given the following, state the domain and range and whether the relation is a function.

a)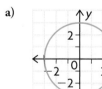

b)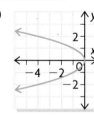

Lesson 1.2

4. The time it takes for a pendulum to make one complete swing and return to the original position is called the period. The period changes according to the length of the pendulum.

Length of Pendulum (cm)	6.2	24.8	55.8	99.2	155.0
Period (s)	0.5	1.0	1.5	2.0	2.5

a) Identify whether the relationship between the length of the pendulum and the period is linear, quadratic, or neither.

b) If a pendulum is 40 cm long, estimate its period.

c) Predict the length of a pendulum if its period is 2.2 s.

5. The distance a car skids depends on the speed of the car just before the brakes are applied. The chart shows the car's speed and the length of the skid.

Speed (km/h)	1	10	20	30	40	50
Length of Skid (m)	0.0	0.7	2.8	6.4	11.4	17.8

Speed (km/h)	60	70	80	90	100
Length of Skid (m)	25.7	35.0	45.7	57.8	71.4

a) Create a scatter plot for the data. Draw a curve of good fit for the data.

b) Estimate the initial speed of the car if the skid mark is 104 m long.

c) Determine whether a linear or a quadratic relation can model the data.

Lesson 1.3

6. A function h is defined by $h(x) = 2x - 5$. Evaluate.

a) $h(-2)$ b) $h(2m)$ c) $h(3) + h(n)$

7. A function g is given by $g(x) = 2x^2 - 3x + 1$. Evaluate.

a) $g(-1)$ b) $g(3m)$ c) $g(0)$

YOU WILL NEED

- graphing calculator or graphing software
- graph paper

GOAL

Understand how the parameters a, h, and k in $f(x) = a(x - h)^2 + k$ affect and change the graph of $f(x) = x^2$.

EXPLORE the Math

transformations

transformations are operations performed on functions to change the position or shape of the associated curves or lines

Transformations of curves, lines, and shapes form the basis for many patterns, like those you see on fabrics, wallpapers, and gift-wrapping paper. They are also used in designing tessellations such as those used in the graphic art of Maurits Cornelis Escher (1898–1972).

? How do the parameters a, h, and k in a quadratic function of the form $f(x) = a(x - h)^2 + k$ change the graph of $f(x) = x^2$?

Tech | **Support**

For help on graphing functions, see Technical Appendix, B-2.

A. Use a graphing calculator or graphing software to graph the quadratic functions shown in the table on the same set of axes. Sketch and label the graph of each function on the same set of axes. Then copy and complete the table.

	Function	Value of k in $f(x) = x^2 + k$	Direction of Opening	Vertex	Axis of Symmetry	Congruent to $f(x) = x^2$?
a)	$f(x) = x^2$	0	up	(0, 0)	$x = 0$	yes
b)	$f(x) = x^2 + 2$	2				
c)	$f(x) = x^2 + 4$					
d)	$f(x) = x^2 - 1$					
e)	$f(x) = x^2 - 3$					

B. Use the graphs and the table from part A to answer the following:
- **i)** What information about the graph does the value of k provide in functions of the form $f(x) = x^2 + k$?
- **ii)** What happens to the graph of the function $f(x) = x^2 + k$ when the value of k is changed? Consider both positive and negative values of k.
- **iii)** What happens to the x-coordinates of all points on $f(x) = x^2$ when the function is changed to $f(x) = x^2 + k$? What happens to the y-coordinates?

C. Clear all previous equations from your calculator or graphing program. Repeat part A for the quadratic functions shown in the table.

	Function	Value of h in $f(x) = (x - h)^2$	Direction of Opening	Vertex	Axis of Symmetry	Congruent to $f(x) = x^2$?
a)	$f(x) = x^2$	0	up	(0, 0)	$x = 0$	yes
b)	$f(x) = (x - 2)^2$	2				
c)	$f(x) = (x - 4)^2$					
d)	$f(x) = (x + 2)^2$					
e)	$f(x) = (x + 4)^2$					

D. Use the graphs and the table from part C to answer the following:
   i) What information about the graph does the value of h provide in functions of the form $f(x) = (x - h)^2$?
   ii) What happens to the graph of the function $f(x) = (x - h)^2$ when the value of h is changed? Consider both positive and negative values of h.
   iii) What happens to the x-coordinates of all points on $f(x) = x^2$ when the function is changed to $f(x) = (x - h)^2$? What happens to the y-coordinates?

E. Clear all previous equations from your calculator or graphing program. Repeat part A for the quadratic functions shown in the table.

	Function	Value of a in $f(x) = ax^2$	Direction of Opening	Vertex	Axis of Symmetry	Congruent to $f(x) = x^2$?
a)	$f(x) = x^2$	1	up	(0, 0)	$x = 0$	yes
b)	$f(x) = 2x^2$					
c)	$f(x) = 0.5x^2$					
d)	$f(x) = -2x^2$					
e)	$f(x) = -0.5x^2$					

F. Use the graphs and the table from part E to answer the following:
   i) What information about the graph does the value of a provide in functions of the form $f(x) = ax^2$? Consider values of a that are greater than 1 and values of a between 0 and 1.
   ii) What happens to the graph of the function $f(x) = ax^2$ when the value of a is changed? Consider both positive and negative values of a.
   iii) What happens to the x-coordinates of all points on $f(x) = x^2$ when the function is changed to $f(x) = ax^2$? What happens to the y-coordinates?

Reflecting

G. Use examples and labelled graphs that compare the transformed functions with the function $f(x) = x^2$. Summarize the effect that a, h, and k have on the parabola's

   a) direction of opening

   b) vertex

   c) axis of symmetry

   d) congruency to $f(x) = x^2$

   e) x-coordinates and y-coordinates

In Summary

Key Ideas

- Quadratic functions can be written in the form $g(x) = a(x - h)^2 + k$.
- Each of the constants a, h, and k changes the position and/or shape of the graph of $f(x) = x^2$.

Need to Know

- Changing the values of h and k changes the position of the parabola and, as a result, the locations of the vertex and the axis of symmetry. The new parabola is congruent to the parabola $f(x) = x^2$.
- Changing the value of a can change the shape of the parabola, as well as the direction in which the parabola opens. The new parabola is not congruent to the parabola $f(x) = x^2$ when $a \neq 1$ or -1.

FURTHER Your Understanding

1. The parabola $f(x) = x^2$ is transformed as described. The equation of the transformed graph has the form $f(x) = a(x - h)^2 + k$. Determine the values of a, h, and k for each of the following transformations.

   a) The parabola moves 4 units to the right.

   b) The parabola moves 5 units up.

   c) The parabola moves 2 units to the left.

   d) The parabola moves 3 unit down.

   e) The parabola is congruent to $f(x) = x^2$ and opens down.

   f) The parabola is narrower and opens upward. The y-coordinates have been multiplied by a factor of 2.

   g) The parabola is wider and opens downward. The y-coordinates have been multiplied by a factor of $\frac{1}{2}$.

Graphing Quadratic Functions by Using Transformations

Use transformations to sketch the graphs of quadratic functions.

LEARN ABOUT the Math

This photograph shows the interior of BCE Place in Toronto. Architects design structures that involve the quadratic model because it combines strength with elegance.

You have seen how changing the values of a, h, and k changes the shape and position of the graph of $f(x)$ in functions of the form $f(x) = x^2 + k$, $f(x) = (x - h)^2$, and $f(x) = ax^2$. This information can be used together with the properties of the quadratic function $f(x) = x^2$ to sketch the graph of the **transformed function**.

? How do you use transformations and the properties of the quadratic function $f(x) = x^2$ to graph the function $g(x) = (x + 2)^2 - 4$?

EXAMPLE 1	Graphing quadratic functions by using a transformation strategy: translating

Use transformations to sketch the graph of $g(x) = (x + 2)^2 - 4$.

Dave's Solution

$g(x) = (x + 2)^2 - 4$

$g(x) = (x - (-2))^2 - 4$ ←

$a = 1, h = -2, k = -4$

> I expressed the relation in the form $g(x) = a(x - h)^2 + k$.
>
> The values of h and k show that I must apply a horizontal and a vertical **translation** to the graph of $f(x) = x^2$ by moving each point on this graph 2 units to the left and 4 units down.

transformed function

the resulting function when the shape and/or position of the original graph of $f(x)$ are changed

translation

two types of translations can be applied to the graph of a function:
- Horizontal translations—all points on the graph move to the right when $h > 0$ and to the left when $h < 0$
- Vertical translations—all points on the graph move up when $k > 0$ and down when $k < 0$

key points

points of any function that define its general shape

Key Points of $f(x) = x^2$

x	$f(x) = x^2$
-3	9
-2	4
-1	1
0	0
1	1
2	4
3	9

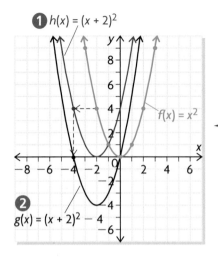

1 $h(x) = (x + 2)^2$

$f(x) = x^2$

2 $g(x) = (x + 2)^2 - 4$

1 I subtracted 2 from the x-coordinates of the **key points**. The base curve $f(x) = x^2$ (in green) moved 2 units to the left. The resulting graph is $h(x) = (x + 2)^2$ (in red).

2 I subtracted 4 from the y-coordinates of the key points of $h(x) = (x + 2)^2$. Each point of the curve $h(x) = (x + 2)^2$ moves down 4 units. The resulting graph is $g(x) = (x + 2)^2 - 4$ (in black).

The vertex changed from $(0, 0)$ to $(-2, -4)$.

The axis of symmetry changed from $x = 0$ to $x = -2$.

The shape of the graph did not change.

Reflecting

A. Consider Dave's solution. Given three points $O(0, 0)$, $A(-2, 4)$, and $B(1, 1)$ on the graph of $f(x) = x^2$, what would the coordinates of the corresponding images of the points on $f(x) = (x + 2)^2 - 4$ be if they were labelled O', A', and B'?

B. Dave translated the graph of $f(x) = x^2$ to the left 2 units and then 4 units down. Had he translated the graph 4 units down and then 2 units to the left, would the resulting graph be the same? Explain.

C. If the graph of $f(x) = x^2$ is only translated up/down and left/right, will the resulting graph always be congruent to the original graph? Explain.

APPLY the Math

EXAMPLE **2**	Graphing quadratic functions by using a transformation strategy: stretching vertically

Use transformations to sketch the graph of $h(x) = 2(x - 4)^2$.

Amanda's Solution

$h(x) = 2(x - 4)^2$

$a = 2, h = 4,$ and $k = 0.$

> $h(x)$ is in the form
> $h(x) = a(x - h)^2 + k.$
> No vertical translation is required, because $k = 0$.

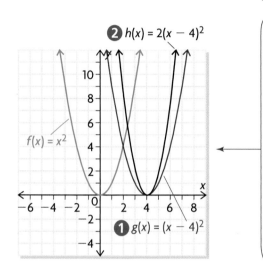

② $h(x) = 2(x - 4)^2$

$f(x) = x^2$

① $g(x) = (x - 4)^2$

> **①** Since $h > 0$, I added 4 to the x-coordinates of the key points of the graph $f(x) = x^2$ (in green). This moved the graph 4 units to the right. The resulting graph is $g(x) = (x - 4)^2$ (in red).
>
> **②** I multiplied the y-coordinates of the key points of $g(x) = (x - 4)^2$ by 2 because $a = 2$. This resulted in a **vertical stretch** and gave me the graph of $h(x) = 2(x - 4)^2$ (in black).

vertical stretch

when $a > 1$, the graph of the function $f(x)$ is stretched vertically

The vertex changed from $(0, 0)$ to $(4, 0)$.

The axis of symmetry changed from $x = 0$ to $x = 4$.

The shape of the parabola also changed—it's narrower than $f(x) = x^2$.

EXAMPLE 3 Graphing quadratic functions by using transformation strategies: reflecting and compressing vertically

Use transformations to sketch the graph of $g(x) = -0.5(x + 2)^2$.

Chantelle's Solution

$g(x) = -0.5(x + 2)^2$

$g(x) = -0.5(x - (-2))^2 + 0$ ←

$a = -0.5$, $h = -2$, and $k = 0$.

> I wrote $g(x)$ in the form $g(x) = a(x - h)^2 + k$. No vertical translation is required because $k = 0$.

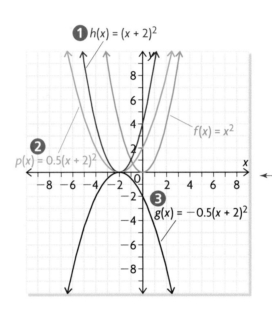

1 $h(x) = (x + 2)^2$

2 $p(x) = 0.5(x + 2)^2$

$f(x) = x^2$

3 $g(x) = -0.5(x + 2)^2$

1 I subtracted 2 from each of the x-coordinates of the key points. The graph of $f(x) = x^2$ (in green) moved 2 units to the left, since $h < 0$. This resulted in the graph of $h(x) = (x + 2)^2$ (in red).

2 I multiplied the y-coordinates of the key points of $h(x) = (x + 2)^2$ by 0.5 to get the graph of $p(x) = 0.5(x + 2)^2$. This resulted in a **vertical compression** and gave me the graph of $p(x) = 0.5(x + 2)^2$ (in blue).

vertical compression

when $0 < a < 1$, the graph is compressed vertically

The vertex changed from $(0, 0)$ to $(-2, 0)$.

The axis of symmetry changed from $x = 0$ to $x = -2$.

The shape of the parabola also changed—it's wider than $f(x) = x^2$.

vertical reflection

when $a < 0$, the graph is reflected in the x-axis

The parabola opens downward.

3 Since a is also negative, I reflected the graph of $p(x) = 0.5(x + 2)^2$ in the x-axis to get the graph $g(x) = -0.5(x + 2)^2$ (in black). This resulted in a **vertical reflection**.

EXAMPLE **4**

Graphing quadratic functions by using transformation strategies

Use transformations to sketch the graph of $m(x) = \frac{1}{3}x^2 + 2$.

Craig's Solution

$m(x) = \frac{1}{3}x^2 + 2$

$m(x) = \frac{1}{3}(x - 0)^2 + 2$ ◀──────

> I wrote $m(x)$ in the form $m(x) = a(x - h)^2 + k$.
> No horizontal translation is required because $h = 0$.

$a = \frac{1}{3}$, $h = 0$, and $k = 2$.

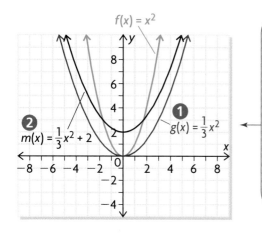

> **1** I multiplied the y-coordinates of $f(x) = x^2$ (in green) by $\frac{1}{3}$. The graph was vertically compressed by a factor of 3 and resulted in the graph of $g(x) = \frac{1}{3}x^2$ (in red).
>
> **2** Then I moved each point on $g(x)$ up 2 units. This gave me the graph of $m(x) = \frac{1}{3}x^2 + 2$ (in black).

The vertex changed from $(0, 0)$ to $(0, 2)$.

The axis of symmetry remained at $x = 0$.

The shape of the graph also changed—it's wider than $f(x) = x^2$.

In Summary

Key Ideas

- Functions of the form $g(x) = a(x - h)^2 + k$ can be graphed by applying transformations, one at a time, to the key points on the graph of $f(x) = x^2$.
- In graphing $g(x)$, the transformations apply to every point on the graph of $f(x)$. However, to sketch the new graph, you only need to apply the transformations to the key points of $f(x) = x^2$.

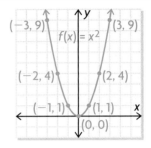

Need to Know

- For a quadratic function in the form $g(x) = a(x - h)^2 + k$,
  - horizontal translations: The graph moves to the right when $h > 0$ and to the left when $h < 0$.

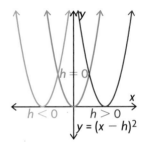

  - vertical translations: The graph moves up when $k > 0$ and down when $k < 0$.

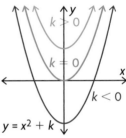

  - vertical stretches: The graph is stretched vertically when $a > 1$.
  - vertical compressions: The graph is compressed vertically when $0 < a < 1$.

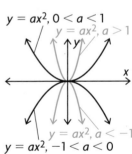

  - vertical reflections: The graph is reflected in the x-axis when $a < 0$.

  - the axis of symmetry is the line $x = h$.
  - the vertex is the point (h, k).

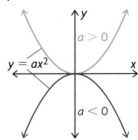

CHECK *Your Understanding*

1. Match each equation with its corresponding graph. Explain how you made your decision.

   a) $y = -(x - 2)^2 - 3$ c) $y = x^2 + 5$ e) $y = (x - 2)^2$

   b) $y = -0.5x^2 - 4$ d) $y = 2(x + 2)^2$ f) $y = -\dfrac{1}{3}(x + 4)^2 + 2$

   i) 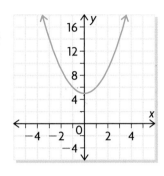   iii) 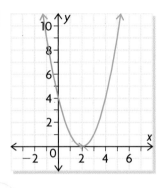   v)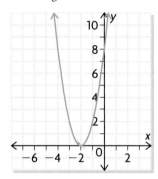

   ii) 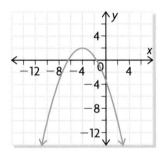   iv) 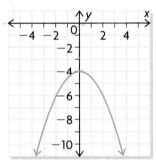   vi)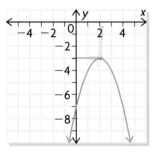

2. For each function,
   i)   identify the values of the parameters a, h, and k
   ii) identify the transformations
   iii) use transformations to graph the function and check that it is correct with a table of values or a graphing calculator

   a) $f(x) = -3x^2$
   b) $f(x) = (x + 3)^2 - 2$
   c) $f(x) = (x - 1)^2 + 1$
   d) $f(x) = -x^2 - 2$
   e) $f(x) = -(x - 2)^2$
   f) $f(x) = \dfrac{1}{2}(x + 3)^2$

PRACTISING

3. Match each graph with the correct equation. The graph of $y = x^2$ is shown in green in each diagram.

a) $y = x^2 + 5$

c) $y = -2x^2 + 5$

b) $y = (x + 5)^2$

d) $y = 2(x + 5)^2$

i)
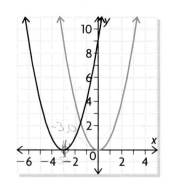

iii)
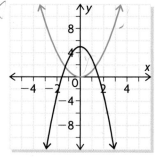

ii)
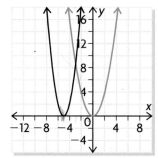

iv)
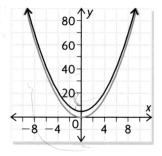

4. Describe the translations applied to the graph of $y = x^2$ to obtain a graph of each quadratic function. Sketch the graph.

a) $y = (x - 5)^2 + 3$

c) $y = 3x^2 - 4$

b) $y = (x + 1)^2 - 2$

d) $y = -\dfrac{1}{3}(x + 4)^2$

5. Consider a parabola P that is congruent to $y = x^2$ and with vertex at $(0, 0)$. Find the equation of a new parabola that results if P is

a) stretched vertically by a factor of 5

b) compressed vertically by a factor of 2

c) translated 2 units to the right and reflected in the x-axis

d) compressed vertically by 3, reflected in the x-axis, and translated 2 units up

6. Consider a parabola P that is congruent to $y = x^2$ and with vertex $(2, -4)$. Find the equation of a new parabola that results if P is

a) translated 2 units down

b) translated 4 units to the left

c) translated 2 units to the left and translated 3 units up

d) translated 3 units to the right and translate 1 unit down

7. Write an equation of a parabola that satisfies each set of conditions.
   a) opens upward, congruent to $y = x^2$, and vertex $(0, 4)$
   b) opens upward, congruent to $y = x^2$, and vertex $(5, 0)$
   c) opens downward, congruent to $y = x^2$, and vertex $(5, 0)$
   d) opens upward, narrower than $y = x^2$, and vertex $(2, 0)$
   e) opens downward, wider than $y = x^2$, and vertex $(-2, 0)$
   f) opens upward, wider than $y = x^2$, and vertex $(1, 0)$

8. Determine the answers to the following questions for each of the given transformed quadratic functions.
   i) How does the shape of the graph compare with the graph of $f(x) = x^2$?
   ii) What are the coordinates of the vertex and the equation of the axis of symmetry?
   iii) Graph the transformed function and $f(x) = x^2$ on the same set of axes.
   iv) Label the points $O(0, 0)$, $A(-2, 4)$, and $B(1, 1)$ on the graph of $f(x) = x^2$. Determine the images of these points on the transformed function. Label the images O', A', and B'.

   a) $f(x) = -(x - 2)^2$
   b) $f(x) = \frac{1}{2}x^2 + 2$
   c) $f(x) = (x + 2)^2 - 2$

9. For each of the following, state the equation of a parabola congruent to $y = x^2$ with the given property.
   a) The graph is 2 units to the right of the graph of $y = x^2$.
   b) The graph is 4 units to the left of the graph of $y = x^2$.
   c) The graph is 4 units to the left and 5 units down from the graph of $y = x^2$.
   d) The graph is vertically compressed by a factor of 4.
   e) The graph is vertically stretched by a factor of 2 and is 4 units to the left of the graph of $y = x^2$.
   f) The graph is vertically stretched by a factor of 3 and is 2 units to the right and 1 unit down from the graph of $y = x^2$.

10. For each of the following, state the condition on a and k such that the
   T parabola $y = a(x - h)^2 + k$ has the given property.
   a) The parabola intersects the x-axis at two distinct points.
   b) The parabola intersects the x-axis at one point.
   c) The parabola does not intersect the x-axis.

9 - demial.

11. The acceleration due to gravity, g, is 9.8 m/s^2 on Earth, 3.7 m/s^2 on Mars, 10.5 m/s^2 on Saturn, and 11.2 m/s^2 on Neptune. The height, $h(t)$, of an object, in metres, dropped from above each surface is given by $h(t) = -0.5gt^2 + k$.

a) Describe how the graphs will differ for an object dropped from a height of 100 m on each of the four planets.

b) On which planet will the object be moving fastest when it hits the surface?

c) On which planet will it be moving slowest?

12. Describe how the x- and y-coordinates of the given quadratic functions differ from the x- and y-coordinates of corresponding points of $y = x^2$.

a) $y = (x + 7)^2$ c) $y = -2(x - 4)^2$

b) $y = x^2 + 7$ d) $y = -\dfrac{1}{2}x^2 - 4$

Extending

13. Predict what the graphs of each group of equations would look like. Check your predictions by using graphing technology.

a) $f(x) = 10x^2$ b) $f(x) = 0.1x^2$
   $f(x) = 100x^2$ $f(x) = 0.01x^2$
   $f(x) = 1000x^2$ $f(x) = 0.001x^2$
   $f(x) = 10000x^2$ $f(x) = 0.0001x^2$

14. a) If $y = x^2$ is the base curve, write the equations of the parabolas that produce the following pattern shown on the calculator screen below. The scale on both axes is 1 unit per tick mark.

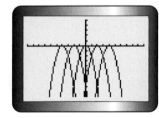

b) Create your own pattern using parabolas, and write the associated equations. Use $y = x^2$ as the base parabola.

1.6 Using Multiple Transformations to Graph Quadratic Functions

GOAL

Apply multiple transformations to $f(x) = x^2$ to graph quadratic functions defined by $f(x) = a(x - h)^2 + k$.

LEARN ABOUT the Math

Dave and Nathan have each graphed the function $f(x) = 3(x - 1)^2 + 2$. Dave's graph is different from Nathan's graph. They both applied the transformations in different orders.

Dave's Graph

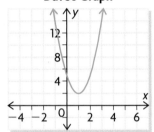

Nathan's Graph

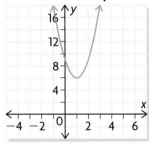

? Does the order in which transformations are performed matter?

EXAMPLE **1**   Reasoning about order: following the order of operations

Dave's Solution: Stretching and then Translating

$f(x) = 3(x - 1)^2 + 2$ ⟵

$a = 3, b = 1, k = 2$

I began with $g(x) = x^2$. I applied the transformations by following the order of operations.

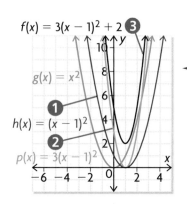

1 I performed the operation inside the bracket first. I translated the graph 1 unit to the right to get the graph $h(x) = (x - 1)^2$ (in red).

2 Next, I performed the multiplication. The translated graph stretches vertically by a factor of 3 to get the graph $p(x) = 3(x - 1)^2$ (in blue).

3 I translated the stretched graph 2 units up to get the graph $f(x) = 3(x - 1)^2 + 2$ (in black).

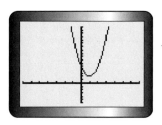

I checked my graph on the graphing calculator. My graph looks the same. The order I used must be correct.

Nathan's Solution: Translating and then Stretching

$f(x) = 3(x - 1)^2 + 2$

$a = 3, h = 1, k = 2$

I began with $g(x) = x^2$. I applied the translations first.

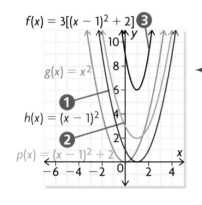

$f(x) = 3[(x - 1)^2 + 2]$ ③

$g(x) = x^2$

①

$h(x) = (x - 1)^2$

②

$p(x) = (x - 1)^2 + 2$

① I shifted the graph 1 unit to the right, since $h = 1$, to get the graph $h(x) = (x - 1)^2$ (in red).

② I translated the graph 2 units up, since $k = 2$, to get the graph $p(x) = (x - 1)^2 + 2$ (in blue).

③ Next, I applied the vertical stretch. The graph stretches vertically by a factor of 3, since $a = 3$, to get the graph $f(x) = 3[(x - 1)^2 + 2]$ (in black).

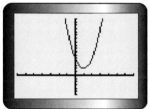

I checked my graph on the graphing calculator. My graph looks different. The order I used can't be correct.

Reflecting

A. List the y-coordinates for $x = -2, -1, 0, 1,$ and 2 for the final graph of each student's solution.

B. List the y-coordinates for $x = -2, -1, 0, 1,$ and 2 for the graph of the function $f(x) = 3(x - 1)^2 + 2$.

C. How do the coordinates from part B compare with those from each student's solution?

D. Explain why one of the solutions is incorrect.

E. List the order of transformations of the correct solution. Does the order of transformations of the correct solution apply to all functions of the form $f(x) = a(x - h)^2 + k$?

APPLY *the Math*

EXAMPLE 2	Applying multiple transformations to sketch the graph of a quadratic function

Sketch $h(x) = 2(x + 3)^2 - 1$ by applying the appropriate transformations to the graph of $f(x) = x^2$.

Mei's Solution

$h(x) = 2(x + 3)^2 + 1$ ←————————————

$h(x) = 2(x - (-3))^2 + 1$

$a = 2, h = -3, k = 1$

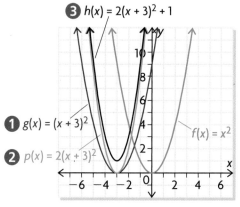

$\quad \textbf{3}\ h(x) = 2(x + 3)^2 + 1$

$\textbf{1}\ g(x) = (x + 3)^2$

$\textbf{2}\ p(x) = 2(x + 3)^2$

⎡ I applied the transformations one at a time, following the order of operations. ⎤

1 I moved the graph of $f(x) = x^2$ (in green) 3 units to the left to get the graph of $g(x) = (x + 3)^2$ (in red).

2 I multiplied the y-coordinates of $g(x) = (x + 3)^2$ by 2 to get the graph of $p(x) = 2(x + 3)^2$ (in blue).

3 I moved the resulting graph 1 unit up to get the graph of $h(x) = 2(x + 3)^2 + 1$ (in black).

The vertex changed from $(0, 0)$ to $(-3, 1)$.
The axis of symmetry changed from $x = 0$ to $x = -3$.
The final graph is narrower than $f(x) = x^2$.

EXAMPLE 3	Identifying transformations from the quadratic function

Describe the transformations you would use to graph the function $f(x) = -(x - 2.5)^2 - 5$.

Deirdre's Solution

$f(x) = -(x - 2.5)^2 - 5$

$a = -1, h = 2.5, k = -5$

Step 1. Horizontal translation 2.5 units to the right $h = 2.5$

Step 2. No vertical stretch or compression $a = -1$

Step 3. Reflection in the x-axis $a < 0$

Step 4. Vertical translation 5 units down $k = -5$

EXAMPLE 4	Applying multiple transformations to sketch the graph of a quadratic function

Graph $g(x) = -7 - (x + 3)^2$ by using transformations.

Jared's Solution

$g(x) = -7 - (x + 3)^2$

$g(x) = -(x + 3)^2 - 7$

$g(x) = -(x - (-3))^2 - 7$ ←

$a = -1, h = -3, k = -7$

> I wrote the function in the form $g(x) = a(x - h)^2 + k$ to help me identify the values of a, h, and k.

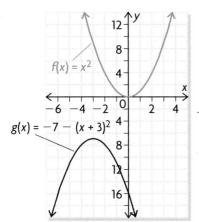

> Since $h < 0$, I moved the graph of $f(x) = x^2$ (in green) 3 units to the left. Next, I reflected it in the x-axis because $a = -1$. Then I moved it 7 units down because $k < 0$. This gave me the graph of $g(x)$ (in black).

The vertex changed from $(0, 0)$ to $(-3, -7)$.
The axis of symmetry changed from $x = 0$ to $x = -3$.
The shape did not change, since no stretching or compressing occurred.

In Summary

Key Idea

- Functions of the form $g(x) = a(x - h)^2 + k$ can be graphed by hand by applying the appropriate transformations, one at a time, to the graph of $f(x) = x^2$.

Need to Know

- Transformations can be applied following the order of operations:
  - horizontal translations
  - vertical stretches or compressions
  - reflections, if necessary
  - vertical translations
- You can use fewer steps if you combine the stretch/compression with the reflection and follow this with the necessary translations. This works because you are multiplying before adding/subtracting like the order of operations

CHECK Your Understanding

1. List the sequence of steps required to graph each function.

   a) $f(x) = 3(x + 2)^2$

   b) $f(x) = -2(x - 3)^2 + 1$

   c) $f(x) = \dfrac{1}{3}x^2 - 3$

   d) $f(x) = -\dfrac{1}{2}(x + 2)^2 + 4$

2. Sketch the final graph for each of the functions in question 1. Verify at least one of the key points, other than the vertex, by substituting its x-value into the equation and solving for y.

PRACTISING

3. Match each function to its graph.

   a) $f(x) = (x + 3)^2 + 1$

   b) $f(x) = -2(x + 4)^2 + 3$

   c) $f(x) = -(x - 3)^2 - 2$

   d) $f(x) = \dfrac{1}{2}(x - 3)^2 - 5$

i)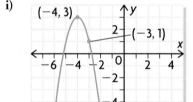

iii)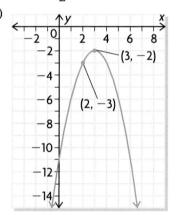

ii)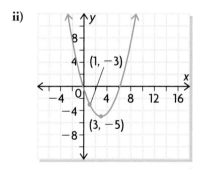

iv)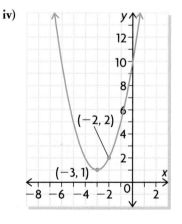

4. Transformations are applied to the graphs of $y = x^2$ to obtain the black parabolas. Describe the transformations that were applied. Write an equation for each black parabola.

a)

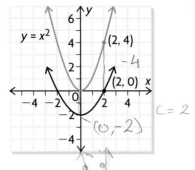

c)

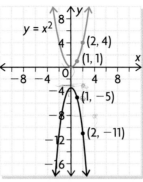

b)

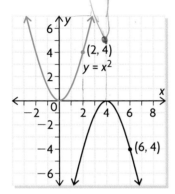

d)

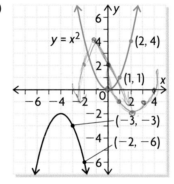

5. Write an equation of a parabola that satisfies each set of conditions.
   a) opens upward, congruent with $y = 2x^2$, vertex $(4, -1)$
   b) opens downward, congruent with $y = \frac{1}{3}x^2$, vertex $(-2, 3)$
   c) opens downward, congruent with $y = \frac{1}{2}x^2$, vertex $(-3, 2)$
   d) vertex $(2, -2)$, x-intercepts of 0 and 4
   e) vertex $(-1, 4)$, y-intercept of 2
   f) vertex $(4, 5)$, passes through $(2, 9)$

6. Consider a parabola P that is congruent to $y = x^2$, opens upward, and has vertex $(2, -4)$. Now find the equation of a new parabola that results if P is
   a) stretched vertically by a factor of 5
   b) compressed by a factor of $\frac{1}{2}$
   c) translated 2 units to the left
   d) translated 3 units up
   e) reflected in the x-axis and translated 2 units to the right and 4 units down

7. Describe the transformations applied to the graph of $y = x^2$ to obtain a graph of each quadratic relation. Sketch the graph by hand. Start with the graph of $y = x^2$ and use the appropriate transformations.

a) $y = -4(x - 5)^2 + 3$

b) $y = 2(x + 1)^2 - 8$

c) $y = \frac{2}{3}(x + 2)^2 + 1$

d) $y = -\frac{1}{2}(x - 1)^2 - 5$

e) $y = -(x - 3)^2 + 2$

f) $y = 2(x + 1)^2 + 4$

8. Sketch the graph of the transformed function $g(x) = -3(x - 2)^2 + 5$.
K Start with the graph of $f(x) = x^2$ and use the appropriate transformations.

9. If $y = x^2$ is the base curve for the graphs shown, what equations could
T be used to produce these screens on a graphing calculator? The scale on both axes is 1 unit per tick mark.

a)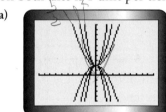

b)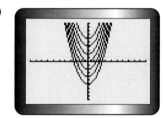

10. The graphs of $y = x^2$ (in red) and another parabola (in black) are
A shown at the left.

a) Determine a combination of transformations that would produce the second parabola from the first.

b) Determine a possible equation for the second parabola.

11. Describe the transformations applied to the graph of $y = x^2$ to obtain
C the graph of each quadratic relation.

a) $y = 2(x + 7)^2 - 3$

b) $y = 2x^2 + 7$

c) $y = -3(x - 4)^2 + 2$

d) $y = -3x^2 - 4$

Extending

12. A graphing calculator was used together with the vertex form $y = a(x - h)^2 + k$ to graph the screens shown. For the set of graphs on each screen, tell which of the variables a, h, and k remained constant and which changed. Give possible values for the variables that remained constant.

a)

b)

c)

The Domain and Range of a Quadratic Function

GOAL

Determine the domain and range of quadratic functions that model situations.

LEARN ABOUT the Math

A flare is shot vertically upward. A motion sensor records its height above ground every 0.2 s. The results are shown in the table.

Time (s)	0.0	0.2	0.4	0.6	0.8	1.0	1.2	1.4	1.6	1.8	2.0
Height (m)	0.0	1.8	3.2	4.2	4.8	5.0	4.8	4.2	3.2	1.8	0.0

The data are plotted. The function $h(t) = -5t^2 + 10t$ models the height of the flare, in metres, as a function of time from the time the flare is first shot into the air to the time that it returns to the ground.

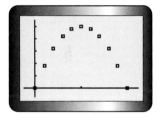

? How does the motion of the flare affect the domain and range of this quadratic function?

EXAMPLE 1	Reasoning about restricting the domain and range of a function

Wanda's Solution

The flare is initially shot at $t = 0$ s. The height of the flare when it is first shot is 0 m.

The flare reaches a maximum height of 5 m at $t = 1.0$ s.

After reaching 5 m, the flare starts to fall back to the ground.

At $t = 2$ s, the flare is on the ground.

(1, 5)

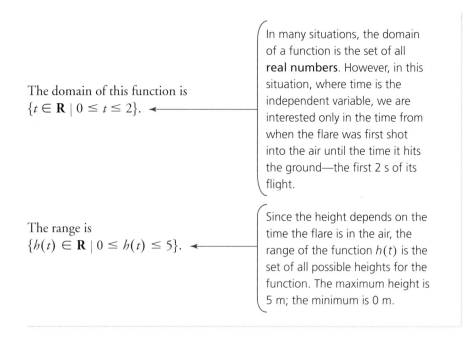

The domain of this function is
$\{t \in \mathbf{R} \mid 0 \le t \le 2\}$.

In many situations, the domain of a function is the set of all **real numbers**. However, in this situation, where time is the independent variable, we are interested only in the time from when the flare was first shot into the air until the time it hits the ground—the first 2 s of its flight.

The range is
$\{h(t) \in \mathbf{R} \mid 0 \le h(t) \le 5\}$.

Since the height depends on the time the flare is in the air, the range of the function $h(t)$ is the set of all possible heights for the function. The maximum height is 5 m; the minimum is 0 m.

Reflecting

A. Use your graphing calculator to graph the quadratic function defined by $f(x) = -5x^2 + 10x$. Then identify the domain and range. Explain why these differ from the domain and range of the function used to model the height of the flare.

B. For each of the following descriptions, identify the independent variable and the dependent variable. Then describe and justify reasonable values for the domain.
   i) The height of a stone that is thrown upward and falls to the ground, as a function of time
   ii) The height of a stone that is thrown upward and falls over a cliff, as a function of time
   iii) The percent of sale prices that a supermarket cashier can remember, as a function of time

APPLY the Math

EXAMPLE **2**

Connecting the domain and range of a function to its graph

Find the domain and range of $y = 3(x - 2)^2 + 3$.

Indira's Solution

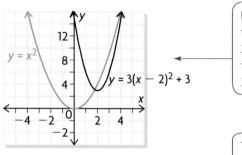

I graphed the function by translating $y = x^2$ to the right 2 units, vertically stretching by a factor of 3, and finally translating 3 units up.

The domain is $\{x \in \mathbf{R}\}$.

The function is defined for all values of x. The domain is the set of all real numbers.

The range is $\{y \in \mathbf{R} \mid y \geq 3\}$.

The vertex is (2, 3).

Since the function is a parabola that opens upward, the y-coordinate of the vertex is a minimum value.

The range is all values of y greater than or equal to 3.

EXAMPLE **3**

Connecting the domain and range to linear functions

Find the domain and range of each linear function.
a) $f(x) = -3x + 4$ **b)** $y = 5$

Paul's Solution

a)

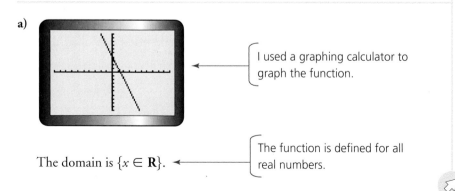

I used a graphing calculator to graph the function.

The domain is $\{x \in \mathbf{R}\}$.

The function is defined for all real numbers.

The range is $\{y \in \mathbf{R}\}$. ← Every real number corresponds to a value in the domain.

b)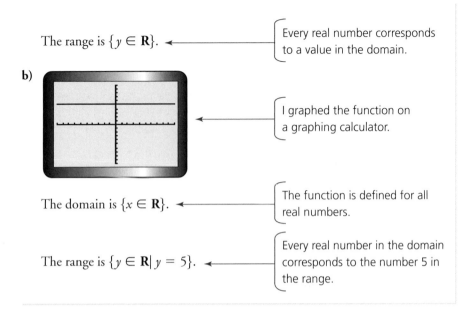

I graphed the function on a graphing calculator.

The domain is $\{x \in \mathbf{R}\}$. ← The function is defined for all real numbers.

The range is $\{y \in \mathbf{R}|\ y = 5\}$. ← Every real number in the domain corresponds to the number 5 in the range.

Any time you work with a function that models a real-world situation, it is necessary to restrict the domain and range.

<table>
<tr><td>EXAMPLE **4**</td><td>**Placing restrictions on the domain and range**</td></tr>
</table>

A baseball thrown from the top of a building falls to the ground below. The path of the ball is modelled by the function $h(t) = -5t^2 + 5t + 30$, where $h(t)$ is the height of the ball above ground, in metres, and t is the elapsed time in seconds. What are the domain and range of this function?

Kendall's Solution

Tech | **Support**

For help on using the table of values, see Technical Appendix, B-6.

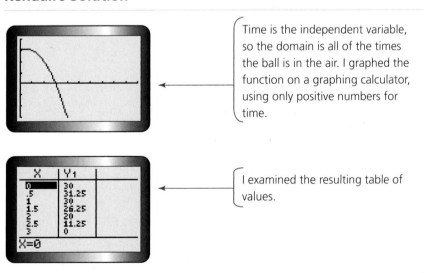

Time is the independent variable, so the domain is all of the times the ball is in the air. I graphed the function on a graphing calculator, using only positive numbers for time.

I examined the resulting table of values.

The domain is $\{t \in \mathbf{R} \mid 0 \le t \le 3\}.$ ◄──

The baseball is thrown at $t = 0$ s and is in the air until it hits the ground at $t = 3$ s.

Height is the dependent variable, so the range is all of the heights the ball reaches during its flight.

The range is $\{h(t) \in \mathbf{R} \mid 0 \le h(t) \le 31.25\}.$ ◄──

The ball reaches its maximum height at $t = 0.5$, so

$$h(0.5) = 5(0.5)^2 + 5(0.5) + 30$$
$$= 31.25 \text{ m}$$

It reaches its minimum height of 0 m when it hits the ground at $t = 3$, so $h(3) = 0$.

Since the height cannot be negative, I disregarded the y-values below 0.

In Summary

Key Ideas

- The domain of a function is the set of values for which the function is defined. As a result, the range of a function depends on the defining equation of the function.
- When quadratic functions model real-world situations, the domain and range are often restricted to values that make sense for the situation. Values that don't make sense are excluded.

Need to Know

- Most linear functions have the set of real numbers as their domain and range. The exceptions are horizontal and vertical lines.
- The range of a horizontal line is the value of y.
- The domain of a vertical line is the value of x.
- Quadratic functions have the set of real numbers as their domain. The range depends on the location of the vertex and whether the parabola opens up or down.

CHECK Your Understanding

1. a) If $f(x)$ is a linear function with a positive or negative slope, what will the domain and range of this function be?

   b) Consider linear relations that represent either horizontal lines, such as $y = 4$, or vertical lines, such as $x = 6$. What are the domain and range of these relations?

2. Find the domain and range of each function. Explain your answers.
   a) $f(x) = -2(x + 3)^2 + 5$ **c)** $f(x) = 3x - 4$
   b) $f(x) = 2x^2 + 4x + 7$ **d)** $x = 5$

3. The height of a flare is a function of the elapsed time since it was fired. An expression for its height is $f(t) = -5t^2 + 100t$. Express the domain and range of this function in set notation. Explain your answers.

PRACTISING

4. For each graph, state the domain and range in set notation. The scale on both axes is 1 unit per tick mark in each calculator screen.

a)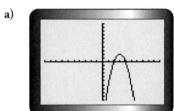

c)

e)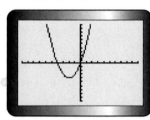

b)

d)

f)

5. State the domain and range of the quadratic function represented by the following table of values.

x	1	2	3	4	5	6	7	8	9	10	11
y	12	20	27	32	35	36	35	32	27	20	12

6. Use a graphing calculator or graphing software to graph each function. State its domain and range.
   a) $f(x) = x^2 - 6$
   b) $g(t) = 2 + t - 10t^2$
   c) $f(x) = -2x^2 + 5$, where $x \geq 0$
   d) $h(x) = 2(x - 1)^2 + 3$
   e) $g(x) = x^2 - 6x + 2$
   f) $h(x) = 3x^2 - 14x - 5$, where $x \geq 0$

7. A pebble is dropped from a bridge into a river. The height of the pebble above the water after it has been released is modelled by the function $h(t) = 80 - 5t^2$, where $h(t)$ is the height in metres and t is time in seconds.
   a) Graph the function for reasonable values of t.
   b) Explain why the values you chose for t in part (a) are reasonable.
   c) How high is the bridge? Explain.
   d) How long does it take the pebble to hit the water? Explain.
   e) Express the domain and range in set notation.

8. The cost of a banquet is $550 for the room rental, plus $18 for each person served.
   a) Create a table of values and construct a scatter plot for this function.
   b) Determine the equation that models the function.
   c) State the domain and range of the function in this situation.

Tech | **Support**
For help with scatter plots, see Technical Appendix, B-10.

9. A submarine at sea level descends 50 m every 5 min.
   T a) Determine the function that models this situation.
   b) If the ocean floor is 3 km beneath the surface of the water, how long will it take the submarine to reach the ocean floor.
   c) State the domain and range of this function in this situation.

10. A baseball is hit from a height of 1 m. The height of the ball is modelled by the function $h(t) = -5t^2 + 10t + 1$, where t is time in seconds.
    a) Graph the function for reasonable values of t.
    b) Explain why the values you chose for t in part (a) are reasonable.
    c) What is the maximum height of the ball?
    d) At what time does the ball reach the maximum height?
    e) For how many seconds is the ball in the air?
    f) For how many seconds is the ball higher than 10 m?
    g) Express the domain and range in set notation.

11. In the problem about selling raffle tickets in Getting Started on page 5, the student council wants to determine the price of the tickets. The council surveyed students to find out how many tickets would be bought at different prices. The council found that
    • if they charge $0.50, they will be able to sell 200 tickets; and
    • if they raise the price to $1.00, they will sell only 50 tickets.
    a) The formula for revenue, $R(x)$, as a function of ticket price, x, is $R(x) = -300x^2 + 350x$. Use a graphing calculator to graph the function.
    b) What ticket price should they charge to generate the maximum amount of revenue?
    c) State the range and domain of the function.

12. Sometimes when functions are used to model real-world situations, the **C** domain and range of the function must be restricted.
    a) Explain what this means.
    b) Explain why restrictions are necessary. Use an example in your explanation.

Extending

13. The picture shows four circles inside a square. Each small circle has a radius r. The area of the shaded region as a function is $A(r) = (16 - 4\pi)r^2$.
    a) What is the domain of the function?
    b) What is the range of the function?

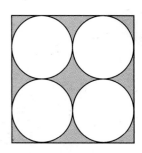

FREQUENTLY ASKED Questions

Q: **What are some of the different kinds of transformations that can be applied to the function $f(x) = x^2$?**

A: The transformations include:

1. horizontal translations: The graph of the function is shifted to the left or right.

2. vertical translations: The graph of the function is shifted up or down.

3. vertical stretches: The graph is stretched vertically. Each y-value on the graph is multiplied by a factor that is greater than 1.

4. vertical compressions: The graph is compressed vertically. Each y-value on the graph is multiplied by a factor that is between 0 and 1.

5. reflections: The graph is reflected in the x-axis when each y-value is multiplied by a negative factor.

> Study | **Aid**
> * See Lesson 1.5, Examples 1 to 4.
> * Try Chapter Review Question 6.

Q: **How is the form $f(x) = a(x - h)^2 + k$ of a quadratic function useful?**

A: Every quadratic function can be written in the form $f(x) = a(x - h)^2 + k$.

The constants a, h, and k each change the location and/or shape of the graph of $f(x) = x^2$.
* Horizontal translations: The graph moves to the right when $h > 0$ and to the left when $h < 0$.
* Vertical stretches: The graph is stretched vertically when $a > 1$.
* Vertical compressions: The graph is compressed vertically when $0 < a < 1$.
* Vertical reflections: The graph is reflected in the x-axis when $a < 0$.
* Vertical translations: The graph moves up when $k > 0$ and down when $k < 0$.

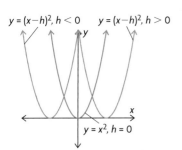

horizontal translations

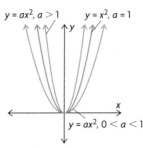

vertical stretch/compressions

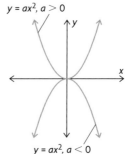

vertical reflections

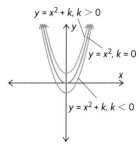

vertical translations

Q: Why is the order in which I apply the transformations when graphing a quadratic function important?

A: Horizontal translations are applied first, vertical stretches and compressions second, reflections third, and then vertical translations. This order of transformations ensures that the y-values of $f(x) = x^2$ are transformed correctly by following the order of operations for every function of the form $f(x) = a(x - h)^2 + k$.

Study | *Aid*
- See Lesson 1.6, Examples 1 to 4.
- Try Chapter Review Questions 7 to 10.

Q: How do you determine the domain and range of a linear and a quadratic function?

A: The domain of a function is the set of values for which the function is defined. As a result, the range of each function depends on the defining equation of the function.

Study | *Aid*
- See Lesson 1.7, Examples 1 to 4.
- Try Chapter Review Questions 11 to 13.

Most linear functions have the set of real numbers as their domain and range. The exceptions are horizontal and vertical lines. When the line is horizontal, the range is the value of y. When the line is vertical, the domain is the value of x.

Quadratic functions have the set of real numbers for their domain. The range depends on the location of the vertex and whether the parabola opens up or down.

When linear and quadratic functions are used to model real-world situations, the domain and range are often restricted to values that make sense for the situation. Values that don't make sense are excluded.

PRACTICE Questions

Lesson 1.1

1. The data in the table show the average mass of a boy as he grows between the ages of 1 and 12. State the following:
   a) domain
   b) range
   c) whether the relation is a function

Age (years)	1	2	3	4	5	6
Mass (kg)	11.5	13.7	16.0	20.5	23.0	23.0

Age (years)	7	8	9	10	11	12
Mass (kg)	30.0	33.0	39.0	38.5	41.0	49.5

Lesson 1.2

2. Determine, without graphing, which type of relationship (linear, quadratic, or neither) best models this table of values. Explain.

x	−1	0	1	2	3
y	1	2	−3	−14	−31

3. State the degree of each function and whether each is linear, or quadratic, or neither.
   a) $f(x) = -8 + 3x$
   b) $g(x) = 4x^2 - 3x + 5$
   c) $y = (x - 4)(4x^2 - 3)$

Lesson 1.3

4. Evaluate the function $f(x) = 3x^2 - 3x + 1$ at the given values.
   a) $f(-1)$
   b) $f(3)$
   c) $f(0.5)$

5. For each of the following, determine $f(3)$.
   a) $f = \{(1, 2), (2, 3), (3, 5), (4, 5)\}$
   b)

x	1	3	5	7
f(x)	2	4	6	8

   c) $f(x) = 4x^2 - 2x + 1$
   d)

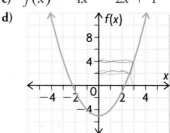

Lesson 1.5

6. Use transformations to determine the vertex, axis of symmetry, and direction of opening of each parabola. Sketch the graph.
   a) $y = x^2 - 7$
   b) $y = -(x + 1)^2 + 10$
   c) $y = -\frac{1}{2}(x + 2)^2 - 3$
   d) $y = 2(x - 5)^2$

Lesson 1.6

7. Describe how the graph of $y = x^2$ can be transformed to the graphs of the relations from question 6.

8. a) Describe how the graph of $y = x^2$ can be transformed into the graph of the given quadratic function.
   i) $y = 5x^2 - 4$
   ii) $y = \frac{1}{4}(x - 5)^2$
   iii) $y = -3(x + 5)^2 - 7$
   b) List the domain and range of each function. Compare these with the original graph of $y = x^2$.

9. a) Describe the transformations to the graph of $y = x^2$ to obtain $y = -2(x + 5)^2 - 3$.

b) Graph $y = x^2$. Then apply the transformations in part (a) to graph $y = -2(x + 5)^2 - 3$.

10. The graphs of $f(x) = x^2$ (in green) and another parabola (in black) are shown.

a) Draw a combination of transformations that would produce the second parabola from the first.

b) Determine a possible equation for the second parabola.

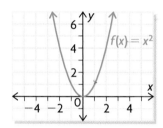

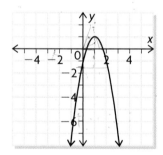

11. On a calculator like the TI-83 Plus, enter these equations: $Y_1 = X^2$, $Y_2 = Y_1 - 4$, and $Y_3 = 2 * Y_2$.

a) Use an appropriate WINDOW setting to graph Y_1, Y_2, and Y_3. Write an equation in vertex form for Y_3.

b) Change Y_2 and Y_3 to $Y_2 = 2 * Y_1$ and $Y_3 = Y_2 - 4$. Find the coordinates of the vertex for Y_3. Write an equation in vertex form for Y_3.

c) Explain why the graph of Y_3 in part (a) is different from the one in part (b).

d) Describe the sequence of transformations needed to transform the graph of $y = x^2$ into the graph of $y = 2x^2 - 4$.

Lesson 1.7

12. A football is thrown into the air. The height, $h(t)$, of the ball, in metres, after t seconds is modelled by $h(t) = -4.9(t - 1.25)^2 + 9$.

a) How high off the ground was the ball when it was thrown?

b) What was the maximum height of the football?

c) How high was the ball at 2.5 s?

d) Is the football in the air after 6 s?

e) When does the ball hit the ground?

13. Clay shooting disks are launched from the ground into the air from a machine 12 m above the ground. The height of each disk, $h(t)$, in metres, is modelled by $h(t) = -5t^2 + 30t + 12$, where t is the time in seconds since it was launched.

a) What is the maximum height the disks reach?

b) At what time do the disks hit the ground?

c) Determine the domain and range of this model.

14. The height, h, in metres, of an object t seconds after it is dropped is $h = -0.5gt^2 + k$, where g is the acceleration due to gravity and k is the height from which the object is released. If an object is released from a height of 400 m, how much longer does it take to fall to a height of 75 m on the Moon compared with falling to the same height on Earth? The acceleration due to gravity is 9.8 m/s² on Earth and 1.6 m/s² on the Moon.

1. For each of the following relations, state
   i) the domain and range
   ii) whether or not it is a function, and justify your answer
   a) $f = \{(1, 2), (3, 1), (4, 2), (7, 2)\}$
   b)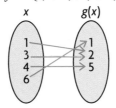
   c)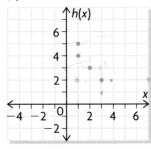

2. Define the term *function* and give an example and a non-example.

Time (s)	Height (m)
0	0
1	30
2	40
3	40
4	30
5	0

3. Use a difference table to determine whether the data in the table at the left represent a linear or quadratic relationship. Justify your decision.

4. If $f(x) = 3x^2 - 2x + 6$, determine
   a) $f(2)$
   b) $f(x - 1)$

5. $f(x) = 3(x - 2)^2 + 1$
   a) Evaluate $f(-1)$.
   b) What does $f(1)$ represent on the graph of f?
   c) State the domain and range of the relation.
   d) How do you know if f is a function from its graph?
   e) How do you know if f is a function from its equation?

6. A function is defined by the equation $d(x) = 5(x - 3)^2 + 1$.
   a) List the transformations to the graph of $f(x) = x^2$ to get $d(x)$.
   b) What is the maximum or minimum value of the transformed function $d(x)$?
   c) State the domain and range of $d(x)$.
   d) Graph the function $d(x)$.

7. A football is kicked from a height of 0.5 m. The height of the football is modelled by the function $h(t) = -5t^2 + 18t + 0.5$, where t is time in seconds and $h(t)$ is height in metres.
   a) Graph the function for reasonable values of t.
   b) Explain why the values you chose for t in part (a) are reasonable.
   c) What is the maximum height of the football?
   d) At what time does the football reach the maximum height?
   e) For how many seconds is the football in the air?
   f) Express the domain and range in set notation.

Using Transformations and Quadratic Function Models

Quadratic functions can be used as mathematical models of many real-life situations. Since the graphs of these functions are parabolas, many familiar examples can be found that appear to utilize this shape. Here are some of them:

❓ How can you use transformations to determine the equation of a quadratic function that models a parabola seen in a picture?

A. Use the Internet to find a picture that appears to have a parabola (or part of one) in it.

B. Copy the picture and paste it into *Geometer's Sketchpad*. Place a grid over top of your picture.

C. Create a quadratic function of the form $f(x) = ax^2$ by estimating an appropriate value for a. Adjust this value and re-graph your function until you are satisfied with the width of your parabola.

D. Create a quadratic function of the form $f(x) = a(x - h)^2 + k$ by estimating appropriate values for h and k. Adjust these values and re-graph your function until you are satisfied that the graph of your function closely matches that of the parabola in the photo. Print your final results.

E. List the transformations you used to create your graphical model as well as the final equation of your quadratic function.

F. State any restrictions on the domain and range that must be applied to ensure that your graph matches only the parabola shown in the photo.

G. Repeat parts A to F using a picture that contains a parabola that opens in the opposite direction to the one you originally found.

Tech | **Support**

For help using *Geometer's Sketchpad* to graph functions, see Technical Appendix, B-19.

Task | **Checklist**

✔ Did you show all your steps?

✔ Did you include a printout of your graph and photo?

✔ Did you support your choice of transformations used?

✔ Did you explain your thinking clearly?

✔ Did you restrict the domain and range?

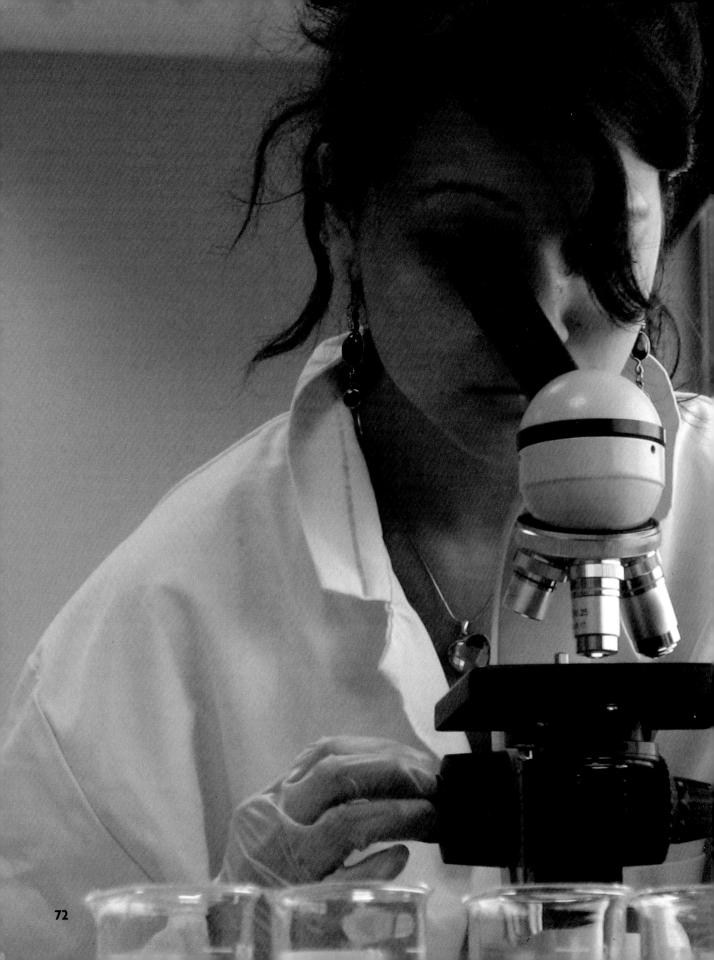

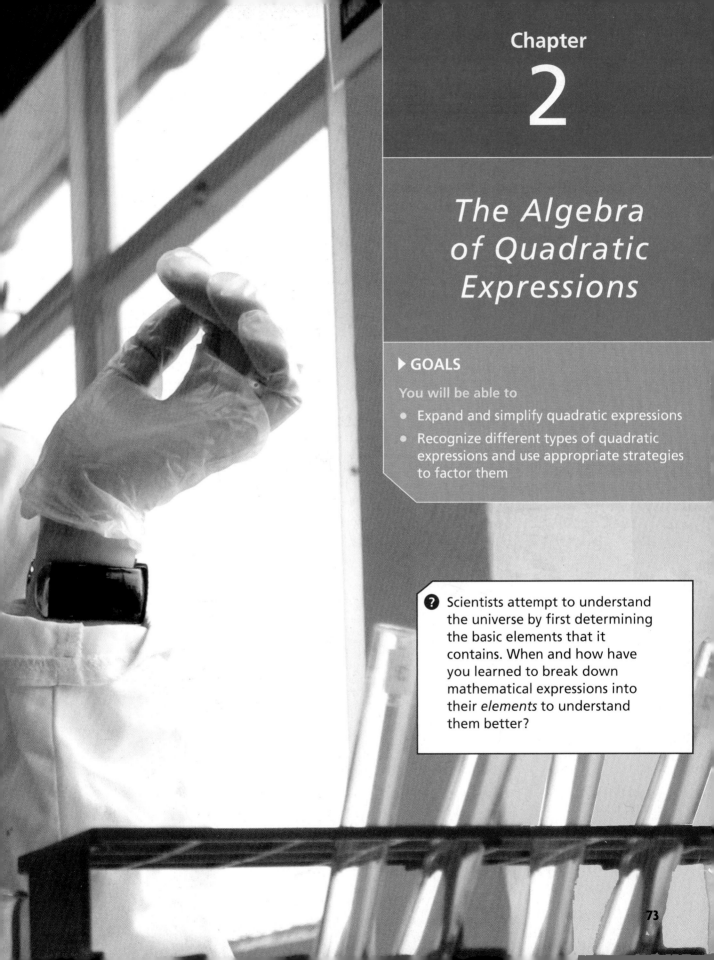

Chapter

2

The Algebra of Quadratic Expressions

▶ **GOALS**

You will be able to

- Expand and simplify quadratic expressions
- Recognize different types of quadratic expressions and use appropriate strategies to factor them

? Scientists attempt to understand the universe by first determining the basic elements that it contains. When and how have you learned to break down mathematical expressions into their *elements* to understand them better?

WORDS *You Need to Know*

1. Match each word with the expression that best illustrates its definition.

a) variable
b) coefficient

c) like terms
d) unlike terms

e) binomial
f) trinomial

g) expanding
h) factoring

i) $3x + 5$

ii) $2(x - 5) = 2x - 10$

iii) $3x^2 + 7x - 1$

iv) $5x^2$

v) $5x + 15 = 5(x + 3)$

vi) $10y$ and $10y^2$

vii) x

viii) $6x$ and $-3x$

SKILLS AND CONCEPTS *You Need*

Study **Aid**

For help, see Essential Skills Appendix, A-9.

Simplifying Algebraic Expressions by Collecting Like Terms

To simplify an algebraic expression, collect like terms by adding or subtracting.

EXAMPLE

Simplify.
$$(2x^2 + 3) + (-4x^2 + 8)$$
$$= 2x^2 + (-4x^2) + 3 + 8$$
$$= -2x^2 + 11$$

2. Simplify each expression.

a) $2x - 5y + 6y - 8x$

b) $7xy - 8x^2 + 6xy - 2x^2 - 12xy + 10x^2$

c) $(4x - 5y) + (6x + 3) - (7x - 2y)$

d) $(2a - 8ab) - (7b + 9a) + (ab - 2a)$

Study **Aid**

For help, see Essential Skills Appendix, A-3.

Exponent Laws for Multiplication and Division			
Rule	**Written Description**	**Algebraic Description**	**Worked Example in Standard Form**
Multiplication	To multiply powers with the same base, add the exponents, leaving the base as is.	$b^m \times b^n = b^{m+n}$	$(x^2)(x^4)$ $= x^{2+4}$ $= x^6$
Division	To divide powers with the same base, subtract the exponents, leaving the base as is.	$b^m \div b^n = b^{m-n}$	$x^5 \div x^3$ $= x^{5-3}$ $= x^2$

3. Simplify.

a) $(x^3)(x^2)$ b) $(2x^2)(5x)$ c) $x^4 \div x^2$ d) $4x^5 \div 2x^3$

4. State an expression for the area of each shape.

a)
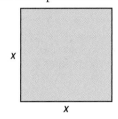
x · x

b)
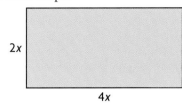
$2x$ · $4x$

Expanding Using the Distributive Property

To expand an algebraic expression, use the distributive property to multiply the expression by a constant or a term.

Study | *Aid*

For help, see Essential Skills Appendix, A-9.

$$a(b + c) = ab + ac$$

EXAMPLE

Expand and simplify.
$2(2a + b) - 3(3a - 2b)$
$= 2(2a + b) - 3(3a - 2b)$
$= 4a + 2b - 9a + 6b$
$= 4a - 9a + 2b + 6b$
$= -5a + 8b$

5. Expand and simplify.

a) $3(3x - 8)$
b) $-4(8x^2 - 2x + 1)$
c) $2(7x^2 + 3x + 5) - 2(8x + 1)$
d) $(3d^3 - 6d + 5d^2) + 4(9 - 2d^3 - 4d^2)$
e) $2x^2(3x + 5)$
f) $-5x^2(x^2 - 3x + 4)$

Dividing Out a Common Factor

To factor an algebraic expression, divide out the greatest number or term that will divide into all terms.

EXAMPLE

Factor each expression.
a) $4x^2 - 12x + 4$
$\quad = 4(x^2 - 3x + 1)$

b) $3x^2 + 6x^5 - 9x^4$
$\quad = 3x^2(1 + 2x^3 - 3x^2)$

6. Factor each expression.

a) $2x - 10$
b) $6x^2 + 24x + 30$
c) $25x^2 + 20x - 100$
d) $7x^4 + 12x^3 - 9x^5$

PRACTICE

Study | **Aid**

For help, see Essential Skills
Appendix.

Question	Appendix
7	A-3

7. Simplify.
a) $x^2 \times x^3$
b) $-5x^3 \times 6x^4$
c) $x \times 3x^2 \times 5x^4$
d) $2 \times (-x) \times (-x^2)$

8. a) Which expressions are monomials?
b) Which are binomials?
c) Which are trinomials?
d) Which are quadratic?
i) 4
ii) $3x - 2$
iii) $4x^2 + 3x - 1$
iv) $2x(3x + 1)$
v) $7x^3$
vi) $-3x^2(2x - 5)$

9. Name the greatest common factor for each pair.
a) 24, 32
b) 56, 80
c) 108, 90
d) $3x, 2x^2$
e) $3x + 2, 6x + 4$
f) $25x^2, 15x$

10. Factor using only prime numbers.
a) 78
b) 63
c) 3025
d) 41

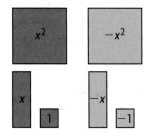

11. Sketch copies of tiles like those shown at the left to represent each expression.
a) $2x + 3$
b) $-4x - 2$
c) $x^2 - 2x + 1$
d) $-x^2 + 1$
e) $3x^2 - 2x - 3$
f) $1 + x - x^2$

12. Match each diagram with the correct expression.
a) $2x + 2$
b) $-2x + 2$
c) $2x - 2$

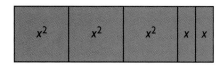

13. This rectangle shows $3x^2 + 2x$.

Sketch rectangles to show each expression.
a) $3x^2 - 2x$
b) $2x + 4$
c) $-2x^2 - x$
d) $x^2 + 3x$

14. Decide whether you agree or disagree with each statement. Explain why.
a) You can factor a number by determining the length and width of a rectangle with that area.
b) One way to factor 12 is as $9 \times \frac{4}{3}$.
c) The only way to factor $2x + 6$ is as $2(x + 3)$.

APPLYING What You Know

YOU WILL NEED

- graph paper or square tiles

Rearranging Tiles

Using 1 × 1 square tiles, Fred builds shapes like those shown at the right. Each shape has a length 2 greater than the width.

Fred notices that he can rearrange the tiles in each of shapes *R*, *S*, and *T* to form a rectangle.

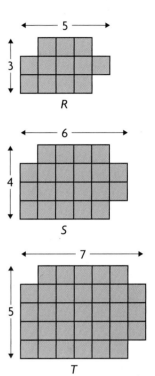

?   Is it always possible to rearrange the tiles to form a rectangle?

A.   Rearrange the tiles in shape *R* to form a rectangle. Sketch all possibilities on graph paper. Repeat, using the tiles from *S* and from *T*.

B.   Select one of your rectangles from part A for shapes *R*, *S*, and *T* so that the three rectangles chosen form a pattern.

C.   How does the length of the rectangle you sketched relate to the original length of each shape?

D.   How does the width of the rectangle you sketched relate to the original width of each shape?

E.   Use square tiles or graph paper to create new shapes *U* and *V* to extend the pattern of shapes *R*, *S*, *T*, ...

F.   Predict how shapes *U* and *V* can be rearranged into rectangles to extend the pattern you created in part B. Check your predictions.

G.   Explain how the general shape represents the shapes in the pattern *R*, *S*, *T*, *U*, *V*, ...

H.   What algebraic expression involving *n* could you use to describe the area of the general shape in part G? Explain.

I.   What algebraic expression involving *n* could you use to describe the length and the width of the rectangle you can make by rearranging the general shape in part G? Explain.

J.   How do you know that any shape like the one in part G can be rearranged into a rectangle as long as $n \geq 3$?

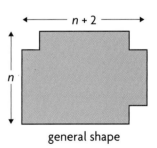

general shape

2.1 Working with Quadratic Expressions

YOU WILL NEED

- algebra tiles

GOAL

Expand and simplify quadratic expressions.

INVESTIGATE *the Math*

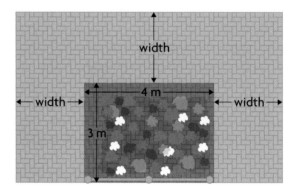

Ali wants to know the area of brick he needs to build a path around his garden.

? **What is a simplified expression for the area of the path?**

A. Use w to represent the width of the path.

B. Think about the large rectangle that includes the garden and the path. What algebraic expressions represent the length and width of this rectangle?

C. What algebraic expression represents the area of the large rectangle in part B? How do you know?

D. What is the area of the garden without the path?

E. Use your results from parts C and D to write an expression for the area of the path.

F. What simplified expression represents the area of the path?

Reflecting

G. Why was a variable introduced in part A?

H. Explain how you expanded and simplified the expression you wrote in part E.

I. Is the algebraic expression representing the area of the path linear or quadratic? Explain how you know.

APPLY the Math

EXAMPLE **1**	Multiplying two binomials by using algebra tiles

Evelyn is sewing a quilt as shown. If the width of the border is x, state the area of the quilt as a function of x.

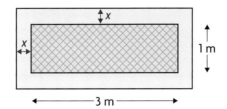

Devin's Solution

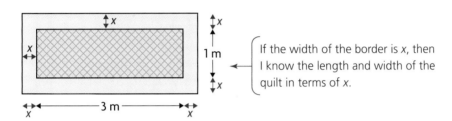

If the width of the border is x, then I know the length and width of the quilt in terms of x.

length: $x + 3 + x = 2x + 3$

width: $x + 1 + x = 2x + 1$

The dimensions are $2x + 3$ and $2x + 1$.

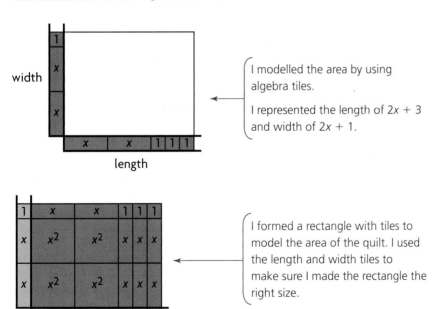

I modelled the area by using algebra tiles.

I represented the length of $2x + 3$ and width of $2x + 1$.

I formed a rectangle with tiles to model the area of the quilt. I used the length and width tiles to make sure I made the rectangle the right size.

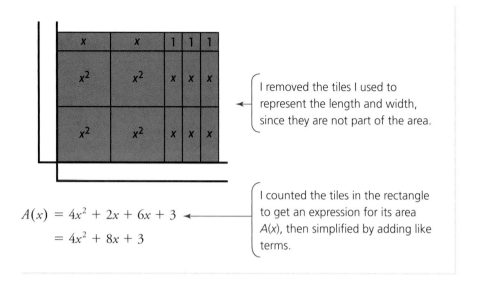

I removed the tiles I used to represent the length and width, since they are not part of the area.

$$A(x) = 4x^2 + 2x + 6x + 3$$
$$= 4x^2 + 8x + 3$$

I counted the tiles in the rectangle to get an expression for its area $A(x)$, then simplified by adding like terms.

EXAMPLE 2 | **Squaring a binomial by using algebra tiles**

Use algebra tiles to simplify $(3x + 1)^2$.

Lisa's Solution

I used tiles to create a model with a length and width of $3x + 1$.

$(3x + 1)^2$ is the area of a square with side length $3x + 1$.

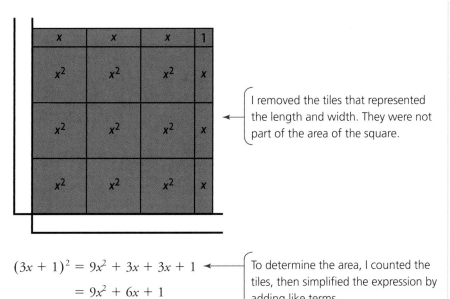

I removed the tiles that represented the length and width. They were not part of the area of the square.

$$(3x + 1)^2 = 9x^2 + 3x + 3x + 1$$
$$= 9x^2 + 6x + 1$$

To determine the area, I counted the tiles, then simplified the expression by adding like terms.

EXAMPLE 3 Determining the product of a sum and difference of two terms using algebra tiles

Dave claims that

$$(2x + 3)(2x - 3) = (2x)^2 - 3^2$$
$$= 4x^2 - 9$$

a) Confirm the relationship by evaluating each expression when $x = 2$ and $x = 3$.
b) Show that Dave is correct no matter what the value of x.

Tracy's Solution

a)

	Left Side	Right Side	
	$(2x + 3)(2x - 3)$	$4x^2 - 9$	
$x = 2$	$(2 \times 2 + 3)(2 \times 2 - 3)$ $= 7 \times 1$ $= 7$	$4 \times 2^2 - 9$ $= 7$	Values are equal.
$x = 3$	$(2 \times 3 + 3)(2 \times 3 - 3)$ $= 9 \times 3$ $= 27$	$4 \times 3^2 - 9$ $= 27$	Values are equal.

I substituted the values 2 and 3 into both sides of the equation.

It appears Dave's claim is true, since both expressions give the same result when $x = 2$ and $x = 3$. However, I can't be certain since I only showed that this worked for two cases.

b)

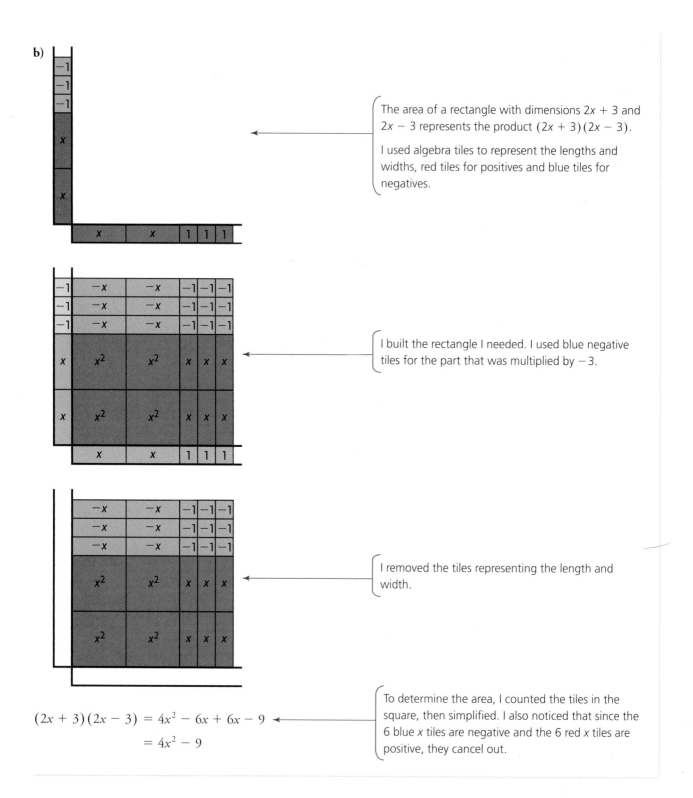

The area of a rectangle with dimensions $2x + 3$ and $2x - 3$ represents the product $(2x + 3)(2x - 3)$.

I used algebra tiles to represent the lengths and widths, red tiles for positives and blue tiles for negatives.

I built the rectangle I needed. I used blue negative tiles for the part that was multiplied by -3.

I removed the tiles representing the length and width.

$(2x + 3)(2x - 3) = 4x^2 - 6x + 6x - 9$
$= 4x^2 - 9$

To determine the area, I counted the tiles in the square, then simplified. I also noticed that since the 6 blue x tiles are negative and the 6 red x tiles are positive, they cancel out.

You can multiply algebraic expressions without using algebra tiles if you use the distributive property.

| EXAMPLE 4 | Expanding and simplifying quadratic expressions symbolically |

Expand and simplify.

a) $x(2x + 1)$

b) $5(2x - 3)^2$

c) $-2(a - 5)(3a - 2)$

d) $4n(n - 3) + (5n + 1)(3n + 2)$

Link's Solutions

a) $x(2x + 1)$

$= x \times 2x + x \times 1$

$= 2x^2 + x$

I used the distributive property to multiply each term in the brackets by x.

b) $5(2x - 3)^2$

$= 5(2x - 3)(2x - 3)$

$= 5[(2x)(2x - 3) + (-3)(2x - 3)]$

$= 5[4x^2 - 6x - 6x + 9]$

$(2x - 3)^2$ means $(2x - 3)(2x - 3)$.

I used the distributive property to expand the two binomials. I multiplied $2x$ by both $2x$ and -3, then multiplied -3 by both $2x$ and -3. I simplified by collecting like terms.

$= 5[4x^2 - 12x - 9]$

$= 20x^2 - 60x + 45$

I multiplied the product by 5, again using the distributive property.

c) $-2(a - 5)(3a - 2)$

$= -2[a(3a - 2) - 5(3a - 2)]$

$= -2(3a^2 - 2a - 15a + 10)$

I used the distributive property to find the product of the two binomials. I multiplied a by both $3a$ and -2, and -5 by both $3a$ and -2.

$= -2(3a^2 - 17a + 10)$

$= -6a^2 + 34a - 20$

Then I collected like terms and multiplied by -2, using the distributive property again.

d) $4n(n - 3) + (5n + 1)(3n + 2)$ ◄──── I used the distributive property again. I multiplied the first product and then the second one.

$$= (4n \times n - 4n \times 3) +$$
$$(5n \times 3n + 5n \times 2 + 1$$
$$\times 3n + 1 \times 2)$$

$$= (4n^2 - 12n) +$$
$$(15n^2 + 10n + 3n + 2)$$

Then I collected like terms. Since I was adding $13n$ and $-12n$, only $1n$ was left.

$$= 4n^2 - 12n + 15n^2 + 13n + 2$$ ◄────

$$= 19n^2 + 1n + 2$$

$$= 19n^2 + n + 2$$ ◄──────── I wrote $1n$ as n.

In Summary

Key Idea

- Quadratic expressions can be expanded by using the distributive property and then simplified by collecting like terms.

Need to Know

- One way to multiply two linear expressions is to use an area model with algebra tiles. If you multiply two expressions, the expressions describe the length and width of a rectangle. The area of the rectangle is the product.

- In these models, x^2 can be represented as the area of a square with side length x.

- x can be represented as the area of a rectangle with side lengths of 1 and x.

- 1 can be represented as a square with side length 1.

- Red is used to represent positive quantities, blue to represent negative quantities.

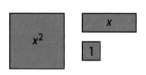

- For example, to multiply $2x(3x + 1)$, build a rectangle with a width of $2x$ and a length of $3x + 1$ and count the tiles in the area as the product.

- For the product of a monomial and a binomial, the distributive property states that

$$a(b + c) = ab + ac$$

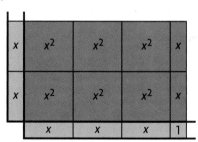

- For the product of a binomial and a binomial, apply the distributive property twice:

$$(a + b)(c + d)$$

$$= a(c + d) + b(c + d)$$
$$= ac + ad + bc + bd$$

- Like terms have the same variables with the same exponents.

- Three special multiplication patterns are

$$(a + b)^2 = (a + b)(a + b) = (a^2 + ab + ba + b^2) = a^2 + 2ab + b^2$$

$$(a - b)^2 = (a - b)(a - b) = (a^2 - ab - ba + b^2) = a^2 - 2ab + b^2$$

$$(a + b)(a - b) = (a^2 - ab + ba - b^2) = a^2 - b^2$$

CHECK *Your Understanding*

1. For each diagram, state the terms representing the length and the width of the rectangle. Then determine the product represented by the area.

   a)

1	x	x	1
1	x	x	1
1	x	x	1
x	x^2	x^2	x
	x	x	1

   b)

-1	-x	-x	-1	-1	-1
-1	-x	-x	-1	-1	-1
x	x^2	x^2	x	x	x
x	x^2	x^2	x	x	x
x	x^2	x^2	x	x	x
	x	x	1	1	1

2. Expand and simplify.

   a) $(x + 7)(x - 3)$

   b) $(a + 6)(a + 6)$

   c) $(2x - 5)^2$

   d) $(m - 9)(m + 9)$

3. Expand and simplify

   a) $3(x - 6)(x + 5)$

   b) $3a(a - 5) - (2a + 1)(a - 7)$

   c) $-2n(2n + 1) + (n + 2)^2$

   d) $3(2x + 1)^2 - 2(3x - 1)^2$

PRACTISING

4. For each diagram, describe the terms representing the length and the width of the rectangle and the product represented by the area.

a)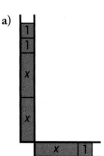

b)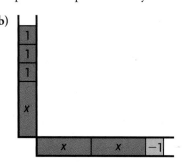

5. Expand and simplify.
   a) $(3x - 2)(4x + 5)$
   b) $5(3x + 2)^2$
   c) $2(x - 3)^2 - (4x + 1)(4x - 1)$
   d) $-a(2a + 3) + 2(a + 4)(3a + 4)$
   e) $2n(n + 3) - 3n(5 - 2n) + 7n(1 - 4n)$
   f) $(2x + 5)^2 + (2x - 5)^2 - (2x - 5)(2x + 5)$

6. a) Evaluate $3 - 2x$ for $x = -4$.
   b) Evaluate $x + 2$ for $x = -4$.
   c) Expand and simplify $(3 - 2x)(x + 2)$.
   d) Evaluate your answer from part (c) for $x = -4$.
   e) How are your answers to parts (a) and (b) related to the answer in part (d)?

7. Expand and simplify.
   K a) $2x(x - 5)$
   b) $(a + 7)(a - 9)$
   c) $(3x + 7)(6x - 5)$
   d) $2m(3m + 1) + (m - 4)(5m + 3)$

8. How do you know that the product will be quadratic when you expand an expression such as $(2x + 3)(3x - 2)$?

9. a) Sketch two different rectangles with an area of $4x^2 + 8x$.
   b) List the dimensions of each rectangle.

Communication | *Tip*

The formula for the area of
- a parallelogram is $A = bh$
- a trapezoid is
  $A = h(b_1 + b_2) \div 2$

10. Write an expression for the area of each shape. Expand and simplify.
    A a)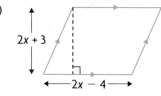

   b)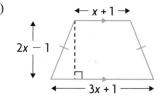

11. A circular pizza has a radius of x cm.

T
a) Write an expression for the area of the pizza.
b) Write an expression for the area of a pizza with a radius that is 5 cm greater.
c) How much greater is the second area? Write the difference as a simplified expression.

> **Communication | Tip**
> The formula for the area of a circle is $A = \pi r^2$.

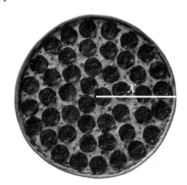

12. When you add $(6x^2 - 8x)$ and $(-15x^2 - 18)$ and collect like terms, you end up with three terms.
a) Give an example to show that the sum of two quadratic binomials could have only two terms.
b) Give an example to show that the sum of two quadratic binomials could have only one term.

13. **a)** Use an example to show that you can multiply two binomials and
C end up with three terms after you simplify.
b) Use other binomials to show that you can end up with two terms after you multiply and simplify.

Extending

14. Expand and simplify.
a) $(2x - y)(3x + y)$
b) $(3a - 5b)^2$
c) $(5m - 7n)(5m + 7n)$
d) $-2(x + 3y)(2x - y)$

15. A Pythagorean triple is a triple of natural numbers a, b, c that satisfies the Pythagorean theorem $a^2 + b^2 = c^2$, or equivalently, $a^2 = c^2 - b^2$. For example, 3, 4, 5 is a Pythagorean triple, since $3^2 + 4^2 = 5^2$, or $3^2 = 5^2 - 4^2$.
a) Show that $(n + 1)^2 - n^2 = 2n + 1$.
b) Determine n if $5^2 = 2n + 1$.
c) Use your answer from part (b) to determine a Pythagorean triple.
d) Determine three more Pythagorean triples.

2.2 Factoring Polynomials: Common Factoring

YOU WILL NEED

- graph paper

GOAL

Factor polynomials by dividing out the greatest common factor.

LEARN ABOUT the Math

Elvira squared the numbers 3, 4, and 5 and then added 1 to get a sum of 51.

$$3^2 + 4^2 + 5^2 + 1 = 9 + 16 + 25 + 1$$
$$= 51$$

She repeated this process with the numbers 8, 9, and 10 and got a sum of 246.

$$8^2 + 9^2 + 10^2 + 1 = 64 + 81 + 100 + 1$$
$$= 246$$

Both of her answers are divisible by 3.

$$\frac{51}{3} = 17 \quad \text{and} \quad \frac{246}{3} = 82$$

? Is one more than the sum of the squares of three consecutive integers always divisible by 3?

EXAMPLE 1   Selecting a strategy to determine the greatest common factor

Ariel's Solution: Using Algebra

$n^2 + (n + 1)^2 + (n + 2)^2 + 1 \longleftarrow$ I let the three consecutive integers be n, $n + 1$, and $n + 2$, where n represents the first integer. I then wrote an expression for the sum of their squares and added 1.

$= n^2 + (n + 1)(n + 1) + (n + 2)(n + 2) + 1 \longleftarrow$ I simplified the resulting expression by squaring and collecting like terms.

$= n^2 + (n^2 + 2n + 1) + (n^2 + 4n + 4) + 1$

$= 3n^2 + 6n + 6$

$= 3(n^2 + 2n + 2) \longleftarrow$ I factored the polynomial by finding the greatest common factor (GCF) of its terms: 3. I divided it out.

Three is a factor of the expression. This shows that one more than the sum of the squares of three consecutive integers is always divisible by 3.

David's Solution: Using Area Models

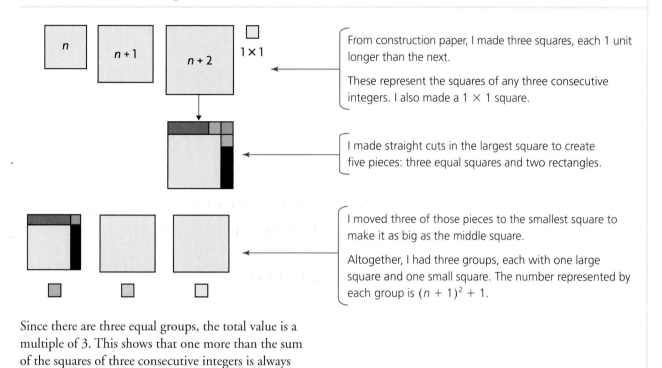

From construction paper, I made three squares, each 1 unit longer than the next.

These represent the squares of any three consecutive integers. I also made a 1 × 1 square.

I made straight cuts in the largest square to create five pieces: three equal squares and two rectangles.

I moved three of those pieces to the smallest square to make it as big as the middle square.

Altogether, I had three groups, each with one large square and one small square. The number represented by each group is $(n + 1)^2 + 1$.

Since there are three equal groups, the total value is a multiple of 3. This shows that one more than the sum of the squares of three consecutive integers is always divisible by 3.

Reflecting

A. Why did Ariel introduce a variable to solve this problem?

B. Why did Ariel want to write the expression with a factor of 3?

C. Ariel used the expression $3(n^2 + 2n + 2)$ to show that the sum was divisible by 3. David used $3[(n + 1)^2 + 1]$. Are the two expressions equivalent? Explain.

D. What are the advantages and disadvantages of each method of showing divisibility?

EXAMPLE **2** | **Selecting a strategy to represent the greatest common factor**

Factor $2x^2 - 6x$.

Chloe's Solution: Using Algebra Tiles

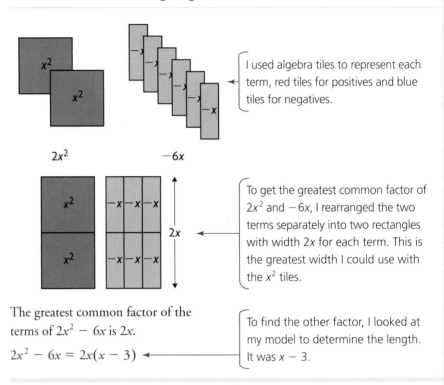

I used algebra tiles to represent each term, red tiles for positives and blue tiles for negatives.

To get the greatest common factor of $2x^2$ and $-6x$, I rearranged the two terms separately into two rectangles with width $2x$ for each term. This is the greatest width I could use with the x^2 tiles.

The greatest common factor of the terms of $2x^2 - 6x$ is $2x$.

$2x^2 - 6x = 2x(x - 3)$

To find the other factor, I looked at my model to determine the length. It was $x - 3$.

Luis's Solution: Using Symbols

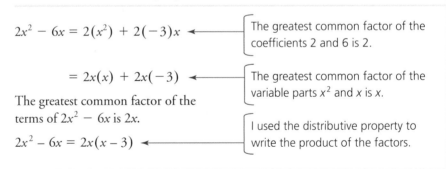

$2x^2 - 6x = 2(x^2) + 2(-3)x$

The greatest common factor of the coefficients 2 and 6 is 2.

$= 2x(x) + 2x(-3)$

The greatest common factor of the variable parts x^2 and x is x.

The greatest common factor of the terms of $2x^2 - 6x$ is $2x$.

$2x^2 - 6x = 2x(x - 3)$

I used the distributive property to write the product of the factors.

Sometimes algebraic expressions have a binomial as a common factor.

EXAMPLE 3   Using reasoning to factor

Factor $3x(x + 1) - 2(x + 1)$.

Ida's Solution

$3x(x + 1) - 2(x + 1) = (3x - 2)(x + 1)$

Both $3x$ and -2 are being multiplied by $x + 1$, so $x + 1$ is a common factor. I wrote an equivalent expression using the distributive property.

EXAMPLE 4   Solving a problem by factoring

A triangle has an area of $12x^2 + 48x$ and a height of $3x$. What is the length of its base?

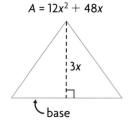

Aaron's Solution

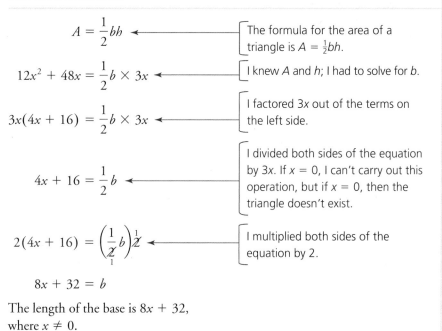

$$A = \frac{1}{2}bh$$

The formula for the area of a triangle is $A = \frac{1}{2}bh$.

$$12x^2 + 48x = \frac{1}{2}b \times 3x$$

I knew A and h; I had to solve for b.

$$3x(4x + 16) = \frac{1}{2}b \times 3x$$

I factored $3x$ out of the terms on the left side.

$$4x + 16 = \frac{1}{2}b$$

I divided both sides of the equation by $3x$. If $x = 0$, I can't carry out this operation, but if $x = 0$, then the triangle doesn't exist.

$$2(4x + 16) = \left(\frac{1}{2}b\right)2$$

I multiplied both sides of the equation by 2.

$$8x + 32 = b$$

The length of the base is $8x + 32$, where $x \neq 0$.

In Summary

Key Ideas

- Factoring algebraic expressions is the opposite of expanding. Expanding involves multiplying, while factoring involves looking for the expressions to multiply. For example:

expanding
$$\xrightarrow{\hspace{3cm}}$$
$$2x(3x - 5) = 6x^2 - 10x$$
$$\xleftarrow{\hspace{3cm}}$$
factoring

- One way to factor a polynomial is to look for the greatest common factor of its terms as one of its factors. For example, $6x^2 + 2x - 4$ can be factored as $2(3x^2 + x - 2)$, since 2 is the greatest common factor of each term.

Need to Know

- It is possible to factor a polynomial by dividing by a common factor that is not the greatest common factor. This will result in another polynomial that still has a common factor. For example:

$$4x + 8 = 2(2x + 4)$$
$$= 2(2)(x + 2)$$
$$= 4(x + 2)$$

- A polynomial is factored fully when only 1 or -1 is a common factor of every term.
- A common factor can have any number of terms. For example, a common factor of $8x^2 + 6x$ is $2x$, a monomial. But a common factor of $(2x - 1)^2 - 4(2x - 1)$ is $(2x - 1)$, a binomial.

CHECK Your Understanding

1. In each diagram, two terms of a polynomial have been rearranged to show their common factor. For each, identify the terms of the polynomial and the common factor.

a)

x^2	x^2	x^2	x	x
x^2	x^2	x^2	x	x

b)

$-x^2$	$-x^2$	$-x^2$	x	x	x	x
$-x^2$	$-x^2$	$-x^2$	x	x	x	x
$-x^2$	$-x^2$	$-x^2$	x	x	x	x
$-x^2$	$-x^2$	$-x^2$	x	x	x	x

2. Name the common factor of the terms of the polynomial.

    a) $3x^2 - 9x + 12$ **b)** $5x^2 + 3x$

3. Factor, using the greatest common factor.

    a) $4x^2 - 6x + 2$ **c)** $5a(a + 7) + 2(a + 7)$

    b) $5x^2 - 20x$ **d)** $4m(3m - 2) - (3m - 2)$

PRACTISING

4. In each diagram, two terms of a polynomial have been rearranged to show their common factor. For each, identify the terms of the polynomial and the common factor.

    a)

    b)

5. Name the greatest common factor of each polynomial.

    a) $6x^2 + 12x - 18$ **c)** $16x^2 - 8x + 10$

    b) $4x^2 + 14x$ **d)** $-15x^2 - 10$

6. Factor.

    a) $27x^2 - 9x$ **d)** $-2a^2 - 4a + 6$

    b) $-8m^2 + 20m$ **e)** $3x(x + 7) - 2(x + 7)$

    c) $10x^2 - 5x + 25$ **f)** $x(3x - 2) + (3x - 2)(x + 1)$

7. The area, A, of each figure is given. Determine the unknown measurement.

    a) $A = 18x^2 - 9x$ **b)** $A = 10m^2 - 20m + 20$

8. The formula for the surface area of a cylinder is $SA = 2\pi r^2 + 2\pi rh$. A cylinder has a height of 10 units and a radius of r units. Determine a factored expression for its total surface area.

9. Show that $(3a - 2)$ is a common factor of $2a(3a - 2) + 7(2 - 3a)$.

10. a) Give three examples of a quadratic binomial with greatest common factor $6x$, and then factor each one.

b) Give three examples of a quadratic trinomial with greatest common factor 7, and then factor each one.

11. Colin says that the greatest common factor of $-8x^2 + 4x - 6$ is 2, but Colleen says that it is -2. Explain why both answers could be considered correct.

12. For what values of k is it possible to divide out a common factor from $6x^2 + kx - 12$, but not from $6x^2 + kx + 4$? Explain.

13. It seems that the sum of the squares of two consecutive even or odd integers is always even. For example:

$$4^2 + 6^2 = 16 + 36$$
$$= 52$$
$$7^2 + 9^2 = 49 + 81$$
$$= 130$$

Let n represent the first integer.

a) What expression represents the second integer?

b) What expression represents the sum of the two squares?

c) Use algebra to show that the result is always even.

14. How does knowing how to determine the greatest common factor of two numbers help you factor polynomials?

Extending

15. Factor.

a) $5x^2y - 10xy^2$

b) $10a^2b^3 + 20a^2b - 15a^2b^2$

c) $3x(x + y) - y(x + y)$

d) $5y(x - 2) - 7(2 - x)$

16. A factor might be common to only some terms of a polynomial, but grouping these terms sometimes allows the polynomial to be factored. For example:

$ax - ay + bx - by$
$= a(x - y) + b(x - y)$ ← Now you see the common factor $(x - y)$.
$= (x - y)(a + b)$

Factor these expressions by grouping.

a) $9xa + 3xb + 6a + 2b$

b) $10x^2 - 5x - 6xy + 3y$

c) $(x + y)^2 + x + y$

d) $1 + xy + x + y$

2.3 Factoring Quadratic Expressions: $x^2 + bx + c$

GOAL

Factor quadratic expressions of the form $ax^2 + bx + c$, where $a = 1$.

YOU WILL NEED

• algebra tiles

INVESTIGATE the Math

Seth claims that for any whole number n, the function $f(n) = n^2 + 8n + 15$ always produces a number that has factors other than 1 and itself.

He tested several examples:

$f(1) = 24 = 4 \times 6$, so $f(1)$ is not prime.

$f(2) = 35 = 5 \times 7$, so $f(2)$ is not prime.

$f(3) = 48 = 6 \times 8$, so $f(3)$ is not prime.

$f(4) = 63 = 7 \times 9$, so $f(4)$ is not prime.

? **Is Seth's claim true?**

A. Identify a pattern relating the value of n to the two factors of $f(n)$.

B. Use this pattern to factor $f(n)$; verify Seth's claim by expanding.

C. Arrange an x^2 tile, 8 x tiles, and 15 unit tiles to form a rectangle.

D. What are the dimensions of the rectangle?

E. Is Seth's claim true? Explain.

Reflecting

F. What are the advantages and disadvantages of the two methods of factoring (using patterns in parts A and B, and using algebra tiles in parts C and D)?

APPLY the Math

EXAMPLE 1	Factoring using algebra tiles

Factor $x^2 + 7x + 12$.

Andy's Solution

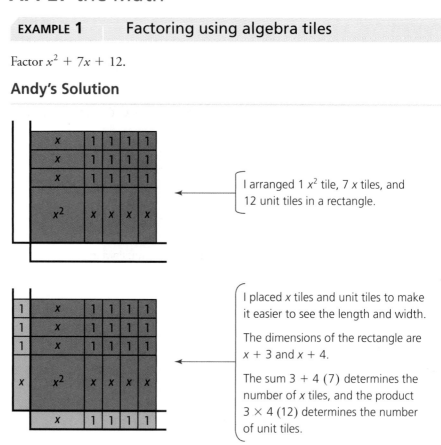

I arranged 1 x^2 tile, 7 x tiles, and 12 unit tiles in a rectangle.

I placed x tiles and unit tiles to make it easier to see the length and width.

The dimensions of the rectangle are $x + 3$ and $x + 4$.

The sum 3 + 4 (7) determines the number of x tiles, and the product 3×4 (12) determines the number of unit tiles.

$$x^2 + 7x + 12 = (x + 3)(x + 4)$$

EXAMPLE 2 Factoring by using the sum-and-product method

Factor $x^2 + 4x - 5$.

Yusef's Solution

$x^2 + 4x - 5 = (x \ ?)(x \ ?)$ ◄——————

The two factors for the quadratic could be binomials that start with x.

I needed two numbers whose sum is 4 and whose product is -5.

$= (x + ?)(x - ?)$ ◄——————

I started with the product. Since -5 is negative, one of the numbers must be negative.

Since the sum is positive, the positive number must be farther from zero than the negative one.

$= (x + 5)(x - 1)$ ◄—— The numbers are 5 and -1.

Check: ◄—————— I checked by multiplying.

$(x + 5)(x - 1) = x^2 - 1x + 5x - 5$
$= x^2 + 4x - 5$

EXAMPLE 3 Factoring quadratic expressions

Factor each expression.
a) $x^2 + 7x + 12$ **b)** $a^2 - 10a + 21$ **c)** $x^2 + 4x + 5$

Ryan's Solution

a) $x^2 + 7x + 12$ ◄——————
$= (x + 4)(x + 3)$

I needed two numbers whose sum is 7 and whose product is 12.

The product is positive, so either both numbers are negative or both are positive.

Since the sum is positive, both numbers must be positive.

The numbers are 4 and 3.

b) $a^2 - 10a + 21$ ←

$$= (a - 3)(a - 7)$$

I needed two numbers whose sum is -10 and whose product is 21.

The product is positive, so either both numbers are negative or both are positive.

Since the sum is negative, both numbers are negative.

The numbers are -3 and -7.

c) $x^2 + 4x + 5$ ←

Cannot be factored

I needed two numbers whose sum is 4 and whose product is 5.

The product is positive, so either both numbers are negative or both are positive.

Since the sum is positive, both numbers are positive.

There are no such numbers, because the only factors of 5 are 1 and 5. But their sum is 6, not 4.

EXAMPLE 4 | **Factoring quadratic expressions by first using a common factor**

Factor $4x^2 + 16x - 48$.

Larry's Solution

$4x^2 + 16x - 48$ ←

First, I divided out the greatest common factor, 4, since all terms are divisible by 4.

$$= 4(x^2 + 4x - 12)$$ ←
$$= 4(x - 2)(x + 6)$$

To factor the trinomial, I needed two numbers whose sum is 4 and whose product is -12.

The numbers are -2 and 6.

In Summary

Key Idea

- If they can be factored, quadratic expressions of the form $x^2 + bx + c$ can be factored into two binomials $(x + r)(x + s)$, where $r + s = b$ and $r \times s = c$, and r and s are integers.

Need to Know

- To factor $x^2 + bx + c$ as $(x + r)(x + s)$, the signs in the trinomial can help you determine the signs of the numbers you are looking for:

Trinomial	Factors
$x^2 + bx + c$	$(x + r)(x + s)$
$x^2 - bx + c$	$(x - r)(x - s)$
$x^2 - bx - c$	$(x - r)(x + s)$, where $r > s$
$x^2 + bx - c$	$(x + r)(x - s)$, where $r > s$

- To factor $x^2 + bx + c$ by using algebra tiles, form a rectangle from the tiles. The factors are given by the dimensions of the rectangle.
- It is easier to factor any polynomial expression if you factor out the greatest common factor first.

CHECK *Your Understanding*

1. The tiles in each diagram represent a polynomial. Identify the polynomial and its factors.

   a)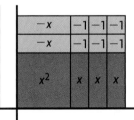

   b)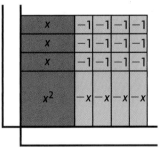

2. One factor is provided and one is missing. What is the missing factor?
   a) $x^2 + 11x + 30 = (x + 6)(?)$
   b) $x^2 - 3x - 28 = (x - 7)(?)$
   c) $x^2 + 3x - 40 = (?)(x - 5)$
   d) $x^2 - 8x - 20 = (?)(x - 10)$

3. Factor.
   a) $x^2 + 9x + 20$
   c) $m^2 + m - 6$
   b) $a^2 - 11a + 30$
   d) $2n^2 - 4n - 70$

PRACTISING

4. The tiles in each diagram represent a polynomial. Identify the polynomial and its factors.

a)
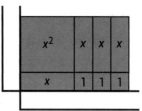

b)
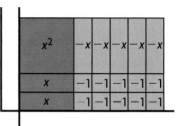

5. One factor is provided and one is missing. What is the missing factor?
   a) $x^2 + 10x + 24 = (x + 4)(?)$
   b) $x^2 - 13x + 42 = (x - 6)(?)$
   c) $x^2 - 3x - 40 = (?)(x - 8)$
   d) $x^2 + 6x - 27 = (?)(x + 9)$

6. Factor.
   a) $x^2 - 7x + 10$ c) $x^2 - 3x - 10$ e) $x^2 - 14x + 33$
   b) $y^2 + 6y - 55$ d) $w^2 - 3w - 18$ f) $n^2 - n - 90$

7. Write three different quadratic trinomials that have $(x + 5)$ as a factor.

8. Tony factored $x^2 - 7x + 10$ as $(x - 2)(x - 5)$. Fred factored it as $(2 - x)(5 - x)$. Why are both answers correct?

9. Factor.
   a) $a^2 + 7a + 10$ c) $z^2 - 10z + 25$ e) $x^2 + 3x - 10$
   b) $-3x^2 - 27x - 54$ d) $x^2 + 11x - 60$ f) $y^2 + 13y + 42$

10. Explain why the function $f(n) = n^2 - 2n - 3$ results in a prime number when $n = 4$, but not when n is any integer greater than 4.

11. How can writing $x^2 - 16$ as $x^2 + 0x - 16$ help you factor it?

12. Choose a pair of integers for b and c that will make each statement true.
    a) $x^2 + bx + c$ can be factored, but $x^2 + cx + b$ cannot.
    b) Both $x^2 + bx + c$ and $x^2 + cx + b$ can be factored.
    c) Neither $x^2 + bx + c$ nor $x^2 + cx + b$ can be factored.

13. For what values of k can the polynomial be factored? Explain.
    a) $x^2 + kx + 4$ b) $x^2 + 4x + k$ c) $x^2 + kx + k$

Extending

14. Factor.
    a) $x^2 + 3xy - 10y^2$ c) $-5m^2 + 15mn - 10n^2$
    b) $a^2 + 4ab + 3b^2$ d) $(x + y)^2 - 5(x + y) + 6$

15. Factor $x + 7 + \dfrac{12}{x}$.

FREQUENTLY ASKED *Questions*

Q: **How do you simplify quadratic expressions?**

A: You can simplify by expanding and collecting like terms. For example, $2x(3 + x) + 4x + 2$ can be simplified by first expanding using the distributive property:

$$2x(3 + x) = 2x \times 3 + 2x \times x$$
$$= 6x + 2x^2$$

Another way to expand is to determine the area of a rectangle with length and width based on the factors. For example, $2x(3 + x)$ is the area of the rectangle shown at the right.

After expanding, collect like terms.

$$2x(3 + x) + 4x + 2 = 6x + 2x^2 + 4x + 2$$
$$= 2x^2 + 10x + 2$$

Study **Aid**
• See Lesson 2.1, Examples 1 to 4.
• Try Mid-Chapter Review Questions 1 to 4.

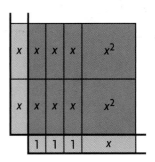

Q: **How do you use the greatest common factor of the terms of a polynomial to factor it?**

A1: You can represent the terms with algebra tiles and rearrange them into rectangles with the same and greatest possible width. That width is the greatest common factor.

For example, the greatest common factor of the terms of $6x^2 - 9x$ is $3x$, since each term can be rearranged into rectangles with width $3x$, which is the largest possible width.

Study **Aid**
• See Lesson 2.2, Examples 1 to 4.
• Try Mid-Chapter Review Questions 5, 6, and 7.

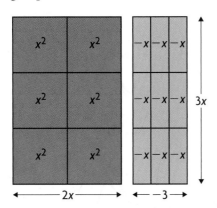

Once you divide out the common factor, the remaining terms represent the other dimension of each rectangle.

$$6x^2 - 9x = 3x(2x - 3)$$

A2: You can determine the greatest common factor of the coefficients and of the variables, and then multiply them together.

For example, for $6x^2 - 9x$, the GCF of 6 and 9 is 3. The GCF of x^2 and x is x. GCF of $6x^2 - 9x$ is $3x$.

Then you can divide out the common factor:

$$6x^2 - 9x = 3x(2x - 3)$$

Study | *Aid*

• See Lesson 2.3, Examples 1 to 4.
• Try Mid-Chapter Review Questions 8 to 11.

Q: How can you factor quadratic expressions of the form $x^2 + bx + cx$?

A1: You can form a rectangle using algebra tiles. The length and width are the factors.

For example, to factor $x^2 + 8x + 12$:

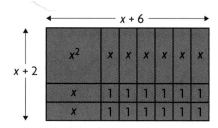

$$x^2 + 8x + 12 = (x + 6)(x + 2)$$

A2: Look for two numbers with a sum of b and a product of c and use them to factor.

For example, to factor $x^2 + 3x - 18$:
• Numbers whose product is -18 are 6 and -3, 9 and -2, 1 and -18, -6 and 3, -9 and 2, and -1 and 18.
• The only pair that adds to $+3$ is 6 and -3.

The factors are $(x + 6)(x - 3)$.

PRACTICE Questions

Lesson 2.1

1. Expand and simplify.
   a) $2x(x - 6) - 3(2x - 5)$
   b) $(3n - 2)^2 + (3n + 2)^2$
   c) $3x(2x - 1) - 4x(3x + 2) - (-x^2 + 4x)$
   d) $-2(3a + b)(3a - b)$

2. The diagram below represents a polynomial multiplication. Which two polynomials are being multiplied and what is the product?

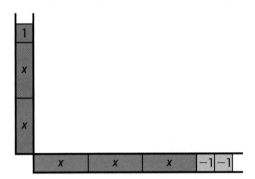

3. Write a simplified expression to represent the area of the triangle shown.

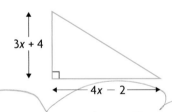

4. A rectangle has dimensions $2x + 1$ and $3x - 2$, where $x > 0$. Determine the increase in its area if each dimension is increased by 1.

Lesson 2.2

5. Factor.
   a) $-8x^2 + 4x$
   b) $3x^2 - 6x + 9$
   c) $5m^2 - 10m - 5$
   d) $3x(2x - 1) + 5(2x - 1)$

6. The tiles shown represent the terms of a polynomial. Identify the polynomial and the common factor of its terms.

$-x^2$	$-x^2$	$-x^2$	$-x^2$	x	x
$-x^2$	$-x^2$	$-x^2$	$-x^2$	x	x

7. Consider the binomials $2x + 4$ and $3x + 6$. The greatest common factor of the first pair of terms is 2 and of the second pair is 3.
   a) Determine the product of the polynomials.
   b) Is the greatest common factor of the terms of their product equal to the product of 2 and 3?
   c) Why might you have expected the answer you got in part (b)?

Lesson 2.3

8. The tiles shown represent a polynomial. Identify the polynomial and its factors.

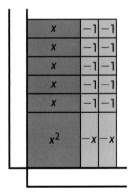

9. Factor.
   a) $x^2 + 2x - 15$ c) $x^2 - 12x + 35$
   b) $n^2 - 8n + 12$ d) $2a^2 - 2a - 24$

10. How do you know that $(x - 4)$ can't be a factor of $x^2 - 18x + 6$?

11. If $x^2 + bx + c$ can be factored, then can $x^2 - bx + c$ be factored? Explain.

2.4 Factoring Quadratic Expressions: $ax^2 + bx + c$

Factor quadratic expressions of the form $ax^2 + bx + c$, where $a \neq 1$.

Martina is asked to factor the expression $3x^2 + 14x + 8$. She is unsure of what to do because the first term of the expression has the coefficient 3, and she hasn't worked with these kinds of polynomials before.

? How do you factor $3x^2 + 14x + 8$?

LEARN ABOUT the Math

EXAMPLE 1	Selecting a strategy to factor a trinomial where $a \neq 1$

Llewelyn's Solution: Using Algebra Tiles

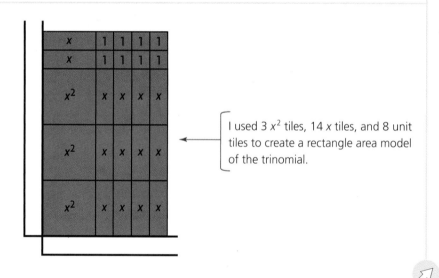

I used 3 x^2 tiles, 14 x tiles, and 8 unit tiles to create a rectangle area model of the trinomial.

I placed the tiles along the length and width of the rectangle to read off the factors. The length was $3x + 2$ and the width was $x + 4$.

$$3x^2 + 14x + 8 = (3x + 2)(x + 4)$$

Albert's Solution: Using Guess and Check

$$3x^2 + 14x + 8 = (x + ?)(3x + ?)$$

To get $3x^2$, I had to multiply x by $3x$, so I set up the equation to show the factors.

$$(x + 8)(3x + 1) = 3x^2 + 25x + 8$$
WRONG FACTORS
$$(x + 1)(3x + 8) = 3x^2 + 11x + 8$$
WRONG FACTORS
$$(x + 4)(3x + 2) = 3x^2 + 14x + 8$$
WORKED!
$$3x^2 + 14x + 8 = (x + 4)(3x + 2)$$

I had to multiply two numbers together to get 8.
The ways to get a product of 8 are 8×1, 4×2, $-8 \times (-1)$, and $-4 \times (-2)$. Since the middle term was positive, I tried the positive values only.
I used guess and check to see which pair of numbers worked.

Murray's Solution: Using Decomposition

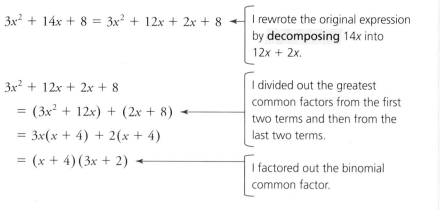

decomposing

breaking a number or expression into parts that make it up

$3x^2 + 14x + 8 = 3x^2 + 12x + 2x + 8$ ◄── I rewrote the original expression by **decomposing** $14x$ into $12x + 2x$.

$3x^2 + 12x + 2x + 8$

$= (3x^2 + 12x) + (2x + 8)$ ◄── I divided out the greatest common factors from the first two terms and then from the last two terms.

$= 3x(x + 4) + 2(x + 4)$

$= (x + 4)(3x + 2)$ ◄── I factored out the binomial common factor.

Reflecting

A. Why did Llewlyn try to form a rectangle with the algebra tiles?

B. What was Murray's goal when he "decomposed" the coefficient of the x-term?

C. How was Albert's factoring method similar to the sum-and-product method?

D. What are the advantages and disadvantages of each of the three factoring methods?

EXAMPLE 2   Factoring a trinomial by using algebra tiles

Use algebra tiles to factor $6x^2 - 7x - 3$.

Ariel's Solution

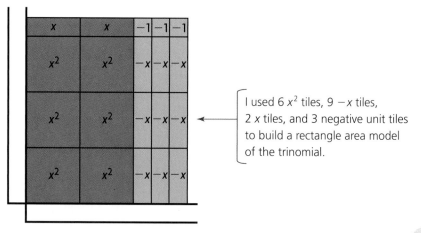

I used 6 x^2 tiles, 9 $-x$ tiles, 2 x tiles, and 3 negative unit tiles to build a rectangle area model of the trinomial.

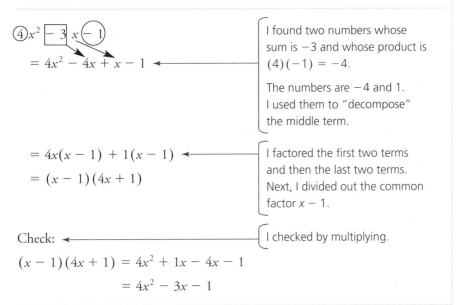

$$6x^2 - 7x - 3 = (3x + 1)(2x - 3)$$

EXAMPLE 3   Factoring a trinomial by decomposition

Factor $4x^2 - 3x - 1$.

Florian's Solution

$\textcircled{4}x^2 \boxed{-3} x \enclose{circle}{-1}$

$= 4x^2 - 4x + x - 1$ I found two numbers whose sum is -3 and whose product is $(4)(-1) = -4$.

The numbers are -4 and 1. I used them to "decompose" the middle term.

$= 4x(x - 1) + 1(x - 1)$ I factored the first two terms and then the last two terms. Next, I divided out the common factor $x - 1$.

$= (x - 1)(4x + 1)$

Check: I checked by multiplying.

$(x - 1)(4x + 1) = 4x^2 + 1x - 4x - 1$

$= 4x^2 - 3x - 1$

EXAMPLE 4 Factoring a trinomial by guess and check

Factor $3x^2 - 5x + 2$.

Nadia's Solution

$3x^2 - 5x + 2$

$(3x - 1)(x - 2) = 3x^2 - 7x + 2$

WRONG FACTORS

$(3x - 2)(x - 1) = 3x^2 - 5x + 2$

WORKED

$$3x^2 - 5x + 2 = (3x - 2)(x - 1)$$

The factors of $3x^2$ are $3x$ and x.

The factors of 2 are 2 and 1 or -2 and -1.

I tried the negative values, since the middle term was negative.

I used guess and check to place the values in each set of brackets.

In Summary

Key Idea

- If the quadratic expression $ax^2 + bx + c$, where $a \neq 1$ can be factored, then the factors are of the form $(px + r)(qx + s)$, where $pq = a$, $rs = c$, and $ps + rq = b$.

Need to Know

- If the quadratic expression $ax^2 + bx + c$, where $a \neq 1$ can be factored, then the factors can be found by a variety of strategies, such as
  - forming a rectangle with algebra tiles
  - decomposition
  - guess and check
- A trinomial of the form $ax^2 + bx + c$ can be factored if two integers can be found whose product is ac and whose sum is b.
- The decomposition method involves decomposing b into a sum of two numbers whose product is ac.

CHECK *Your Understanding*

1. Each diagram represents a polynomial. Identify the polynomial and its factors.

a)

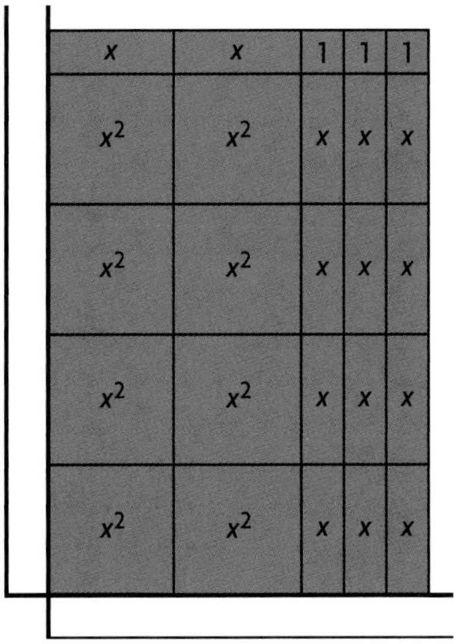

b)

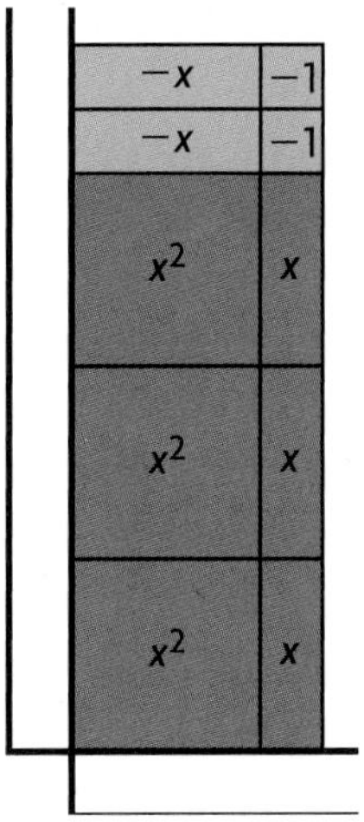

2. State the missing factor.

 a) $6a^2 + 5a - 4 = (3a + 4)(?)$
 b) $5x^2 - 22x + 8 = (?)(x - 4)$
 c) $3x^2 + 7x + 2 = (3x + 1)(?)$
 d) $4n^2 + 8n - 60 = (?)(n + 5)$

PRACTISING

3. The diagram below represents a polynomial. Identify the polynomial and its factors.

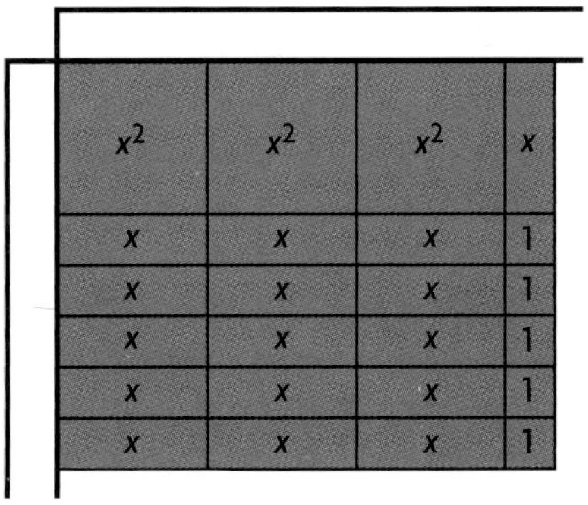

4. Factor. You may first need to determine a common factor.
   a) $2x^2 - 7x - 4$
   b) $3x^2 + 18x + 15$
   c) $5x^2 + 17x + 6$
   d) $2x^2 + 10x + 8$
   e) $3x^2 + 12x - 63$
   f) $2x^2 - 15x + 7$

5. Factor.
   a) $8x^2 + 10x + 3$
   b) $6m^2 - 3m - 3$
   c) $2a^2 - 11a + 12$
   d) $15x^2 - 4x - 4$
   e) $6n^2 + 26n - 20$
   f) $16x^2 + 4x - 6$

6. Write three different quadratic trinomials that have $(2x - 5)$ as a factor.

7. For each expression, name an integer, k, such that the quadratic trinomial can be factored.
   a) $kx^2 + 4x + 1$
   b) $4x^2 + kx - 10$
   c) $8x^2 - 14x + k$

8. Can the guess and check method for factoring trinomials of the form $ax^2 + bx + c$, where $a \neq 1$, be applied when $a = 1$? Explain.

9. Factor.
K
   a) $6x^2 - x - 12$
   b) $8k^2 + 43k + 15$
   c) $30r^2 - 85r - 70$
   d) $12n^3 - 75n^2 + 108n$
   e) $3k^2 - 6k - 24$
   f) $24y^2 - 10y - 25$

10. Factor.
   a) $x^2 + 5x + 6$
   b) $x^2 - 36$
   c) $5a^2 - 13a - 6$
   d) $a^2 - a - 12$
   e) $4x^2 + 16x - 48$
   f) $6x^2 + 7x - 3$

11. Is there an integer, n, such that $6n^2 + 10n + 4$ is divisible by 50?
T Explain.

12. How does knowing that factoring is the opposite of expanding help
C you factor a polynomial such as $-4x^2 + 38x - 48$?

Extending

13. Factor.
   a) $6x^2 + 11xy + 3y^2$
   b) $5a^2 - 7ab - 6b^2$
   c) $8x^2 - 14xy + 3y^2$
   d) $12a^2 + 52a - 40$

14. Can $ax^2 + bx + c$ be factored if a, b, and c are odd? Explain.

2.5 Factoring Quadratic Expressions: Special Cases

Factor perfect-square trinomials and differences of squares.

LEARN ABOUT the Math

The area of a square is given by $A(x) = 9x^2 + 12x + 4$, where x is a natural number.

? What are the dimensions of the square?

EXAMPLE 1 | Factoring a perfect-square trinomial

Factor $9x^2 + 12x + 4$.

Fred's Solution: Using Algebra Tiles

x	x	x	1	1
x	x	x	1	1
x^2	x^2	x^2	x	x
x^2	x^2	x^2	x	x
x^2	x^2	x^2	x	x

I used 9 x^2 tiles, 12 x tiles, and 4 unit tiles to create an area model of the trinomial.

The only arrangement of tiles that seemed to work was a square.

I lined up tiles along two edges to determine the length (or width) of the square.

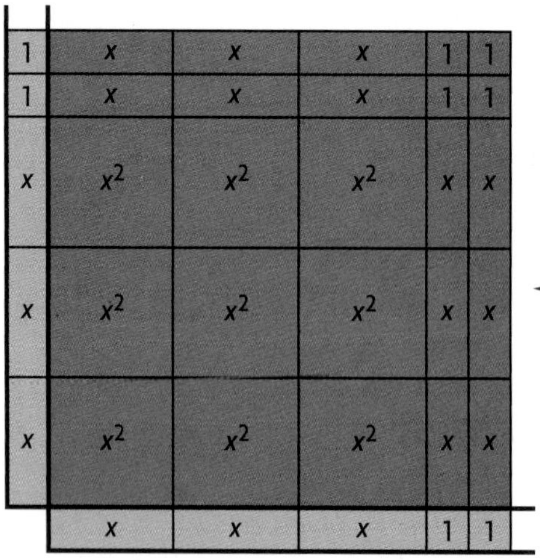

Each edge is $(3x + 2)$ long.
$9x^2 + 12x + 4 = (3x + 2)^2$

Haroun's Solution: Using a Pattern

$A(x) = 9x^2 + 12x + 4$

$9 = 3^2$

$4 = 2^2$

9 and 4 are both perfect squares.

Maybe $9x^2 + 12x + 4 = (3x + 2)^2$.

I tested to see if the polynomial was a perfect square.

$(3x + 2)^2 = (3x + 2)(3x + 2)$

$= 9x^2 + 6x + 6x + 4$

$= 9x^2 + 12x + 4$

WORKED!

This worked, since I got the correct coefficient of x.

$9x^2 + 12x + 4 = (3x + 2)^2$

Reflecting

A. Why is $9x^2 + 6x + 4$ not a perfect square?

B. How can you recognize a trinomial that might be a perfect square? Explain.

APPLY the Math

EXAMPLE 2 | Factoring a difference of squares by forming a square

Factor $4x^2 - 9$.

Lisa's Solution

I started with $4\,x^2$ tiles and 9 negative unit tiles. I made a square out of each term. When using algebra tiles to factor, the ones tiles are always diagonal from the x^2 tiles, so I put them that way.

The rectangle was not complete, so I added 6 horizontal $-x$ tiles and 6 vertical x tiles. This is adding zero to the polynomial, because $6x + (-6x) = 0x$.

The rectangle turned out to be a square.

The resulting dimensions, $2x - 3$ and $2x + 3$, are the factors of $4x^2 - 9$.

$$4x^2 - 9 = (2x - 3)(2x + 3)$$

EXAMPLE **3**

Factoring a perfect-square trinomial by decomposition

Factor $16x^2 + 24x + 9$.

Tolbert's Solution

$16 \times 9 = 144$

$24 = 12 + 12$

and $\quad 12 \times 12 = 144$

> I needed to decompose 24, the coefficient of x, as the sum of two numbers whose product is 144.

$16x^2 + 12x + 12x + 9$

> I wrote the x-term in decomposed form.

$= 4x(4x + 3) + 3(4x + 3)$

> I divided out the GCF from the first two terms and the GCF from the last two terms.

$= (4x + 3)(4x + 3)$

$= (4x + 3)^2$

> Then I divided out the binomial common factor.

Check:

> I checked by multiplying.

$(4x + 3)^2 = (4x + 3)(4x + 3)$

$\qquad = 16x^2 + 12x + 12x + 9$

$\qquad = 16x^2 + 24x + 9$

$16x^2 + 24x + 9 = (4x + 3)^2$

In Summary

Key Ideas

- A polynomial of the form $a^2x^2 \pm 2abx + b^2$ is a perfect-square trinomial and can be factored as $(ax \pm b)^2$.
- A polynomial of the form $a^2x^2 - b^2$ is a difference of squares and can be factored as $(ax - b)(ax + b)$.

Need to Know

- A perfect-square trinomial and a difference of squares can be factored by
  - forming a square using algebra tiles
  - decomposition
  - guess and check or sum and product

CHECK *Your Understanding*

1. Each diagram represents a polynomial. Identify the polynomial and its factors.

a)

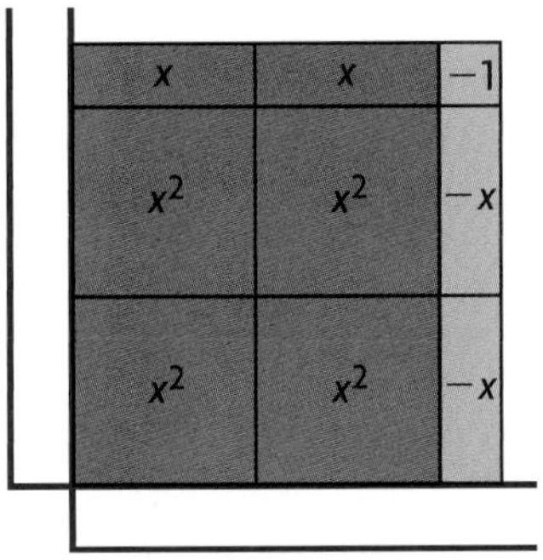

b)

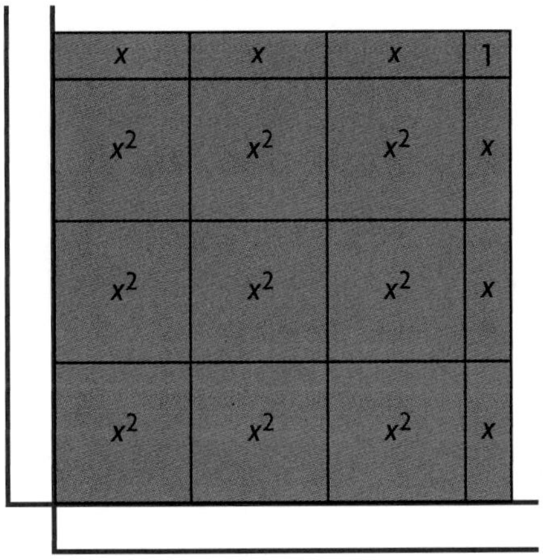

2. State the missing factor.
   a) $x^2 - 25 = (x + 5)(?)$
   b) $n^2 + 8n + 16 = (?)(n + 4)$
   c) $25a^2 - 36 = (?)(5a - 6)$
   d) $28x^2 - 7 = (?)(2x - 1)(2x + 1)$
   e) $4m^2 - 12m + 9 = (?)^2$
   f) $18x^2 + 12x + 2 = 2(?)^2$

PRACTISING

3. Factor.
   a) $x^2 - 36$
   b) $x^2 + 10x + 25$
   c) $x^2 - 64$
   d) $x^2 - 24x + 144$
   e) $x^2 - 100$
   f) $x^2 + 4x + 4$

4. Factor, if possible.
   a) $49a^2 + 42a + 9$
   b) $x^2 - 121$
   c) $-8x^2 + 24x - 18$
   d) $20a^2 - 180$
   e) $16 - 36x^2$
   f) $(x + 1)^2 + 4(x + 1) + 4$

5. Some shortcuts in mental arithmetic are based on factoring.

A For example, $21^2 - 19^2$ can be easily calculated mentally with the difference-of-squares method.

$$21^2 - 19^2 = (21 - 19)(21 + 19)$$
$$= 2(40) = 80$$

Calculate mentally.

a) $52^2 - 48^2$ b) $34^2 - 24^2$

6. Explain how you know that $5x^2 + 20x + 9$ cannot be factored in the

K indicated way.

a) as $(ax + b)^2$ b) as $(ax + b)(ax - b)$

7. Factor $x^4 - 13x^2 + 36$.

8. Determine all integers, m and n, such that $m^2 - n^2 = 24$.

T

9. Fred claims that the difference between the squares of any two consecutive odd numbers is 4 times their median. For example,

$$9^2 - 7^2 = 81 - 49$$
$$= 32 = 4(8)$$
$$15^2 - 13^2 = 225 - 169$$
$$= 56 = 4(14)$$

Use variables to explain why Fred is correct.

10. Explain how you would recognize a polynomial as a perfect-square

C trinomial or as a difference of squares. Then explain how you would factor each one.

Extending

11. Factor.

a) $100x^2 - 9y^2$ c) $(2x - y)^2 - 9$

b) $4x^2 + 4xy + y^2$ d) $90x^2 - 120xy + 40y^2$

12. The method of grouping can sometimes be applied to factor polynomials that contain a perfect-square trinomial. For example:

$x^2 + 6x + 9 - y^2$ ← The first three terms form a perfect-square trinomial, which is factored.

$= (x + 3)^2 - y^2$

$= (x + 3 - y)(x + 3 + y)$ ← This produces a difference of squares, which is factored to complete the factoring.

Factor.

a) $4x^2 - 20xy + 25y^2 - 4z^2$ b) $81 - x^2 + 14x - 49$

Factoring Using Number Patterns

Many people are more comfortable working with numbers than algebraic expressions. Number patterns can be used to help factor algebraic expressions.

For example, to factor $4x^2 - 1$, you can substitute numbers for x in a systematic way and try to identify a pattern.

$$
\begin{aligned}
\text{Let} \quad x = 1 &\rightarrow 4(1)^2 - 1 = 3 & \rightarrow & \quad 1 \times 3 \\
x = 2 &\rightarrow 4(2)^2 - 1 = 15 & \rightarrow & \quad 3 \times 5 \\
x = 3 &\rightarrow 4(3)^2 - 1 = 35 & \rightarrow & \quad 5 \times 7 \\
x = 4 &\rightarrow 4(4)^2 - 1 = 63 & \rightarrow & \quad 7 \times 9 \\
x = 5 &\rightarrow 4(5)^2 - 1 = 99 & \rightarrow & \quad 9 \times 11
\end{aligned}
$$

One possible pattern you might observe is that the factors are two apart from each other. Also, the factors are 1 greater and 1 less than double the number that was substituted. So, $(2x + 1)(2x - 1)$ are the factors.

Can you see any other patterns that might help you factor this expression?

Here is another example. To factor $9x^2 + 6x + 1$:

$$
\begin{aligned}
\text{Let} \quad x = 1 &\rightarrow 9(1)^2 + 6(1) + 1 = 9 + 6 + 1 = 16 & \rightarrow & \; 4 \times 4 \\
x = 2 &\rightarrow 9(2)^2 + 6(2) + 1 = 36 + 12 + 1 = 49 & \rightarrow & \; 7 \times 7 \\
x = 3 &\rightarrow 9(3)^2 + 6(3) + 1 = 81 + 18 + 1 = 100 & \rightarrow & \; 10 \times 10 \\
x = 4 &\rightarrow 9(4)^2 + 6(4) + 1 = 144 + 24 + 1 = 169 & \rightarrow & \; 13 \times 13 \\
x = 5 &\rightarrow 9(5)^2 + 6(5) + 1 = 225 + 30 + 1 = 256 & \rightarrow & \; 16 \times 16
\end{aligned}
$$

Here the pattern shows the factors are identical, so this must be a perfect square. Each factor is 1 greater than 3 times the number substituted. So, $(3x + 1)(3x + 1)$ are the factors.

1. Use number patterns to factor each expression.

a) $x^2 + 3x + 2$ d) $4x^2 + 10x + 6$

b) $2x^2 + 3x - 2$ e) $9x^2 - 1$

c) $4x^2 + 4x + 1$ f) $x^2 + 10x + 25$

FREQUENTLY ASKED Questions

Study Aid

- See Lesson 2.4, Examples 1 to 4.
- Try Chapter Review Questions 11, 12, and 13.

Q: How can you factor quadratic expressions of the form $ax^2 + bx + c$, where $a \neq 1$?

A: Some polynomials of this form can be factored, but others cannot. Try to factor the expression by using one of the methods below. If none of them works, the polynomial cannot be factored.

Method 1:

Arrange algebra tiles to form a rectangle, and read off the length and width as the factors. For example, for $6x^2 + 7x - 3$, the rectangle wasn't filled, so add $+2x + (-2x)$ to make it work.

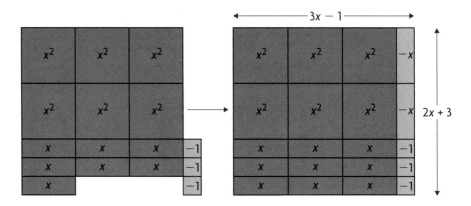

$$6x^2 + 7x - 3 = (3x - 1)(2x + 3)$$

Method 2:

Use guess and check.

If there are factors, the factored expression looks like $(\_\_x + \_\_)(\_\_x - \_\_)$, since the constant is -3.

The coefficients of x must multiply to 6, so they could be 6 and 1, -6 and -1, 3 and 2, or -3 and -2.

The constants must multiply to -3. They could be 1 and -3, or -3 and 1.

Try different combinations until one works.

$$6x^2 + 7x - 3 = (3x - 1)(2x + 3)$$

Method 3:

Use decomposition. Decompose $+7$ as the sum of two numbers that multiply to $-3 \times 6 = -18$. Use -2 and 9.

$$6x^2 + 7x - 3 = 6x^2 - 2x + 9x - 3$$
$$= 2x(3x - 1) + 3(3x - 1)$$
$$= (3x - 1)(2x + 3)$$

Q: **How do you recognize a perfect-square trinomial and how do you factor it?**

A: If the coefficient of x^2 is a perfect square and the constant is a perfect square, test to see if the middle coefficient is twice the product of the two square roots.

For example, $16x^2 + 40x + 25$ is a perfect square, since $16 = 4^2$, $25 = 5^2$, and $40 = 2 \times 4 \times 5$. So, $16x^2 + 40x + 25 = (4x + 5)^2$.

Q: **How do you recognize a difference of squares and how do you factor it?**

A: If a polynomial is made up of two perfect-square terms that are subtracted, it can be factored.

For example, $36x^2 - 49$ is a difference of squares, since $36x^2 = (6x)^2$ and $49 = 7^2$.

So, $36x^2 - 49 = (6x + 7)(6x - 7)$.

Study *Aid*

• See Lesson 2.5, Examples 1, 2, and 3.
• Try Chapter Review Questions 14 to 19.

PRACTICE Questions

Lesson 2.1

1. Expand and simplify.
   a) $3x(2x - 3) + 9(x - 1) - x(-x - 11)$
   b) $-9(4a - 5)(4a + 5)$
   c) $2(x^2 - 5) - 7x(8x - 9)$
   d) $-5(2n - 5)^2$

2. Which two polynomials are being multiplied and what is the product?

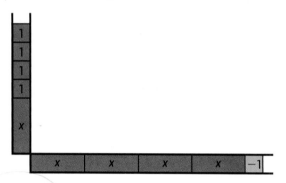

Lesson 2.2

3. Determine the missing factor.
   a) $7x^2 - 14x = 7x(?)$
   b) $3a^2 + 15a - 9 = (?)(a^2 + 5a - 3)$
   c) $10b^4 - 20b^2 = 10b^2(?)$
   d) $4x(x - 3) + 5(x - 3) = (?)(x - 3)$

4. Factor.
   a) $10x^2 - 5x$
   b) $12n^2 - 24n + 48$
   c) $-2x^2 - 6x + 8$
   d) $3a(5 - 7a) - 2(7a - 5)$

5. A rectangle has an area of $6x^2 - 8$.
   a) Determine the dimensions of the rectangle.
   b) Is there more than one possibility? Explain.

6. a) Give three examples of polynomials that have a greatest common factor of $7x$.
   b) Factor each polynomial from part (a).

Lesson 2.3

7. What are the factors of each polynomial being modelled?
   a)
   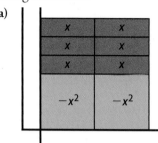

   b)
   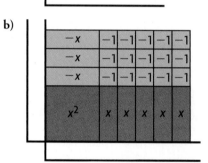

8. Determine the missing factor.
   a) $x^2 + 9x + 14 = (x + 2)(?)$
   b) $a^2 + 3a - 28 = (?)(a + 7)$
   c) $b^2 - b - 20 = (b - 5)(?)$
   d) $-8x + x^2 + 15 = (?)(x - 3)$

9. Factor.
   a) $x^2 + 7x + 10$ c) $x^2 + x - 42$
   b) $x^2 - 12x + 27$ d) $x^2 - x - 90$

10. Determine consecutive integers b and c, and also m and n, such that
   $$x^2 + bx + c = (x + m)(x + n)$$

Lesson 2.4

11. How would you decompose the x-term to factor each polynomial?
   a) $6x^2 + x - 1$ c) $7x^2 - 50x - 48$
   b) $12x^2 + 9x - 30$ d) $30x^2 - 9x - 3$

12. Determine the missing factor.
  a) $2x^2 + 7x + 5 = (x + 1)(?)$
  b) $3a^2 + 10a - 8 = (?)(3a - 2)$
  c) $4b^2 - 4b - 15 = (2b - 5)(?)$
  d) $20 + 27x + 9x^2 = (?)(3x + 5)$

13. Factor.
  a) $6x^2 - 19x + 10$ c) $20x^2 + 9x - 18$
  b) $10a^2 - 11a - 6$ d) $6n^2 + 13n + 7$

Lesson 2.5

14. Each diagram represents a polynomial. Identify the polynomial and its factors.
  a)

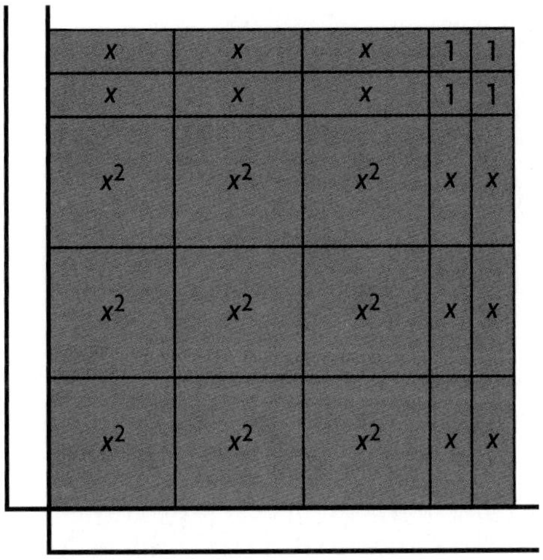

  b)

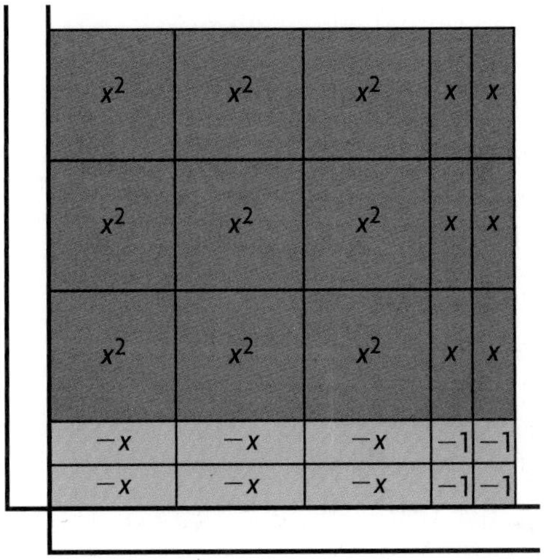

15. Determine the missing factor.
  a) $x^2 - 25 = (x + 5)(?)$
  b) $9a^2 + 6a + 1 = (?)(3a + 1)$
  c) $4b^2 - 20b + 25 = (2b - 5)(?)$
  d) $9x^2 - 64 = (?)(3x + 8)$

16. Factor.
  a) $4x^2 - 9$
  b) $16a^2 - 24a + 9$
  c) $x^8 - 256$
  d) $(x - 2)^2 + 6(x - 2) + 9$

17. The polynomial $x^2 - 1$ can be factored. Can the polynomial $x^2 + 1$ be factored? Explain.

18. Factor each expression. Remember to divide out all common factors first.
  a) $x^2 + 2x - 15$
  b) $5m^2 + 15m - 20$
  c) $2x^2 - 18$
  d) $18x^2 + 15x - 3$
  e) $36x^2 + 48x + 16$
  f) $15c^3 + 25c^2$

19. How is factoring a polynomial related to expanding a polynomial? Use an example in your explanation.

1. Expand and simplify.

    a) $-2x(3x - 4) - x(x + 6)$

    b) $-3(5n - 4)^2 - 5(5n + 4)^2$

    c) $-8(x^2 - 5x + 7) + 5(2x - 5)(3x - 7)$

    d) $-3(5a - 4)(5a + 4) - 3a(a - 7)$

2. What two binomials are being multiplied and what is the product?

3. A rectangle has a width of $2x - 3$ and a length of $3x + 1$.

    a) Write its area as a simplified polynomial.

    b) Write expressions for the dimensions if the width is doubled and the length is increased by 2.

    c) Write the new area as a simplified polynomial.

4. Use pictures and words to show how to factor $-2x^2 + 8x$.

5. Factor.

    a) $x^2 + x - 12$ **c)** $-5x^2 + 75x - 280$

    b) $a^2 + 16a + 63$ **d)** $y^2 + 3y - 54$

6. Factor.

    a) $2x^2 - 9x - 5$ **c)** $6x^2 - 15x + 6$

    b) $12n^2 - 67n + 16$ **d)** $8a^2 - 14a - 15$

7. What dimensions can a rectangle with an area of $12x^2 - 3x - 15$ have?

8. State all the integers, m, such that $x^2 + mx - 13$ can be factored.

9. Factor.

    a) $121x^2 - 25$ **c)** $x^4 - 81$

    b) $36a^2 - 60a + 25$ **d)** $(3 - n)^2 - 12(3 - n) + 36$

10. Determine all integers, m and n, such that $m^2 - n^2 = 45$.

The Algebra Challenge

Part 1:

Make each statement true by replacing each ▪ with one of these digits:

| 1 | 2 | 3 | 4 | 5 | 6 | 7 | 8 | 9 |
| 1 | 2 | 3 | 4 | 5 | 6 | 7 | 8 | 9 |

Use each of the 18 digits once.

$$(3x + 2)^2 = 9x^2 + \blacksquare\, \blacksquare\, x + 4$$
$$\blacksquare x^2 - \blacksquare = (2x + \blacksquare)(2x - 3)$$
$$\blacksquare x^2 + 14x - 2\blacksquare = \blacksquare(x^2 + 2x - 3)$$
$$(\blacksquare x + \blacksquare)(\blacksquare x + \blacksquare) = 40x^2 + 93x + \blacksquare 4$$
$$\blacksquare(\blacksquare x^2 - \blacksquare) = \blacksquare x^2 - \blacksquare$$

Part 2:

Make up a puzzle of your own, like the one you just solved, for a partner to solve.

> **Task | Checklist**
>
> ✔ Did you use each of the 18 digits listed once?
>
> ✔ Did you verify that each statement is correct?
>
> ✔ How do you know the puzzle you created works?

Working with Quadratic Functions: Standard and Factored Forms

▶ GOALS

You will be able to

- Expand and simplify quadratic expressions, solve quadratic equations, and relate the roots of a quadratic equation to the corresponding graph

- Make connections between the numeric, graphical, and algebraic representations of quadratic functions

- Model situations and solve problems involving quadratic functions

❓ When will the gymnast reach his maximum height? How does this height relate to the start of his vault and the end of his vault?

WORDS *You Need to Know*

1. Match each term with its picture or example.

a) linear equation   c) function   e) range
b) zeros   d) domain   f) function notation

i) $f(x) = 2x^2 - 5x + 3$ iii)   v) $y = 3x + 4$

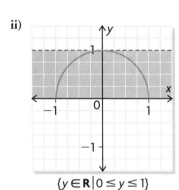

ii)

$\{y \in \mathbf{R} \mid 0 \le y \le 1\}$

iv)

$\{x \in \mathbf{R} \mid -1 \le x \le 1\}$

vi)

$x = -2$ and $x = 3$

Study | Aid
- For help, see Essential Skills Appendix, A-11.

SKILLS AND CONCEPTS *You Need*

Solving Linear Equations

- To solve a linear equation, isolate the variable.
- Whatever opearation you do to one side, you must do to the other.
- A linear equation has only one solution.

EXAMPLE

Solve $7y - 12 = 30$.

Solution

$7y - 12 + 12 = 30 + 12$ ← To isolate the variable y, add +12 to both sides. Simplify.

$\dfrac{7y}{7} = \dfrac{42}{7}$ ← Divide both sides by the coefficient of the variable.

$y = 6$

2. Solve the following linear equations.

   a) $2y + 3 = 15$ **c)** $-2a + 3 = -6$

   b) $3x - 5 = 11$ **d)** $4c + 6 = 13$

Determining the Characteristics of a Quadratic Function from Its Graph

State the vertex, equation of the axis of symmetry, domain, and range of this function.

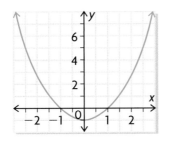

Solution

vertex: $(0, -1)$

equation of the axis of symmetry: $x = 0$

domain: $\{x \in \mathbf{R}\}$

range: $\{y \in \mathbf{R} \mid y \geq -1\}$

3. For each of the following, state the vertex, equation of the axis of symmetry, domain, and range.

   a) $y = 2x^2 + 2$ **b)** $y = -(x - 3)^2 - 5$

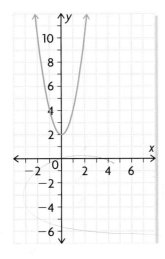

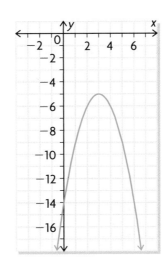

PRACTICE

Study | Aid

- For help, see Essential Skills Appendix.

Question	Appendix
4	A-9
6	A-10
7	A-7 and 8
9	A-12

4. Simplify the following.

a) $2x + 3y - 7x + 2y$

b) $3(m + 2) + 5(m - 2)$

c) $2x^2(3x^2 + 5x)$

d) $2x(x + 2) - 3(x^2 + 2x - 4)$

5. Determine the x- and y-intercepts.

a) $y = 3x - 7$

b) $2x - 6y = 12$

c) $5x = 10 - 2y$

d) $3x + 4y - 12 = 0$

6. Factor the following.

a) $5x^2 + 15 - 5x$

b) $x^2 - 11x + 10$

c) $2x^2 + 7x - 15$

d) $6x^2 + 7x - 3$

e) $2x^2 + 4x - 6$

f) $x^2 - 121$

7. Graph the following.

a) $y = 3x - 2$

b) $2x - 4y - 8 = 0$

c) $y = x^2 + x - 3$

d) $y = x^2 - 4$

8. For each graph, identify the x- and y-intercepts and the maximum or minimum value. State why that value is a maximum or minimum and what it tells you about a, the **coefficient** of the x^2-term of the quadratic function.

a)

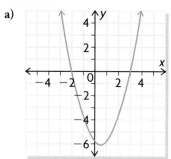

b)

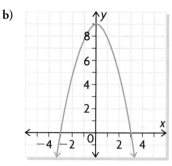

9. Decide whether the functions in tables (a) and (b) are linear or nonlinear. Justify your answer.

a)

x	y
0	2
1	1
2	0
3	-1
4	-2
5	-3
6	-4
7	-5
8	-6

b)

x	y
-3	15
-2	5
-1	-1
0	-3
1	-1
2	5
3	15
4	29
5	47

10. Complete the chart by writing down what you know about how and when to **factor** quadratic expressions.

Factoring strategies:		
	Factoring Quadratics	
Examples:		Non-examples:

APPLYING *What You Know*

Properties of the Quadratic Function

Talia is babysitting Marc.

Talia notices that when Marc rolls a large marble on a flat surface, it travels in a straight line. Talia and Marc think that the path will look different if they roll the marble on a slanted table.

To see the path the marble travels, they dip the marble in paint and roll it on graph paper taped to a slanted table. The resulting graph is shown.

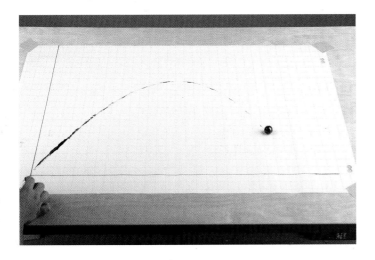

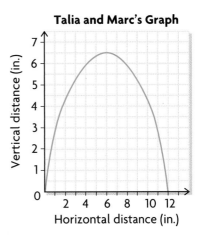

Talia and Marc's Graph

? How could you describe the graph?

A. What shape resulted? Was Talia and Marc's hypothesis correct?

B. What are the *x*-intercepts, or zeros, of the graph?

C. What was the maximum distance travelled by the marble from the edge of the table?

D. When was this maximum distance reached?

E. What is the equation of the axis of symmetry of the curve? Explain how this relates to the *x*-intercepts (zeros) of the function.

F. What is the vertex of the graph? Explain how it is related to the location of the maximum.

G. What factors do you think could affect the size and shape of the parabola made by the rolling marble?

Exploring a Situation Using a Quadratic Function

- chart graph paper
- two different-coloured counters

GOAL

Use quadratic functions to model real-world applications.

EXPLORE the Math

In the game of leapfrog, you move red and blue counters along a line of spaces in a board. Equal numbers of red and blue counters are placed in the spaces opposite each other and separated by an empty space in the middle.

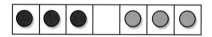

The object of the game is to move the red counters into the spaces occupied by the blue counters, and vice versa.

The rules allow two types of moves:
- a move to an adjacent space
- a jump over a counter of the other colour

? What is the minimum number of moves needed to switch seven blue and seven red counters?

A. Play the game by drawing a grid on paper with at least 11 boxes. The grid must contain an odd number of boxes, and you must use equal numbers of red and blue counters.

B. In a table, relate the number of counters of each colour, N, to the minimum number of moves, M, required to switch the counters.

Number of Counters of each Colour, N	1	2	3	4	5
Minimum Number of Moves, M					

C. Consult other groups to see if your minimum number of moves is the same as theirs.

D. Create a scatter plot of the data you collected. Identify the independent and dependent variables.

E. Determine the type of graph represented by your data.

F. What type of equation could be used to model your data?

Reflecting

G. Predict the minimum number of moves needed to switch seven red and seven blue counters.

H. Holly thinks that the function $f(x) = x^2 + 2x$ models the minimum number of moves required to switch equal numbers of counters. What do x and $f(x)$ represent in this situation? Is Holly correct? Explain.

I. What are the domain and range of $f(x)$? Is it necessary to restrict the domain? Explain.

In Summary

Key Ideas

- A scatter plot that appears to have a parabolic shape or part of a parabolic shape might be modelled by a quadratic function.
- The domain and range of the quadratic model may need to be restricted for the situation you are dealing with.

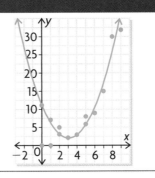

Need to Know

- The domain and range may need to be limited to whole numbers if the data are **discrete**; that is, if fractions, decimals, and negative numbers are not valid for the situation.

FURTHER Your Understanding

1. Play leapfrog with six counters of each colour, and determine the minimum number of moves. Does the algebraic model predict the minimum number of moves? Explain.

2. Play the game again, increasing N, the number of counters used each time you play. This time, for each move, keep track of the colour of the counter and the type of move made.

   For example, with one red and one blue counter, the moves could be red slide (RS), blue jump (BJ), and then red slide (RS). Set up a table as shown to record your results. Identify any patterns you see in your table.

N			
1	RS	BJ	RS
2			
3			

3. Play the game with one side having one more counter than the other side. Can a quadratic model be used to relate the minimum number of moves to the lower number of counters? Explain.

Relating the Standard and Factored Forms

YOU WILL NEED

- graph paper
- graphing calculator (optional)

GOAL

Compare the factored form of a quadratic function with its standard form.

Communication | **Tip**

A company's revenue is its money earned from sales. A revenue function, $R(x)$, is the product of the number of items sold and price.

Revenue = (number of items sold)(price)

LEARN ABOUT the Math

To raise money, some students sell T-shirts. Based on last year's sales, they know that

- they can sell 40 T-shirts a week at $10 each
- if they raise the price by $1, they will sell one less T-shirt each week

They picked some prices, estimated the number of shirts they might sell, and calculated the revenue.

Price ($)	T-shirts Sold	Revenue ($)
10	40	10 × 40 = 400
10 + 15 = 25	40 − 15 = 25	25 × 25 = 625
10 + 30 = 40	40 − 30 = 10	40 × 10 = 400

Rachel and Andrew noticed that as they increased the price, the revenue increased and then decreased. Based on this pattern, they suggested quadratic functions to model revenue, where x is the number of $1 increases.

- Rachel's function is $R(x) = (40 - x)(10 + x)$.
- Andrew's function is $R(x) = -x^2 + 30x + 400$.

❓ What is the maximum revenue they can earn on T-shirt sales?

EXAMPLE 1	Connecting the standard and factored forms of quadratic functions

Compare the two functions:

$R(x) = (40 - x)(10 + x)$ and $R(x) = -x^2 + 30x + 400$

Enrique's Solution: Comparing and Analyzing the Graphs

Rachel's Function	
x	$R(x) = (40 - x)(10 + x)$
0	$= (40 - 0)(10 + 0)$ $= (40)(10)$ $= 400$
4	504
8	576
12	616
16	624
20	600
24	544

Andrew's Function	
x	$R(x) = -x^2 + 30x + 400$
0	$= -0^2 + 30(0) + 400$ $= 0 + 0 + 400$ $= 400$
4	504
8	576
12	616
16	624
20	600
24	544

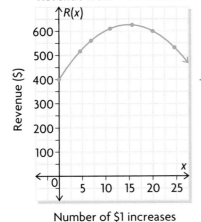

Revenue from T-shirt Sales

Number of $1 increases

The tables of values are the same for both functions, so the graphs will be the same. I need only one graph.

I used Rachel's table of values to make a graph.

The two functions are equivalent.

The maximum revenue is about $625.

To get the maximum revenue, I looked at my graph and found the y-coordinate of the vertex. Since the vertex occurs at $(15, 625)$, $R(15) = 625$.

factored form

a quadratic function in the form $f(x) = a(x - r)(x - s)$

standard form

a quadratic function in the form $f(x) = ax^2 + bx + c$

Rachel's revenue function is in **factored form**. If you expand and simplify it to express it in **standard form**, then you can compare both functions to see if they are the same.

Beth's Solution: Comparing and Analyzing the Functions

$R(x) = (40 - x)(10 + x)$

$R(x) = 400 + 40x - 10x - x^2$ ◄—— I multiplied the binomials and collected like terms. Rachel's revenue function in standard form is the same as Andrew's revenue function, but the terms are in a different order.

$R(x) = 400 + 30x - x^2$

$R(x) = -x^2 + 30x + 400$ ◄—— I rearranged my function to match Andrew's.

zeros of a relation

the values of x for which a relation has the value zero. The zeros of a relation correspond to the x-intercepts of its graph

$0 = -x^2 + 30x + 400$ or

$0 = (40 - x)(10 + x)$ ◄—— The x-intercepts, or **zeros**, occur when the revenue is 0, so $R(x) = 0$. I used Rachel's function to find the x-intercepts by setting each factor equal to zero and solving for x.

$40 - x = 0 \quad$ and $\quad 10 + x = 0$

$40 = 0 + x \quad$ and $\quad x = 0 - 10$

$40 = x \quad$ and $\quad x = -10$

$x = \dfrac{(40 + (-10))}{2}$ ◄—— The maximum value is the y-coordinate of the vertex, and the vertex lies on the axis of symmetry.

$x = \dfrac{30}{2}$ ◄—— To find the axis of symmetry, I added the zeros and divided by 2.

$x = 15$

$R(15) = (40 - 15)(10 + 15)$ ◄—— To find the maximum revenue I substituted the x-value into Rachel's equation.

$R(15) = (25)(25)$

$R(15) = 625$

The maximum revenue is $625.

Reflecting

A. If there is no increase in the cost of the T-shirts, what will the revenue be? Whose revenue model can be used to see this value most directly?

B. Which revenue function would you use to determine the maximum value? Explain.

APPLY the Math

If you are given a quadratic function in standard form and it is possible to write it in factored form, you can use the factored form to determine the coordinates of the vertex.

EXAMPLE 2	Selecting a factoring strategy to determine the vertex of a quadratic function

Determine the coordinates of the vertex of $f(x) = 2x^2 - 5x - 12$, and sketch the graph of $f(x)$.

Talia's Solution

$f(x) = 2x^2 - 5x - 12$

$f(x) = 2x^2 - 8x + 3x - 12$ ◄──────

> First, I factored the trinomial by decomposition. I found two numbers that multiply to give -24 (from 2×-12) and add to give -5. The numbers are -8 and 3. I replaced the middle term with $-8x$ and $3x$.

$f(x) = (2x^2 - 8x) + (3x - 12)$ ◄──────

$f(x) = 2x(x - 4) + 3(x - 4)$

$f(x) = (x - 4)(2x + 3)$

$0 = (x - 4)(2x + 3)$

> I grouped the first two terms and the second two terms. I divided out the common factor $(x - 4)$ from both groups. I used the factors to find the zeros.

$x - 4 = 0$ and $2x + 3 = 0$

$x = 0 + 4$ and $2x = 0 + (-3)$

$x = 4$ and $x = -\dfrac{3}{2}$ or -1.5

$$x = \frac{4 + (-1.5)}{2}$$

$$x = \frac{2.5}{2}$$

$$x = 1.25$$

> I used the zeros to find the axis of symmetry. I know that the axis of symmetry is halfway between the zeros. I added the two zeros and divided by 2.

The equation of the axis of symmetry is $x = 1.25$.

$$f(1.25) = 2(1.25)^2 - 5(1.25) - 12$$
$$= 3.125 - 6.25 - 12$$
$$= -15.125$$

> I put $x = 1.25$ into the original function and solved to get the y-coordinate of the vertex.

The vertex is $(1.25, -15.125)$.

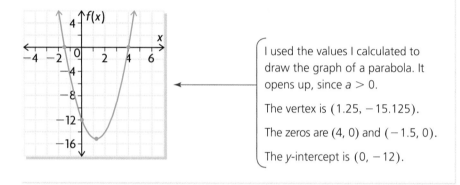

> I used the values I calculated to draw the graph of a parabola. It opens up, since $a > 0$.
>
> The vertex is $(1.25, -15.125)$.
>
> The zeros are $(4, 0)$ and $(-1.5, 0)$.
>
> The y-intercept is $(0, -12)$.

EXAMPLE 3   **Solving problems using a quadratic function model**

The height of a football kicked from the ground is given by the function $h(t) = -5t^2 + 20t$, where $h(t)$ is the height in metres and t is the time in seconds from its release.

a) Write the function in factored form.
b) When will the football hit the ground?
c) When will the football reach its maximum height?
d) What is the maximum height the football reaches?
e) Graph the height of the football in terms of time without using a table of values.

Matt's Solution

a) $h(t) = -5t^2 + 20t$

$h(t) = -5t(t - 4)$

> I wrote the function in factored form by dividing out the common factor $-5t$.

b) $0 = -5t(t - 4)$

$-5t = 0$ and $t - 4 = 0$

$t = 0$ and $t = 4$

When the football hits the ground, the height is zero. $h(t) = 0$ when each factor is 0. The zeros are 0 and 4. The football is on the ground to start and then returns 4 s later.

c) $t = 0$ and $t = 4$ are the zeros.

$t = \dfrac{0 + 4}{2}$

$t = 2$

The ball reaches its maximum height when $t = 2$ s.

I know that the maximum value is the y-coordinate of the vertex, and it lies on the axis of symmetry, which is halfway between the zeros. The coefficient of x^2 is -5, which means that the parabola opens down. Therefore, there is a maximum value. The equation $t = 2$ is the axis of symmetry.

d) $h(t) = -5t^2 + 20t$

$h(2) = -5(2)^2 + 20(2)$

$= -20 + 40$

$= 20$

The maximum height is 20 m.

I know the ball reaches the maximum height at 2 s. So I calculated $h(2)$, the maximum height.

e)

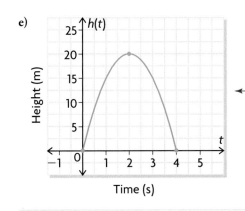

I marked the two zeros and the vertex that I found on graph paper. Since $a < 0$, those three points are part of a parabola that opens down, so I joined them by sketching a parabola. I can see from the graph that the y-intercept is 0. This is also seen in the function when it's written in standard form: $h(t) = -5t^2 + 20t + 0$.

If you are given the graph of a quadratic function and you can determine the zeros and the coordinates of another point on the parabola, you can determine the equation of the function in factored and standard forms.

EXAMPLE **4**

Representing a quadratic function in factored and standard forms from a graph

From the graph of this quadratic function, determine the function's factored and standard forms.

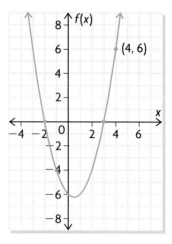

Randy's Solution

$f(x) = a(x - r)(x - s)$

$f(x) = a(x - (-2))(x - (+3))$ ◄——

$f(x) = a(x + 2)(x - 3)$

> A quadratic function can be written in the form $f(x) = a(x - r)(x - s)$, where r and s are the zeros of the function and a indicates the shape of the parabola. From the graph, the zeros are $x = -2$ and $x = 3$. I substituted -2 for r and 3 for s, then simplified.

Let $(x, y) = (4, 6)$

$6 = a(4 + 2)(4 - 3)$ ◄——

$6 = a(6)(1)$

$a = 1$

> To get a, I picked any point on the curve, substituted its coordinates into the function, and then solved for a.

$f(x) = 1(x + 2)(x - 3)$

$f(x) = 1(x^2 - 3x + 2x - 6)$ ◄——

$f(x) = x^2 - x - 6$

> I used my value of a to rewrite the function in factored form.
> I expanded to get the standard form.

> From the graph the y-intercept of the parabola is -6. This is also seen in the function when it's written in standard form.

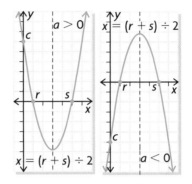
CHECK *Your Understanding*

1. The parabola shown is congruent to $y = x^2$.
   a) What are the zeros of the function?
   b) Write the equation in factored form.

2. Express each quadratic function in factored form. Then determine the zeros, the equation of the axis of symmetry, and the coordinates of the vertex.
   a) $f(x) = 2x^2 + 12x$
   b) $f(x) = x^2 - 7x + 12$
   c) $f(x) = -x^2 + 100$
   d) $f(x) = 2x^2 + 5x - 3$

3. Express each quadratic function in standard form. Determine the y-intercept.
   a) $f(x) = 3x(x - 4)$
   b) $f(x) = (x - 5)(x + 7)$
   c) $f(x) = 2(x - 4)(3x + 2)$
   d) $f(x) = (3x - 4)(2x + 5)$

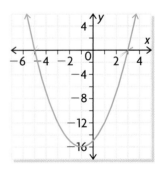

Communication | Tip

Two parabolas that are congruent have exactly the same size and shape.

PRACTISING

4. For each quadratic function, determine the zeros, the equation of the axis of symmetry, and the coordinates of the vertex without graphing.
   a) $g(x) = 2x(x + 6)$
   b) $g(x) = (x - 8)(x + 4)$
   c) $g(x) = (x - 10)(2 - x)$
   d) $g(x) = (2x + 5)(9 - 2x)$
   e) $g(x) = (2x + 3)(x - 2)$
   f) $g(x) = (5 - x)(5 + x)$

5. Express each function in factored form. Then determine the zeros, the equation of the axis of symmetry, and the coordinates of the vertex without graphing.
   a) $g(x) = 3x^2 - 6x$
   b) $g(x) = x^2 + 10x + 21$
   c) $g(x) = x^2 - x - 6$
   d) $g(x) = 3x^2 + 12x - 15$
   e) $g(x) = 2x^2 - 13x - 7$
   f) $g(x) = -6x^2 + 24$

6. Match the factored form on the left with the correct standard form on the right. How did you decide on your answer?
   a) $y = (2x + 3)(x - 4)$
   b) $y = (4 - 3x)(x + 3)$
   c) $y = (3x - 4)(x - 3)$
   d) $y = (3 - 4x)(4 - x)$
   e) $y = (x + 3)(3x - 4)$

   i) $y = 4x^2 - 19x + 12$
   ii) $y = -3x^2 - 5x + 12$
   iii) $y = 2x^2 - 5x - 12$
   iv) $y = 3x^2 - 13x + 12$
   v) $y = 3x^2 + 5x - 12$

7. Determine the maximum or minimum value for each quadratic function.
   a) $f(x) = (7 - x)(x + 2)$
   b) $f(x) = (x + 5)(x - 9)$
   c) $f(x) = (2x + 3)(8 - x)$
   d) $g(x) = x^2 + 7x + 10$
   e) $g(x) = -x^2 + 25$
   f) $g(x) = 4x^2 + 4x - 3$

8. **K** Graph each quadratic function by hand by determining the zeros, vertex, axis of symmetry, and y-intercept.
   a) $g(x) = (2x - 1)(x - 4)$
   b) $f(x) = (3x - 1)(2x - 5)$
   c) $f(x) = x^2 - x - 20$
   d) $g(x) = 2x^2 + 2x - 12$
   e) $f(x) = -x^2 - 2x + 24$
   f) $f(x) = -4x^2 - 16x + 33$

9. a) When a quadratic function is in standard form, what information about the graph can be easily determined? Provide an example.
   b) When a quadratic function is in factored form, what information about the graph can be easily determined? Provide an example.

10. Graph each function, and complete the table.

	Factored Form	Standard Form	Axis of Symmetry	Zeros	y-intercept	Vertex	Maximum or Minimum Value
a)	$R(x) = (40 - x)(10 + x)$	$R(x) = -x^2 + 30x + 400$					
b)	$f(x) = (x - 4)(x + 2)$						
c)		$g(x) = -x^2 + 2x + 8$					
d)	$p(x) = (3 - x)(x + 1)$						
e)		$j(x) = 4x^2 - 121$					

11. The height of water, $h(t)$, in metres, from a garden hose is given by the function $h(t) = -5t^2 + 15t$, where t is time in seconds. Express the function in factored form, and then use the zeros to determine the maximum height reached by the water.

12. For each graph, write the equation in both factored and standard forms.

a)

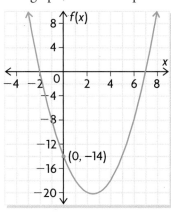

c)

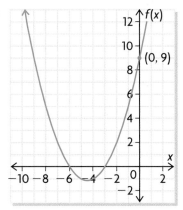

b)

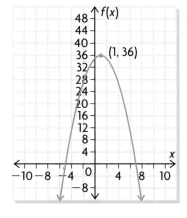

d)

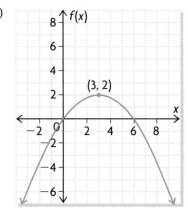

13. Complete the table.

	Zeros	Axis of Symmetry	Maximum or Minimum Value	Vertex	Function in Factored Form	Function in Standard Form
a)	2 and 8		6			
b)	−7 and −2		−2			
c)	−1 and 9		5			
d)	−8 and 0		−5			

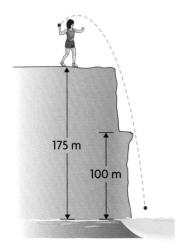

175 m

100 m

14. A ball is thrown into water from a cliff that is 175 m high. The height of the ball above the water after it is thrown is modelled by the function $h(t) = -5t^2 + 10t + 175$, where $h(t)$ is the height in metres and t is time in seconds.
  a) When will the ball reach the water below the cliff?
  b) When will the ball reach a ledge that is 100 m above the water?

15. The safe stopping distance for a boat travelling at a constant speed in calm water is given by $d(v) = 0.002(2v^2 + 10v + 3000)$, where $d(v)$ is the distance in metres and v is the speed in kilometres per hour. What is the initial speed of the boat if it takes 30 m to stop?

16. Describe how you would sketch the graph of $f(x) = 2x^2 - 4x - 30$ without using a table of values.

Extending

17. The stainless-steel Gateway Arch in St. Louis, Missouri, is almost parabolic in shape. It is 192 m from the base of the left leg to the base of the right leg. The arch is 192 m high. Determine a function, in standard form, that models the shape of the arch.

18. A model rocket is shot straight up into the air. The table shows its height, $h(t)$, at time t. Determine a function, in factored form, that estimates the height of the rocket at any given time.

Time (s)	0	1	2	3	4	5	6
Height (m)	0.0	25.1	40.4	45.9	41.6	27.5	3.6

19. The path of a shot put is given by $h(d) = 0.0502(d^2 - 20.7d - 26.28)$, where $h(d)$ is the height and d is the horizontal distance, both in metres.
  a) Rewrite the relation in the form $h(d) = a(d - r)(d - s)$, where r and s are the zeros of the relation.
  b) What is the significance of r and s in this question?

3.3 Solving Quadratic Equations by Graphing

YOU WILL NEED
- graph paper
- graphing calculator

GOAL

Use graphs to solve quadratic equations.

INVESTIGATE the Math

A model rocket is launched from the roof of a building.

The height, $h(t)$, in metres, at any time, t, in seconds, is modelled by the function $h(t) = -5t^2 + 15t + 20$.

❓ When will the rocket hit the ground?

A. Use a graphing calculator to graph the height function. Determine the zeros, the axis of symmetry, and the vertex of this function.

B. What is the rocket's height when it hits the ground? Use this value to write a **quadratic equation** you could solve to determine when the rocket hits the ground.

C. Substitute one of the zeros you determined in part A into the quadratic equation you wrote in part B. Repeat this for the other zero. What do you notice?

D. What are the **roots** of the quadratic equation you wrote in part B?

E. State the domain and range of the function in the context of this problem.

F. What is the starting height of the rocket? Where is it on the graph? Where is it in the function?

G. When will the rocket hit the ground?

quadratic equation

an equation that contains a polynomial whose highest degree is 2; for example, $x^2 + 7x + 10 = 0$

root of an equation

a number that when substituted for the unknown, makes the equation a true statement

for example, $x = 2$ is a root of the equation $x^2 - x - 2 = 0$ because $2^2 - 2 - 2 = 0$

the root of an equation is also known as a *solution* to that equation

Reflecting

H. How are the zeros of the function $h(t)$ related to the roots of the quadratic equation $0 = -5t^2 + 15t + 20$?

I. The quadratic equation $0 = -5t^2 + 15t + 20$ has two roots. Explain why only one root is an acceptable solution in this situation. How does your explanation relate to the domain of the function $h(t)$?

J. How can the graph of the function $f(x) = ax^2 + bx + c$ help you solve the quadratic equation $ax^2 + bx + c = 0$?

APPLY *the Math*

EXAMPLE 1	Connecting graphs to the solutions of a quadratic equation

Determine the solutions of the quadratic equation $x^2 - 8x + 12 = -3$ by graphing. Check your solutions.

Nash's Solution: Rearranging the Equation to Determine the Zeros on the Corresponding Graph

$x^2 - 8x + 12 = -3$

$x^2 - 8x + 15 = 0$

$g(x) = x^2 - 8x + 15$ ←

> I rearranged the equation so that it was equal to zero. I used the corresponding function, $g(x) = x^2 - 8x + 15$, to solve the equation by finding its zeros.

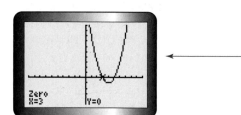

> I graphed $g(x) = x^2 - 8x + 15$. I used the zero operation on my calculator to see where the graph crosses the x-axis.

One solution is $x = 3$.

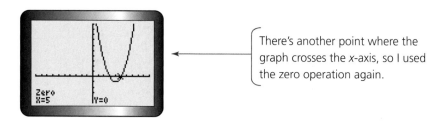

> There's another point where the graph crosses the x-axis, so I used the zero operation again.

The other solution is $x = 5$.

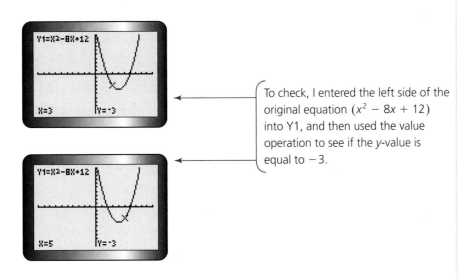

To check, I entered the left side of the original equation ($x^2 - 8x + 12$) into Y1, and then used the value operation to see if the y-value is equal to -3.

Tech | Support

For help determining values and zeros using a graphing calculator, see Technical Appendix, B-3 and B-8.

Both solutions give a value of -3, so I know they're correct.

Rearranging a quadratic equation into the form $ax^2 + bx + c$ and then graphing $f(x) = ax^2 + bx + c$ is one method you can use to solve a quadratic equation. Another method is to treat each side of the equation as a function and then graph both to determine the point(s) of intersection.

Lisa's Solution: Determining the Points of Intersection of Two Corresponding Functions

$x^2 - 8x + 12 = -3$

$f(x) = x^2 - 8x + 12$ and $g(x) = -3$

I separated the equation into two functions, using each side of the equation. I called the left side $f(x)$ and the right side $g(x)$.

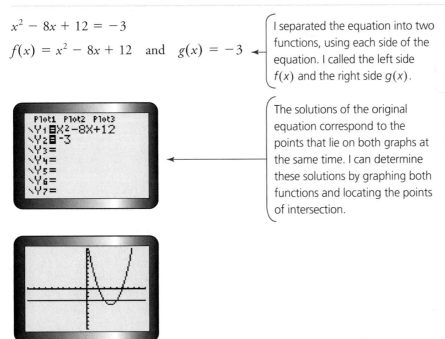

The solutions of the original equation correspond to the points that lie on both graphs at the same time. I can determine these solutions by graphing both functions and locating the points of intersection.

Tech | Support

For help determining points of intersection using a graphing calculator, see Technical Appendix, B-11.

Chapter 3 Working with Quadratic Functions: Standard and Factored Forms   **145**

I found the points of intersection by using the intersect operation on the calculator. The x-coordinates of these points are the solutions of the original equation.

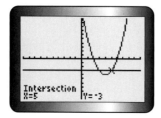

The two solutions are $x = 3$ and $x = 5$.

Check:

L.S. $= x^2 - 8x + 12$ R.S. $= -3$

When $x = 3$,

L.S. $= (3)^2 - 8(3) + 12$

$= 9 - 24 + 12$

$= -3$

$=$ R.S.

When $x = 5$,

L.S. $= (5)^2 - 8(5) + 12$

$= 25 - 40 + 12$

$= -3$

$=$ R.S.

I substituted each of my answers into the original equation to see if I had the correct solutions. They both work, so I know my solutions are correct.

The population of an Ontario city is modelled by the function $P(t) = 0.5t^2 + 10t + 300$, where $P(t)$ is the population in thousands and t is the time in years. *Note:* $t = 0$ corresponds to the year 2000.

a) What was the population in 2000?
b) What will the population be in 2010?
c) When is the population expected to be 1 050 000?

Guillaume's Solution

a) $P(0) = 0.5(0)^2 + 10(0) + 300$

$\qquad = 0 + 0 + 300$ ◄───────

$\qquad = 300$

> The year 2000 corresponds to $t = 0$, so I set $t = 0$ in the equation and evaluated. I remembered to multiply the solution by 1000 because P is in thousands.

The population was 300 000 in the year 2000.

b) $P(10) = 0.5(10)^2 + 10(10) + 300$

$\qquad = 0.5(100) + 100 + 300$ ◄───────

$\qquad = 50 + 100 + 300$

$\qquad = 450$

> The year 2010 corresponds to $t = 10$, so I set $t = 10$ in the equation. Again, I had to remember to multiply the solution by 1000 because P is in thousands.

The population will be 450 000 in the year 2010.

c) $1050 = 0.5t^2 + 10t + 300$ ◄───────

$\quad 0 = 0.5t^2 + 10t - 750$

> To see when the population will be 1 050 000, I let P be 1050 because P is in thousands, and $\dfrac{1\,050\,000}{1000} = 1050$.

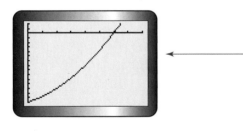

> I rearranged the equation to get zero on the left side. I used my graphing calculator to graph $P(t) = 0.5t^2 + 10t - 750$, and then I looked for the zeros. I decided to use $t \geq 0$ as the domain, so I'll have only one zero.

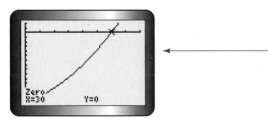

> Using the zero operation on my graphing calculator, I got 30 for the zero.

The population should reach 1 050 000 in the ◄─────── year 2030.

> Since $t = 0$ is used to represent the year 2000, I added 30 to get 2030.

EXAMPLE 3	Solving problems involving a quadratic equation

The function $h(t) = 2 + 50t - 1.862t^2$, where $h(t)$ is the height in metres and t is time in seconds, models the height of a golf ball above the planet Mercury's surface during its flight.

a) What is the maximum height reached by the ball?
b) How long will the ball be above the surface of Mercury?
c) When will it reach a height of 200 m on the way down?

Candace's Solution

a)

The maximum height of the golf ball is about 337.7 m.

> I graphed the equation $y = 2 + 50t - 1.862t^2$ on my graphing calculator. I used the maximum operation on my calculator to get the maximum value.

b)

(graph showing Zero X=26.892787 Y=0)

The ball will be above the planet's surface for about 26.9 s.

> When the ball reaches a height of 0 m, it's no longer above the planet. So I need to identify the zeros.
>
> The first zero is before the golf ball has been hit. I want the second zero.

c) $2 + 50t - 1.862t^2 = 200$

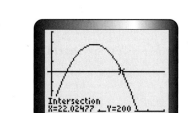

The ball will reach a height of 200 m on its way down at about 22.0 s.

> To find the time when the ball is at a height of 200 m, I set $h(t)$ equal to 200. To solve this equation, I graphed $y = 2 + 50t - 1.862t^2$ and $y = 200$ on the same axes. From the graph, I can see that there are two points of intersection. The second point is the height on the way down, so I found it using the intersection operation on the calculator.

Tech | Support

For help determining maximum and minimum values using a graphing calculator, see Technical Appendix, B-9.

Tech | Support

For help determining points of intersection using a graphing calculator, see Technical Appendix, B-11.

CHECK *Your Understanding*

1. Graph each function by hand. Then use the graph to solve the corresponding quadratic equation.
   a) $g(x) = -x^2 + 5x + 14$ and $-x^2 + 5x + 14 = 0$
   b) $f(x) = x^2 + 8x + 15$ and $x^2 + 8x + 15 = 0$

2. Graph each function using a graphing calculator. Then use the graph to solve the corresponding quadratic equation.
   a) $h(x) = x^2 - x - 20$ and $x^2 - x - 20 = 0$
   b) $f(x) = x^2 - 5x - 9$ and $x^2 - 5x - 9 = 0$

PRACTISING

3. For each function, write the corresponding quadratic equation whose solutions are also the zeros of the function.

a)

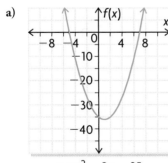

$y = x^2 - 2x - 35$

b)

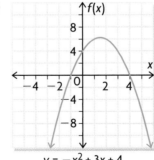

$y = -x^2 + 3x + 4$

c)

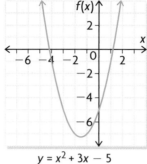

$$y = x^2 + 3x - 5$$

d)

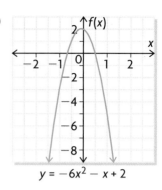

$$y = -6x^2 - x + 2$$

4. Graph the corresponding function to determine the roots of each equation. Verify your solutions.

a) $x^2 - 8x = -16$

b) $2x^2 + 3x - 20 = 0$

c) $-5x^2 + 15x = 10$

d) $x^2 + 4x = 8$

e) $x^2 + 5 = 0$

f) $4x^2 - 64 = 0$

5. Graph each function. Then use the graph to solve the quadratic equation.

a) $p(x) = 3x^2 + 5x - 2$ and $3x^2 + 5x - 2 = 0$

b) $f(x) = 2x^2 - 11x - 21$ and $2x^2 - 11x - 21 = 0$

c) $p(x) = 8x^2 + 2x - 3$ and $8x^2 + 2x - 3 = 0$

d) $f(x) = 3x^2 + x + 1$ and $3x^2 + x + 1 = 0$

6. The population, $P(t)$, of an Ontario city is modelled by the function $P(t) = 14t^2 + 650t + 32\,000$. *Note: t = 0 corresponds to the year 2000.*

a) What will the population be in 2035?

b) When will the population reach 50 000?

c) When was the population 25 000?

7. The function $h(t) = 2.3 + 50t - 1.86t^2$ models the height of an arrow shot from a bow on Mars, where $h(t)$ is the height in metres and t is time in seconds. How long does the arrow stay in flight?

8. The height of an arrow shot by an archer is given by the function $h(t) = -5t^2 + 18t - 0.25$, where $h(t)$ is the height in metres and t is time in seconds. The centre of the target is in the path of the arrow and is 1 m above the ground. When will the arrow hit the centre of the target?

9. The student council is selling cases of gift cards as a fundraiser. The revenue, $R(x)$, in dollars, can be modelled by the function $R(x) = -25x^2 + 100x + 1500$, where x is the number of cases of gift cards sold. How many cases must the students sell to maximize their revenue?

10. The Wheely Fast Co. makes custom skateboards for professional riders. The company models its profit with the function $P(b) = -2b^2 + 14b - 20$, where b is the number of skateboards produced, in thousands, and $P(b)$ is the company's profit, in hundreds of thousands of dollars.

Communication | Tip

A company's break-even point is the point at which the company shows neither a profit nor a loss. This occurs when the profit is zero.

  a) How many skateboards must be produced for the company to break even?

  b) How many skateboards does Wheely Fast Co. need to produce to maximize profit?

11. A ball is tossed upward from a cliff that is 40 m above water. The height of the ball above the water is modelled by $h(t) = -5t^2 + 10t + 40$, where $h(t)$ is the height in metres and t is the time in seconds. Use a graph to answer the following questions.

  a) What is the maximum height reached by the ball?

  b) When will the ball hit the water?

12. The cost, $C(n)$, in dollars, of operating a concrete-cutting machine is modelled by $C(n) = 2.2n^2 - 66n + 655$, where n is the number of minutes the machine is in use.

  a) How long must the machine be in use for the operating cost to be a minimum?

  b) What is the minimum cost?

13. a) For each condition, determine an equation in standard form of a quadratic function that

  i) has two zeros

  ii) has one zero

  iii) has no zeros

  b) What is the maximum number of zeros that a quadratic function can have? Explain.

14. a) What quadratic function could be used to determine the solution of the quadratic equation $3x^2 - 2x + 5 = 4$?

  b) Explain how you could use the function in part (a) to determine the solutions of the equation.

Extending

2 m

15. A parabolic arch is used to support a bridge. Vertical cables are every 2 m along its length. The table of values shows the length of the cables with respect to their placement relative to the centre of the arch. Negative values are to the left of the centre point. Write an algebraic model that relates the length of each cable to its horizontal placement.

Distance from Centre of Arch (m)	Length of Cable (m)
−10	120.0
−8	130.8
−6	139.2
−4	145.2
−2	148.8
0	150.0
2	148.8
4	145.2
6	139.2
8	130.8
10	120.0

16. Determine the points of intersection of
a) $y = 4x - 1$ and $y = 2x^2 + 5x - 7$
b) $y = x^2 - 2x - 9$ and $y = -x^2 + 5x + 6$

17. Show that the function $y = 2x^2 - 3x + 4$ cannot have a y-value less than 1.5.

FREQUENTLY ASKED Questions

Q: **How are the factored and standard forms of a quadratic function related?**

A: The factored form and the standard form are different algebraic representations of the same function. They have the same zeros, axis of symmetry, and maximum or minimum values. As a result, they have the same graph.

Study | **Aid**
- See Lesson 3.2, Example 1.
- Try Mid-Chapter Review Question 5.

EXAMPLE

$$f(x) = (x - 4)(x + 7) \quad \text{and} \quad g(x) = x^2 + 3x - 28$$

Expand the factored form to get the standard form.
$$f(x) = (x - 4)(x + 7)$$
$$= x^2 + 7x - 4x - 28$$
$$= x^2 + 3x - 28$$
$$= g(x)$$

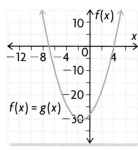

Factor the standard form to get the factored form.
$$g(x) = x^2 + 3x - 28$$
$$= (x + 7)(x - 4)$$
$$= f(x)$$

Q: **What information can you determine about a parabola from the factored and standard forms of a quadratic function?**

A: From the factored form, $f(x) = a(x - r)(x - s)$, you can determine
- the zeros, or x-intercepts, which are r and s
- the equation of the axis of symmetry, which is $x = (r + s) \div 2$
- the coordinates of the vertex by substituting the value of the axis of symmetry for x in the function

From the standard form, $f(x) = ax^2 + bx + c$, you can determine
- the y-intercept, which is c

From both forms you can determine
- the direction in which the parabola opens: up when $a > 0$ and down when $a < 0$

Q: How do you solve a quadratic equation by graphing?

A1: If the quadratic equation is in the form $ax^2 + bx + c = 0$, then graph the function $f(x) = ax^2 + bx + c$ to see where the graph crosses the x-axis. The roots, or solutions, of the equation are the zeros, or x-intercepts, of the function.

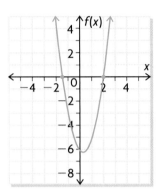

A2: If the quadratic equation is not in the form $ax^2 + bx + c = 0$, for example, $ax^2 + bx + c = d$, then either

graph the functions
$f(x) = ax^2 + bx + c$
and $g(x) = d$ and
see where they
intersect,

or

rewrite $ax^2 + bx + c = d$
in the form $ax^2 + bx +$
$c - d = 0$ and then
graph the function
$f(x) = ax^2 + bx + c - d$ to
see where it crosses the x-axis.

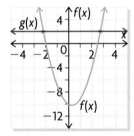

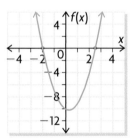

In this case, the
solutions to the equation
$ax^2 + bx + c = d$
are the points of
intersection of the
graphs.

In this case, the
solutions to the equation
$ax^2 + bx + c = d$
are the zeros of the
graph.

PRACTICE Questions

Lesson 3.2

1. Write each function in standard form.
   a) $f(x) = (x + 7)(2x + 3)$
   b) $g(x) = (6 - x)(3x + 2)$
   c) $f(x) = -(2x + 3)(4x - 5)$
   d) $g(x) = -(5 - 3x)(-2x + 1)$

2. Match each function to its graph.
   a) $f(x) = x^2 + 7x + 10$
   b) $f(x) = (5 - x)(x + 2)$
   c) $f(x) = -x^2 + 3x + 10$
   d) $f(x) = (x + 2)(x + 5)$
   e) $f(x) = x^2 - 3x - 10$

   i)
   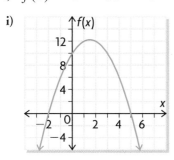

   ii)
   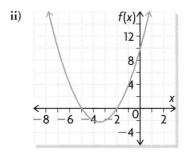

   iii)
   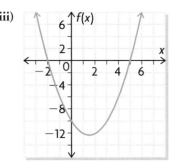

3. Determine the maximum or minimum of each function.
   a) $f(x) = x^2 + 2x - 35$
   b) $g(x) = -2x^2 - 6x + 36$
   c) $f(x) = 2x^2 - 9x - 18$
   d) $g(x) = -2x^2 + x + 15$

4. Which form of a quadratic function do you find most useful? Use an example in your explanation.

Lesson 3.3

5. Graph each function, and then use the graph to locate the zeros.
   a) $f(x) = x^2 + 2x - 8$
   b) $g(x) = 15x^2 - 2x - 1$
   c) $f(x) = 8x^2 + 6x + 1$
   d) $g(x) = 2x^2 - 3x - 5$

6. Solve by graphing.
   a) $x^2 + 2x - 15 = 0$
   b) $(x + 3)(2x + 5) = 0$
   c) $2x^2 - x - 6 = 0$
   d) $x^2 + 7x = -12$

7. A field-hockey ball must stay below waist height, approximately 1 m, when shot; otherwise, it is a dangerous ball. Sally hits the ball. The function $h(t) = -5t^2 + 10t$, where $h(t)$ is in metres and t is in seconds, models the height of the ball. Has she shot a dangerous ball? Explain.

8. Are $x = 3$ and $x = -4$ the solutions to the equation $x^2 - 7x + 12 = 0$? Explain how you know.

3.4

Solving Quadratic Equations by Factoring

GOAL

Use different factoring strategies to solve quadratic equations.

LEARN ABOUT the Math

The profit of a skateboard company can be modelled by the function $P(x) = -63 + 133x - 14x^2$, where $P(x)$ is the profit in thousands of dollars and x is the number of skateboards sold, also in thousands.

? When will the company break even, and when will it be profitable?

EXAMPLE **1**	**Selecting a factoring strategy to determine the break-even points and profitability**

Determine when the company is profitable by calculating the break-even points.

Tim's Solution

$-63 + 133x - 14x^2 = 0$ ◄——————
> The company will break even when the profit is 0, so I set the profit function equal to zero. I can also write the function in factored form, then I can determine the zeros without having to graph the function.

$-14x^2 + 133x - 63 = 0$ ◄——————
> I rewrote the equation in the form $ax^2 + bx + c = 0$. It's easier to see if I can factor it in standard form.

$-7(2x^2 - 19x + 9) = 0$ ◄——————
> I divided out the common factor of -7 because this makes it easier to factor the trinomial.

$$-7(2x^2 - 18x - 1x + 9) = 0$$
$$-7[(2x(x - 9) - 1(x - 9)] = 0$$
$$-7(x - 9)(2x - 1) = 0$$

> I used decomposition to factor the trinomial. I need two numbers that multiply to give 18 (from 2×9) and add to give -19. The two numbers are -18 and -1. I rewrote the middle term as two terms: $-18x$ and $-1x$. I grouped the terms and divided out the common factor of $(x - 9)$.

$$x - 9 = 0 \quad \text{or} \quad 2x - 1 = 0$$

> Since the zeros of the function $f(x) = -14x^2 + 133x - 63$ are the roots of the equation $-14x^2 + 133x - 63 = 0$, I determined these numbers by setting each factor equal to zero and solving for x without having to make a graph first.

$$x = 9 \quad \text{or} \quad 2x = 1$$
$$x = \frac{1}{2}$$

> The solutions of the equations are 9 and $\frac{1}{2}$. These numbers represent the number of skateboards sold in the thousands, so I multiplied the solutions by 1000.

The company will break even when it sells 500 or 9000 skateboards. The company will be profitable if it sells between 500 and 9000 skateboards.

Reflecting

A. If a quadratic equation is expressed in factored form, how can you determine the solutions?

B. Why does equating each of the factors of a quadratic equation to zero enable you to determine its solutions?

APPLY *the* Math

Since some quadratic equations can be solved by factoring, you will need to use a suitable factoring strategy based on the type of quadratic in each equation.

| EXAMPLE **2** | Selecting factoring strategies to solve quadratic equations |

Solve by factoring.
a) $x^2 - 2x - 24 = 0$ b) $9x^2 - 25 = 0$ c) $4x^2 + 20x = -25$

Damir's Solution

a) $x^2 - 2x - 24 = 0$ ◄─────────────── This equation involves a trinomial where $a = 1$. First, I found two numbers that multiply to give -24 and add to give -2. The two numbers are -6 and 4.

$\quad (x - 6)(x + 4) = 0$

$\quad\quad x - 6 = 0 \quad$ and $\quad x + 4 = 0$ ◄─────── I set each factor equal to zero and solved.

$\quad\quad\quad x = 6 \quad$ and $\quad x = -4$

b) $9x^2 - 25 = 0$ ◄─────────────── This equation involves a **difference of squares** because $9x^2 = (3x)^2$, $25 = 5^2$, and the terms are separated by a minus sign. A difference of squares in the form $a^2 - b^2$ factors to $(a - b)(a + b)$.

$\quad (3x - 5)(3x + 5) = 0$

$\quad\quad 3x - 5 = 0 \quad$ and $\quad 3x + 5 = 0$ ◄─────── I set each factor equal to zero and solved.

$\quad\quad\quad 3x = 5 \quad$ and $\quad 3x = -5$

$\quad\quad\quad x = \dfrac{5}{3} \quad$ and $\quad x = -\dfrac{5}{3}$

c) $4x^2 + 20x = -25$

First, I wrote the equation in the form $ax^2 + bx + c = 0$.

This equation involves a trinomial where $a \neq 1$.

I factored it by decomposition, so I found two numbers that multiply to give 100 (from 4×25) and add to give 20. The numbers are 10 and 10. Then I rewrote the $20x$-term as $10x + 10x$.

$\quad 4x^2 + 20x + 25 = 0$ ◄───────────

$\quad 4x^2 + 10x + 10x + 25 = 0$

$\quad 2x(2x + 5) + 5(2x + 5) = 0$ ◄─────── Now I can factor by grouping in pairs and divide out the common factor of $(2x + 5)$.

$\quad\quad (2x + 5)(2x + 5) = 0$

$\quad\quad\quad\quad 2x + 5 = 0$

$\quad\quad\quad\quad\quad 2x = -5$

$\quad\quad\quad\quad\quad x = -\dfrac{5}{2}$ ◄─────── Since both factors are the same, I set one factor equal to zero and solved. This equation has two solutions that are the same, so it really has only one solution.

Some quadratic equations may require some algebraic manipulation before you can solve by factoring.

EXAMPLE 3   Reasoning to solve a more complicated quadratic equation

Solve $2(x + 3)^2 = 5(x + 3)$. Verify your solution.

Beth's Solution: Expanding and Rearranging the Equation

$$2(x + 3)^2 = 5(x + 3)$$
$$2(x + 3)(x + 3) = 5(x + 3)$$
$$2(x^2 + 3x + 3x + 9) = 5x + 15$$
$$2(x^2 + 6x + 9) = 5x + 15$$
$$2x^2 + 12x + 18 = 5x + 15$$
$$2x^2 + 12x + 18 - 5x - 15 = 0$$
$$2x^2 + 7x + 3 = 0$$
$$2x^2 + 6x + 1x + 3 = 0$$
$$2x(x + 3) + 1(x + 3) = 0$$
$$(x + 3)(2x + 1) = 0$$

$$x + 3 = 0 \quad \text{and} \quad 2x + 1 = 0$$
$$x = -3 \quad \text{and} \quad 2x = -1$$
$$x = -\frac{1}{2}$$

I expanded both sides by multiplying. I wrote $(x + 3)^2$ as $(x + 3)(x + 3)$ to make it easier to expand. Next, I multiplied each expression using the distributive property, then collected like terms and rewrote the equation so that the right side is equal to zero. It's easier to find the factors when one side of the equation is equal to zero.

I used decomposition to factor because the coefficient of x^2 is not 1. I need two numbers that multiply to give 6 (from 2×3) and add to give 7. The two numbers are 6 and 1, so I rewrote $7x$ as $6x + 1x$. I grouped the terms and divided out the common factor $(x + 3)$.

I set each factor equal to zero and solved.

Then I checked my solutions.

To see if my solutions are correct, I substituted each x-value into the original equation. I checked to see if the left side of the equation and the right side of the equation give the same answer.

Check:

When $x = -3$,

L.S. $= 2(x + 3)^2$ R.S. $= 5(x + 3)$
$\quad = 2(-3 + 3)^2$ $\quad = 5(-3 + 3)$
$\quad = 2(0)^2$ $\quad = 5(0)$
$\quad = 2(0) = 0$ $\quad = 0$
$\quad = $ R.S.

Both my solutions are correct.

When $x = -\dfrac{1}{2}$,

L.S. $= 2\left(-\dfrac{1}{2} + 3\right)^2$ R.S. $= 5\left(-\dfrac{1}{2} + 3\right)$
$\quad = 2\left(-\dfrac{1}{2} + \dfrac{6}{2}\right)^2$ $\quad = 5\left(-\dfrac{1}{2} + \dfrac{6}{2}\right)$
$\quad = 2\left(\dfrac{5}{2}\right)^2$ $\quad = 5\left(\dfrac{5}{2}\right)$
$\quad = 2\left(\dfrac{25}{4}\right)$ $\quad = \dfrac{25}{2}$
$\quad = \dfrac{25}{2}$
$\quad = $ R.S.

Steve's Solution: Dividing out the Common Factor

$$2(x + 3)^2 = 5(x + 3)$$

$$2(x + 3)^2 - 5(x + 3) = 0$$

$$(x + 3)[2(x + 3) - 5] = 0$$

$$(x + 3)(2x + 6 - 5) = 0$$

$$(x + 3)(2x + 1) = 0$$

$$x + 3 = 0 \quad \text{and} \quad 2x + 1 = 0$$

$$x = 0 - 3 \quad \text{and} \quad 2x = 0 - 1$$

$$x = -3 \quad \text{and} \quad 2x = -1$$

$$x = -\frac{1}{2}$$

Since an equation is easier to solve when one side is equal to zero, I moved $5(x + 3)$ to the left side.

I noticed that $(x + 3)$ is common to both terms, so I divided out this common factor.

I simplified the expression in the second brackets by expanding and collecting like terms.

To solve, I set each factor equal to zero.

EXAMPLE 4   Solving a problem involving a quadratic equation

The path a dolphin travels when it rises above the ocean's surface can be modelled by the function $h(d) = -0.2d^2 + 2d$, where $h(d)$ is the height of the dolphin above the water's surface and d is the horizontal distance from the point where the dolphin broke the water's surface, both in feet. When will the dolphin reach a height of 1.8 feet?

Tyson's Solution

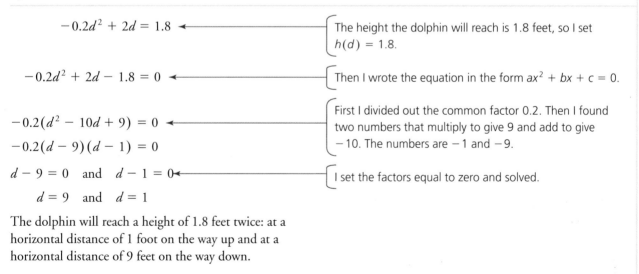

$$-0.2d^2 + 2d = 1.8$$

The height the dolphin will reach is 1.8 feet, so I set $h(d) = 1.8$.

$$-0.2d^2 + 2d - 1.8 = 0$$

Then I wrote the equation in the form $ax^2 + bx + c = 0$.

$$-0.2(d^2 - 10d + 9) = 0$$

$$-0.2(d - 9)(d - 1) = 0$$

First I divided out the common factor 0.2. Then I found two numbers that multiply to give 9 and add to give -10. The numbers are -1 and -9.

$$d - 9 = 0 \quad \text{and} \quad d - 1 = 0$$

$$d = 9 \quad \text{and} \quad d = 1$$

I set the factors equal to zero and solved.

The dolphin will reach a height of 1.8 feet twice: at a horizontal distance of 1 foot on the way up and at a horizontal distance of 9 feet on the way down.

In Summary

Key Idea

- If a quadratic equation of the form $ax^2 + bx + c = 0$ can be written in factored form, then the solutions of the quadratic equation can be determined by setting each of the factors to zero and solving the resulting equations.

Need to Know

- All quadratic equations of the form $ax^2 + bx + c = d$ must be expressed in the form $ax^2 + bx + (c - d) = 0$ before factoring. Doing this is necessary because the zeros of the corresponding function, $f(x) = ax^2 + bx + (c - d)$, are the roots of the equation $ax^2 + bx + c = d$.
- When factoring quadratic equations in the form $ax^2 + bx + c = 0$, apply the same strategies you used to factor quadratic expressions. These strategies include looking for
  - a common factor
  - a pair of numbers r and s, where $rs = ac$ and $r + s = b$
  - familiar patterns such as a difference of squares or perfect squares
- Not all quadratic expressions are factorable. As a result, not all quadratic equations can be solved by factoring. To determine whether $ax^2 + bx + c = 0$ is factorable, multiply a and c. If two numbers can be found that multiply to give the product ac and also add to give b, then the equation can be solved by factoring. If not, then factoring cannot be used. If this is the case, then other methods must be used.

CHECK Your Understanding

1. Solve.

a) $(x + 3)(x - 5) = 0$

b) $5(x - 6)(x - 9) = 0$

c) $(2x + 1)(3x - 5) = 0$

d) $2x(x - 3) = 0$

2. Solve by factoring.

a) $x^2 + x - 20 = 0$

b) $x^2 = 36$

c) $x^2 + 12x = -36$

d) $x^2 = 10x$

3. Determine whether the given number is a root of the quadratic equation.

a) $x = 2;\ x^2 + 6x - 16 = 0$

b) $x = -4;\ 2x^2 - 5x - 35 = 0$

c) $x = 1;\ 6x^2 + 7x = x^2 + 12$

d) $x = -1;\ 5x^2 + 7x = 2x^2 - 6$

PRACTISING

4. Solve by factoring. Verify your solutions.

a) $x^2 - 3x - 54 = 0$ d) $x^2 - 17x + 42 = 0$

b) $x^2 - 169 = 0$ e) $2x^2 - 9x - 5 = 0$

c) $x^2 + 14x + 49 = 0$ f) $3x^2 + 11x - 4 = 0$

5. Solve by factoring. Verify your solutions.

a) $x^2 = 289$ d) $2x^2 + 3x = 16x + 7$

b) $9x^2 - 30x = -25$ e) $4x^2 - 5x = 2x^2 - x + 30$

c) $x^2 - 5x = -3x + 15$ f) $x^2 + 3x + 10 = 3x^2 - 4x - 5$

6. Solve by factoring. Verify your solutions.

a) $3x(x - 2) = 4x(x + 1)$

b) $2x^2(x + 3) = -4x^2(x - 1)$

c) $(x + 5)^2 - 6 = (x + 5)$

d) $(x + 3)(x - 1) = 2(x - 5)(x + 3)$

e) $3(x - 5)^2 = x - 5$

f) $x^3 + 4x^2 = x^3 - 2x^2 - 17x - 5$

7. Solve by factoring. Verify your solutions.

a) $x^2 + 8x = 9x + 42$

b) $2(x + 1)(x - 4) = 4(x - 2)(x + 2)$

c) $x^2 + 5x - 36 = 0$

d) $3(x - 2)^2 = 2(x - 2)$

e) $8x^2 - 3x = -17x - 3$

f) $-8x^2 + 5x = 2x^2 - 8x - 3$

8. A model airplane is shot into the air. Its path is approximated by the function $h(t) = -5t^2 + 25t$, where $h(t)$ is the height in metres and t is the time in seconds. When will the airplane hit the ground?

9. The area of a rectangular enclosure is given by the function $A(w) = -2w^2 + 48w$, where $A(w)$ is the area in square metres and w is the width of the rectangle in metres.

a) What values of w give an area of 0?

b) What is the maximum area of the enclosure?

10. Snowy's Snowboard Co. manufactures snowboards. The company uses the function $P(x) = 324x - 54x^2$ to model its profit, where $P(x)$ is the profit in thousands of dollars and x is the number of snowboards sold, in thousands.

a) How many snowboards must be sold for the company to break even?

b) How many snowboards must be sold for the company to be profitable?

11. A rock is thrown down from a cliff that is 180 m high. The function
    A $h(t) = -5t^2 - 10t + 180$ gives the approximate height of the rock
    above the water, where $h(t)$ is the height in metres and t is the time
    in seconds. When will the rock reach a ledge that is 105 m above
    the water?

12. A helicopter drops an aid package. The height of the package
    above the ground at any time is modelled by the function
    $h(t) = -5t^2 - 30x + 675$, where $h(t)$ is the height in metres and
    t is the time in seconds. How long will it take the package to hit
    the ground?

13. The manager of a hardware store knows that the weekly
    revenue function for batteries sold can be modelled with
    $R(x) = -x^2 + 10x + 30\ 000$, where both the revenue, $R(x)$, and
    the cost, x, of a package of batteries are in dollars. According to
    the model, what is the maximum revenue the store will earn?

14. Kool Klothes has determined that the revenue function for selling
    T x thousand pairs of shorts is $R(x) = -5x^2 + 21x$. The cost function
    $C(x) = 2x + 10$ is the cost of producing the shorts.
    a) Write a profit function.
    b) How many pairs of shorts must the company sell in order to
       break even?

15. Can factoring always be used to solve quadratic equations? Explain.

16. What are the advantages and disadvantages of solving quadratic
    C equations by factoring?

Extending

17. A hot-air balloon drops a sandbag. The table shows the height of the
    sandbag at different times. When will the sandbag reach the ground?

Time (s)	0	1	2	3	4	5	6	7	8	9	10
Height (m)	1000	995	980	955	920	875	820	755	680	595	500

18. What numbers could t be if t^2 must be less than $12t - 20$?

3.5 Solving Problems Involving Quadratic Functions

YOU WILL NEED

- graphing calculator

GOAL

Select and apply factoring and graphing strategies to solve applications involving quadratics.

LEARN ABOUT the Math

A computer software company models the profit on its latest video game using the function $P(x) = -2x^2 + 32x - 110$, where x is the number of games, in thousands, that the company produces and $P(x)$ is the profit, in millions of dollars.

? How can you determine the maximum profit the company can earn?

EXAMPLE 1 | **Selecting a strategy to solve the problem**

Choose a strategy to determine the maximum profit possible.

Matt's Solution: Using a Table of Values

x	P(x)
0	− 110
2	− 54
4	− 14
5	0
6	10
7	16
8	18
9	16

I used different values for x and then picked the highest number in the $P(x)$ column for the maximum profit.

The highest value of $P(x)$ is 18, and it occurs when $x = 8$.

The company must sell 8000 games to earn a profit of $18 million.

Tony's Solution: Using a Graphing Calculator

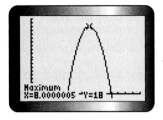

The number of games they must produce to make a maximum profit is 8000. The profit will be $18 million.

I used a graphing calculator. I entered $-2x^2 + 32x - 110$ into Y1 and then used the maximum operation.

According to the graphing calculator, the maximum is 18 when $x = 8$. At first I thought that meant that the company must sell 8 games to earn a profit of $18, and that's not much profit. Then I remembered that the function is for x thousand games and $P(x)$ is in millions of dollars.

Tech | *Support*
For help determining the maximum or minimum value using a graphing calculator, see Technical Appendix, B-9.

Donica's Solution: Factoring

$$P(x) = -2x^2 + 32x - 110$$

This is an equation of a parabola that opens downward, since the coefficient of x^2 is negative.

$$P(x) = -2(x - 5)(x - 11)$$
$$x - 5 = 0 \quad \text{and} \quad x - 11 = 0$$
$$x = 5 \quad \text{and} \quad x = 11$$

The maximum value is at the vertex, which is halfway between the function's two zeros. So I factored $P(x)$ to find the zeros. I set each factor equal to zero to solve.

The maximum occurs at

$$x = \frac{5 + 11}{2}$$

$$x = 8$$

The maximum is halfway between the zeros. I added the zeros and divided by 2.

$$P(8) = -2(8^2) + 32(8) - 110$$
$$= 18$$

I put 8 into the profit function to get $P(8) = 18$.

The company must sell 8000 games to earn a profit of $18 million.

Reflecting

A. Can Matt always be certain he has determined the maximum value using his method?

B. Will Donica always be able to use her method to determine the maximum (or minimum) value of a function? Explain.

C. How do you know that each student has determined the maximum profit and that no other maximum could exist?

D. Why is finding the domain and range important for quadratic equations that model real-world situations?

E. How do you choose your strategy? What factors will affect the method you choose to solve a problem?

APPLY the Math

The strategy you use to solve a problem involving a quadratic function depends on what you are asked to determine, the quadratic function you are given, and the degree of accuracy required.

EXAMPLE 2 **Selecting a tool or strategy to solve a quadratic equation**

Sally is standing on the top of a river slope and throws a ball. The height of the ball at a given time is modelled by the function $h(t) = -5t^2 - 10t + 250$, where $h(t)$ is the height in metres and t is time in seconds. When will the ball be 10 m above the ground?

Rachel's Solution: Factoring

$-5t^2 - 10t + 250 = 10$ ← I set the original function equal to 10 because I wanted to know when the height is 10 m.

$-5t^2 - 10t + 240 = 0$ ← Then I rearranged the equation to get zero on the right side.

$-5(t^2 + 2t - 48) = 0$ ← I factored and set each factor equal to zero and solved for t to find the roots.

$-5(t + 8)(t - 6) = 0$

$t + 8 = 0$ and $t - 6 = 0$

$t = -8$ and $t = 6$

The ball will be 10 m above the ground 6 s into its flight.

Only one of the roots makes sense. The domain for the function is $t \geq 0$, since it doesn't make sense for time to be negative. This means that the only solution is $t = 6$.

Stephanie's Solution: Using a Graphing Calculator

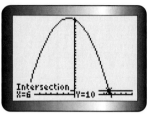

I entered $-5x^2 - 10x + 250$ into Y1 and 10 into Y2. I remembered to change the window setting so that I could see the whole parabola.

I noticed that the line $y = 10$ crosses the parabola twice. This means that I have to find two intersection points.

I used the intersection operation on the calculator to find the points of intersection. The intersections occur when $x = 6$ and when $x = -8$. I can use only the positive solution because the domain is $t \geq 0$.

6 s after the ball is initially thrown, it has a height of 10 m.

In Summary

Key Ideas

- Problems involving quadratic functions can be solved with different strategies, such as
  - a table of values
  - graphing (with or without graphing technology)
  - an algebraic approach involving factoring
- The strategy used depends on what you need to determine and whether or not an estimate or a more accurate answer is required.

Need to Know

- The value of a in the quadratic equation determines the direction in which the parabola opens and whether there is a maximum or a minimum. If $a > 0$, the parabola opens up and there will be a minimum. If $a < 0$, the parabola opens down and there will be a maximum.

- The quadratic function can be used to determine the maximum or minimum value and/or the zeros of the quadratic model. These can then be used as needed to interpret or solve the situation presented.

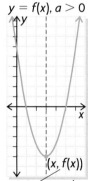

$y = f(x), a > 0$

minimum value

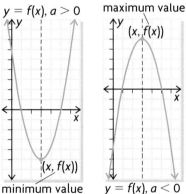

maximum value

$y = f(x), a < 0$

CHECK *Your Understanding*

1. Solve by a table of values, a graphing calculator, and factoring:
A computer software company models its profit with the function
$P(x) = -x^2 + 13x - 36$, where x is the number of games, in
hundreds, that the company sells and $P(x)$ is the profit, in thousands
of dollars.
   a) How many games must the company sell to be profitable?
   b) Which method do you prefer? Why?

2. Use factoring to solve the following problem: The height of a ball
above the ground is given by the function $h(t) = -5t^2 + 45t + 50$,
where $h(t)$ is the height in metres and t is time in seconds. When will
the ball hit the ground?

PRACTISING

3. The function $d(s) = 0.0056s^2 + 0.14s$ models the stopping distance
of a car, $d(s)$, in metres, and the speed, s, in kilometres per hour. What
is the speed when the stopping distance is 7 m? Use a graph to solve.

4. The population of a city is modelled by
$P(t) = 14t^2 + 820t + 52\ 000$, where t is time in years. *Note: $t = 0$*
corresponds to the year 2000. According to the model, what will the
population be in the year 2020? Here is Beverly's solution:

Beverly's Solution

$P(2020) = 14(2020)^2 + 820(2020) + 52\ 000$ | I substituted 2020 into the function for t because that's the year for which I want to know the population. Then I solved.

$= 57\ 125\ 600 + 1\ 656\ 400 + 52\ 000$

$= 58\ 834\ 000$

The population will be 58 834 000.

Are Beverly's solution and reasoning correct? Explain.

5. Solve $4x^2 - 10x - 24 = 0$ using three different methods.
K

6. The population of a city is modelled by the function
$P(t) = 0.5t^2 + 10t + 200$, where $P(t)$ is the population in thousands
and t is time in years. *Note: $t = 0$* corresponds to the year 2000.
According to the model, when will the population reach 312 000?

7. Which methods can be used to solve $-4.9t^2 + 9.8t + 73.5 = 0$?
T Explain why each method works.

8. Water from a hose is sprayed on a fire burning at a height of 10 m up the side of a wall. If the function $h(x) = -0.15x^2 + 3x$, where x is the horizontal distance from the fire, in metres, models the height of the water, $h(x)$, also in metres, how far back does the firefighter have to stand in order to put out the fire?

9. The president of a company that manufactures toy cars thinks that the function $P(c) = -2c^2 + 14c - 20$ represents the company's profit, where c is the number of cars produced, in thousands, and $P(c)$ is the company's profit, in hundreds of thousands of dollars. Determine the maximum profit the company can earn.

10. **A** A ball is thrown vertically upward from the top of a cliff. The height of the ball is modelled by the function $h(t) = 65 + 10t - 5t^2$, where $h(t)$ is the height in metres and t is time in seconds. Determine when the ball reaches its maximum height.

11. **C** A computer software company models the profit on its latest video game with the function $P(x) = -2x^2 + 32x - 110$, where x is the number of games the company produces, in thousands, and $P(x)$ is the profit, in thousands of dollars. How many games must the company sell to make a profit of $16 000?
    a) Write a solution to the problem. Indicate why you chose the strategy you did.
    b) Discuss your solution and reasoning with a partner. Be ready to share your ideas with the class.

Extending

12. The population of a city, $P(t)$, is modelled by the quadratic function $P(t) = 50t^2 + 1000t + 20\,000$, where t is time in years. *Note*: $t = 0$ corresponds to the year 2000. Peg says that the population was 35 000 in 1970. Explain her reasoning for choosing that year.

13. Which pair of numbers that add to 10 will multiply to give the greatest product?

14. Jasmine and Raj have 24 m of fencing to enclose a rectangular garden. What are the dimensions of the largest rectangular garden they can enclose with that length of fencing?

Creating a Quadratic Model from Data

YOU WILL NEED

- graphing calculator
- graph paper

Determine the equation of a curve of good fit using the factored form.

INVESTIGATE the Math

A ball is thrown into the air from the top of a building. The table of values gives the height of the ball at different times during the flight.

Time (s)	0	1	2	3	4	5
Height (m)	30	50	60	60	50	30

? **What is a function that will model the data?**

A. Create a scatter plot, with an appropriate scale, from the data.

curve of good fit

a curve that approximates or is close to the distribution of points in a scatter plot

B. What shape best describes the graph? Draw a **curve of good fit**.

C. Extend the graph to estimate the location of the zeros.

D. Use the zeros to write an equation in factored form.

E. In what direction does the parabola open? What does this tell you?

F. Using one of the points in the table, calculate the coefficient of x^2. Write the equation for the data in factored form and in standard form.

quadratic regression

a process that fits the second-degree polynomial $ax^2 + bx + c$ to the data

G. Using a graphing calculator and **quadratic regression**, determine the quadratic function model.

H. How does your model compare with the graphing calculator's model?

Reflecting

I. How does the factored form of an equation help you determine the curve of good fit?

Tech | Support

For help determining the equation of a curve by quadratic regression, see Technical Appendix, B-10.

J. How will you know whether the equation is a good representation of your data?

K. How would your model change if it has only one zero? What if the model has no zeros?

APPLY the Math

> **EXAMPLE 1** | **Representing a quadratic function from its graph**

Determine an equation in factored form that best represents the graph shown. Express your equation in standard form as well.

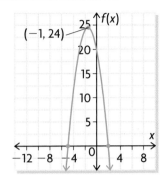

Angela's Solution: Using the Zeros and the Factored Form of a Quadratic Function

$y = a(x + 5)(x - 2)$ ← I used the graph to determine that the zeros of the function are -5 and 2. So I wrote the factored form of the equation.

$24 = a(-1 + 5)(-1 - 2)$ ← The graph passes through the point $(-1, 24)$, so I substituted $x = -1$ and $y = 24$. Then I solved for a.

$24 = a(4)(-3)$

$24 = -12a$

$-2 = a$

The equation in factored form is
$y = -2(x + 5)(x - 2)$ ← I substituted the value for a into the original equation. I expanded to get the standard form.

$= -2(x^2 - 2x + 5x - 10)$

$= -2(x^2 + 3x - 10)$

$= -2x^2 - 6x + 20$

The equation in standard form is
$y = -2x^2 - 6x + 20$.

If the curve of good fit is a parabola and it crosses the x-axis, then you can estimate the zeros and use the factored form of a quadratic function to determine its equation.

EXAMPLE 2 — Representing a quadratic function from data

The track for the main hill of a roller coaster forms a parabolic arch. Vertical support columns are set in the ground to reinforce the arch every 2 m along its length. The table of values shows the length of the columns in terms of their placement relative to the centre of the arch. Negative values are to the left of the centre point. Write an algebraic model in factored form that relates the length of each column to its horizontal placement. Check your answer with a graphing calculator.

Distance from Centre of Arch (m)	Length of Support Column (m)
−10	35.0
−8	40.0
−6	44.6
−4	47.6
−2	49.4
0	50.0
2	49.4
4	47.6
6	44.4
8	40.4
10	35.0

Giovanni's Solution

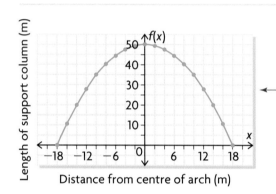

I graphed the data from the table by hand. It looks quadratic. I drew a curve of good fit and then extended the graph to identify the zeros. They're $x = 18$ and $x = -18$. I substituted those values into the factored form of the equation.

$$y = a(x - 18)(x + 18)$$
$$35 = a(10 - 18)(10 + 18)$$
$$35 = a(-8)(28)$$
$$35 = -224a$$

The graph must pass through one of the pairs of data given in the table. I chose a point to get a. I used (10, 35). I put 10 in for x and 35 in for y in the equation, then solved for a.

$$-0.15625 = a$$
$$y = -0.15625(x - 18)(x + 18)$$
$$y = -0.15625(x^2 - 324)$$
$$y = -0.15625x^2 + 50.625$$

I substituted a into my original equation and expanded into the standard form, so that I could compare my solution with the one the graphing calculator gives.

I compared my equation to the answer the calculator gave. I ignored the value of b since it was so small.

From the calculator, rounding to two decimal places, $y = -0.15x^2 + 49.97$, and from my scatter plot, $y = -0.15625x^2 + 50.625$.

I think I did a pretty good job of finding an equation, but the graphing calculator found it much faster than I could by hand.

Tech | Support

For help creating a scatter plot and determining a curve of good fit using regression, see Technical Appendix, B-10.

You can use a graphing calculator to create a scatter plot, and then use guess and check to graph the equation that fits the data.

Victoria's Solution: Using the Zeros and Adjusting *a* with a Graphing Calculator

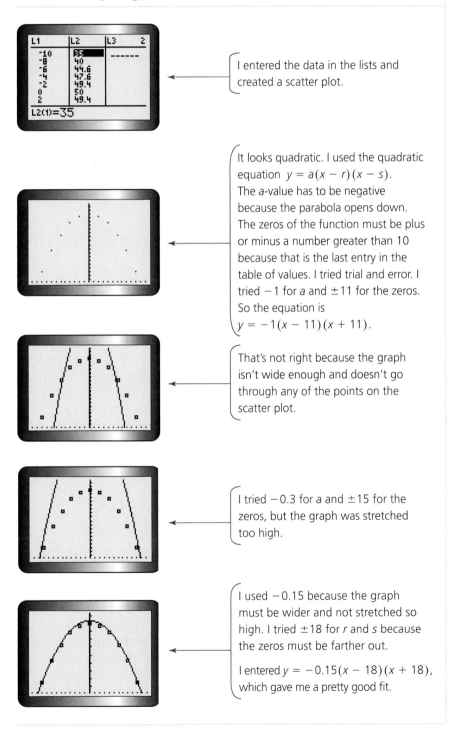

I entered the data in the lists and created a scatter plot.

It looks quadratic. I used the quadratic equation $y = a(x - r)(x - s)$. The *a*-value has to be negative because the parabola opens down. The zeros of the function must be plus or minus a number greater than 10 because that is the last entry in the table of values. I tried trial and error. I tried -1 for *a* and ± 11 for the zeros. So the equation is $y = -1(x - 11)(x + 11)$.

That's not right because the graph isn't wide enough and doesn't go through any of the points on the scatter plot.

I tried -0.3 for *a* and ± 15 for the zeros, but the graph was stretched too high.

I used -0.15 because the graph must be wider and not stretched so high. I tried ± 18 for *r* and *s* because the zeros must be farther out.

I entered $y = -0.15(x - 18)(x + 18)$, which gave me a pretty good fit.

With some graphing programs you can make sliders. Sliders make it easier to change the parameters in the equation so that you can adjust and fit the curve to your data.

<table>
<tr><td>EXAMPLE 3</td><td>Representing a quadratic function with graphing software</td></tr>
</table>

A stone is dropped from a bridge 20 m above the water. The table of values shows the time, in seconds, and the height of the stone above the water, in metres. Write an algebraic model for the height of the stone, and use it to estimate when the stone will be 10 m above the water.

Collection 1

	1	2	3	4	5
Time (s)	0.0	0.5	1.0	1.5	2.0
Position (m)	20.00	18.75	15.00	8.75	0.00

Kommy's Solution

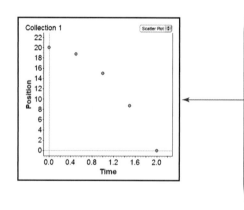

I used graphing software to create a scatter plot and see the shape.

It looks quadratic because it is not a straight line but does curve downward. The maximum height happens when $x = 0$, the time the stone is dropped. One zero happens when $x = 2$, and the maximum happens halfway between the zeros. This means that the other zero is $x = -2$.

$$f(x) = a(x - 2)(x + 2)$$

I substituted the zeros into the factored form of the equation.

I added a slider and function to my graph. I used a slider to find the value for a. I tried to find a good fit by changing the value of a.

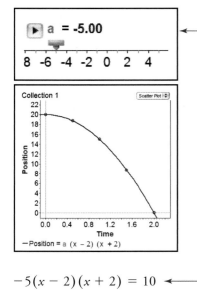

When $a = -5$ for the slider, the fit is pretty good.

So the equation is $f(x) = -5(x - 2)(x + 2)$. I have to state the domain because the model will make sense only when $x \geq 0$, since time can't be negative.

Tech | **Support**

For help with graphing software, see Technical Appendix, B-23.

I used my equation to find when the stone will have a height of 10 m. I set $f(x)$ equal to 10.

$$-5(x - 2)(x + 2) = 10$$
$$-5(x^2 - 4) = 10$$
$$-5x^2 + 20 = 10$$
$$-5x^2 = -10$$
$$x^2 = 2$$
$$x = \pm\sqrt{2}, x \geq 0$$
$$x \doteq 1.4$$

I expanded the left side of the equation. Since the equation has only an x^2-term and no x-term, I got the x^2-term by itself. Then I solved for x.

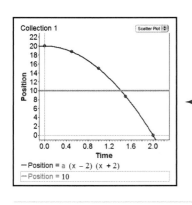

I checked my answer by graphing the line $y = 10$.

In Summary

Key Ideas

- If a scatter plot of data has a parabolic shape and its curve of good fit passes through the x-axis, then the factored form of the quadratic function can be used to determine an algebraic model for the relationship.
- Once the algebraic model has been determined, it can be used to solve problems involving the relationship.

Need to Know

- The x-intercepts, or zeros, of the curve of good fit represent the values of r and s in the factored form of a quadratic function $f(x) = a(x - r)(x - s)$.
- The value of a can be determined
  - algebraically: substitute the coordinates of a point that lies on or close to the curve of good fit into $f(x)$ and solve for a.
  - graphically: estimate the value of a and graph the resulting parabola with graphing technology. By observing the graph, you can adjust your estimate of a and regraph until the parabola passes through or close to a large number of points of the scatter plot.
- Graphing technology can be used to determine the algebraic model of the curve of good fit. You can use quadratic regression if the data have a parabolic pattern.

CHECK *Your Understanding*

1. Determine the equation of each parabola by
   - determining the zeros of the function
   - writing the function in factored form
   - determining the value of a using a point on the curve
   - expressing your answer in standard form

a)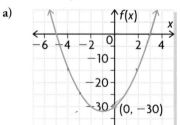

b)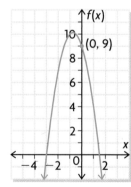

2. Create a scatter plot from the data in the table, and decide whether the data appear to be quadratic. Justify your decision.

a)

Time (h)	0	1	2	3	4	5
Population	1200	2400	4800	9600	19 200	38 400

b)

Distance (m)	0	1	2	3	4	5	6	7	8
Height (m)	0	0.9	1.6	2.1	2.4	2.5	2.4	2.0	1.4

PRACTISING

3. Determine the equation of each parabola.

a)
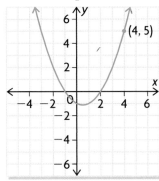

b)
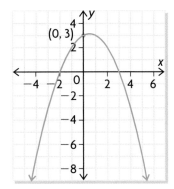

4. Write the standard form of the quadratic equation for each case.
K

	x-intercepts	*y*-intercept
a)	-3 and 4	24
b)	-2 and -5	10
c)	5 and -7	-105
d)	4 and 2	-24

5. A garden hose sprays a stream of water across a lawn. The table shows the approximate height of the stream at various distances from the nozzle. Determine an equation of the curve of good fit. Use your algebraic model to see how far away you need to stand to water a potted plant that is 1 m above the ground.

Distance from Nozzle (m)	0	0.5	1.0	1.5	2.0	2.5	3.0	3.5	4.0
Height above Lawn (m)	0.5	1.2	1.6	1.9	2.0	1.9	1.6	1.2	0.5

6. While on vacation, Steve filmed a cliff diver. He analyzed the video and recorded the time and height of the diver in the table below. Determine an algebraic model to fit the data. Then use the model to predict when the diver will be 5 m above the water.

Time (s)	0	0.5	1.0	1.5	2.0
Height (m)	35.00	33.75	30.00	23.75	15.00

Time (s)	Height (m)
0	1.0
0.5	4.5
1.0	6.0
1.5	4.5
2.0	1.0

7. The height, at a given time, of a child above the ground when the child is on a trampoline is shown in the table. Determine an algebraic model for the data. Then use the model to predict when the child will reach a height of 3 m.

8. The height of an arrow shot by an archer is given in the table. Determine the equation of a curve of good fit. Use it to predict when the arrow will hit the ground.

t (s)	0	0.5	1.0	1.5	2.0	2.5
h (m)	0.5	8.2	13.4	16.2	16.5	14.3

9. The data in the table describe the flight of a plastic glider launched from a tower on a hilltop. The height values are negative whenever the glider is below the height of the hilltop.

Time (s)	Height (m)
0	7.2
1	4.4
2	2.0
3	0
4	−1.6
5	−2.8
6	−3.6
7	−4.0
8	−4.0
9	−3.6
10	−2.8

Time (s)	Height (m)
11	−1.6
12	0
13	2.0
14	4.4
15	7.2
16	10.4
17	14.0
18	18.0
19	22.4
20	27.2

a) Write an equation to model the flight of the glider.
b) What is the lowest point in the glider's flight?

10. The Paymore Shoe Company introduced a new line of neon-green high-heeled running shoes. The table shows the number of pairs of shoes sold at one store over an 11-month period.
   a) What is the equation of a curve of good fit?
   b) According to your equation, when will the store sell no neon-green high-heeled running shoes?

Month	Pairs of Shoes Sold
1	56
2	60
3	62
4	62
5	60
6	56
7	50
8	42
9	32
10	20
11	6

11. The diagrams show points joined by all possible segments.

T

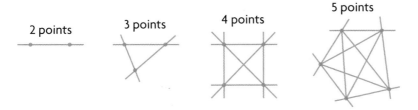

2 points   3 points   4 points   5 points

   a) Extend the pattern to include a figure with six points.
   b) Write an algebraic equation for the number of line segments in terms of the number of points. Assume that the number of line segments for 0 points and 1 point is zero, since you cannot draw a line in these situations.

Number of Points	Number of Lines
0	0
1	0
2	
3	
4	
5	
6	

12. Explain how to determine the equation of a quadratic function if the zeros of the function can be estimated from a scatter plot.

C

Extending

13. The Golden Gate Bridge, in San Francisco, is a suspension bridge supported by a pair of cables that appear to form parabolas. The cables are attached at either end to a pair of towers at points 152 m above the roadway. The towers are 1280 m apart, and the cable reaches its lowest point when it is 4 m above the roadway. Determine an algebraic expression that models the cable as it hangs between the towers. (*Hint:* Transfer the data to a graph such that the parabola lies below the *x*-axis.)

14. For a school experiment, Marcus had to record the height of a model rocket during its flight. However, during the experiment, he discovered that the motion detector he was using had stopped working. Before the detector quit, it collected the data in the table.
   a) The trajectory is quadratic. Complete the table up to the time when the rocket hit the ground.
   b) Determine an equation that models the height of the rocket.
   c) What is the maximum height of the rocket?

Time (s)	Height (m)
0	1.500
0.5	12.525
1.0	21.100
1.5	27.225
2.0	30.900

Curious Counting

How many handshakes will there be if 8 people must shake hands with each other only once?

How many lines can be drawn that connect each pair of points if there are 8 points in total?

How many angles are there in this diagram?

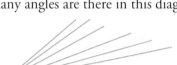

YOU WILL NEED
- graphing calculator (optional)

How are these three problems related to each other?

1. For each problem, examine the simplest case. Determine
   - the number of handshakes if 2 people shake hands with each other only once
   - the number of lines that can be drawn that connect 2 points
   - the number of angles formed from 2 lines

2. For each problem, examine the next case by increasing the number of people, points, and lines to 3. Increase the number of people, points, and lines systematically by 1 each time, and examine each case. Record your findings in the table. What do you notice?

n, Number of People/ Points/Lines	2	3	4	5	6	7	8
Number of Handshakes							
Number of Lines							
Number of Angles							

3. Create a single scatter plot using the number of people/points/lines as the independent variable and the number of handshakes/lines/angles as the dependent variable. Determine the first and second differences of the dependent variable.

4. What type of function can be used to model each situation? Explain how you know.

5. Determine the function that represents each relationship.

6. Predict how many handshakes/lines/angles there will be if 15 people/points/lines are used.

7. How are these problems related to each other?

FREQUENTLY ASKED Questions

Q: **How do you solve a quadratic equation by factoring?**

A: First, express the equation in standard form, $ax^2 + bx + c = 0$. Then use an appropriate factoring strategy to factor the quadratic expression. To determine the roots, or solutions, set each factor equal to zero and solve the resulting equations.

Study | Aid
- See Lesson 3.4, Examples 1, 2, and 4.
- Try Chapter Review Questions 6 and 7.

EXAMPLE

$$x^2 - 24 = x + 6$$
$$x^2 - x - 24 - 6 = 0$$
$$x^2 - x - 30 = 0$$
$$(x - 6)(x + 5) = 0$$
$$x - 6 = 0 \quad \text{and} \quad x + 5 = 0$$
$$x = 6 \quad \text{and} \quad x = -5$$

Q: **What strategies can be used to solve problems involving quadratic functions?**

A: A table of values might work, but it is time-consuming. Graphing by hand is also time-consuming. Using a graphing calculator is faster and is very helpful if the answer is not an integer. Factoring works only if the equation is written so that one side is zero and the equation can be factored. It may be difficult to tell whether the equation can be factored.

Study | Aid
- See Lesson 3.5, Examples 1 and 2.
- Try Chapter Review Questions 8 and 9.

Q: **What strategies can be used to create a quadratic function from data?**

A: Graph the data either by hand or using a graphing calculator to see if the curve looks quadratic. If the function has zeros that are easily identifiable, use them to write the function in the form $f(x) = a(x - r)(x - s)$. Then substitute one of the points from the graph for x and $f(x)$, and solve for a. Rewrite the function, including the zeros and a. You can leave it in factored form, or you can expand the factored form to put the equation in standard form. Verify that your equation matches the data.

Study | Aid
- See Lesson 3.6, Examples 1 and 2.
- Try Chapter Review Questions 10 and 11.

PRACTICE Questions

Lesson 3.2

1. Match each factored form with the correct standard form.

a) $f(x) = (x + 3)(2x - 7)$
b) $f(x) = 2(x + 7)(x - 3)$
c) $f(x) = (2x + 1)(x + 7)$
d) $f(x) = (x + 7)(x - 3)$

i) $f(x) = x^2 + 4x - 21$
ii) $f(x) = 2x^2 - x - 21$
iii) $f(x) = 2x^2 + 8x - 42$
iv) $f(x) = 2x^2 + 15x + 7$

2. Determine the maximum or minimum of each function.

a) $f(x) = x^2 - 2x - 35$
b) $f(x) = 2x^2 + 7x + 3$
c) $g(x) = -2x^2 + x + 15$

Lesson 3.3

3. Determine the zeros and the maximum or minimum value for each function.

a) $f(x) = x^2 + 2x - 15$
b) $f(x) = -x^2 + 8x - 7$
c) $f(x) = 2x^2 + 18x + 16$
d) $f(x) = 2x^2 + 7x + 3$
e) $f(x) = 6x^2 + 7x - 3$
f) $f(x) = -x^2 + 49$

4. The function $h(t) = 1 + 4t - 1.86t^2$ models the height of a rock thrown upward on the planet Mars, where $h(t)$ is height in metres and t is time in seconds. Use a graph to determine

a) the maximum height the rock reaches
b) how long the rock will be above the surface of Mars

5. Determine the zeros, the coordinates of the vertex, and the y-intercept for each function.

a)
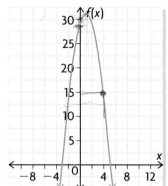

b)
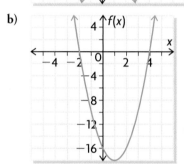

Lesson 3.4

6. Solve by factoring.

a) $x^2 + 2x - 35 = 0$
b) $-x^2 - 5x = -24$
c) $9x^2 = 6x - 1$
d) $6x^2 = 7x + 5$

7. A firecracker is fired from the ground. The height of the firecracker at a given time is modelled by the function $h(t) = -5t^2 + 50t$, where $h(t)$ is the height in metres and t is time in seconds. When will the firecracker reach a height of 45 m?

8. The population of a city, $P(t)$, is given by the function $P(t) = 14t^2 + 820t + 42\,000$, where t is time in years. *Note: $t = 0$ corresponds to the year 2000.*

   a) When will the population reach 56 224? Provide your reasoning.

   b) What will the population be in 2035? Provide your reasoning.

9. Fred wants to install a wooden deck around his rectangular swimming pool. The function $C(x) = 120x^2 + 1800x + 5400$ represents the cost of installation, where x is the width of the deck in metres and $C(x)$ is the cost in dollars. What will the width be if Fred spends $9480 for the deck? Here is Steve's solution.

Steve's Solution

I used a graphing calculator to solve this problem. I entered $120x^2 + 1800x + 5400$ into Y1 and 9480 into Y2 to see where they intersect.

They intersect at two places: $x = 2$ and $x = -17$. Since both answers must be positive, use $x = 2$ and $x = 17$. Because you will get more deck with a higher number, use only $x = 17$.

Do you agree with his reasoning? Why or why not?

10. A toy rocket sitting on a tower is launched vertically upward. Its height y at time t is given in the table.

t Time (s)	y Height (m)
0	16
1	49
2	60
3	85
4	88
5	81
6	64
7	37
8	0

   a) What is an equation of a curve of good fit?

   b) How do you know that the equation in part (a) is a good fit?

11. Determine the equation of a curve of good fit for the scatter plot shown.

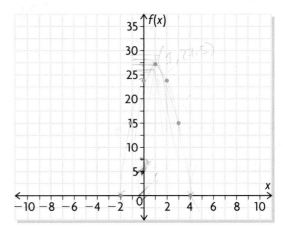

1. Write in standard form.
   a) $f(x) = (3x - 5)(2x - 9)$ b) $f(x) = (6 - 5x)(x + 2)$

2. Write in factored form.
   a) $f(x) = x^2 - 81$ b) $f(x) = 6x^2 + 5x - 4$

3. Determine the zeros, the axis of symmetry, and the maximum or minimum value for each function.
   a) $f(x) = x^2 + 2x - 35$ b) $f(x) = -4x^2 - 12x + 7$

4. Can all quadratic equations be solved by factoring? Explain.

5. Solve by graphing.
   a) $x^2 + 6x - 3 = -3$ b) $2x^2 - 5x = 7$

6. Solve by factoring.
   a) $2x^2 + 11x - 6 = 0$ b) $x^2 = 4x + 21$

7. The population of a town, $P(t)$, is modelled by the function $P(t) = 6t^2 + 110t + 3000$, where t is time in years. *Note:* $t = 0$ represents the year 2000.
   a) When will the population reach 6000?
   b) What will the population be in 2030?

8. Target-shooting disks are launched into the air from a machine 12 m above the ground. The height, $h(t)$, in metres, of the disk after launch is modelled by the function $h(t) = -5t^2 + 30t + 12$, where t is time in seconds.
   a) When will the disk reach the ground?
   b) What is the maximum height the disk reaches?

9. Students at an agricultural school collected data showing the effect of different annual amounts of rainfall, x, in hectare-metres (ha·m), on the yield of broccoli, y, in hundreds of kilograms per hectare (100 kg/ha). The table lists the data.
   a) What is an equation of the curve of good fit?
   b) How do you know whether the equation in part (a) is a good fit?
   c) Use your equation to calculate the yield when there is 1.85 ha·m of annual rainfall.

10. Why is it useful to have a curve of good fit?

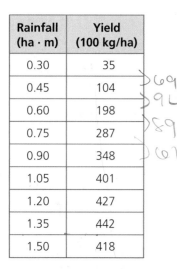

Rainfall (ha · m)	Yield (100 kg/ha)
0.30	35
0.45	104
0.60	198
0.75	287
0.90	348
1.05	401
1.20	427
1.35	442
1.50	418

Quadratic Cases

Quadratic functions are used as mathematical models for many real-life situations, such as designing architectural supports, decorative fountains, and satellite dishes. Quadratic models can show trends and provide predictions about relationships among data. They can also be used to solve problems involving maximizing profit, minimizing the amount of material used to manufacture something, and calculating the location of a projectile, such as a ball.

? **What can you model with a quadratic function?**

A. Collect some data for which the zeros can be determined easily, from either a secondary source, such as the Internet, or an experiment that can be modelled with a quadratic function.

B. Determine a model in factored form that can be used to represent your data.

C. Create several questions that can be answered by your model.

D. In a report, indicate
  - how you decided that a quadratic model would be appropriate
  - how you manually determined the factored form of the equation that fits your data
  - how well you think the equation fits your data
  - the questions you created and the answers you determined by using the model

Task | Checklist

✔ Did you show all your steps?

✔ Did you include a graph?

✔ Did you support your choice of data?

✔ Did you explain your thinking clearly?

Multiple Choice

1. If $f(x) = 4x^2 - 3x + 8$, then $f(-2) = $ ▢.

   a) 30

   b) 18

   c) -6

   d) 8

2. Which of the following functions is quadratic?

   a) $f(x) = 4 + 2x$

   b) $y = -x^3 + 4$

   c) $f(x) = 6x^2 - 3x + 12$

   d) $y = (2x^2 - 4)(x + 4)$

3. Which of the following is the graph of $y = (x + 5)^2$?

   a)

   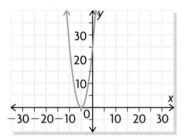

   b)

   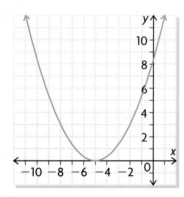

c)

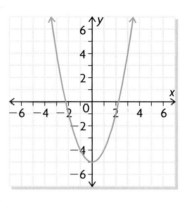

d)

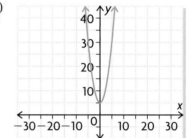

4. Which function has the following domain and range?

$$D = \{x \in \mathbf{R}\}$$
$$R = \{y \in \mathbf{R} \mid y \le 10\}$$

   a)

   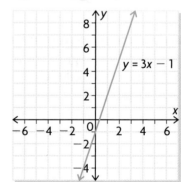

   $y = 3x - 1$

b)

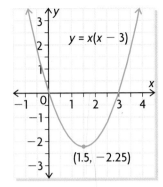

$y = x(x - 3)$

$(1.5, -2.25)$

c)

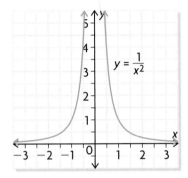

$y = \dfrac{1}{x^2}$

d)

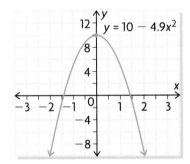

$y = 10 - 4.9x^2$

5. Identify the equation that corresponds to the following transformations applied to the graph of $y = x^2$: reflected through the x-axis, then translated down 7 units.

a) $y = \dfrac{1}{7}x^2 + 7$

b) $y = 7(x + 7)^2$

c) $y = -x^2 - 7$

d) $y = -7(x - 7)^2 - 7$

6. Identify the equation that corresponds to the following transformations applied to the graph of $y = x^2$: stretched vertically by a factor of 7, then translated left 7 units.

a) $y = \dfrac{1}{7}x^2 + 7$

b) $y = 7(x + 7)^2$

c) $y = -x^2 - 7$

d) $y = -7(x - 7)^2 - 7$

7. Identify the equation that corresponds to the following transformations applied to the graph of $y = x^2$: reflected through the x-axis, stretched vertically by a factor of 7, then translated to the right 7 units and down 7 units.

a) $y = \dfrac{1}{7}x^2 + 7$

b) $y = 7(x + 7)^2$

c) $y = -x^2 - 7$

d) $y = -7(x - 7)^2 - 7$

8. Identify the relation that is not a function.

a)

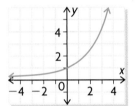

b)

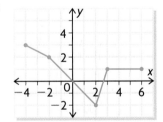

c)

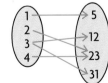

d) $(8, 9), (3, 2), (5, 7), (1, 0), (4, 6)$

9. Expand and simplify $3x(2x - 5) - (2x + 1)^2$.
   a) $2x^2 - 4$
   b) $6x^2 - 19x + 1$
   c) $2x^2 - 11x - 14$
   d) $2x^2 - 19x - 1$

10. Identify the missing factor:
   $6x^2 - 11x - 10 = (3x + 2)$ ▨
   a) $2x + 5$
   b) $2x - 5$
   c) $5x + 2$
   d) $5x - 2$

11. For the expression $kx^2 + 6x + 8$, identify the values of k that make the trinomial unfactorable.
   a) $k = 1$
   b) $k = -2$
   c) $k = 2$
   d) $k = 3$

12. A model rocket is launched straight upward with an initial velocity of 22 m/s. The height of the rocket, h, in metres, can be modelled by $h(t) = -5t^2 + 22t$, where t is the elapsed time in seconds. What is the maximum height the rocket reaches?
   a) 19.5 m
   b) 10.2 m
   c) 24.2 m
   d) 29.6 m

13. Identify the expressions that cannot be factored.
   a) $4x^2 - 12x + 9$
   b) $x^2 + 3x + 2$
   c) $x^2 - 3x + 5$
   d) $100 - x^2$

14. A rectangular enclosure has an area in square metres given by $A(W) = -2w^2 + 36w$, where w is the width of the rectangle in metres. Determine the width that would create a rectangular enclosure of 130 m².
   a) 5 m
   b) 13 m
   c) 10 m
   d) 7 m

15. Which of the following functions is equivalent to $f(x) = -2(x - 5)^2 + 3$?
   a) $g(x) = -2x^2 - 5x + 3$
   b) $g(x) = x^2 - 10x + 28$
   c) $g(x) = -2x^2 + 20x - 47$
   d) $g(x) = -2x^2 - 47$

16. Which of the following expressions is a perfect square?
   a) $4x^2 - 9$
   b) $4x^2 + 37x + 9$
   c) $4x^2 + 12x + 9$
   d) $4x^2 + 36x + 9$

17. Which are the correct zeros of $f(x) = x^2 - x - 12$?
   a) $x = 4$ and $x = -3$
   b) $x = 4$ and $x = 3$
   c) $x = -4$ and $x = 3$
   d) $x = -4$ and $x = -3$

18. Which are the coordinates of the vertex of $f(x) = 2(x - 4)(x - 10)$?
   a) $(4, 10)$
   b) $(7, -18)$
   c) $(-7, 18)$
   d) $(-4, -10)$

Investigations

19. Analyzing Quadratic Functions

Analyze the function $f(x) = 3(x - 4)^2 + 5$ in-depth. Include

a) the domain and range

b) the relationship to the function $f(x) = x^2$, including all applied transformations

c) a sketch of the function

20. Coast Guard Rescue

Over the ocean, an inflatable raft is dropped from a coast guard helicopter to a sinking ship below. The table shows the height of the raft above the water at different times as it falls.

t (s)	0	1	2	3	4	5
h (m)	320	315	300	275	240	195

a) Draw a scatter plot of the data.

b) What type of model represents the relationship between the height of the raft and time? Explain how you know.

c) Use first and second differences to extend the table of values until the raft reaches the water.

d) Draw a curve of good fit. Is the height of the raft a function of time? Explain.

e) Use your graph to determine the location of the vertex, the axis of symmetry, and the zeros. Use this information to help you determine a function that models this relationship.

f) State the domain and range of your function in this context.

g) Use your function to determine

    i) the height of the raft at 7.5 s

    ii) the time it takes the raft to reach a height of 50 m

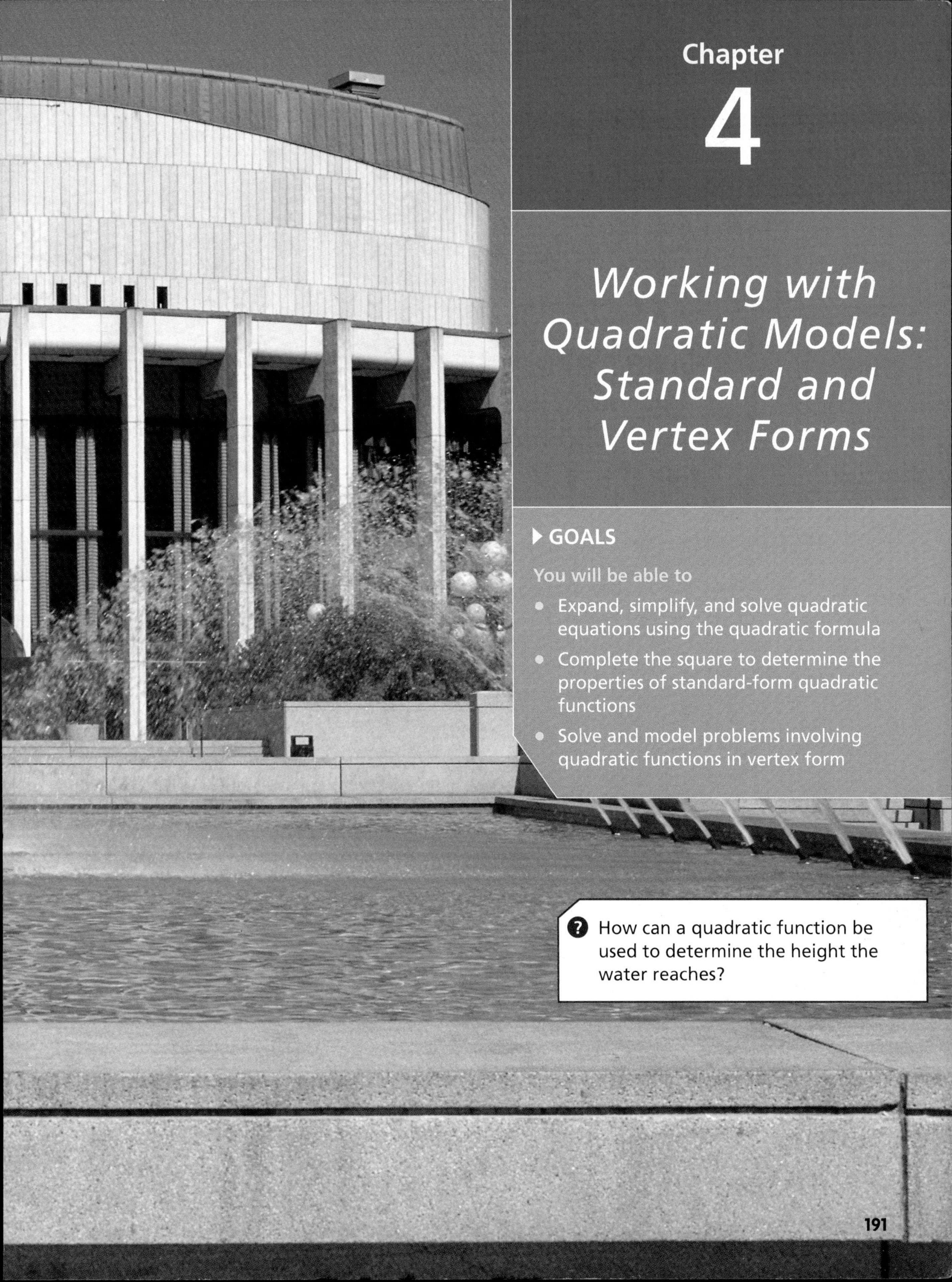

Working with Quadratic Models: Standard and Vertex Forms

▶ **GOALS**

You will be able to

- Expand, simplify, and solve quadratic equations using the quadratic formula

- Complete the square to determine the properties of standard-form quadratic functions

- Solve and model problems involving quadratic functions in vertex form

❓ How can a quadratic function be used to determine the height the water reaches?

WORDS *You Need to Know*

1. Match the term with the picture or example that best illustrates its definition.

a) quadratic equation
b) solutions to a quadratic equation
c) quadratic function

d) perfect square trinomial
e) curve of good fit
f) vertex

i)

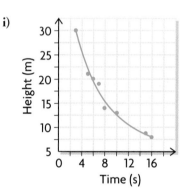

iv) $-12x^2 - 31x + 27 = 0$
v) $f(x) = x^2 - 3x + 9$
vi)

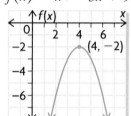

ii) $x^2 + 10x + 25$

iii) $x^2 - 5x - 24 = 0$
$(x - 8)(x + 3) = 0$
$x = 8$ and $x = -3$

SKILLS AND CONCEPTS *You Need*

Study Aid

For help, see Essential Skills Appendix, A-8.

Solving Quadratic Equations by Graphing and Factoring

If the value of y is known for the function $y = ax^2 + bx + c$, then the corresponding values of x can be determined either graphically or algebraically.

EXAMPLE

Solve $x^2 - 2x - 15 = 0$ by a) factoring and b) graphing.

Solution

a) Factoring
$$x^2 - 2x - 15 = 0$$
$$(x - 5)(x + 3) = 0$$
$$(x - 5) = 0 \quad \text{and} \quad (x + 3) = 0$$
$$x = 5 \quad \text{and} \quad x = -3$$

Factor the trinomial.
Set each factor equal to zero.
Solve.

b) Graphing

$y = x^2 - 2x - 15$

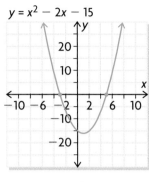

The x-intercepts, or zeros, of the graph are the solutions to the equation. In this case, the zeros are $x = -3$ and $x = 5$.

2. Solve by factoring. Confirm your results by graphing.

a) $x^2 + 7x - 30 = 0$ **c)** $x^2 - x - 6 = 0$

b) $x^2 + 8x + 15 = 0$ **d)** $x^2 - 5x + 6 = 0$

Identifying and Factoring Perfect Square Trinomials

A trinomial of the form $a^2 + 2ab + b^2$ or $a^2 - 2ab + b^2$ is a perfect square. It has two identical factors of the form $(a + b)^2$ or $(a - b)^2$, respectively.

Study *Aid*

For help, see Essential Skills Appendix, A-10.

EXAMPLE

Factor the following perfect square trinomials.

a) $x^2 - 10x + 25$

b) $16x^2 + 24x + 9$

Solution

a) $a = 1x, b = 5$

$\qquad = (x - 5)(x - 5)$

$\qquad$ or $\quad (x - 5)^2$

b) $a = 4x, b = 3$

$\qquad = (4x + 3)(4x + 3)$

$\qquad$ or $\quad (4x + 3)^2$

3. Factor the following perfect square trinomials.

a) $x^2 + 6x + 9$ **c)** $9x^2 + 6x + 1$

b) $x^2 - 8x + 16$ **d)** $4x^2 - 12x + 9$

PRACTICE

Study Aid

For help, see Essential Skills Appendix.

Question	Appendix
4	A-9
5, 7	A-10
9	A-8

4. Simplify.
- **a)** $(2x + 5 - 6x^2) + (5x - 9x^2 + 3)$
- **b)** $(4x^2 - 3x + 7) - (2x^2 - 3x + 1)$
- **c)** $(3x + 5)(4x - 7)$
- **d)** $(2x - 1)(2x + 1)$

5. Factor.
- **a)** $x^2 + 3x - 40$
- **b)** $6x^2 + 5x - 6$
- **c)** $81x^2 - 49$
- **d)** $9x^2 + 6x + 1$

6. What term will make each expression a perfect trinomial square?
- **a)** $x^2 + 6x + \blacksquare$
- **b)** $x^2 + \blacksquare + 25$
- **c)** $4x^2 + \blacksquare + 49$
- **d)** $9x^2 - 24x + \blacksquare$

7. Write in factored form.
- **a)** $f(x) = x^2 - 7x - 18$
- **b)** $g(x) = -2x^2 + 17x - 8$
- **c)** $h(x) = 4x^2 - 25$
- **d)** $y = 6x^2 + 13x - 5$

8. Determine the vertex of each quadratic function, and state the domain and range of each.
- **a)** $y = x^2 + 6x + 5$
- **b)** $f(x) = 2x^2 - 5x - 12$
- **c)** $g(x) = -6x^2 - 7x + 3$
- **d)** $h(x) = -3x^2 + 9x + 30$

9. Sketch the graph of each function by hand. Start with $y = x^2$ and use the appropriate transformations.
- **a)** $f(x) = x^2 - 5$
- **b)** $y = (x - 2)^2 + 1$
- **c)** $f(x) = 2(x + 3)^2 - 4$
- **d)** $y = -\frac{1}{2}(x - 4)^2 + 2$

10. The height of a diver above the water is given by the quadratic function $h(t) = -5t^2 + 5t + 10$, with t in seconds and $h(t)$ in metres. When will the diver reach the maximum height?

11. Use what you know about quadratic equations to complete the chart.

Essential characteristics:		Non-essential characteristics:
	Quadratic Equation	
Examples:		Non-examples:

APPLYING *What You Know*

Cutting the Cake

Kommy wants to share his birthday cake with as many people as possible.

YOU WILL NEED

• graph paper or graphing calculator
• ruler

? If Kommy makes 12 cuts in the cake, what is the maximum number of people he can serve?

A. Copy and complete the table by drawing a diagram for each case.

Number of Cuts	0	1	2	3	4	5	6
Diagram of Cake with Cuts							
Maximum Number of Pieces of Cake							

B. Compare your table with that of other classmates. Do you have the same answers? Discuss and make any necessary changes.

C. How should the cake be cut to get the maximum number of pieces?

D. Graph the data from your table, with number of cuts along the horizontal axis and pieces of cake along the vertical axis. What relationship is there between the number of cuts made and the maximum number of pieces of cake? Explain how you know.

E. With a partner, repeat part A with a circular cake and then a triangular cake. (You do one type of cake, your partner another.) Does the shape of the cake affect the relationship between the number of cuts and the maximum number of pieces? Explain.

F. If Kommy makes 12 cuts in the cake, what is the maximum number of people he can serve? Explain your reasoning.

4.1

The Vertex Form of a Quadratic Function

YOU WILL NEED

- graph paper
- graphing calculator

GOAL

Compare the standard and vertex forms of a quadratic function.

LEARN ABOUT the Math

The school environment club will use 80 square concrete slabs 1 m in length to surround a rectangular garden. The largest possible area must be enclosed.

The math team has suggested two quadratic functions to model the enclosed area:

$$f(w) = -w^2 + 40w \quad \text{and}$$
$$g(w) = -(w - 20)^2 + 400$$

Here, w is the width in metres, and $f(w)$ and $g(w)$ are the areas in square metres.

? Which function should the environment club use?

EXAMPLE 1	Connecting functions in standard and vertex forms

Compare the two functions suggested by the math team.

Kirsten's Solution: Using Algebra

$g(w) = -(w - 20)^2 + 400$ ⟵

$g(w) = -(w - 20)(w - 20) + 400$

$g(w) = -(w^2 - 20w - 20w + 400) + 400$

$g(w) = -(w^2 - 40w + 400) + 400$

$g(w) = -w^2 + 40w - 400 + 400$

$g(w) = -w^2 + 40w$ ⟵

> If the functions have the same form, it's easier to compare them. I expanded $g(w)$ to get it into standard form, like $f(w)$.

> This is the same as $f(w)$.

The environment club can use either equation.

Marc's Solution: Using a Table of Values and Graphing

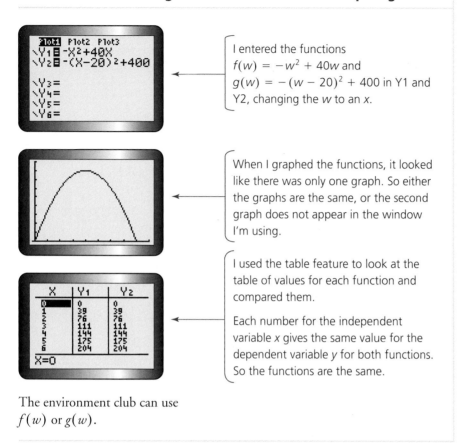

I entered the functions
$f(w) = -w^2 + 40w$ and
$g(w) = -(w - 20)^2 + 400$ in Y1 and
Y2, changing the w to an x.

When I graphed the functions, it looked like there was only one graph. So either the graphs are the same, or the second graph does not appear in the window I'm using.

I used the table feature to look at the table of values for each function and compared them.

Each number for the independent variable x gives the same value for the dependent variable y for both functions. So the functions are the same.

The environment club can use
$f(w)$ or $g(w)$.

The vertex form of a quadratic function provides useful information about some of the characteristics of its graph.

EXAMPLE 2	Connecting information about the parabola to the vertex form of the function

What is the maximum area of a garden defined by $f(w) = -w^2 + 40w$?
How does this relate to the function $g(w) = -(w - 20)^2 + 400$?

Kirsten's Solution: Using Algebra

$f(w) = -w^2 + 40w$

$f(w) = -w(w - 40)$

zeros are $w = 0$ and $w = 40$

$f(w)$ is a quadratic function, and the leading coefficient is less than zero. The graph is a parabola that opens down. I factored the function. The two zeros occur when $w = 0$ or $w = 40$, and the maximum will be halfway between them.

x-coordinate of vertex

$$x = \frac{0 + 40}{2}$$

$$= 20$$

$$f(20) = -(20)^2 + 40(20)$$

$$= -400 + 800$$

$$= 400$$

> I used the zeros to find the axis of symmetry by adding them and dividing by 2. This corresponds to the x-coordinate of the vertex.

> I substituted 20 into $f(w)$ to get the y-coordinate, which is the maximum value.

The maximum area is 400 m² when the width is 20 m. Both numbers appear in the function $g(w)$, which is written in **vertex form**.

vertex form

a quadratic function in the form $f(x) = a(x - h)^2 + k$, where the vertex is (h, k)

Marc's Solution: Using a Graph

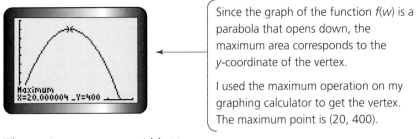

> Since the graph of the function $f(w)$ is a parabola that opens down, the maximum area corresponds to the y-coordinate of the vertex.

> I used the maximum operation on my graphing calculator to get the vertex. The maximum point is (20, 400).

The maximum occurs at width 20 m and the area is 400 m². Both numbers appear in $g(w)$.

Reflecting

A. How can a quadratic function in vertex form be written in standard form?

B. What is the significance of a when the quadratic function is written in vertex form? Does a change when you change from vertex to standard form?

C. Can you always easily identify the zeros from vertex form? Explain.

D. What is the significance of h and k in vertex form?

APPLY the Math

EXAMPLE 3 — Identifying features of the parabola from a quadratic function in vertex form

a) Determine the direction of opening, the axis of symmetry, the minimum value, and the vertex of the quadratic function $f(x) = 3(x - 5)^2 + 7$.

b) Without using graphing technology, use the information you determined to sketch a graph of $f(x)$.

c) State the domain and range of $f(x)$.

Steve's Solution

a) $f(x) = a(x - h)^2 + k$

$f(x) = 3(x - 5)^2 + 7$ ◄──── The function is written in vertex form. Since $a = 3$, a positive number, the parabola opens up.

opens up

$x = 5$ ◄──── The equation of the axis of symmetry equals the value of h because a parabola is symmetric about the vertical line that passes through its vertex.

minimum of 7 ◄──── The parabola opens up, so the minimum value of 7 is at the vertex, when $x = 5$.

The vertex is $(5, 7)$.

b) When $x = 0$,

$y = 3(0 - 5)^2 + 7$

$y = 3(-5)^2 + 7$ ◄──── I need some other points to graph the parabola. I substituted $x = 0$ into the quadratic function and solved to find the y-intercept.

$y = 3(25) + 7$

$y = 75 + 7$

$y = 82$

A point on the curve is $(0, 82)$.

Another x-coordinate that has ◄──── a y-value of 82 is $x = 5 + 5$.

$x = 10$

Another point on the curve is $(10, 82)$.

I know that there's another point at $(x, 82)$ on the opposite branch of the parabola. Since 0 is 5 units to the left of the axis of symmetry, $x = 5$, another x-value that will have 82 as its y-coordinate must be 5 units to the right of the axis of symmetry.

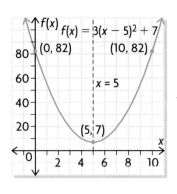

$f(x)$

$f(x) = 3(x - 5)^2 + 7$

80 (0, 82) (10, 82)

60

$x = 5$

40

20

(5, 7)

0   2   4   6   8   10

I plotted the vertex and my two points and connected them with a parabola.

c) domain: $\{x \in \mathbf{R}\}$

The function represents a parabola and is defined for any real number.

range: $\{y \in \mathbf{R} | y \geq 7\}$

The range depends on where the vertex is and which way the parabola opens. Since the vertex is at (5, 7) and the parabola opens up, the range is all values of y greater than or equal to 7.

EXAMPLE 4

Selecting a strategy to determine the zeros from a quadratic function in vertex form

Determine the zeros of the function $f(x) = (x - 5)^2 - 36$.

Marc's Solution: Expanding and Factoring

$$(x - 5)^2 - 36 = 0$$

To find the zeros, I let y or $f(x)$ equal zero.

$$(x - 5)(x - 5) - 36 = 0$$

I wrote the squared term as $(x - 5)(x - 5)$ to help me expand.

$$x^2 - 5x - 5x + 25 - 36 = 0$$

I multiplied the binomials and collected like terms.

$$x^2 - 10x - 11 = 0$$
$$(x - 11)(x + 1) = 0$$

I factored by finding two numbers that multiply to -11 and add to -10. They are -11 and 1.

$$x - 11 = 0 \quad \text{and} \quad x + 1 = 0$$
$$x = 11 \quad \text{and} \quad x = -1$$

I set each factor equal to zero and solved.

are the zeros of this function.

Beth's Solution: Using Inverse Operations

$$(x - 5)^2 - 36 = 0$$
$$(x - 5)^2 - 36 + 36 = 0 + 36$$
$$(x - 5)^2 = 36$$

To get the zeros, I set the function equal to zero.

I used inverse operations to isolate the term containing x.

$$(x - 5) = \pm\sqrt{36}$$

I took the square root of both sides because that's the inverse of squaring. I remembered that the square root can be positive or negative.

$$x - 5 = \pm 6$$

I took the square root of 36.

$$x - 5 + 5 = \pm 6 + 5$$
$$x = 5 \pm 6$$
$$x = 5 + 6 \quad \text{and} \quad x = 5 - 6$$
$$x = 11 \quad \text{and} \quad x = -1$$

are the zeros of $f(x)$.

I isolated x by adding 5 to both sides.

I split up the expression and solved for x in each.

Rachel's Solution: Using a Graphing Calculator

I entered the function in Y1.

I adjusted the window to see both zeros.

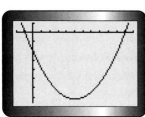

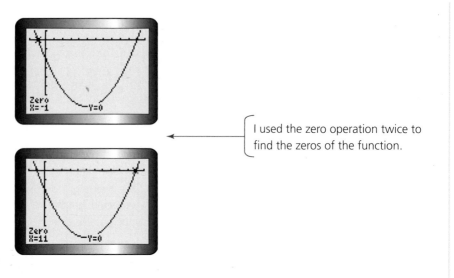

I used the zero operation twice to find the zeros of the function.

The zeros are at $x = -1$ and $x = 11$.

If you are given the graph of a quadratic function and you know the coordinates of the vertex and another point on the parabola, you can use the vertex form to determine the function's equation.

| EXAMPLE 5 | Using the vertex form to write the equation of a quadratic function from its graph |

Determine the equation in vertex form of the quadratic function shown.

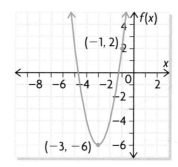

Veena's Solution

$f(x) = a(x - h)^2 + k$

$f(x) = a(x - (-3))^2 + (-6)$ ← The vertex is $(-3, -6)$, so I replaced h with -3 and k with -6.

$f(x) = a(x + 3)^2 - 6$

$$2 = a(-1 + 3)^2 - 6$$

$$2 = a(2)^2 - 6$$

$$2 = 4a - 6$$

$$2 + 6 = 4a$$

$$8 = 4a$$

$$2 = a$$

$$f(x) = 2(x + 3)^2 - 6$$

To get the value of a, I needed another point on the parabola. I used $(-1, 2)$. I replaced $f(x)$ with 2 and x with -1 and solved for a.

I rewrote the quadratic function in vertex form with my values for a, h, and k.

In Summary

Key Idea

- A quadratic function in vertex form, $f(x) = a(x - h)^2 + k$, can be expressed in standard form, $f(x) = ax^2 + bx + c$, by expanding and simplifying. The two forms are equivalent.

Need to Know

- If a quadratic function is expressed in vertex form, $f(x) = a(x - h)^2 + k$, then
  - the vertex is located at (h, k)
  - the equation of the axis of symmetry is $x = h$
  - the function has a maximum value of k when $a < 0$
  - the function has a minimum value of k when $a > 0$

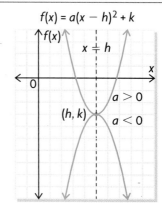

CHECK Your Understanding

1. For each function, state the vertex and whether the function has a maximum or minimum value. Explain how you decided.
   a) $f(x) = 2(x - 3)^2 - 5$
   b) $f(x) = -3(x - 5)^2 - 1$
   c) $f(x) = -(x + 1)^2 + 6$
   d) $f(x) = (x + 5)^2 - 3$

2. For each function in question 1, identify the equation of the axis of symmetry, and determine the domain and range.

3. Graph and then write in standard form.
   a) $f(x) = 2(x - 3)^2 + 1$
   b) $f(x) = -(x + 1)^2 - 3$

PRACTISING

4. Complete the table.

K

	Function	Vertex	Axis of Symmetry	Opens Up/Down	Range	Sketch
a)	$f(x) = (x - 3)^2 + 1$					
b)	$f(x) = -(x + 1)^2 - 5$					
c)	$y = 4(x + 2)^2 - 3$					
d)	$y = -3(x + 5)^2 + 2$					
e)	$f(x) = -2(x - 4)^2 + 1$					
f)	$y = \frac{1}{2}(x - 4)^2 + 3$					

Tech | Support

For help determining the zeros using a graphing calculator, see Technical Appendix, B-8.

5. Use a graphing calculator to determine the zeros for each function.

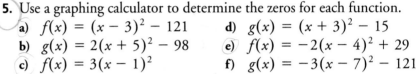

a) $f(x) = (x - 3)^2 - 121$ d) $g(x) = (x + 3)^2 - 15$

b) $g(x) = 2(x + 5)^2 - 98$ e) $f(x) = -2(x - 4)^2 + 29$

c) $f(x) = 3(x - 1)^2$ f) $g(x) = -3(x - 7)^2 - 121$

6. The same quadratic function $f(x)$ can be expressed in three different forms:

$$f(x) = (x - 7)^2 - 25$$
$$f(x) = x^2 - 14x + 24$$
$$f(x) = (x - 12)(x - 2)$$

What information about the parabola does each form provide?

7. The height above the ground of a bungee jumper is modelled by the

A quadratic function $h(t) = -5(t - 0.3)^2 + 110$, where height, $h(t)$, is in metres and time, t, is in seconds.

a) When does the bungee jumper reach maximum height? Why is it a maximum?

b) What is the maximum height reached by the jumper?

c) Determine the height of the platform from which the bungee jumper jumps.

8. Write the equation of the quadratic function, first in vertex form and then in standard form.

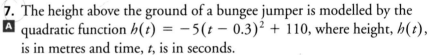

a) vertex $(-4, 8)$ and passing through $(2, -4)$

b) vertex $(3, 5)$ and passing through $(1, 1)$

c) vertex $(1, -7)$ and passing through $(-2, 29)$

d) vertex $(-6, -5)$ and passing through $(-3, 4)$

9. For each graph, write the quadratic equation in vertex form.

a)
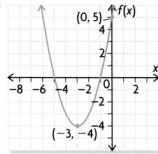

c)
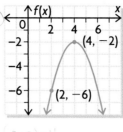

b)
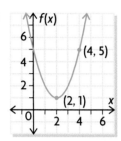

d)
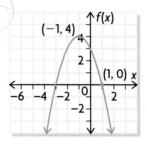

10. The path of a ball is modelled by the quadratic function
$h(t) = -5(t - 2)^2 + 23$, where height, $h(t)$, is in metres and
time, t, is in seconds.
a) What is the maximum height the ball reaches?
b) When does it reach the maximum height?
c) When will the ball reach a height of 18 m?

11. For each function, determine the vertex and two points that satisfy the
equation. Use the information to sketch the graph of each.
a) $y = (x - 4)^2$
c) $y = 2(x - 1)^2 - 3$
b) $y = x^2 + 4$
d) $y = -2(x + 3)^2 + 5$

12. A quadratic function has zeros at 1 and -3 and passes through the
T point (2, 10). Write the equation in vertex form.

13. An equation is given in vertex form. Explain how to write it in
standard form. Illustrate with an example.

14. What information about the graph do you know immediately when a
C quadratic function is written in
a) standard form?
b) vertex form?

Extending

15. The table at the right shows the number of cigarettes sold from 1994
to 2005. Determine the equation of a curve of good fit, in vertex form,
that can be used to model the data.

16. Write the function $f(x) = (x - 7)(x + 5)$ in vertex form.

Year	Cigarettes (millions of units)
1994	18.2
1995	18.9
1996	19.5
1997	18.6
1998	19.0
1999	18.8
2000	18.2
2001	17.5
2002	16.3
2003	15.7
2004	14.4
2005	13.2

Relating the Standard and Vertex Forms: Completing the Square

GOAL

Write a quadratic function, given in standard form, in vertex form.

LEARN ABOUT the Math

A model rocket is shot from a platform 1 m above the ground. The height of the rocket above the ground is recorded in the table.

Time (s)	0	0.5	1.0	1.5	2.0	2.5	3.0	3.5	4.0
Height (m)	1.00	14.75	26.00	34.75	41.00	44.75	46.00	44.75	41.00

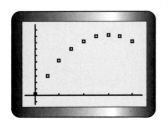

Sally and Jake used a graphing calculator to create a scatter plot from the data.

They decided that a quadratic function models the data. They used different methods to determine the function and got different results. The method each used, along with their reasoning, is shown.

Sally's Solution: Using a Table of Values and a Graph

$$y = a(x - h)^2 + k$$
$$y = a(x - 3)^2 + 46$$

From the table of values and my graph, it looks like the vertex is (3, 46), so I replaced h with 3 and k with 46.

$$1 = a(0 - 3)^2 + 46$$
$$1 = a(-3)^2 + 46$$
$$1 = 9a + 46$$

To find the value of a, I picked the point (0, 1) and substituted 0 for x and 1 for y. Then I solved for a.

$$1 - 46 = 9a$$
$$-5 = a$$
$$y = -5(x - 3)^2 + 46$$

I replaced a with -5 and wrote the equation.

This equation gives the height of the rocket during its flight.

Jake's Solution: Using Quadratic Regression

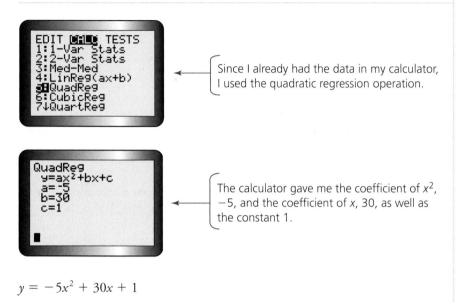

Since I already had the data in my calculator, I used the quadratic regression operation.

The calculator gave me the coefficient of x^2, -5, and the coefficient of x, 30, as well as the constant 1.

$y = -5x^2 + 30x + 1$

This is the equation that gives the height of the rocket during its flight.

Tech | *Support*

For help using the quadratic regression operation on a graphing calculator, see Technical Appendix, B-10.

? Do both equations represent the same function?

EXAMPLE 1   Comparing the two equations

Tim's Solution

$h(x) = -5(x - 3)^2 + 46$

$h(x) = -5(x - 3)(x - 3) + 46$

If I expand Sally's equation and get Jake's equation, then I know they're equal.

I rewrote the squared term and then multiplied the two binomials together.

$h(x) = -5(x^2 - 3x - 3x + 9) + 46$

$h(x) = -5(x^2 - 6x + 9) + 46$

$h(x) = -5x^2 + 30x - 45 + 46$

$h(x) = -5x^2 + 30x + 1$

I collected like terms and multiplied by -5.

Then I simplified.

Sally's equation is the same as Jake's equation.

$$h(x) = -5x^2 + 30x + 1$$
$$h(x) = -5x^2 + 30x - 45 + 46$$
$$h(x) = -5(x^2 - 6x + 9) + 46$$
$$h(x) = -5(x - 3)(x - 3) + 46$$
$$h(x) = -5(x - 3)^2 + 46$$

> I wrote the solution in reverse to help me see how to go from standard form to vertex form.

Reflecting

A. When changing from standard form to vertex form, why do the brackets appear?

B. What type of trinomial resulted when the common factor was divided out?

C. In the trinomial $x^2 - 6x + 9$, how is the coefficient 6 related to 9?

D. Why must we use a perfect-square trinomial to write the equation of a quadratic function in vertex form?

APPLY the Math

EXAMPLE **2**	Selecting a strategy to write a quadratic function in vertex form

completing the square

the process of adding a constant to a given quadratic expression to form a perfect trinomial square

for example, $x^2 + 6x + 2$ is not a perfect square, but if 7 is added to it, it becomes $x^2 + 6x + 9$, which is $(x + 3)^2$

Write each quadratic function in vertex form by **completing the square.** Then use your graphing calculator to verify that the standard form and your vertex form are equivalent.

a) $y = x^2 + 6x + 5$
b) $f(x) = -2x^2 + 16x + 1$

Tommy's Solution

a) $y = x^2 + 6x + 5$

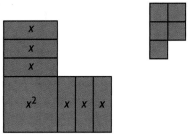

> I used algebra tiles to model the expression. The vertex form has two identical factors, so I had to arrange my tiles to form a square. To do this, I divided the x tiles in half. $6 \div 2 = 3$.
>
> This arrangement is not a perfect square because the 5 unit tiles can't be used to create a square. But if I add 9 tiles to the model, I can create a perfect square. Dividing the coefficient of x and squaring it gave me that number. $(6 \div 2)^2 = 3^2 = 9$

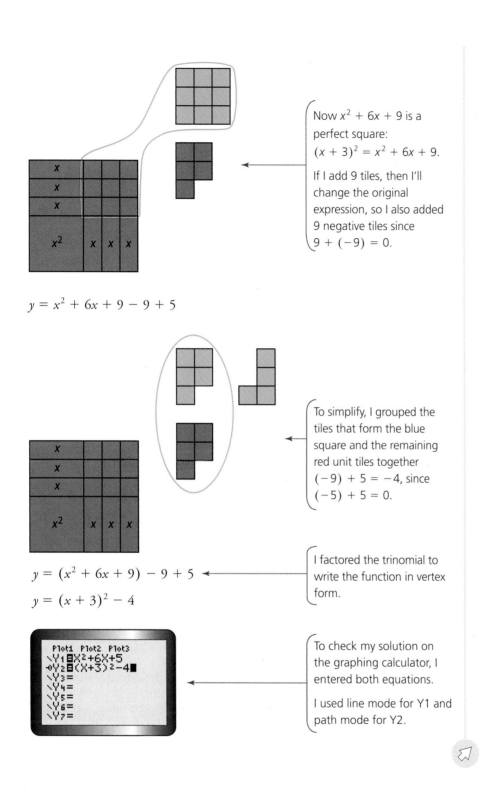

Now $x^2 + 6x + 9$ is a perfect square: $(x + 3)^2 = x^2 + 6x + 9$.

If I add 9 tiles, then I'll change the original expression, so I also added 9 negative tiles since $9 + (-9) = 0$.

$y = x^2 + 6x + 9 - 9 + 5$

To simplify, I grouped the tiles that form the blue square and the remaining red unit tiles together $(-9) + 5 = -4$, since $(-5) + 5 = 0$.

$y = (x^2 + 6x + 9) - 9 + 5$

$y = (x + 3)^2 - 4$

I factored the trinomial to write the function in vertex form.

To check my solution on the graphing calculator, I entered both equations.

I used line mode for Y1 and path mode for Y2.

Since the cursor traces the parabola they're equal.

b) $f(x) = -2x^2 + 16x + 1$

$f(x) = -2x^2 + 16x + 1$

$f(x) = -2(x^2 - 8x) + 1$

Since $a \neq 1$, I divided the first two terms by the common factor -2. I used only the first two terms because they're all I need to complete the square.

$f(x) = -2(x^2 - 8x + 16 - 16) + 1$

I completed the square for the terms in the brackets. Half of -8 is -4, and -4 squared is 16, so I added 16 and subtracted 16 inside the brackets.

$f(x) = -2(x^2 - 8x + 16) + 32 + 1$

The first three terms inside the brackets form a perfect square, so I multiplied the last term in the brackets by -2 to group the three terms together.

$f(x) = -2(x - 4)(x - 4) + 32 + 1$

I factored the perfect square and added the constants.

$f(x) = -2(x - 4)^2 + 33$

To see if the two equations were equal, I entered the standard form in Y1 and the vertex form in Y2. Then I looked at the table the calculator creates and compared the y-values for each x-value.

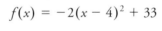

Since the y-values are equal, the equations are equivalent.

EXAMPLE **3**

Completing the square to write a quadratic function in vertex form

Write the quadratic function $y = 2x^2 - 3x - 7$ in vertex form.

Marc's Solution: Using Fractions

$y = 2x^2 - 3x - 7$

$y = 2\left(x^2 - \dfrac{3}{2}x\right) - 7$ ← I factored 2 out of the first two terms of the expression because it's easier to complete the square if the coefficient of x^2 is 1. Then I completed the square.

$y = 2\left(x^2 - \dfrac{3}{2}x + \dfrac{9}{16} - \dfrac{9}{16}\right) - 7$ ← Half of $-\dfrac{3}{2}$ is $-\dfrac{3}{4}$. If I square $-\dfrac{3}{4}$ I get $\dfrac{9}{16}$, so I added and subtracted $\dfrac{9}{16}$ inside the brackets.

$y = 2\left(x^2 - \dfrac{3}{2}x + \dfrac{9}{16}\right) - \dfrac{9}{8} - 7$ ← The perfect square consists of only the first three terms, so I multiplied $-\dfrac{9}{16}$ by 2 to get $-\dfrac{9}{8}$ and moved it outside of the brackets.

$y = 2\left(x - \dfrac{3}{4}\right)^2 - \dfrac{9}{8} - 7$

I factored the expression in the brackets.

$y = 2\left(x - \dfrac{3}{4}\right)^2 - \dfrac{9}{8} - \dfrac{56}{8}$

$y = 2\left(x - \dfrac{3}{4}\right)^2 - \dfrac{65}{8}$ ← To add the remaining constants, I used the common denominator 8 and left the sum as an improper fraction.

Rachel's Solution: Using Decimals

$y = 2x^2 - 3x - 7$

$y = 2(x^2 - 1.5x) - 7$ ← I factored 2 from the first two terms so that the coefficient of x^2 would be 1. I decided to write -3 divided by 2 as the decimal -1.5.

$y = 2(x^2 - 1.5x + 0.5625 - 0.5625) - 7$ ← Half the coefficient of x, -1.5, is -0.75, so I squared -0.75 and got 0.5625. I added and subtracted this to keep my equation the same.

$$y = 2(x^2 - 1.5x + 0.5625) - 1.125 - 7$$

Only the first three terms in the brackets form a perfect square, so I multiplied -0.5625 by 2 and moved it out of the brackets.

$$y = 2(x - 0.75)^2 - 1.125 - 7$$
$$y = 2(x - 0.75)^2 - 8.125$$

The perfect square factored to $(x - 0.75)^2$. Then I collected like terms.

EXAMPLE 4   Solving a problem by completing the square

Judy wants to fence three sides of the yard in front of her house. She bought 60 m of fence and wants the maximum area she can fence in. The quadratic function $f(x) = 60x - 2x^2$, where x is the width of the yard in metres, represents the area to be enclosed. Write an equation in vertex form that gives the maximum area that can be enclosed.

Asif's Solution

$$f(x) = -2x^2 + 60x$$

I rearranged the equation.

$$f(x) = -2(x^2 - 30x)$$

I factored -2 from both terms.

$$f(x) = -2(x^2 - 30x + 225 - 225)$$

Half the coefficient of x, -30, is -15, so I squared -15 and got 225. I added and subtracted this to keep my equation the same.

$$f(x) = -2(x^2 - 30x + 225) + 450$$

Only the first three terms in brackets form a perfect square, so I multiplied -225 by -2 and moved it out of the brackets.

$$f(x) = -2(x - 15)^2 + 450$$

A maximum area of 450 m^2 can be fenced in.

The perfect square factored to $(x - 15)^2$. The vertex is $(15, 450)$.

In Summary

Key Ideas

- All quadratic functions in standard form can be written in vertex form by completing the square. The equations are equivalent.
- Both the standard form and the vertex form provide useful information for graphing the parabola.

___

Need to Know

- To complete the square, follow these steps:

$f(x) = ax^2 + bx + c$

$f(x) = a\left(x^2 + \dfrac{b}{a}x\right) + c$ ⟵ Factor the coefficient of x^2 from the first two terms.

$f(x) = a\left[x^2 + \dfrac{b}{a}x + \left(\dfrac{b}{2a}\right)^2 - \left(\dfrac{b}{2a}\right)^2\right] + c$ ⟵ Add and subtract the square of half the coefficient of x inside the brackets.

$f(x) = a\left[x^2 + \dfrac{b}{a}x + \left(\dfrac{b}{2a}\right)^2\right] - a\left(\dfrac{b}{2a}\right)^2 + c$ ⟵ Group the three terms that form the perfect square. Multiply the fourth term by a, and move it outside the brackets.

$f(x) = a\left(x + \dfrac{b}{2a}\right)^2 - \dfrac{b^2}{4a} + c$ ⟵ Factor the perfect square and simplify.

- Any quadratic function in standard form with $a \neq 1$ can be expressed in vertex form when fractions are used to complete the square. Decimals can be used only if the coefficient of x results in a terminating decimal when a is factored from the x^2- and x-terms of the quadratic.

CHECK *Your Understanding*

1. What number must you add to the following to create a perfect square?

a)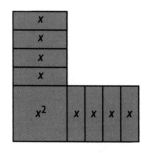

b)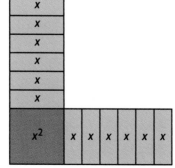

c) $x^2 + 10x$

d) $x^2 - 5x$

2. Determine the values of m and n required to create a perfect-square trinomial.

a) $x^2 - 10x = x^2 - mx + n - n$

b) $x^2 + 6x = x^2 + mx + n - n$

c) $5x^2 + 60x = 5(x^2 + mx + n - n)$

d) $2x^2 - 7x = 2(x^2 - mx + n - n)$

3. Factor.

a) $x^2 + 14x + 49$

b) $x^2 - 18x + 81$

c) $x^2 - 20x + 100$

d) $x^2 + 6x + 9$

4. Complete the square.

a) $y = x^2 + 12x + 40$

b) $y = x^2 - 6x + 2$

c) $y = x^2 - 10x + 29$

d) $y = x^2 - x - 3$

PRACTISING

5. Determine the values of a, h, and k that make the equation true.

a) $3x^2 - 12x + 17 = a(x - h)^2 + k$

b) $-2x^2 - 20x - 53 = a(x - h)^2 + k$

c) $2x^2 - 12x + 23 = a(x - h)^2 + k$

d) $\frac{1}{2}x^2 + 3x - \frac{1}{2} = a(x - h)^2 + k$

6. Write the function in vertex form.

a) $f(x) = x^2 + 8x + 3$

b) $f(x) = x^2 - 12x + 35$

c) $f(x) = 2x^2 + 12x + 7$

d) $f(x) = -x^2 + 6x + 7$

e) $f(x) = -x^2 + 3x - 2$

f) $f(x) = 2x^2 + 3x + 1$

7. Complete the square to express each function in vertex form. Then graph each, and state the domain and range.

a) $f(x) = x^2 - 4x + 5$

b) $f(x) = x^2 + 8x + 13$

c) $f(x) = 2x^2 + 12x + 19$

d) $f(x) = -x^2 + 2x - 7$

e) $f(x) = -3x^2 - 12x - 11$

f) $f(x) = \frac{1}{2}x^2 + 3x + 4$

8. For the quadratic function $g(x) = 4x^2 - 24x + 31$:

K a) Write the equation in vertex form.

b) Write the equation of the axis of symmetry.

c) Write the coordinates of the vertex.

d) Determine the maximum or minimum value of $g(x)$. State a reason for your choice.

e) Determine the domain of $g(x)$.

f) Determine the range of $g(x)$.

g) Graph the function.

9. Colin completed the square to write $y = 2x^2 - 6x + 5$ in vertex form. Is his solution correct or incorrect? If incorrect, identify the error and show the correct solution.

$$y = 2(x^2 - 3x) + 5$$
$$y = 2\left(x^2 - 3x + \frac{9}{4} - \frac{9}{4}\right) + 5$$
$$y = 2\left(x - \frac{3}{4}\right)^2 - \frac{9}{2} + 5$$
$$y = 2\left(x - \frac{3}{4}\right)^2 + \frac{1}{2}$$

10. A lifeguard wants to rope off a rectangular area for swimmers to swim in. She has 700 m of rope. The area, $A(x)$, that is to be enclosed can be modelled by the function $A(x) = 700x - 2x^2$, where x is the width of the rectangle. What is the maximum area that can be enclosed?

11. **A** A theatre company's profit, $P(x)$, on a production is modelled by $P(x) = -60x^2 + 1800x + 16\,500$, where x is the cost of a ticket in dollars. According to the model, what should the company charge per ticket to make the maximum profit?

12. **a)** Write the quadratic equation $y = 3x^2 - 30x + 73$ in vertex form.
   b) What information does the vertex form give that is not obvious from the standard form?

13. **T** What transformations must be applied to the graph of $y = x^2$ to produce the graph of $y = -2x^2 + 16x - 29$? Justify your reasoning.

14. Why express quadratic equations in several different equivalent forms?

15. **C** List the steps followed to change a quadratic function in standard form with $a = 1$ to vertex form. Illustrate with an example.

Extending

16. Betty shoots an arrow into the air. The height of the arrow is recorded in the table. What is the equation of a curve of good fit for the height of the arrow in terms of time?

Time (s)	0	0.5	1.0	1.5	2.0
Height (m)	1.00	4.75	6.00	4.75	1.00

17. The points $(-2, -12)$ and $(2, 4)$ lie on the parabola $y = a(x - 1)^2 + k$. What is the vertex of this parabola?

Solving Quadratic Equations Using the Quadratic Formula

YOU WILL NEED

• graphing calculator

GOAL

Understand and apply the quadratic formula.

LEARN ABOUT the Math

A quarter is thrown from a bridge 15 m above a pool of water. The height of the quarter above the water at time t is given by the quadratic function $h(t) = -5t^2 + 10t + 15$, where time, t, is in seconds and height, $h(t)$, is in metres.

? How can you determine when the quarter hits the water?

EXAMPLE 1	Selecting a strategy to determine the zeros of a quadratic function

Determine when the quarter will hit the water.

Ben's Solution: Using a Table of Values

t	$h(t)$
0	15
1	20
2	15
3	0

When the quarter hits the water, its height will be 0. I made a table of values. I substituted different values of t into the original function. Time can't be negative, so I chose numbers greater than or equal to zero. I stopped when I found an $h(t)$ value equal to zero.

The quarter will hit the water at 3 s.

Talia's Solution: Graphing

When the quarter hits the water, its height will be 0. I graphed the function to get its zeros. I looked at the part of the graph where $t \geq 0$ because time can't be negative.

The graph crosses the horizontal axis at $t = 3$.

The quarter will hit the water at 3 s.

Jim's Solution: Using Factoring

$-5t^2 + 10t + 15 = 0$ ◄───────

> The quarter will hit the water when the height is zero, so I let $h(t) = 0$.

$-5(t^2 - 2t - 3) = 0$ ◄───────

$-5(t - 3)(t + 1) = 0$

> I factored -5 from all terms in the equation. I need two numbers that multiply to give -3 and add to -2. The numbers are -3 and 1.

$t - 3 = 0$ or $t + 1 = 0$ ◄───────

$t = 3$ or $t = -1$

> I set each of the factors equal to zero and solved.

$t = 3$ because $t \geq 0$ ◄───────

The quarter hits the water at 3 s.

> Since t represents time, it must be greater than or equal to zero. So $t = 3$ is the only acceptable solution.

Elva's Solution: Using the Vertex Form

$-5t^2 + 10t + 15 = 0$ ◄───────

> The height of the quarter will be zero when it hits the water, so I let $h(t) = 0$.

$-5(t^2 - 2t) + 15 = 0$ ◄───────

> I factored -5 from the first two terms.

$-5(t^2 - 2t + 1 - 1) + 15 = 0$ ◄───────

> I completed the square by adding and subtracting half the coefficient of the term containing t.

$-5(t^2 - 2t + 1) + 5 + 15 = 0$ ◄───────

$-5(t^2 - 2t + 1) + 20 = 0$

> I rewrote the expression so that the terms of the perfect square were in the brackets. Then I collected like terms.

$-5(t - 1)^2 + 20 = 0$ ◄───────

$-5(t - 1)^2 = -20$

> I factored the perfect square and rewrote it with the term involving t on one side.

$(t - 1)^2 = 4$ ◄───────

$(t - 1) = \pm\sqrt{4}$

> Then I divided both sides by -5. I took the square root, which can be positive or negative, of both sides.

$(t - 1) = \pm 2$ ◄───────

> I simplified by taking the square root of 4.

$$t = 1 \pm 2$$

I isolated the variable.

$$t = 1 + 2 \quad \text{or} \quad t = 1 - 2$$
$$t = 3 \quad \text{or} \quad t = -1$$

I separated the right side to make two equations.

The quarter hits the water at 3 s.

Since time can't be negative, $t = 3$.

EXAMPLE 2 **Developing a formula to determine the solutions to a quadratic equation**

Given the standard form of the equation $ax^2 + bx + c = 0$, follow Elva's solution in Example 1 to help complete the square of the general quadratic equation $ax^2 + bx + c = 0$ in order to express it in vertex form. Then use it to determine the roots.

Terri's Solution

$$ax^2 + bx + c = 0$$

I factored the coefficient of x^2 from the first two terms.

$$a\left(x^2 + \frac{b}{a}x\right) + c = 0$$

$$a\left(x^2 + \frac{b}{a}x + \frac{b^2}{4a^2} - \frac{b^2}{4a^2}\right) + c = 0$$

I completed the square by adding and subtracting the square of half the coefficient of the x-term.

$$a\left(x^2 + \frac{b}{a}x + \frac{b^2}{4a^2}\right) - \frac{b^2}{4a} + c = 0$$

I rewrote the expression so that only three terms were in the brackets.

$$a\left(x^2 + \frac{b}{a}x + \frac{b^2}{4a^2}\right) - \frac{b^2}{4a} + \frac{4ac}{4a} = 0$$

I gathered together like terms. I used a common denominator before I added the like terms.

$$a\left(x + \frac{b}{2a}\right)^2 - \left(\frac{b^2}{4a} - \frac{4ac}{4a}\right) = 0$$

I factored the perfect square in the brackets.

$$a\left(x + \frac{b}{2a}\right)^2 = \left(\frac{b^2}{4a} - \frac{4ac}{4a}\right)$$

I used inverse operations so that the variable was on one side.

$$\left(x + \frac{b}{2a}\right)^2 = \left(\frac{b^2 - 4ac}{4a^2}\right)$$

I divided both sides by a.

$$\left(x + \frac{b}{2a}\right) = \pm\sqrt{\frac{b^2 - 4ac}{4a^2}}$$

I took the square root of both sides. I remembered that it could be the positive or negative square root.

$$\left(x + \frac{b}{2a}\right) = \pm\frac{\sqrt{b^2 - 4ac}}{2a}$$

$$x = -\frac{b}{2a} \pm \frac{\sqrt{b^2 - 4ac}}{2a}$$

I simplified and isolated the variable.
I determined the zeros.

$$x = \frac{-b \pm \sqrt{b^2 - 4ac}}{2a}$$

$$x = \frac{-b + \sqrt{b^2 - 4ac}}{2a} \quad \text{or} \quad x = \frac{-b - \sqrt{b^2 - 4ac}}{2a}$$

Reflecting

A. What is the maximum number of solutions the **quadratic formula** gives?

B. Why is it easier to use the quadratic formula if the quadratic equation you are solving is in the form $ax^2 + bx + c = 0$?

C. What do the solutions determined using the quadratic formula represent in the original function?

D. How did solving the specific quadratic equation in Example 1 help you understand the development of the quadratic formula in Example 2?

quadratic formula

a formula for determining the roots of a quadratic equation of the form $ax^2 + bx + c = 0$. The formula uses the coefficients of the terms in the quadratic equation:

$$x = \frac{-b \pm \sqrt{b^2 - 4ac}}{2a}$$

APPLY the Math

Unlike factoring, you can use the quadratic formula to solve any quadratic equation. The number of solutions depends on the coefficients of the terms in the equation.

EXAMPLE 3	Solving a quadratic equation using the quadratic formula

Use the quadratic formula to solve each equation.
a) $x^2 - 30x + 225 = 0$ **b)** $3x^2 + 2x + 15 = 0$ **c)** $2x^2 - 5x = 1$

Steve's Solution

a) $x^2 - 30x + 225 = 0$

$$x = \frac{-(-30) \pm \sqrt{(-30)^2 - 4(1)(225)}}{2(1)}$$

I substituted $a = 1$, $b = -30$, and $c = 225$ into the quadratic formula and solved for x.

$$x = \frac{30 \pm \sqrt{900 - 900}}{2}$$

$$x = \frac{30 \pm \sqrt{0}}{2}$$

$$x = \frac{30}{2}$$

$$x = 15$$

There is only one solution.

b) $3x^2 + 2x + 15 = 0$ ←

$$x = \frac{-(2) \pm \sqrt{(2)^2 - 4(3)(15)}}{2(3)}$$

$$x = \frac{-2 \pm \sqrt{4 - 180}}{6}$$

$$x = \frac{-2 \pm \sqrt{-176}}{6}$$ ←

> I substituted $a = 3$, $b = 2$, and $c = 15$ into the quadratic formula and solved for x.

> I can't take the square root of a negative number.

There's no real solution.

c) $2x^2 - 5x - 1 = 0$

$$x = \frac{-(-5) \pm \sqrt{(-5)^2 - 4(2)(-1)}}{2(2)}$$ ←

$$x = \frac{5 \pm \sqrt{25 + 8}}{4}$$

$$x = \frac{5 \pm \sqrt{33}}{4}$$

$$x = \frac{5 + \sqrt{33}}{4} \quad \text{or} \quad x = \frac{5 - \sqrt{33}}{4}$$

$$x \doteq 2.69 \quad \text{or} \quad x \doteq -0.19$$

There are two solutions.

> I rearranged the equation to get zero on the right side.
>
> Then I substituted $a = 2$, $b = -5$, and $c = -1$ into the quadratic formula and solved for x.

EXAMPLE 4   Applying the quadratic formula to solve a problem

The profit on a school drama production is modelled by the quadratic equation $P(x) = -60x^2 + 790x - 1000$, where $P(x)$ is the profit in dollars and x is the price of the ticket, also in dollars.

a) Use the quadratic formula to determine the break-even price for the tickets.

b) At what price should the drama department set the tickets to maximize their profit?

Julia's Solution

a) $-60x^2 + 790x - 1000 = 0$ ◄─── At the break-even price there is no profit. So I set the profit function equal to zero.

$$x = \frac{-790 \pm \sqrt{790^2 - 4(-60)(-1000)}}{2(-60)}$$ ◄─── I substituted $a = -60$, $b = 790$, and $c = -1000$ into the quadratic formula and solved for x.

$$x = \frac{-790 \pm \sqrt{624\ 100 - 240\ 000}}{-120}$$

$$x = \frac{-790 \pm \sqrt{384\ 100}}{-120}$$

$$x = \frac{-790 \pm 619.758}{-120}$$

$$x = \frac{-790 + 619.758}{-120} \quad \text{or} \quad x = \frac{-790 - 619.758}{-120}$$

$$x \doteq 1.42 \quad \text{or} \quad x \doteq 11.75$$

The price of the tickets could be either $1.42 or $11.75.

b) $x = \dfrac{1.42 + 11.75}{2}$ ◄─── The maximum profit is halfway between the break-even prices. I added the two prices and then divided by 2 to get the price that should be charged.

$$x = \frac{13.17}{2} = 6.585$$

The price will be $6.59. ◄─── I can't have three decimal places because I'm dealing with money, so I rounded to $6.59.

In Summary

Key Ideas

- All quadratic equations of the form $ax^2 + bx + c = 0$ can be solved using the quadratic formula,

$$x = \frac{-b \pm \sqrt{b^2 - 4ac}}{2a}$$

- The quadratic formula is derived by completing the square for $ax^2 + bx + c = 0$ and solving for x. It is a direct way of calculating roots without graphing or algebraic manipulation.

Need to Know

- A quadratic equation can have 2, 1, or 0 real solutions, depending on the values of a, b, and c.
- The solutions generated by the quadratic formula for the equation $ax^2 + bx + c = 0$ correspond to the zeros, or x-intercepts, of the function $f(x) = ax^2 + bx + c$.

CHECK *Your Understanding*

1. Identify the values of a, b, and c you would substitute into the quadratic formula to solve each of the following.

   a) $3x^2 - 5x + 2 = 0$ c) $16x^2 + 7 = -24x - 2$

   b) $7 - 3x + 5x^2 = 0$ d) $(x - 3)(2x + 1) = 5(x - 2)$

2. a) Set up the quadratic formula for each equation in question 1.

   b) Solve the equations in part (a).

PRACTISING

3. Use the quadratic formula to solve each quadratic real equation. Round your answers to two decimal places. If there is no real solution, say so.

   a) $x^2 - 5x + 11 = 0$ d) $4x^2 - 5x - 7 = 0$

   b) $-2x^2 - 7x + 15 = 0$ e) $-3x^2 - 5x - 11 = 0$

   c) $4x^2 - 44x + 121 = 0$ f) $-8x^2 + 5x + 2 = 0$

4. Use a graphing calculator to graph the corresponding function for each equation in question 3. Determine the zeros for each to verify that your solutions are correct.

Tech | **Support**

For help using the graphing calculator to determine zeros, see Technical Appendix, B-8.

5. Identify a method that could be used to determine the roots of the given equations. Then use it to determine the roots.

   a) $x^2 = 15x$ d) $(x - 5)(2x - 3) = (x + 4)(x - 3)$

   b) $x^2 = 115$ e) $-2(x - 3)^2 + 50 = 0$

   c) $2x^2 = 19x - 24$ f) $1.5x^2 - 26.7x + 2.4 = 0$

6. On Mars, if you hit a baseball, the height of the ball at time t would be modelled by the quadratic function $h(t) = -1.85t^2 + 20t + 1$, where t is in seconds and $h(t)$ is in metres.

   a) When will the ball hit the ground?

   b) How long will the ball be above 17 m?

7. A gardener wants to fence three sides of the yard in front of her house.

A She bought 60 m of fence and wants an area of about 400 m^2. The quadratic equation $f(x) = 60x - 2x^2$, where x is the width of the yard in metres and $f(x)$ is the area in square metres, gives the area that can be enclosed. Determine the dimensions that will give the desired area.

8. The height of an arrow shot on Neptune can be modelled by the quadratic function $h(t) = 2.3 + 50t - 5.57t^2$, where time, t, is in seconds and height, $h(t)$, is in metres. Use the quadratic formula to determine when the arrow will hit the surface.

9. The quadratic function $d(s) = 0.0056s^2 + 0.14s$ models the relationship between stopping distance, d, in metres and speed, s, in kilometres per hour in driving a car. What is the fastest you can drive and still be able to stop within 60 m?

10. Create five quadratic equations by selecting integer values for a, b, and

T c in $ax^2 + bx + c = 0$. Choose values so that

1. N/F   2. a,b = same

3. 1+2 = 2 different

- 2 of your equations have two solutions
- 2 of your equations have one solution
- 1 of your equations has no solution

Use the quadratic formula to verify.

11. a) Determine the roots of $3(x - 4)^2 - 17 = 0$ to two decimal

C places by isolating $(x - 4)^2$ and then taking the square root of both sides.

b) Solve the equation $3(x - 4)^2 - 17 = 0$ by expanding $(x - 4)^2$ and then using the quadratic formula.

c) Which method was better for you? Explain.

Extending

12. Determine the intersection points of $y - 3x^2 - 4x - 9$ and $y = 2x$.

13. Determine the intersection points of $y = -2x^2 + 5x + 3$ and $y = x^2 + 2x - 5$.

14. Solve each of the following.

a) $x^3 + 5x^2 - 2x = 0$ **b)** $x^4 - 15x^2 + 54 = 0$

Exact Solution Patterns

The congruent parabolas that correspond to the equations below share the same axis of symmetry. Because they share a characteristic, they are called a family of quadratic equations.

The family of quadratic equations on the left has the solutions on the right.

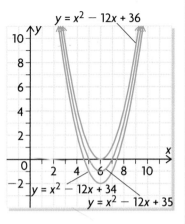

$y = x^2 - 12x + 36$

$y = x^2 - 12x + 34$

$y = x^2 - 12x + 35$

$x^2 - 12x + 36 = 0$	$x = 6$
$x^2 - 12x + 35 = 0$	$x = 6 + \sqrt{1}$ and $x = 6 - \sqrt{1}$
$x^2 - 12x + 34 = 0$	$x = 6 + \sqrt{2}$ and $x = 6 - \sqrt{2}$
$x^2 - 12x + 33 = 0$	$x = 6 + \sqrt{3}$ and $x = 6 - \sqrt{3}$
$x^2 - 12x + 32 = 0$	$x = 6 + \sqrt{4}$ and $x = 6 - \sqrt{4}$
$x^2 - 12x + 31 = 0$	$x = 6 + \sqrt{5}$ and $x = 6 - \sqrt{5}$
$x^2 - 12x + 26 = 0$	$x = 6 + \sqrt{10}$ and $x = 6 - \sqrt{10}$

1. How do the solutions relate to the original equations?

2. Using the above pattern, determine the exact solution of the following family of quadratic equations:

$$x^2 - 22x + 121 = 0$$
$$x^2 - 22x + 120 = 0$$
$$x^2 - 22x + 119 = 0$$
$$x^2 - 22x + 118 = 0$$
$$x^2 - 22x + 115 = 0$$
$$x^2 - 22x + 100 = 0$$
$$x^2 - 22x + 99 = 0$$
$$x^2 - 22x + 122 = 0$$

3. What do you need to start with to build your own family of quadratic equations?

4. Create your own family of quadratic equations from your base function.

FREQUENTLY ASKED Questions

Q: **What information about a parabola is easily determined from the vertex form of a quadratic function?**

A: The vertex, axis of symmetry, direction of opening, domain, and range are easily determined.

Study | **Aid**
• See Lesson 4.1, Example 3.
• Try Mid-Chapter Review Question 2.

EXAMPLE

$f(x) = 2(x - 3)^2 + 7$
vertex $(3, 7)$
axis of symmetry: $x = 3$
opens up
domain: $\{x \in \mathbf{R}\}$
range: $\{y \in \mathbf{R} \,|\, y \geq 7\}$

$g(x) = -3(x + 5)^2 + 2$
vertex $(-5, 2)$
axis of symmetry: $x = -5$
opens down
domain: $\{x \in \mathbf{R}\}$
range: $\{y \in \mathbf{R} \,|\, y \leq 2\}$

Q: **How can you change a quadratic function from standard to vertex form and vice versa?**

A: To change from standard to vertex form, you complete the square.

Study | **Aid**
• See Lesson 4.1, Example 1, Kristen's Solution, and Lesson 4.3, Examples 2, 3, and 4.
• Try Mid-Chapter Review Questions 1, 4, and 5.

EXAMPLE

$g(x) = ax^2 + bx + c, a \neq 1, a \neq 0$
$g(x) = 2x^2 + 12x - 7$
$g(x) = 2(x^2 + 6x) - 7$ ← Factor the coefficient in front of the x^2-term out of the first two terms.

$g(x) = 2(x^2 + 6x + 9 - 9) - 7$
$g(x) = 2(x^2 + 6x + 9) - 18 - 7$ ← Add and subtract the square of half the coefficient of x. Only three terms are required to complete the square, so multiply the last term by the coefficient in front of the brackets.

$g(x) = 2(x + 3)^2 - 25$ ← Factor the perfect trinomial square and collect like terms

To change from vertex to standard form, expand the function and simplify.

Q: **What formula can be used to solve any quadratic equation of the form $ax^2 + bx + c = 0$?**

A: The quadratic formula, which is $x = \dfrac{-b \pm \sqrt{b^2 - 4ac}}{2a}$.

Study | **Aid**
• See Lesson 4.3, Examples 3 and 4.
• Try Mid-Chapter Review Questions 7, 8, and 9.

PRACTICE Questions

Lesson 4.1

1. Write each function in standard form.
   a) $f(x) = (x - 8)^2 + 4$
   b) $g(x) = -(x - 3)^2 - 8$
   c) $f(x) = 4(x - 5)^2 + 9$
   d) $g(x) = -0.5(x - 4)^2 + 2$

2. State the vertex, axis of symmetry, maximum or minimum value, domain, and range of each function. Then graph each function.
   a) $f(x) = (x - 3)^2 + 6$
   b) $f(x) = -(x + 5)^2 - 7$

3. Which form of a quadratic function do you like using? Explain, using an example.

4. Determine the equation of each parabola.
   a)
   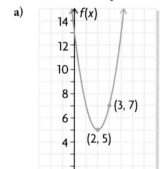
   b)
   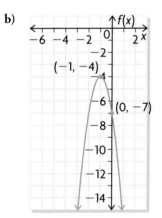

Lesson 4.2

5. Write each function in vertex form, and then sketch its graph.
   a) $f(x) = x^2 + 10x + 12$
   b) $f(x) = 2x^2 + 12x - 3$
   c) $f(x) = -x^2 - 8x - 10$
   d) $g(x) = 2x^2 - 2x + 7.5$

6. The cost, $C(n)$, of operating a cement-mixing truck is modelled by the function $C(n) = 2.2n^2 - 66n + 700$, where n is the number of minutes the truck is running. What is the minimum cost of operating the truck?

7. A police officer has 400 m of yellow tape to seal off the area of a crime scene. What is the maximum area that can be enclosed?

Lesson 4.3

8. Solve using the quadratic formula.
   a) $x^2 + 2x - 15 = 0$
   b) $9x^2 - 6x + 1 = 0$
   c) $2(x - 7)^2 - 6 = 0$
   d) $x^2 + 7x = -24$

9. The height of a ball at a given time can be modelled with the quadratic function $h(t) = -5t^2 + 20t + 1$, where height, $h(t)$, is in metres and time, t, is in seconds. How long is the ball in the air?

10. A theatre company's profit can be modelled by the function $P(x) = -60x^2 + 700x - 1000$, where x is the price of a ticket in dollars. What is the break-even price of the tickets?

11. A model rocket is launched into the air. Its height, $h(t)$, in metres after t seconds is $h(t) = -5t^2 + 40t + 2$.
    a) What is the height of the rocket after 2 s?
    b) When does the rocket hit the ground?
    c) When is the rocket at a height of 77 m?

4.4 Investigating the Nature of the Roots

YOU WILL NEED
- graphing calculator (optional)

GOAL

Determine how many real roots a quadratic equation has without actually locating them.

INVESTIGATE the Math

Quadratic equations can have 2, 1, or 0 real solutions.

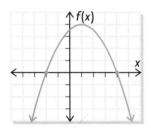 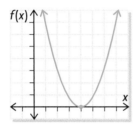 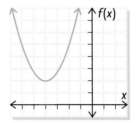

This is because, graphically, the zeros of any parabola $f(x) = ax^2 + bx + c$ are also the solutions of the corresponding quadratic equation $0 = ax^2 + bx + c$.

Sergio wonders whether he can tell how many zeros a quadratic function has without factoring or graphing.

? How can you predict the number of real solutions a quadratic equation has without graphing or determining the solution(s)?

A. Use the quadratic formula to determine the solution(s) of each quadratic equation.
- **a)** $x^2 + 3x + 1 = 0$
- **b)** $2x^2 - x + 1 = 0$
- **c)** $x^2 + 6x + 9 = 0$

B. Create three more quadratic equations and determine their solutions.

C. Based on your results in part A, sketch a parabola that could represent each of the given functions. Confirm your sketch by graphing each function on a graphing calculator.

D. What part of the quadratic formula is directly related to the number of real solutions each quadratic equation has? Explain.

Reflecting

E. Use the graph of the corresponding function to determine when a quadratic equation has
   a) two distinct real solutions
   b) one real solution
   c) no real solution

F. Use the quadratic formula to determine when a quadratic equation has
   a) two real distinct solutions
   b) one real solution
   c) no real solution

discriminant

the expression $b^2 - 4ac$ in the quadratic formula

G. Explain why the **discriminant** determines the number of real solutions of a quadratic equation.

APPLY the Math

> **EXAMPLE 1** — Connecting the number of real roots of a quadratic equation to the value of the discriminant

Use the discriminant to determine the number of roots of each quadratic equation.
a) $-2x^2 - 3x - 5 = 0$
b) $2x^2 - x - 6 = 0$
c) $x^2 - 10x = -25$

Dave's Solution

a) $b^2 - 4ac$

$(-3)^2 - 4(-2)(-5)$ ◄——— I substituted $a = -2$, $b = -3$, and $c = -5$ into the discriminant.

$= 9 - 40$

$= -31$

No real roots.

Since the answer is less than 0, there are no real roots.

b) $b^2 - 4ac$

$(-1)^2 - 4(2)(-6)$ ◄——— I substituted $a = 2$, $b = -1$, and $c = -6$ into the discriminant.

$= 1 + 48$

$= 49$

Two distinct real roots.

Since the answer is greater than 0, there are two distinct real roots.

c) $b^2 - 4ac$

$(-10)^2 - 4(1)(25)$ ← I substituted $a = 1$, $b = -10$, and $c = 25$ into the discriminant.

$= 100 - 100$

$= 0$

Since the answer is equal to 0, there is one real root.

One real root.

EXAMPLE **2**	Connecting the number of zeros of a quadratic function to the value of the discriminant

Without drawing the graph, state whether the quadratic function intersects the x-axis at one point, two points, or not at all.

a) $f(x) = 5x^2 + 3x - 7$

b) $g(x) = 4x^2 - x + 3$

Andrew's Solution

a) $b^2 - 4ac$

$(3)^2 - 4(5)(-7)$ ← I substituted $a = 5$, $b = 3$, and $c = -7$ into the discriminant.

$= 9 + 140$

$= 149$

Since the answer is greater than 0, the graph will intersect the x-axis at two points.

The parabola will intersect the x-axis at two points.

b) $b^2 - 4ac$

$(-1)^2 - 4(4)(3)$ ← I substituted $a = 4$, $b = -1$, and $c = 3$ into the discriminant.

$= 1 - 48$

$= -47$

Since the answer is less than 0, the graph will not intersect the x-axis.

The parabola will not intersect the x-axis.

If a quadratic function is expressed in vertex form, the location of the vertex and the direction in which the parabola opens can be used to identify how many zeros it has.

EXAMPLE 3 | Using reasoning to determine the number of zeros without graphing

Without drawing the graph, state whether the quadratic function intersects the x-axis at one point, two points, or not at all.

a) $f(x) = 2.3(x - 5)^2 + 4.5$

b) $g(x) = -3.7(x + 2)^2 + 3.5$

Talia's Solution

a) $f(x) = 2.3(x - 5)^2 + 4.5$ has no zeros.

> The function is in vertex form. The vertex is (5, 4.5). The value of *a* is 2.3, a positive number. The parabola opens up. The vertex is above the x-axis. Since the parabola opens up, the graph will never cross the x-axis, so this function has no zeros.

b) $g(x) = -3.7(x + 2)^2 + 3.5$ has two zeros.

> The function is in vertex form. The vertex is (−2, 3.5). The value of *a* is −3.7, a negative number. The parabola opens down. The vertex is above the x-axis. Since the parabola opens down, the graph will cross the x-axis twice.

EXAMPLE 4 | Solving a problem using the discriminant

For what value(s) of k does the equation $kx - 10 = 5x^2$ have

a) one real solution?

b) two distinct real solutions?

c) no real solution?

Kelly's Solution

a)
$$kx - 10 = 5x^2$$
$$-5x^2 + kx - 10 = 0$$

> I put the equation into standard form to see the discriminant better.

$$b^2 - 4ac$$
$$= k^2 - \boxed{4(-5)(-10)} \quad {\scriptstyle =200}$$
$$= k^2 - 100$$

> I substituted $a = -5$, $b = k$, and $c = -10$ into the discriminant.

$$k^2 - 100 > 0$$

> For there to be two distinct solutions, the discriminant must be greater than zero.

$$k^2 - 100 = 0$$

> For there to be one solution, the discriminant must equal zero.

230   4.4 Investigating the Nature of the Roots

$k^2 - 100 < 0$ ◄———————— For there to be no solution, the discriminant must be less than zero.

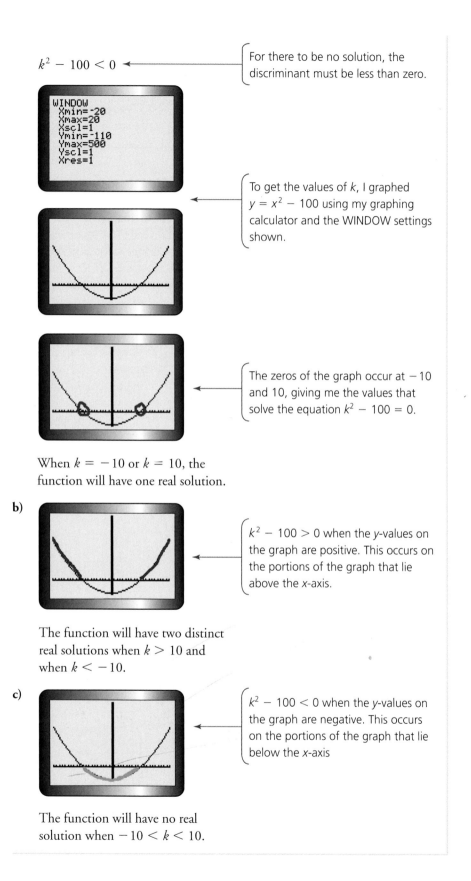

To get the values of k, I graphed $y = x^2 - 100$ using my graphing calculator and the WINDOW settings shown.

The zeros of the graph occur at -10 and 10, giving me the values that solve the equation $k^2 - 100 = 0$.

When $k = -10$ or $k = 10$, the function will have one real solution.

b)

$k^2 - 100 > 0$ when the y-values on the graph are positive. This occurs on the portions of the graph that lie above the x-axis.

The function will have two distinct real solutions when $k > 10$ and when $k < -10$.

c)

$k^2 - 100 < 0$ when the y-values on the graph are negative. This occurs on the portions of the graph that lie below the x-axis

The function will have no real solution when $-10 < k < 10$.

In Summary

Key Idea

- The value of the discriminant, $b^2 - 4ac$, tells you how many real solutions a quadratic equation has and how many x-intercepts the corresponding function has.

Need to Know

- For the quadratic equation $ax^2 + bx + c = 0$ and its corresponding function $f(x) = ax^2 + bx + c$, if
  - $b^2 - 4ac > 0$, then the equation has two distinct real solutions and the function has two x-intercepts
  - $b^2 - 4ac = 0$, then the equation has one real solution and the function has one x-intercept
  - $b^2 - 4ac < 0$, then the equation has no real solution and the function has no x-intercepts

CHECK *Your Understanding*

1. Write the discriminant. Do not evaluate.
   a) $x^2 - 5x + 7 = 0$ c) $3(x + 2)^2 - 19 = 0$
   b) $x^2 + 11x = 6x^2 - 17$ d) $(x + 3)(2x - 1) = 4(x + 2)$

2. Determine the number of real solutions of each quadratic equation. Do not solve.
   a) $x^2 + 3x - 4 = 0$ d) $3(x + 5)^2 + 7 = 0$
   b) $2x^2 - x + 5 = 0$ e) $-2(x - 5)^2 + 3 = 0$
   c) $4x^2 - 8x = 4x - 9$ f) $4(x + 1)^2 = 0$

3. Andrew thinks that the quadratic function $f(x) = x^2 - 5x + 2$ does not intersect the x-axis because the discriminant is negative. Do you agree? Explain.

PRACTISING

4. Determine whether each quadratic function intersects the x-axis at one
   K point, two points, or not at all. Do not draw the graph.
   a) $f(x) = 3x^2 + 6x - 1$ d) $g(x) = -3(x - 5)^2 + 2$
   b) $g(x) = 4(x - 6)^2 + 2$ e) $f(x) = 2x^2 + 3x + 5$
   c) $f(x) = 9x^2 - 30x + 25$ f) $g(x) = -3(x + 2)^2$

5. For what value(s) of k does the function $f(x) = kx^2 - 8x + k$ have no zeros?

6. For what value of m does $g(x) = 49x^2 - 28x + m$ have exactly one zero?

7. For what value of k does $8x^2 + 4x + k = 0$ have two distinct real solutions? one solution? no solution?

8. a) Explain how you would solve this problem: For what value of k does the function $f(x) = 3x^2 - 5x + k$ have only one zero?
b) Use your strategy to find the value of k.

9. The function $f(x) = x^2 + kx + k + 8$ touches the x-axis once. What value(s) could k be?

10. For what values of k does the line $y = x + k$ pass through the circle
T defined by $x^2 + y^2 = 25$ at
    a) 2 points?   **b)** 1 point?   **c)** 0 points?

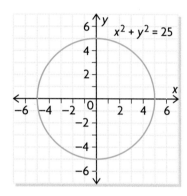

11. Generate 10 quadratic equations by randomly selecting integer values for a, b, and c in $ax^2 + bx + c = 0$. Use the discriminant to identify how many real solutions each equation has.

12. The profit, $P(x)$, of a video company, in thousands of dollars, is given
A by $P(x) = -5x^2 + 550x - 5000$, where x is the amount spent on advertising, in thousands of dollars. Can the company make a profit of $50\ 000$? Explain.

13. State two different ways to determine the number of zeros of the
C function $f(x) = 2(x + 1)^2 - 6$.

Extending

14. Can $p^2 - 13$ equal $-12p$? Explain your answer.

15. Using different values of k, determine the number of zeros of the function $f(x) = (k + 1)x^2 + 2kx + k - 1$.

Using Quadratic Function Models to Solve Problems

YOU WILL NEED

- graphing calculator (optional)

GOAL

Solve problems involving quadratic functions and equations arising from standard and vertex forms.

LEARN ABOUT the Math

The graph shows the height of a rock launched from a slingshot, where time, t, is in seconds and height, $h(t)$, is in metres.

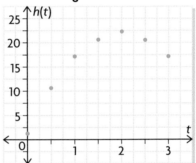

Height of a Rock

? How can you use the vertex form of a quadratic function to model the height of the rock over time and to determine when the rock hits the ground?

EXAMPLE 1   Connecting the vertex to an algebraic model for height

a) Use the graph to determine an algebraic model.
b) Use the model to determine when the rock hits the ground.

Nadia's Solution

a)
$$y = a(x - h)^2 + k$$
$$h(t) = a(t - 2)^2 + 22$$
$$2 = a(0 - 2)^2 + 22$$
$$2 = a(-2)^2 + 22$$
$$2 = 4a + 22$$
$$2 - 22 = 4a$$

> The vertex appears to be (2, 22), so I replaced h with 2 and k with 22.
>
> Then I substituted (0, 2), a point on the parabola, to solve for a.

$$-20 = 4a$$

$$\frac{-20}{4} = a$$

$$-5 = a$$

The function that models the rock's height is $h(t) = -5(t - 2)^2 + 22$.

> I put the value of a in the vertex form.

b) $-5(t - 2)^2 + 22 = 0$

> The height of the ground is zero, so I set $h(t) = 0$ and solved for t.

$$-5(t - 2)^2 = -22$$

$$(t - 2)^2 = \frac{-22}{-5}$$

$$(t - 2)^2 = \frac{22}{5}$$

$$t - 2 = \pm\sqrt{\frac{22}{5}}$$

$$t = 2 \pm \sqrt{\frac{22}{5}}$$

$$t \doteq -0.1 \quad \text{and} \quad t \doteq 4.1$$

Time, t, must be positive, so the domain is $\{t \in \mathbf{R} \mid 0 \leq t \leq 4.1\}$. The rocket hits the ground at about 4.1 s.

Reflecting

A. Why was using the vertex form to determine the function a good strategy for solving this problem?

B. Why is determining the domain important when modelling quadratic functions?

C. What factors can affect the form of the quadratic function you choose to model a problem?

APPLY the Math

EXAMPLE **2** **Selecting a strategy to determine when a quadratic function reaches its maximum value**

Mr. McIntosh has 90 apple trees. He earns an annual revenue of $120 per tree. If he plants more trees, they have less room to grow, resulting in fewer apples per tree. As a result, the annual revenue per tree is reduced by $1 for each additional tree. The revenue, in dollars, is modelled by the function $R(x) = (90 + x)(120 - x)$, where x is the number of additional trees planted.

Regardless of the number of trees planted, the cost of maintaining each tree is $8. The cost, in dollars, is modelled by the function $C(x) = 8(90 + x)$, where x is the number of additional trees planted. How many trees must Mr. McIntosh plant to maximize profit?

Talia's Solution

$P(x) = R(x) - C(x)$ ◄──────── To get the profit function, I subtracted the cost function from the revenue function.

$P(x) = (90 + x)(120 - x) - 8(90 + x)$ ◄─── I expanded and then simplified the profit function.

$P(x) = 10\,800 - 90x + 120x - x^2 - 720 - 8x$

$P(x) = -x^2 + 22x + 10\,080$ ◄──── If the function is written in vertex form, I can just read when the maximum happens and what it will be.

I completed the square to get the function in vertex form.

$P(x) = -(x^2 - 22x) + 10\,080$ ◄──── I factored -1 from the first two terms. I added and subtracted the square of half of the coefficient of the x-term.

$P(x) = -(x^2 - 22x + 121 - 121) + 10\,080$

$P(x) = -(x^2 - 22x + 121) + 121 + 10\,080$ ◄─── I grouped the three terms of the perfect square and I multiplied the last term by -1 to move it outside the brackets.

I factored the perfect square and added the constant terms together.

$P(x) = -(x - 11)^2 + 10\,201$ ◄──── The vertex is (11, 10 201). The parabola opens down, so the maximum is 10 201 when $x = 11$.

Number of trees $= 90 + 11 = 101$

Mr. McIntosh needs to plant 11 extra trees, so he should have a total of 101 trees to maximize his revenue.

EXAMPLE 3

EXAMPLE 3   Selecting a strategy to determine when a quadratic function reaches its minimum value

The cost of running an assembly line is a function of the number of items produced per hour. The cost function is $C(x) = 0.28x^2 - 1.12x + 2$, where $C(x)$ is the cost per hour in thousands of dollars, and x is the number of items produced per hour in thousands. Determine the most economical production level.

Andrew's Solution

$$y = a(x - h)^2 + k$$
$$C(x) = 0.28x^2 - 1.12x + 2$$

"The most economical" means "the minimum cost." I wrote the function in vertex form so I could see the minimum value and when it happens.

$$C(x) = 0.28(x^2 - 4x) + 2$$

I factored 0.28 from the first two terms.

$$C(x) = 0.28(x^2 - 4x + 4 - 4) + 2$$

I added and subtracted the square of half of the coefficient of the x-term.

$$C(x) = 0.28(x^2 - 4x + 4) - 1.12 + 2$$

I grouped the first three terms of the perfect square and multiplied the last term by 0.28 to move it outside the brackets.

$$C(x) = 0.28(x - 2)^2 + 0.88$$

I factored the perfect square and added the constant terms together.

The most economical production level is 2000/h.

The vertex is (2, 0.88). The parabola opens up, so a minimum value of 0.88 happens when $x = 2$.

I multiplied by 1000 because x is in thousands.

| EXAMPLE **4** | Selecting a strategy to determine when a quadratic function reaches a given value |

A bus company usually charges \$2 per ticket, but wants to raise the price by 10¢ per ticket. The revenue that could be generated is modelled by the function $R(x) = -40(x - 5)^2 + 25\,000$, where x is the number of 10¢ increases and the revenue, $R(x)$, is in dollars. What should the price of the tickets be if the company wants to earn \$21 000?

Rachel's Solution: Solving Algebraically

$$-40(x - 5)^2 + 25\,000 = 21\,000$$ ← I set the revenue function equal to 21 000.

$$-40(x - 5)^2 = 21\,000 - 25\,000$$ ← To isolate the squared term, I subtracted 25 000 from both sides and simplified.

$$-40(x - 5)^2 = -4000$$

$$(x - 5)^2 = 100$$ ← I divided both sides by -40 and simplified.

$$(x - 5) = \pm 10$$ — Then I took the square root, which can be positive or negative, of both sides.

$x - 5 = 10$ or $x - 5 = -10$ ← I rewrote the resulting equations so that I could see the two answers.

$x = 10 + 5$ or $x = -10 + 5$

$x = 15$ or $x = -5$

The price of the ticket is $\$2 + 15(0.10) = \$3.50$. ← I can only use the positive value, so there will be 15 ten-cent increases, resulting in a ticket price of \$3.50.

Joyce's Solution: Expanding the Equation and Using the Quadratic Formula

$$-40(x - 5)^2 + 25\,000 = 21\,000$$ ← I set the revenue function equal to 21 000.

$$-40(x - 5)(x - 5) + 25\,000 - 21\,000 = 0$$ — I put the equation into standard form so that I could use the quadratic formula.

$$-40(x^2 - 5x - 5x + 25) + 4000 = 0$$

$$-40(x^2 - 10x + 25) + 4000 = 0$$

$$-40x^2 + 400x - 1000 + 4000 = 0$$

$$-40x^2 + 400x + 3000 = 0$$

$$x = \frac{-b \pm \sqrt{b^2 - 4ac}}{2a}$$ ← I substituted $a = -40$, $b = 400$, and $c = 3000$ into the quadratic formula and solved.

$$x = \frac{-(400) \pm \sqrt{(400)^2 - 4(-40)(3000)}}{2(-40)}$$

$$x = \frac{-400 \pm \sqrt{160\,000 + 480\,000}}{-80}$$

$$x = \frac{-400 \pm \sqrt{640\,000}}{-80}$$

$$x = \frac{-400 \pm 800}{-80}$$

$$x = \frac{-400 + 800}{-80} \quad \text{or} \quad x = \frac{-400 - 800}{-80}$$

$$x = \frac{400}{-80} \quad \text{or} \quad x = \frac{-1200}{-80}$$

$$x = -5 \quad \text{or} \quad x = 15$$

The price of the ticket will be $2 + 15(0.10) = $3.50. ◀── I can only use the positive value for x. There will be 15 ten-cent increases, resulting in a ticket price of $3.50.

In Summary

Key Idea

- The vertex form of a quadratic function, $f(x) = a(x - h)^2 + k$, can be used to solve a variety of problems, such as determining the maximum or minimum value of the quadratic model. This value can then be used as needed to interpret the situation presented.

Need to Know

- If a problem requires you to determine the value of the independent variable, x, for a given value of the dependent variable, $f(x)$, for a quadratic function model, then substitute the number in $f(x)$. This will result in a quadratic equation that can be solved by graphing, factoring, or using the quadratic formula.

CHECK Your Understanding

1. The manager of a hardware store sells batteries for $5 a package. She wants to see how much money she will earn if she increases the price in 10¢ increments. A model of the price change is the revenue function $R(x) = -x^2 + 10x + 3000$, where x is the number of 10¢ increments and $R(x)$ is in dollars. Explain how to determine the maximum revenue.

2. Determine the maximum revenue generated by the manager in question 1.

3. A cliff diver dives from about 17 m above the water. The diver's height above the water, $h(t)$, in metres, after t seconds is modelled by $h(t) = -4.9t^2 + 1.5t + 17$. Explain how to determine when the diver is 5 m above the water.

4. Determine when the diver in question 3 is 5 m above the water.

PRACTISING

5. The function $P(x) = -30x^2 + 360x + 785$ models the profit, $P(x)$, **K** earned by a theatre owner on the basis of a ticket price, x. Both the profit and ticket price are in dollars. What is the maximum profit, and how much should the tickets cost?

6. The population of a town is modelled by the function **A** $P(t) = 6t^2 + 110t + 4000$, where $P(t)$ is the population and t is the time in years since 2000.
   a) What will the population be in 2020?
   b) When will the population be 6000?
   c) Will the population ever be 0? Explain your answer.

7. The profit of a shoe company is modelled by the quadratic function $P(x) = -5(x - 4)^2 + 45$, where x is the number of pairs of shoes produced, in thousands, and $P(x)$ is the profit, in thousands of dollars. How many thousands of pairs of shoes will the company need to sell to earn a profit?

8. Beth wants to plant a garden at the back of her house. She has 32 m of fencing. The area that can be enclosed is modelled by the function $A(x) = -2x^2 + 32x$, where x is the width of the garden in metres and $A(x)$ is the area in square metres. What is the maximum area that can be enclosed?

9. The stopping distance for a boat in calm water is modelled by the function $d(v) = 0.004v^2 + 0.2v + 6$, where $d(v)$ is in metres and v is in kilometres per hour.
   a) What is the stopping distance if the speed is 10 km/h?
   b) What is the initial speed of the boat if it takes 11.6 m to stop?

10. Mario wants to install a wooden deck around a rectangular swimming pool. The function $C(w) = 120w^2 + 1800w$ models the cost, where the cost, $C(w)$, is in dollars and width, w, is in metres. How wide will the deck be if he has \$4080 to spend?

11. The population of a rural town can be modelled by the function $P(x) = 3x^2 - 102x + 25\,000$, where x is the number of years since 2000. According to the model, when will the population be lowest?

12. A bowling alley has a $5 cover charge on Friday nights. The manager is considering increasing the cover charge in 50¢ increments. The revenue is modelled by the function $R(x) = -12.5x^2 + 75x + 2000$, where revenue $R(x)$ is in dollars and x is the number of 50¢ increments.

   a) What cover charge will yield the maximum revenue?
   b) What will the cover charge be if the revenue is $2000?

13. The height of a soccer ball kicked in the air is given by the quadratic equation $h(t) = -4.9(t - 2.1)^2 + 23$, where time, t, is in seconds and height, $h(t)$, is in metres.
   a) What was the height of the ball when it was kicked?
   b) What is the maximum height of the ball?
   c) Is the ball still in the air after 6 s? Explain.
   d) When is the ball at a height of 10 m?

14. The student council is deciding how much to charge for a ticket to the school dance to make the most money. The data collected are listed in the table.

Ticket Cost ($)	3.50	4.00	5.00	6.00	6.50	7.00	8.00
Revenue ($)	1260	1365	1485	1485	1440	1365	1125

   a) Determine the function, in vertex form, that represents these data.
   b) What should the ticket price be to earn $765 in revenue?

15. What are some advantages and disadvantages in using the vertex form to solve questions about quadratic functions?

Extending

16. A rectangle is 7 cm longer than it is wide. The diagonal is 13 cm. What are the rectangle's dimensions?

17. A photo framer wants to place a matte of uniform width all around a photo. The area of the matte should be equal to the area of the photo. The photo measures 40 cm by 60 cm. How wide should the matte be?

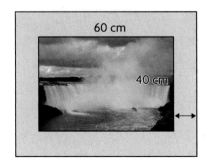

4.6

Using the Vertex Form to Create Quadratic Function Models from Data

YOU WILL NEED

- graphing calculator
- graph paper
- dynamic geometry software
- graphing software

GOAL

Determine the equation of a curve of good fit from data.

INVESTIGATE *the Math*

The table shows data on the percent of 15- to 19-year-old male Canadians who smoke.

Year	1981	1983	1985	1986	1989	1991	1994	1995	1996
Percent	43.4	39.6	26.7	25.2	22.6	22.6	27.3	28.5	29.1

❓ **What is a function that will model the data?**

A. Create a scatter plot of the data with an appropriate scale.

B. What shape best describes the graph? Draw a curve of good fit.

C. Estimate the coordinates of the vertex.

D. Use the vertex to write an equation in vertex form.

E. In what direction does the parabola open? What does this tell you?

F. Using one of the points in the table, calculate the *a*-value. Write the equation for the data in vertex form and in standard form.

G. Determine the domain and range of your model.

H. Using a graphing calculator and quadratic regression, determine a quadratic function that will model the data.

Tech | **Support**

For help using quadratic regression to determine the equation of a curve, see Technical Appendix, B-10.

Reflecting

I. How does your model compare with the graphing calculator's model?

J. How does the vertex form of an equation help you determine an equation for a curve of good fit?

K. How will you know whether the equation is a good representation of your data?

APPLY *the Math*

EXAMPLE **1** **Selecting a strategy to determine the equation of a curve of good fit: dynamic geometry software**

Determine a quadratic equation in vertex form that best represents the arch between the towers in the suspension bridge photo. Express your equation in standard form as well.

Veronica's Solution

I imported the picture into dynamic geometry software and superimposed a grid over the picture.

$$y = a(x - 1)^2 + 2$$

It looks like the vertex is at (1, 2), so I used the vertex form for the function.

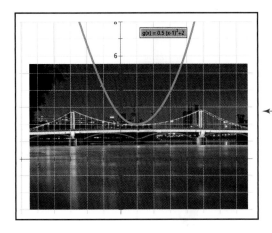

I guessed that 0.5 might work for the value of *a* since the shape is wide. I chose a positive value because the parabola opens up.

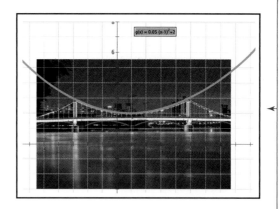

The value of *a* that I chose didn't work, so I tried something less than 0.1. I know that making the value of *a* closer to 0 creates a wider parabola. I tried 0.06. This fit better.

The parabola extends past the towers, so I need to restrict the domain. The towers appear to be located at $x = -3$ and $x = 5$. If so, the range will be restricted to between $y = 2$ and $y = 2.96$.

The equation is $y = 0.06(x - 1)^2 + 2$.

domain: $\{x \in \mathbf{R} \mid -3 \leq x \leq 5\}$

range: $\{y \in \mathbf{R} \mid 2 \leq y \leq 2.96\}$

$y = 0.06(x^2 - 2x + 1) + 2$ ← I expanded to get the standard form.

The standard form is
$y = 0.06x^2 - 0.12x + 2.06$.

| EXAMPLE 2 | Representing a quadratic function from data |

A hose sprays a stream of water across a lawn. The table shows the approximate height of the stream above the lawn at various distances from the person holding the nozzle. Write an algebraic model in vertex form that relates the height of the water to the distance from the person. Check your answer using a graphing calculator.

Distance from Nozzle (m)	0	1	2	3	4	5	6	7	8
Height above Lawn (m)	0.5	1.4	2.1	2.6	2.9	3.0	2.9	2.5	1.9

Betty's Solution: Using the Vertex Form

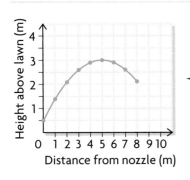

I graphed the data from the table by hand.

It looks quadratic. I estimated the vertex as (5, 3) and substituted these numbers into the vertex form.

$$y = a(x - 5)^2 + 3$$
$$0.5 = a(0 - 5)^2 + 3$$
$$0.5 = a(-5)^2 + 3$$
$$0.5 = a(25) + 3$$
$$0.5 - 3 = 25a$$
$$-2.5 = 25a$$
$$-0.1 = a$$

The graph must pass through one of the pairs of data from the table. I chose the point 0, 0.5) to get a.

$$y = -0.1(x - 5)^2 + 3$$
$$y = -0.1(x - 5)(x - 5) + 3$$
$$y = -0.1(x^2 - 10x + 25) + 3$$
$$y = -0.1x^2 + 1x - 2.5 + 3$$
$$y = -0.1x^2 + 1x + 0.5$$

I substituted a into the original equation and expanded it into standard form.

I checked to see how close my equation was to the calculator's.

I did well, but the graphing calculator was much faster than my hand calculation.

From the calculator,
$y = -0.1068x^2 + 1.0362x - 0.4763$;
from my scatter plot,
$y = -0.1x^2 + x + 0.5$.

Tech | *Support*

For help creating a scatter plot and determining the equation of the curve of best fit, see Technical Appendix, B-10.

Sheila's Solution: Using the Vertex Form and a Graphing Calculator

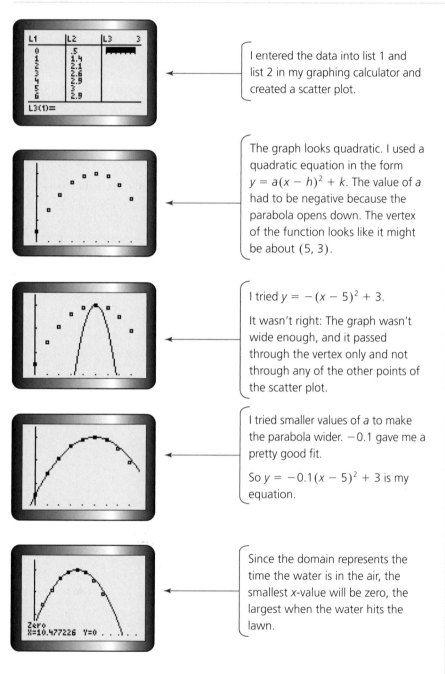

I entered the data into list 1 and list 2 in my graphing calculator and created a scatter plot.

The graph looks quadratic. I used a quadratic equation in the form $y = a(x - h)^2 + k$. The value of a had to be negative because the parabola opens down. The vertex of the function looks like it might be about $(5, 3)$.

I tried $y = -(x - 5)^2 + 3$.

It wasn't right: The graph wasn't wide enough, and it passed through the vertex only and not through any of the other points of the scatter plot.

I tried smaller values of a to make the parabola wider. -0.1 gave me a pretty good fit.

So $y = -0.1(x - 5)^2 + 3$ is my equation.

Since the domain represents the time the water is in the air, the smallest x-value will be zero, the largest when the water hits the lawn.

domain: $\{x \in \mathbf{R} \mid 0 \le x \le 10.5\}$

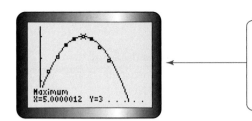

The range represents the water's height, which could be anywhere from 0 m above the ground to the maximum height of 3 m above the ground.

Tech | *Support*

For help creating a scatter plot, inserting a function, and using sliders in Fathom, see Technical Appendix, B-23.

range: $\{y \in \mathbf{R} \mid 0 \le y \le 3\}$

EXAMPLE **3**

Selecting a strategy to determine the equation of a curve of good fit: Graphing software

A plastic glider is launched from a hilltop. The height of the glider above the ground at a given time is recorded in the table. When will it reach a height of 45 m?

Collection 1

Time (s)	0	1	2	3	4	5	6	7	8	9	10
Height (m)	16.0	12.5	9.5	7.0	5.0	3.5	2.5	2.0	2.0	2.5	3.5

Time (s)	11	12	13	14	15	16	17	18	19	20
Height (m)	5.0	7.0	9.5	12.5	16.0	20.0	24.5	29.5	35.0	41.0

Griffen's Solution

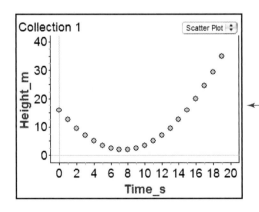

From the trend in the table of values, I predicted that the glider would reach a height of 45 m just after 20 s.

To get a more precise answer, I used graphing software to create a scatter plot. I observed the shape of the plot.

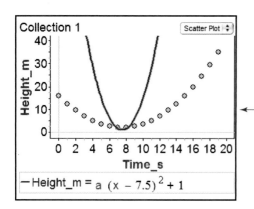

Height_m = a (x − 7.5)² + 1

It looks quadratic because the curve decreases, then increases, forming a parabola. The minimum height occurs when $x = 7.5$ s and appears to be 1.

I substituted the vertex (7.5, 1) into the vertex form of the equation and got $f(x) = a(x − 7.5)^2 + 1$.

I added a function to my graph and used a slider to get the value of a. I tried to get a good fit by changing the value of a.

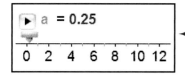

When $a = 0.25$, the fit is pretty good.

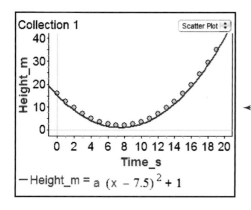

Height_m = a (x − 7.5)² + 1

So the equation that models the data is $f(x) = 0.25(x − 7.5)^2 + 1$.

The domain represents the time the glider is in the air. So time must begin at $x = 0$.

The range is the height reached by the glider. The lowest height is 1 m.

domain: $\{x \in \mathbf{R} \mid x \geq 0\}$

range: $\{y \in \mathbf{R} \mid y \geq 1\}$

$$0.25(x − 7.5)^2 + 1 = 45$$
$$0.25(x − 7.5)^2 = 44$$
$$(x − 7.5)^2 = 176$$
$$x − 7.5 = \pm\sqrt{176}$$
$$x = 7.5 \pm \sqrt{176}$$
$$x \doteq 7.5 + 13.27, x \geq 0$$
$$x = 20.77$$

To find when the glider reaches a height of 45 m, I substituted 45 in $f(x)$ and used inverse operations to solve the resulting quadratic equation.

Since x represents time, I can only use positive values.

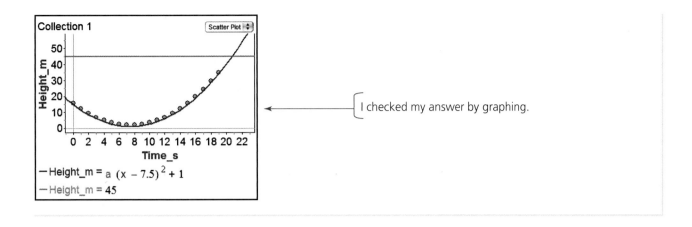

Collection 1

Height_m

Scatter Plot

Time_s

— Height_m = $a(x - 7.5)^2 + 1$

— Height_m = 45

I checked my answer by graphing.

In Summary

Key Ideas

- If a scatter plot has a parabolic shape and its curve of good fit passes through or near the vertex, then the vertex form of the quadratic function can determine an algebraic model of the relationship.
- Once the algebraic model has been determined, it can be used to solve problems involving the relationship.

Need to Know

- Some graphing calculators and graphing and data programs can determine an algebraic model of a scatter plot by regression. This produces the "best" possible model for the situation. If the data form a nonlinear pattern, the graph produced by the regression equation is called the "curve of best fit."
- Curves of good fit are useful for interpolating. They are not necessarily useful for extrapolating because they assume that the trend in the data will continue. Many factors can affect the relationship between the independent and dependent variable and change the trend, resulting in a different curve and a different algebraic model.
- It is often necessary to restrict the domain and range of the function to represent a realistic situation. For example, when a ball is in the air, the domain is between zero and the time the ball hits the ground; the range is between zero and the maximum height of the ball.

CHECK Your Understanding

1. Create a scatter plot, and decide whether the data appear to be quadratic. Justify your decision.

a)

Time (s)	0	1	2	3	4	5
Number	1800	5400	16 200	48 600	145 800	437 400

b)

Distance (m)	0	1	2	3	4	5	6	7	8
Height (m)	3	1	1	3	7	13	21	31	43

2. The following are data on the percent of 15- to 19-year-old female Canadians who smoke.

Year	1981	1983	1985	1986	1989	1991	1994	1995	1996
Percent	41.7	40.5	27.7	27	23.5	25.6	28.9	29.5	31

a) Use your graphing calculator to create a scatter plot.
b) Estimate the coordinates of the vertex.
c) Is the parabola opening up or down? Use your answer to estimate a value for a.
d) Adjust your estimate of a until you are satisfied with the fit.
e) Write the function that defines your curve of good fit.
f) State any restrictions on the domain and range of your model.

PRACTISING

3. Determine the equation of each parabola.

K a)

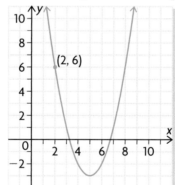

(2, 6)

b)

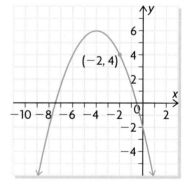

(−2, 4)

4. Write the standard form of the quadratic equation.

	Vertex	y-intercept
a)	(2, 3)	11
b)	(−1, 5)	3
c)	(3, −7)	−43
d)	(−2, −5)	19

5. A car skids in an accident. The investigating police officer knows that
A the distance a car skids depends on the speed of the car just before the brakes are applied.

Speed (km/h)	0	10	20	30	40	50	60	70	80	90	100
Length of Skid (m)	0.0	0.7	2.8	6.4	11.4	17.8	25.7	35.0	45.7	57.8	71.4

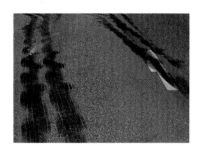

a) Create a scatter plot of the data in the table, and draw a curve of good fit.

b) Determine an equation of the curve of good fit. Assume that there is only one zero, located at the origin.

c) Use the curve to determine the length of the skid mark to the nearest tenth of a kilometre if the initial speed was 120 km/h.

d) State any restrictions on the domain and range of your model.

6. The amount of gas used by a car per kilometre depends on the car's speed.

Speed (km/h)	20	40	60	80	100	120
Cost of Gas (¢/km)	19.1	17.8	17.1	17.1	17.8	19.1

a) Use the data in the table to determine an equation for a curve of good fit.

b) Use the curve to determine the cost of gas if the car's speed is 140 km/h.

c) State any restrictions on the domain and range of your model.

7. The number of new cars sold in Canada from 1982 to 1992 is shown in the table.

Year	1982	1983	1984	1985	1986	1987	1988	1989	1990	1991	1992
New Cars Sold (000s)	718	841	971	1135	1102	1061	1056	985	885	873	798

a) Determine the equation of a curve of good fit.

b) How well does the equation predict car sales after 2000?

c) State any restrictions on the domain and range of your model.

8. The height of an arrow shot by an archer is given in the table. Determine the equation for a curve of good fit. State any restrictions on the domain and range of your model. Use it to predict when the arrow will hit the ground.

Time (s)	0	0.5	1.0	1.5	2.0	2.5
Height (m)	0.5	9.2	15.5	19.3	20.5	19.3

9. A farming cooperative collected data on the effect of different amounts of fertilizer, x, in hundreds of kilograms per hectare (kg/ha) on the yield of carrots, y, in tonnes (t).

Fertilizer (hundreds kg/ha)	0	0.25	0.50	0.75	1.00	1.25	1.50	1.75	2.00
Yield (t)	0.15	0.45	0.65	0.90	0.95	1.10	1.05	0.90	0.80

a) Write an equation to predict the yield of carrots based on fertilizer used.

b) State any restrictions on the domain and range of your model.

c) How much fertilizer is used for a yield of 0.75 t?

10. The data in the table at the left show the average miles per gallon for various cars and their top speeds in miles per hour.

a) Determine the equation of the curve of good fit in vertex form.

b) Use your equation from part (a) to estimate the top speed reached by a car that gets 150 miles per gallon. Does this make sense?

Average Miles per Gallon	Top Speed (miles/h)
96.0	17.5
56.0	97.0
97.0	20.0
105.0	20.0
45.4	97.0
38.8	111.0
35.4	111.0
121.0	45.0
18.1	165.0
17.0	147.0
16.7	157.0
130.0	55.0

11. A quadratic function passes through the points $(-2, 6)$, $(0, 6)$, and **T** $(2, 22)$. Determine its equation algebraically without using quadratic regression.

12. If data appear to be quadratic, explain how an equation of the curve of **C** good fit could be obtained even if the vertex for the function does not appear in the data.

Extending

13. A sprinkler waters a circular area of lawn of radius 9 m. Another sprinkler waters an area that is 100% larger. What is the radius of lawn reached by the second sprinkler?

14. Determine two numbers that add to 39 and multiply to 360. Use a method other than guess and check.

15. A school has decided to sell T-shirts as a fundraiser. Research shows that 800 students will buy one if they cost $5 each. For every 50¢ increase in the price, 20 fewer students will buy T-shirts. What is the maximum revenue, and for what price should the shirts sell?

FREQUENTLY ASKED *Questions*

Q: **How can you determine the number of zeros and the number of solutions of a quadratic equation?**

A1: If the function $f(x) = ax^2 + bx + c$ or the equation $0 = ax^2 + bx + c$ is in standard form, substitute the values of a, b, and c into the discriminant, $b^2 - 4ac$, and evaluate.
- If $b^2 - 4ac > 0$, there are two distinct zeros/real solutions.
- If $b^2 - 4ac = 0$, there is one zero/real solution.
- If $b^2 - 4ac < 0$, there are no zeros/real solutions.

A2: Write the function in vertex form. Then you can identify the vertex and whether the parabola opens up or down. For example, if the vertex is above the x-axis and opens down, there will be two distinct zeros.

Q: **What strategies can you use to solve problems involving quadratic functions and equations?**

A: To determine the maximum or minimum value, you can use a graphing calculator or put the function into vertex form by completing the square.

To solve a quadratic equation:
- A table of values might work but is time-consuming.
- Graphing by hand is also time-consuming.
- Using a graphing calculator is faster and more reliable, especially if the answer is not an integer.
- Factoring works only some of the time and becomes difficult when the numbers in the equation are large.
- The quadratic formula always works, but the quadratic equation must be in standard form.

Q: **What strategies can you use to determine the equation of a curve of good fit of a quadratic function in vertex form from data?**

A: Graph the data either by hand or using a graphing calculator to see if the curve looks quadratic. Locate or estimate the coordinates of the vertex. Replace h and k in the vertex form of the quadratic function, $f(x) = a(x - h)^2 + k$. Pick a point that is on the curve to determine the value of a. Verify that your equation matches the data.

Study | *Aid*
- See Lesson 4.4, Examples 1 and 2.
- Try Chapter Review Question 7.

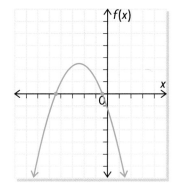

Study | *Aid*
- See Lesson 4.5, Examples 1 to 4.
- Try Chapter Review Questions 9, 10, and 11.

Study | *Aid*
- See Lesson 4.6, Example 2.
- Try Chapter Review Questions 2, 12, and 13.

PRACTICE Questions

Lesson 4.1

1. Write in standard form.
   a) $f(x) = (x + 3)^2 - 7$
   b) $f(x) = -(x + 7)^2 + 3$
   c) $f(x) = 2(x - 1)^2 + 5$
   d) $f(x) = -3(x - 2)^2 - 4$

2. Write the equation of each graph in vertex form.

a)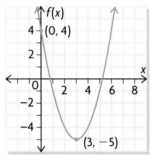

b)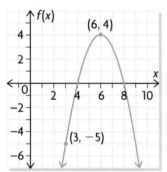

c)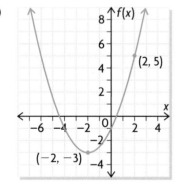

d)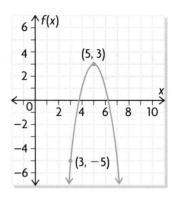

Lesson 4.2

3. Write in vertex form by completing the square.
   a) $f(x) = x^2 + 2x - 15$
   b) $f(x) = -x^2 + 8x - 7$
   c) $f(x) = 2x^2 + 20x + 16$
   d) $f(x) = 3x^2 + 12x + 19$
   e) $f(x) = \dfrac{1}{2}x^2 - 6x + 26$
   f) $f(x) = 2x^2 + 2x + 4$

4. Determine the vertex, the axis of symmetry, the direction the parabola opens, and the number of zeros for each quadratic function. Sketch a graph of each.
   a) $f(x) = 3(x - 5)^2 - 2$
   b) $g(x) = -2(x + 3)^2 - 1$
   c) $f(x) = 2x^2 + 4x + 7$
   d) $g(x) = -x^2 + 16x - 64$

Lesson 4.3

5. Use the quadratic formula to determine the solutions.
   a) $2x^2 - 15x - 8 = 0$
   b) $3x^2 + x + 7 = 0$
   c) $9x^2 = 6x - 1$
   d) $2.5x^2 = -3.1x + 7$

6. A T-ball player hits a ball from a tee that is 0.6 m tall. The height of the ball at a given time is modelled by the function $h(t) = -4.9t^2 + 7t + 0.6$, where height, $h(t)$, is in metres and time, t, is in seconds.
  a) What will the height be after 1 s?
  b) When will the ball hit the ground? → Find 0s

Lesson 4.4

7. Without solving, determine the number of real solutions of each equation.
  a) $x^2 - 5x + 9 = 0$
  b) $3x^2 - 5x - 9 = 0$
  c) $16x^2 - 8x + 1 = 0$

8. For the function $f(x) = kx^2 + 8x + 5$, what value(s) of k will have
  a) two distinct real solutions?
  b) one real solution?
  c) no real solution?

Lesson 4.5

9. The daily production cost, C, of a special-edition toy car is given by the function $C(t) = 0.2t^2 - 10t + 650$, where $C(t)$ is in dollars and t is the number of cars made.
  a) How many cars must be made to minimize the production cost?
  b) Using the number of cars from part (a), determine the cost.

10. The function $A(w) = 576w - 2w^2$ models the area of a pasture enclosed by a rectangular fence, where w is width in metres.
  a) What is the maximum area that can be enclosed?
  b) Determine the area that can be enclosed using a width of 20 m.

c) Determine the width of the rectangular pasture that has an area of 18 144 m².

Lesson 4.6

11. The vertical height of an arrow at a given time, t, is shown in the table.

Time (s)	0	0.5	1.0	1.5	2.0	2.5
Height (m)	0.5	6.3	9.6	10.5	8.9	4.9

  a) Determine an equation of a curve of good fit in vertex form.
  b) State any restrictions on the domain and range of your function.
  c) Use your equation from part (a) to determine when the arrow will hit a target that is 2 m above the ground.

12. The table shows how many injuries resulted from motor vehicle accidents in Canada from 1984 to 1998.

Year	Injuries
1984	237 455
1986	264 481
1988	278 618
1990	262 680
1992	249 821
1994	245 110
1996	230 890
1998	217 614

  a) Create a scatter plot, and draw a curve of good fit.
  b) Determine an equation of a curve of good fit.
  c) Check the accuracy of your model using quadratic regression.
  d) Use one of your models to predict how many accidents will result in injury in 2002. Explain why you chose the model you did.
  e) According to the quadratic regression model, when were accidents that resulted in injury at their maximum levels?

1. Write in standard form.
   a) $f(x) = (x + 3)^2 - 7$ b) $f(x) = -3(x + 5)^2 + 2$

2. Write in vertex form by completing the square.
   a) $f(x) = x^2 - 10x + 33$ b) $f(x) = -5x^2 + 20x - 12$

3. Determine the vertex, the equation of the axis of symmetry, and the maximum or minimum value for each function. Sketch a graph of each.
   a) $f(x) = 2(x - 8)^2 + 3$ b) $f(x) = -3x^2 + 42x - 141$

4. Can all quadratic equations be solved by the quadratic formula? Explain.

5. Solve using the quadratic formula.
   a) $x^2 - 6x - 8 = 0$ b) $3x^2 + x = -9$

6. Use the discriminant to determine the number of real solutions of each quadratic equation.
   a) $5x^2 - 4x + 11 = 0$ b) $x^2 = 8x - 16$

7. The height of a soccer ball is modelled by $h(t) = -4.9t^2 + 19.6t + 0.5$, where height, $h(t)$, is in metres and time, t, is in seconds.
   a) What is the maximum height the ball reaches?
   b) What is the height of the ball after 1 s?

8. The profit, $P(t)$, made at a fair depends on the price of the ticket, t. The profit is modelled by the function $P(t) = -37t^2 + 1776t - 7500$.
   a) What is the maximum profit?
   b) What is the price of a ticket that gives the maximum profit?

9. The price of an ice cream cone and the revenue generated for each price are listed in the table.
   a) What is an equation for a curve of good fit?

Price ($)	0.50	1.00	1.50	2.00	2.50	3.00	3.50
Revenue ($)	2092	3340	4060	4180	3700	2620	940

   b) How do you know whether it is a good fit?
   c) State any restrictions on the domain and range of your model.
   d) Use your equation to calculate the revenue if the price of an ice cream cone is $2.25.

10. Why might someone choose to write a curve of good fit in vertex form rather than factored form?

Water Fountain

The water flowing from a fountain forms a parabola. If you adjust the pressure, the shape of the flow changes.

YOU WILL NEED
- pipe cleaners or flexible wire, such as copper
- graph paper

❓ How does the change in pressure affect the equation of a curve of good fit?

A. Turn the water in the fountain on at about half pressure. Shape a piece of wire so that it matches the shape produced by the flow of water.

B. Measure the height from the bowl to the point where the water leaves the spout of the fountain.

C. On graph paper mark a point on the *y*-axis above the origin that corresponds to the height of the fountain's spout. Using this point as the starting point of your graph, copy the shape of your wire onto the graph paper.

D. Repeat the process for two different water pressures. In the first trial, increase the pressure over the original pressure. In the second trial, decrease the pressure.

E. Determine a curve of good fit, in vertex form, for all of your trials.

F. In a report, indicate
- how you manually determined the vertex form of the equation that fits your data
- the characteristics of each of your curves of good fit (vertex, axis of symmetry, domain, and range)
- how well you think the equation fits your data
- how the change in water pressure affected the equation of the curve of good fit

Task | *Checklist*

✔ Did you show all your steps?

✔ Did you include a graph?

✔ Did you support your choice of data?

✔ Did you explain your thinking clearly?

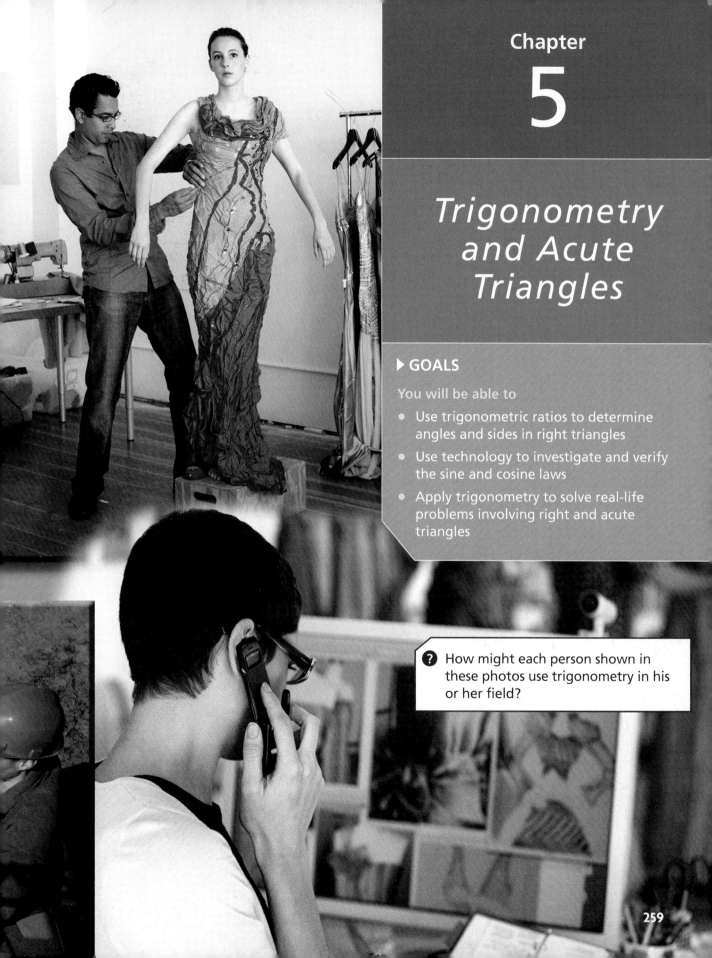

Chapter
5

Trigonometry and Acute Triangles

▶ **GOALS**

You will be able to

- Use trigonometric ratios to determine angles and sides in right triangles
- Use technology to investigate and verify the sine and cosine laws
- Apply trigonometry to solve real-life problems involving right and acute triangles

? How might each person shown in these photos use trigonometry in his or her field?

WORDS You Need to Know

1. Match the term with the picture or example that best illustrates its definition.

a) acute angle

b) right angle

c) Pythagorean theorem

d) hypotenuse

e) opposite side

f) adjacent side

i)

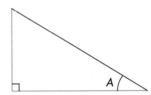

ii)

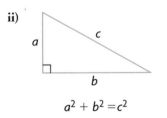

$a^2 + b^2 = c^2$

iii)

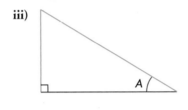

iv)

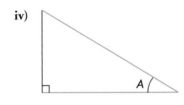

v)

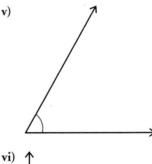

vi)

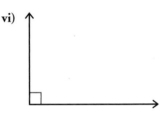

SKILLS AND CONCEPTS You Need

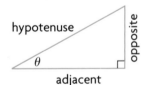

Determining and Using the Primary Trigonometric Ratios in Right Triangles

For a right angle triangle, the three primary trigonometric ratios are:

$$\sin \theta = \frac{\text{opposite}}{\text{hypotenuse}} \qquad \cos \theta = \frac{\text{adjacent}}{\text{hypotenuse}} \qquad \tan \theta = \frac{\text{opposite}}{\text{adjacent}}$$

The opposite side, adjacent side, and hypotenuse are determined by their positions relative to a particular angle in the triangle.

EXAMPLE

State the primary trigonometric ratios for $\angle C$.

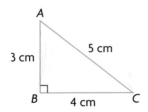

Solution

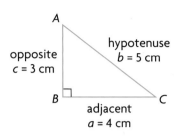

$$\sin C = \frac{\text{opposite}}{\text{hypotenuse}} = \frac{3}{5}$$

$$\cos C = \frac{\text{adjacent}}{\text{hypotenuse}} = \frac{4}{5}$$

$$\tan C = \frac{\text{opposite}}{\text{hypotenuse}} = \frac{3}{4}$$

Communication | Tip

It is common practice to label the vertices of a triangle with upper case letters. The side opposite each angle is labelled with the corresponding lower case letter.

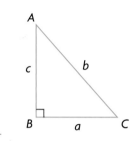

2. State the primary trigonometric ratios for $\angle A$.

a)

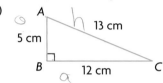

b)

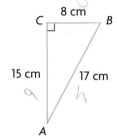

3. Calculate the measure of the indicated side or angle to the nearest unit.

a)

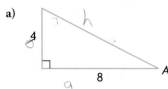

c)

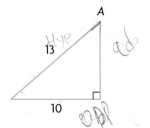

Study | Aid

For help, see Essential Skills Appendix, A-15.

b)

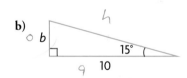

d)

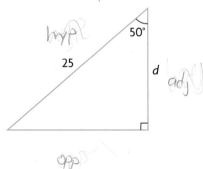

PRACTICE

Study | Aid

For help, see Essential Skills Appendix.

Question	Appendix
4	A-4
5, 6, 7, 10, 11	A-15

4. Use the Pythagorean theorem to determine the value of x to the nearest unit.

a)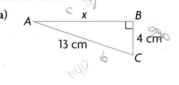

b)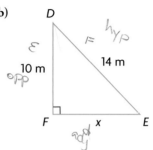

5. Using the triangles in question 4, determine the primary trigonometric ratios for each given angle. Then determine the angle measure to the nearest degree.

a) $\angle A$ b) $\angle D$ c) $\angle C$

6. Use a calculator to evaluate to four decimal places.

a) $\sin 50°$ b) $\cos 11°$ c) $\tan 72°$

7. Use a calculator to determine θ to the nearest degree.

a) $\cos \theta = 0.6820$ b) $\tan \theta = 0.1944$ c) $\sin \theta = 0.9848$

8. Determine each unknown angle to the nearest degree.

a)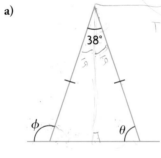

b)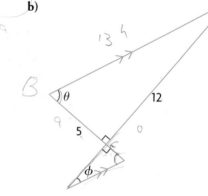

Tech | Support

For help setting a graphing calculator to degree mode and determining angles using the inverse trigonometric keys, see Technical Appendix, B-12.

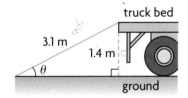

9. Rudy is unloading a piano from his truck. The truck bed is 1.4 m above ground, and he extended the ramp to a length of 3.1 m.

a) At what angle does the ramp meet the ground? Round your answer to the nearest degree.

b) Rudy needs the angle to be 15° or less so that he can control the piano safely. How much more should he extend the ramp? Round your answer to the nearest tenth of a metre.

10. Explain, using examples, how you could use trigonometry to calculate

a) the measure of a side in a right triangle when you know one side and one angle

b) the measure of an angle when you know two sides

APPLYING *What You Know*

Landing at an Airport

For an airplane to land safely, the base of the clouds, or *ceiling*, above the airport must be at least 600 m. Grimsby Airport has a spotlight that shines perpendicular to the ground, onto the cloud above. At 1100 m from the spotlight, Cory measures the angle between the spotlight's vertical beam of light, himself, and the illuminated spot on the base of the cloud to be 55°. Cory's eyes are 1.8 m above ground.

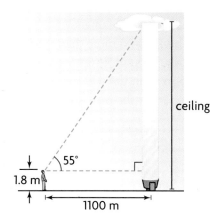

?   With this cloud ceiling, is it safe for a plane to land at Grimsby Airport?

A.   Sketch the right triangle that models this problem. Represent the given information on your sketch and any other information you know.

B.   Label the hypotenuse and the sides that are opposite and adjacent to the 55° angle.

C.   Which trigonometric ratio (sine, cosine, or tangent) would you use to solve the problem? Justify your choice.

D.   Use the trigonometric ratio that you chose in part C to write an equation to calculate the height of the cloud ceiling. Solve the equation and round your answer to the nearest metre.

E.   Add 1.8 m to account for Cory's eyes being above ground. Determine whether it is safe for the plane to land.

5.1

Applying the Primary Trigonometric Ratios

GOAL

Use primary trigonometric ratios to solve real-life problems.

LEARN ABOUT *the Math*

angle of elevation

the angle between the horizontal and the line of sight when looking up at an object

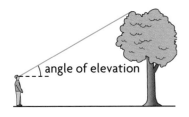

angle of elevation

Eric's car alarm will sound if his car is disturbed, but it is designed to shut off if the car is being towed at an **angle of elevation** of more than 15°. Mike's tow truck can lift a bumper no more than 0.88 m higher than the bumper's original height above ground. Eric's car has these measurements:

• The front bumper is 3.6 m from the rear axle.
• The rear bumper is 2.8 m from the front axle.

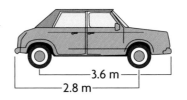

3.6 m
2.8 m

? **Will Mike be able to tow Eric's car without the alarm sounding?**

EXAMPLE 1	**Selecting a strategy to solve a problem involving a right triangle**

Determine whether the car alarm will sound.

Jason's Solution: Calculating the Angle of Elevation

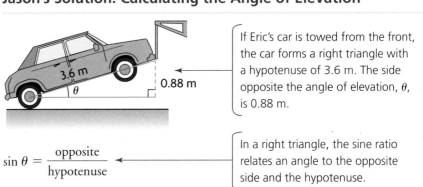

3.6 m
θ
0.88 m

If Eric's car is towed from the front, the car forms a right triangle with a hypotenuse of 3.6 m. The side opposite the angle of elevation, θ, is 0.88 m.

$$\sin \theta = \frac{\text{opposite}}{\text{hypotenuse}}$$

In a right triangle, the sine ratio relates an angle to the opposite side and the hypotenuse.

$$\sin \theta = \frac{0.88}{3.6}$$

$$\theta = \sin^{-1}\left(\frac{0.88}{3.6}\right)$$ ← I used the inverse sine ratio to calculate the angle.

```
sin⁻¹(0.88/3.6)
        14.14900451
```

$\theta \doteq 14°$ ← I rounded to the nearest degree.

Since the angle is less than 15°, the car has not been lifted enough to shut off the alarm.

If the car is towed from the rear, the car still forms a right triangle. The opposite side is still 0.88, but the hypotenuse is now 2.8 m.

$$\sin \theta = \frac{\text{opposite}}{\text{hypotenuse}}$$ ← To determine the angle of elevation, θ, I used the sine ratio, since I knew the lengths of the opposite side and the hypotenuse.

$$\sin \theta = \frac{0.88}{2.8}$$

$$\theta = \sin^{-1}\left(\frac{0.88}{2.8}\right)$$ ← I used the inverse sine ratio to calculate the angle.

```
sin⁻¹(0.88/2.8)
        18.31769841
```

$\theta \doteq 18°$ ← I rounded to the nearest degree.

Since the angle is greater than 15°, the car alarm will shut off.

For the car alarm to shut off, Mike should tow Eric's car from the back.

Monica's Solution: Calculating the Minimum Height

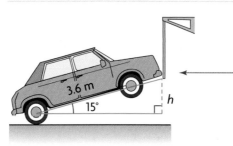

For an angle of elevation of 15°, I wanted to know the height the car must be lifted to shut off the alarm. I started with the front of the car being lifted. I drew a right triangle and labelled the height as h. I knew an angle and the hypotenuse.

$$\sin \theta = \frac{\text{opposite}}{\text{hypotenuse}}$$

Side h is opposite the 15° angle, so I used the sine ratio to calculate h.

$$\sin 15° = \frac{h}{3.6}$$

$$3.6 \times \sin 15° = \overset{1}{\cancel{3.6}} \times \frac{h}{\underset{1}{\cancel{3.6}}}$$

I solved for h by multiplying both sides of the equation by 3.6.

$$h = 3.6 \times \sin 15°$$

```
3.6*sin(15)
        .9317485624
```

I used a calculator to evaluate and rounded to the nearest hundredth.

$$h \doteq 0.93 \text{ m}$$

When the front of the car is lifted 15°, the front bumper is about 0.93 m above its original height.

Mike can't raise the car high enough from the front to shut off the alarm.

This won't work because Mike can lift the bumper only 0.88 m.

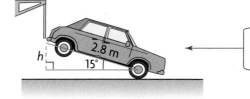

I used the same method, but with the rear of the car being lifted.

$$\sin 15° = \frac{h}{2.8}$$

I used the sine ratio because I knew the opposite side and the hypotenuse.

$$2.8 \times \sin 15° = \overset{1}{\cancel{2.8}} \times \dfrac{h}{\underset{1}{\cancel{2.8}}}$$

I solved for h by multiplying both sides of the equation by 2.8.

$$h = 2.8 \times \sin 15°$$

```
2.8*sin(15)
       .7246933263
```

I used a calculator to evaluate and rounded to the nearest hundredth.

When the rear of the car is lifted 15°, the rear bumper is about 0.72 m above its original height.

$$h \doteq 0.72 \text{ m}$$

For the alarm to shut off, Mike should tow Eric's car from the rear.

This will work, since Mike can lift a car more than that.

Reflecting

A. Compare the two solutions. How are they the same and how are they different?

B. Which solution do you prefer? Why?

C. Could the cosine or tangent ratios be used instead of the sine ratio to solve this problem? Explain.

APPLY the Math

EXAMPLE 2	Selecting the appropriate trigonometric ratios to determine unknown sides

A hot-air balloon on the end of a taut 95 m rope rises from its platform. Sam, who is in the basket, estimates that the **angle of depression** to the rope is about 55°.

a) How far, to the nearest metre, did the balloon drift horizontally?

b) How high, to the nearest metre, is the balloon above ground?

c) Viewed from the platform, what is the angle of elevation, to nearest degree, to the balloon?

angle of depression

the angle between the horizontal and the line of sight when looking down at an object

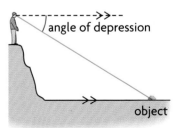

Kumar's Solution

a)

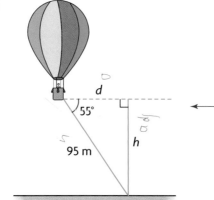

I drew a sketch and labelled the angle of depression as 55°. I labelled the horizontal distance the balloon drifted as d and the height of the balloon as h. Relative to the 55° angle, d is the adjacent side and h is the opposite side.

$$\cos \theta = \frac{\text{adjacent}}{\text{hypotenuse}}$$

I used the cosine ratio to calculate d because in a right triangle, cosine relates an angle to the adjacent side and the hypotenuse.

$$\cos 55° = \frac{d}{95}$$

$$95 \times \cos 55° = \overset{1}{\cancel{95}} \times \frac{d}{\underset{1}{\cancel{95}}}$$

I solved for d by multiplying both sides of the equation by 95.

$$d = 95 \times \cos 55°$$

I used a calculator to evaluate.

$$d \doteq 54 \text{ m}$$

The balloon drifted about 54 m horizontally.

b)
$$\sin 55° = \frac{h}{95}$$

I used the sine ratio to calculate h because in a right triangle, sine relates an angle to the opposite side and the hypotenuse.

$$95 \times \sin 55° = \overset{1}{\cancel{95}} \times \frac{h}{\underset{1}{\cancel{95}}}$$

I solved for h by multiplying both sides of the equation by 95.

$$h = 95 \times \sin 55°$$

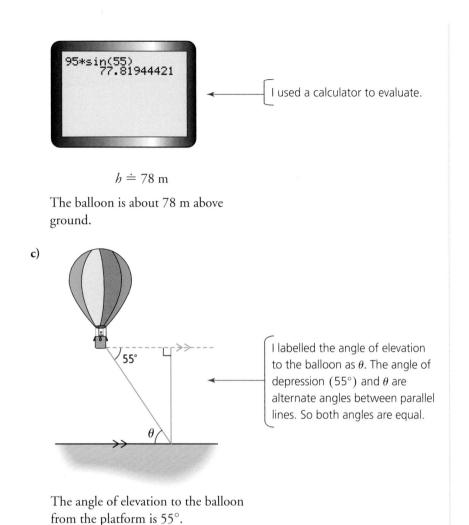

```
95*sin(55)
        77.81944421
```

I used a calculator to evaluate.

$h \doteq 78$ m

The balloon is about 78 m above ground.

c)

I labelled the angle of elevation to the balloon as θ. The angle of depression (55°) and θ are alternate angles between parallel lines. So both angles are equal.

The angle of elevation to the balloon from the platform is 55°.

EXAMPLE 3   Solving a problem by using a trigonometric ratio

A wheelchair ramp is safe to use if it has a minimum slope of $\frac{1}{12}$ and a maximum slope of $\frac{1}{5}$. What are the minimum and maximum angles of elevation to the top of such a ramp? Round your answers to the nearest degree.

Nazir's Solution

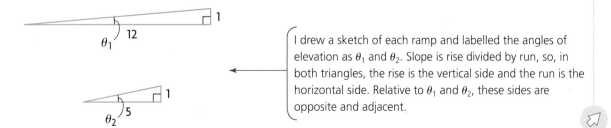

I drew a sketch of each ramp and labelled the angles of elevation as θ_1 and θ_2. Slope is rise divided by run, so, in both triangles, the rise is the vertical side and the run is the horizontal side. Relative to θ_1 and θ_2, these sides are opposite and adjacent.

Chapter 5 Trigonometry and Acute Triangles   **269**

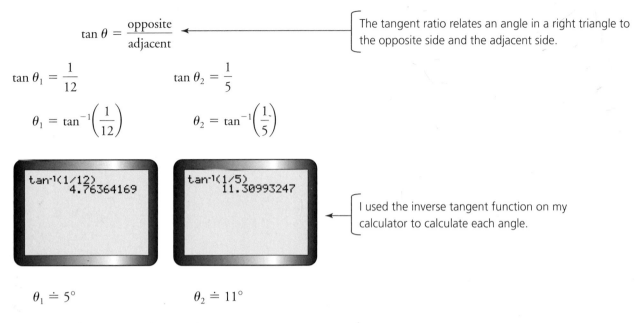

$$\tan \theta = \frac{\text{opposite}}{\text{adjacent}}$$

The tangent ratio relates an angle in a right triangle to the opposite side and the adjacent side.

$$\tan \theta_1 = \frac{1}{12} \qquad\qquad \tan \theta_2 = \frac{1}{5}$$

$$\theta_1 = \tan^{-1}\left(\frac{1}{12}\right) \qquad\qquad \theta_2 = \tan^{-1}\left(\frac{1}{5}\right)$$

```
tan-1(1/12)
        4.76364169
```

```
tan-1(1/5)
        11.30993247
```

I used the inverse tangent function on my calculator to calculate each angle.

$$\theta_1 \doteq 5° \qquad\qquad \theta_2 \doteq 11°$$

The angle of elevation to the top of the wheelchair ramp must be between 5° and 11° to be safe.

In Summary

Key Idea

- The primary trigonometric ratios can be used to determine side lengths and angles in right triangles. Which ratio you use depends on what you want to calculate and what you know about the triangle.

Need to Know

- For any right $\triangle ABC$, the primary trigonometric ratios for $\angle A$ are:

$$\sin A = \frac{\text{opposite}}{\text{hypotenuse}}$$

$$\cos A = \frac{\text{adjacent}}{\text{hypotenuse}}$$

$$\tan A = \frac{\text{opposite}}{\text{adjacent}}$$

- Opposite and adjacent sides are named relative to their positions to a particular angle in a triangle.
- The hypotenuse is the longest side in a right triangle and is always opposite the 90° angle.

CHECK Your Understanding

1. Use a calculator to evaluate to four decimal places.
a) $\sin 15°$
b) $\cos 55°$

2. Use a calculator to determine the angle to the nearest degree.
a) $\tan^{-1}\left(\dfrac{2}{5}\right)$
b) $\sin^{-1}(0.7071)$

3. State all the primary trigonometric ratios for $\angle A$ and $\angle D$. Then determine $\angle A$ and $\angle D$ to the nearest degree.

a)
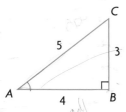

b)
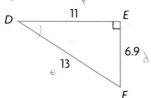

4. For each triangle, calculate x to the nearest centimetre.

a)
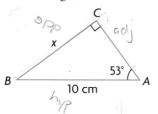

b)
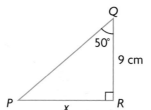

PRACTISING

5. Determine all unknown sides to the nearest unit and all unknown interior angles to the nearest degree.

a)
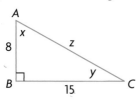

c)
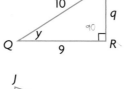

b)
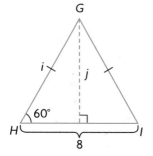

d)

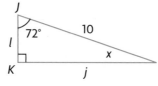

Eiffel Tower

Leaning Tower of Pisa

Empire State Building

Big Ben's clock tower

6. Predict the order, from tallest to shortest, of these famous landmarks. Then use the given information to determine the actual heights to the nearest metre.
a) Eiffel Tower, Paris, France
   68 m from the base, the angle of elevation to the top is 78°.
b) Empire State Building, New York, New York
   267 m from the base, the angle of elevation to the top is 55°.
c) Leaning Tower of Pisa, Pisa, Italy
   The distance from a point on the ground to the tallest tip of the tower is 81 m with an angle of elevation of 44°.
d) Big Ben's clock tower, London, England
   81 m from the base, the angle of elevation is 50°.

7. Manpreet is standing 8.1 m from a flagpole. His eyes are 1.7 m above ground. The top of the flagpole has an angle of elevation of 35°. How tall, to the nearest tenth of a metre, is the flagpole?

8. The CN Tower is 553 m tall. From a position on one of the Toronto Islands 1.13 km away from the base of the tower, determine the angle of elevation, to the nearest degree, to the top of the tower. Assume that your eyes are 1.5 m above ground.

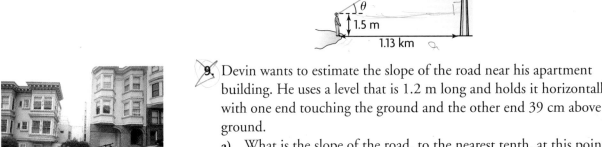

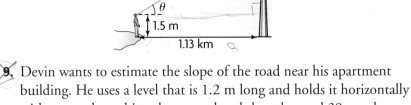

9. Devin wants to estimate the slope of the road near his apartment building. He uses a level that is 1.2 m long and holds it horizontally with one end touching the ground and the other end 39 cm above ground.
a) What is the slope of the road, to the nearest tenth, at this point?
b) What angle, to the nearest degree, would represent the slant of the road?

10. A 200 m cable attached to the top of an antenna makes an angle of 37° with the ground. How tall is the antenna to the nearest metre?

11. An underground parking lot is being constructed 8.00 m below ground level.
a) If the exit ramp is to rise at an angle of 15°, how long will the ramp be? Round your answer to the nearest hundredth of a metre.
b) What horizontal distance, to the nearest hundredth of a metre, is needed for the ramp?

12. The pitch of a roof is the rise divided by the run. If the pitch is greater than 1.0 but less than 1.6, roofers use planks fastened to the roof to stand on when shingling it. If the pitch is greater than 1.6, scaffolding is needed. For each roof angle, what equipment (scaffolding or planks), if any, would roofers need?

a) $\theta = 34°$ b) $\theta = 60°$ c) $\theta = 51°$

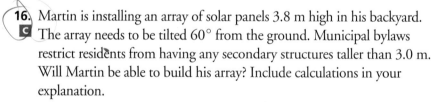

13. To estimate the width of a river near their school, Charlotte and Mavis have a device to measure angles and a pole of known length. Describe how they might calculate the width of the river with this equipment.

14. Ainsley's and Caleb's apartment buildings are exactly the same height. Ainsley measures the distance between the buildings as 51 m and observes that the angle of depression from the roof of her building to the bottom of Caleb's is about 64°. How tall, to the nearest metre, is each building?

15. To use an extension ladder safely, the base must be 1 m out from the wall for every 2 m of vertical height.
a) What is the maximum angle of elevation, to the nearest degree, to the top of the ladder?
b) If the ladder is extended to 4.72 m in length, how high can it safely reach? Round your answer to the nearest hundredth of a metre.
c) How far out from the wall does a 5.9 m ladder need to be? Round your answer to the nearest tenth of a metre.

16. Martin is installing an array of solar panels 3.8 m high in his backyard. The array needs to be tilted 60° from the ground. Municipal bylaws restrict residents from having any secondary structures taller than 3.0 m. Will Martin be able to build his array? Include calculations in your explanation.

Extending

17. When poured into a pile, gravel will naturally form a cylindrical cone with a slope of approximately 34°. If a construction foreman has room only for a pile that is 85 m in diameter, how tall, to the nearest metre, will the pile be?

18. Nalini and Jodi are looking at the top of the same flagpole. They are standing in a line on the same side of the flagpole, 50.0 m apart. The angle of elevation to the top of the pole is 11° from Jodi's position and 7° from Nalini's position. The girls' eyes are 1.7 m above ground. For each question, round your answer to the nearest tenth of a metre.
a) How tall is the flagpole?
b) How far is each person from the base of the flagpole?
c) If Nalini and Jodi were standing in a line on opposite sides of the pole, how tall would the flagpole be? How far would each person be from its base?

5.2 Solving Problems by Using Right-Triangle Models

GOAL

Solve real-life problems by using combinations of primary trigonometric ratios.

LEARN about the Math

A surveyor stands at point *S* and uses a laser transit and marking pole to sight two corners, *A* and *B*, on one side of a rectangular lot. The surveyor marks a reference point *P* on the line *AB* so that the line from *P* to *S* is perpendicular to *AB*. Corner *A* is 150 m away from *S* and 34° east of *P*. Corner *B* is 347 m away from *S* and 69° west of *P*.

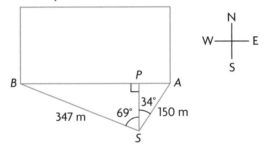

Communication | Tip

For the highest degree of accuracy, save intermediate answers by using the memory keys of your calculator. Round only after you do the very last calculation.

? How long, to the nearest metre, is the lot?

EXAMPLE 1 | Selecting a strategy to determine a distance

Calculate the length of the building lot to the nearest metre.

Leila's Solution: Using the Sine Ratio

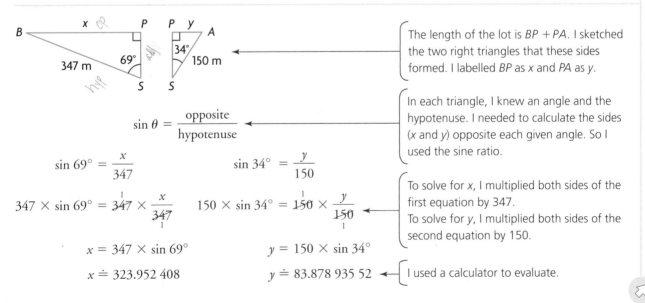

The length of the lot is $BP + PA$. I sketched the two right triangles that these sides formed. I labelled *BP* as *x* and *PA* as *y*.

$$\sin \theta = \frac{\text{opposite}}{\text{hypotenuse}}$$

In each triangle, I knew an angle and the hypotenuse. I needed to calculate the sides (*x* and *y*) opposite each given angle. So I used the sine ratio.

$$\sin 69° = \frac{x}{347} \qquad \sin 34° = \frac{y}{150}$$

$$347 \times \sin 69° = 347 \times \frac{x}{347} \qquad 150 \times \sin 34° = 150 \times \frac{y}{150}$$

To solve for *x*, I multiplied both sides of the first equation by 347.
To solve for *y*, I multiplied both sides of the second equation by 150.

$$x = 347 \times \sin 69° \qquad y = 150 \times \sin 34°$$

$$x \doteq 323.952\ 408 \qquad y \doteq 83.878\ 935\ 52$$

I used a calculator to evaluate.

$AB = BP + PA$

$\quad = x + y$ ────────────── I added x and y to determine the length of the lot.

$\quad = 323.952\ 408 + 83.878\ 935\ 52$

$\quad \doteq 408$ m

The lot is about 408 m long.

Tony's Solution: Using the Cosine Ratio and the Pythagorean Theorem

The length of the lot is $BP + PA$. PS is a shared side of $\triangle BPS$ and $\triangle APS$. If I knew PS, then I could use the Pythagorean theorem to determine BP and PA.

In $\triangle APS$:

$$\cos \theta = \frac{\text{adjacent}}{\text{hypotenuse}}$$ ────────── AS is the hypotenuse and PS is adjacent to the 34° angle. So, to determine PS, I used the cosine ratio.

$$\cos 34° = \frac{PS}{150}$$

$$150 \times \cos 34° = \overset{1}{\cancel{150}} \times \frac{PS}{\underset{1}{\cancel{150}}}$$ ────────── To solve for PS, I multiplied both sides of the equation by 150 and used a calculator to evaluate.

$$PS = 150 \times \cos 34°$$

$$PS \doteq 124.356 \text{ m}$$

In $\triangle APS$:   In $\triangle BPS$:

$AS^2 = PS^2 + PA^2$ $BS^2 = PS^2 + BP^2$ ──── Next, I used the Pythagorean theorem to determine PA and BP.

$150^2 = 124.356^2 + PA^2$ $347^2 = 124.356^2 + BP^2$

$PA^2 = 150^2 - 124.356^2$ $BP^2 = 347^2 - 124.356^2$

$PA^2 = 7036$ $BP^2 = 104\ 945$

$PA = \sqrt{7036}$ $BP = \sqrt{104\ 945}$

$PA \doteq 83.880\ 867\ 9$ m $BP \doteq 323.952\ 157$ m

$AB = BP + PA$ ────────────── I added BP and PA to determine the length of the lot.

$\quad = 83.880\ 867\ 9 + 323.952\ 157$

$\quad \doteq 408$ m

The lot is about 408 m long.

Reflecting

A. Explain why the length of the lot cannot be determined directly using either the Pythagorean theorem or the primary trigonometric ratios.

B. How are Leila's and Tony's solutions the same? How are they different?

C. Why did the surveyor choose point *P* so that *PS* would be perpendicular to *AB*?

APPLY the Math

EXAMPLE 2	Using trigonometric ratios to calculate angles

Karen is a photographer taking pictures of the Burlington Skyway bridge. She is in a helicopter hovering 720 m above the bridge, exactly 1 km horizontally from the west end of the bridge. The Skyway spans a distance of 2650 m from east to west.

a) From Karen's position, what are the angles of depression, to the nearest degree, of the east and west ends of the bridge?

b) If Karen's camera has a wide-angle lens that can capture 150°, can she get the whole bridge in one shot?

Ali's Solution

a)

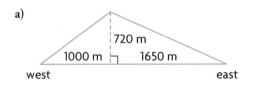

I sketched the position of the helicopter relative to the bridge. The horizontal distance from the helicopter to the west end of the bridge is 1000 m. So I subtracted 1000 from 2650 to determine the horizontal distance from the helicopter to the east end.

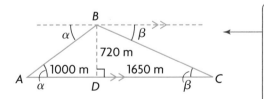

I drew a dashed line parallel to the base of the triangle and labelled the angles of depression as α and β. In $\triangle ABD$, I knew that $\angle BAD = \alpha$ because $\angle BAD$ and α are alternate angles between parallel lines. By the same reasoning, $\angle BCD = \beta$.

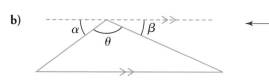

$$\tan \theta = \frac{\text{opposite}}{\text{adjacent}}$$

$$\tan \alpha = \frac{720}{1000} \qquad\qquad \tan \beta = \frac{720}{1650}$$

> I knew the opposite and adjacent sides in each right triangle. So, to determine α and β, I used the tangent ratio.

$$\alpha = \tan^{-1}\left(\frac{720}{1000}\right) \qquad \beta = \tan^{-1}\left(\frac{720}{1650}\right)$$

$$\alpha \doteq 35.8° \qquad\qquad \beta \doteq 23.6°$$

> I used the inverse tangent function on my calculator to evaluate each angle.

The angles of depression are about $36°$ (west end) and about $24°$ (east end).

b)

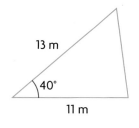

> To determine whether Karen can get the whole bridge in one shot, I needed to know θ, the angle between the east and west ends of the bridge from Karen's perspective.

$$\theta = 180° - \alpha - \beta$$
$$= 180° - 35.8° - 23.6°$$
$$= 120.6°$$

> I subtracted α and β from $180°$.

Since the angle from Karen's perspective is less than $150°$, she will be able to get the whole bridge in one shot.

EXAMPLE 3 | **Using trigonometry to determine the area of a triangle**

Elise's parents have a house with a triangular front lawn as shown. They want to cover the lawn with sod rather than plant grass seed. How much would it cost to put sod if it costs $13.75 per square metre?

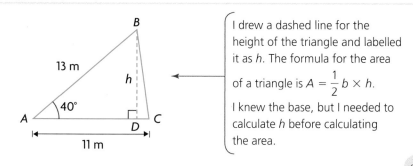

Tina's Solution

> I drew a dashed line for the height of the triangle and labelled it as h. The formula for the area of a triangle is $A = \frac{1}{2} b \times h$.
>
> I knew the base, but I needed to calculate h before calculating the area.

$$\sin \theta = \frac{\text{opposite}}{\text{hypotenuse}}$$

$$\sin 40° = \frac{h}{13}$$

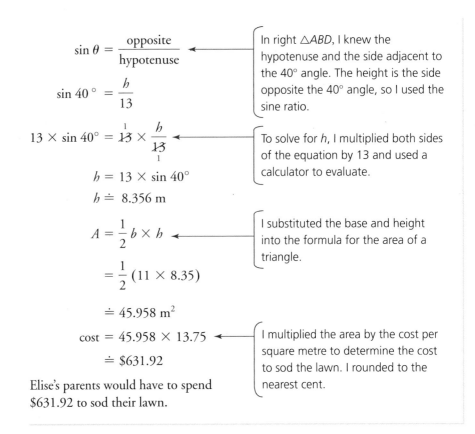

In right $\triangle ABD$, I knew the hypotenuse and the side adjacent to the 40° angle. The height is the side opposite the 40° angle, so I used the sine ratio.

$$13 \times \sin 40° = \overset{1}{\cancel{13}} \times \frac{h}{\underset{1}{\cancel{13}}}$$

$$h = 13 \times \sin 40°$$

$$h \doteq 8.356 \text{ m}$$

To solve for h, I multiplied both sides of the equation by 13 and used a calculator to evaluate.

$$A = \frac{1}{2} b \times h$$

I substituted the base and height into the formula for the area of a triangle.

$$= \frac{1}{2} (11 \times 8.35)$$

$$\doteq 45.958 \text{ m}^2$$

$$\text{cost} = 45.958 \times 13.75$$

$$\doteq \$631.92$$

I multiplied the area by the cost per square metre to determine the cost to sod the lawn. I rounded to the nearest cent.

Elise's parents would have to spend $631.92 to sod their lawn.

EXAMPLE 4 | Solving a problem by using trigonometric ratios

A communications tower is some distance from the base of a 70 m high building. From the roof of the building, the angle of elevation to the top of the tower is 11.2°. From the base of the building, the angle of elevation to the top of the tower is 33.4°. Determine the height of the tower and how far it is from the base of the building. Round your answers to the nearest metre.

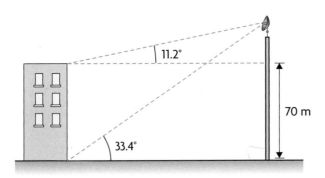

Pedro's Solution

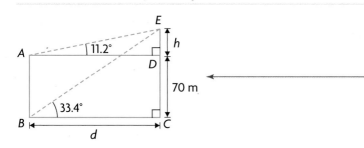

I drew a sketch and labelled the distance between the tower and the building as *d*. The tower is more than 70 m tall, so I labelled the extra height above 70 m as *h*.

In △*BEC*:   In △*AED*:

$$\tan \theta = \frac{\text{opposite}}{\text{adjacent}}$$

$$\tan 33.4° = \frac{h + 70}{d} \qquad \tan 11.2° = \frac{h}{d}$$

In △*BEC*, *d* is adjacent to the 33.4° angle and the opposite side is 70 + *h*. In △*AED*, *d* is adjacent to the 11.2° angle and the opposite side is *h*. Since I knew the adjacent and opposite sides in each right triangle, I used the tangent ratio.

$$d \times \tan 33.4° = \overset{1}{d} \times \frac{h + 70}{\underset{1}{d}} \qquad d \times \tan 11.2° = \overset{1}{d} \times \frac{h}{\underset{1}{d}}$$

$$d \times \tan 33.4° = h + 70 \qquad d \times \tan 11.2° = h$$

$$d = \frac{h + 70}{\tan 33.4°}$$

To solve for *d* in the first equation, I multiplied both sides by *d* and then divided both sides by tan 33.4°. To solve for *h* in the second equation, I multiplied both sides by *d*.

$$d = \frac{d \times \tan 11.2° + 70}{\tan 33.4°}$$

$$\tan 33.4° \times d = d \times \tan 11.2° + 70$$

$$d(\tan 33.4° - \tan 11.2°) = 70$$

$$0.461\ 373\ 177\ 9d \doteq 70$$

$$d = \frac{70}{0.461\ 373\ 177\ 9}$$

$$\doteq 151.721\ 000\ 2 \text{ m}$$

I substituted the second equation into the first and solved for *d*. I multiplied both sides of the equation by tan 33.4°. Then I simplified.

$$h = 151.721\ 000\ 2 \times \tan 11.2°$$

$$\doteq 30 \text{ m}$$

To determine *h*, I substituted the value of *d* into the second equation.

$$70 + 30 = 100 \text{ m}$$

To determine the height of the tower, I added 70 to *h*.

The tower is about 100 m tall and about 152 m from the base of the building.

In Summary

Key Idea

- If a situation involves calculating a length or an angle, try to represent the problem with a right-triangle model. If you can, solve the problem by using the primary trigonometric ratios.

Need to Know

- To calculate the area of a triangle, you can use the sine ratio to determine the height. For example, if you know a, b, and $\angle C$, then

$$\sin C = \frac{h}{a}$$

$$h = a \sin C$$

The area of a triangle is

$$A = \frac{1}{2} b \times h$$

$$A = \frac{1}{2} b (a \sin C)$$

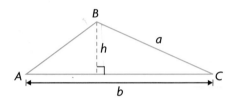

CHECK Your Understanding

1. Calculate the area of each triangle to the nearest tenth of a square centimetre.

a)

5.5 cm, 30°, 6.0 cm, triangle ABC

b)

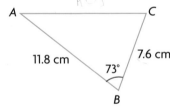

11.8 cm, 7.6 cm, 73°, triangle ABC

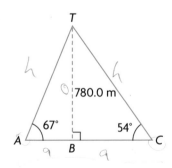

2. A mountain is 780.0 m high. From points A and C, the angles of elevation to the top of the mountain are 67° and 54° as shown at the left. Explain how to calculate the length of a tunnel from A to C.

3. Karen and Anna are standing 23 m away from the base of a 23 m high house. Karen's eyes are 1.5 m above ground and Anna's eyes are 1.8 m above ground. Both girls observe the top of the house and measure its angle of elevation. Which girl will measure the greater angle of elevation? Justify your answer.

PRACTISING

4. The angle of elevation from the roof of a 10 m high building to the top
K of another building is 41°. The two buildings are 18 m apart at the base.
   a) Which trigonometric ratio would you use to solve for the height of
   the taller building? Why?
   b) How tall, to the nearest metre, is the taller building?

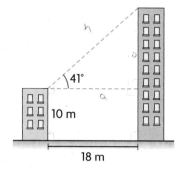

5. If the angle of elevation to the top of the pyramid of Cheops in Giza,
Egypt, is 17.5°, measured 348 m from its base, can you calculate the
height of the pyramid accurately? Explain your reasoning.

6. Darrin wants to lean planks of wood that are 1.5 m, 1.8 m, and 2.1 m
long against the wall inside his garage. If the top of a plank forms an
angle of less than 10° with the wall, the plank might fall over. If
the bottom of a plank sticks out more than 30 cm from the wall,
Darrin won't have room to park his car. Will Darrin be able to store
all three planks in the garage? Justify your answer with calculations.

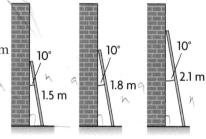

7. Jordan is standing on a bridge over the Welland Canal. His eyes are
5.1 m above the surface of the water. He sees a cargo ship heading
straight toward him. From his position, the bow appears at an angle of
depression of 5° and the stern appears at an angle of depression of 1°.
For each question, round your answer to the nearest tenth of a metre.
   a) What is the straight-line distance from the bow to Jordan?
   b) What is the straight-line distance from the stern to Jordan?
   c) What is the length of the ship from bow to stern?

8. A searchlight is mounted at the front of a helicopter flying 125 m
T above ground. The angle of depression of the light beam is 70°.
An observer on the ground notices that the beam of light measures 5°.
How wide, to the nearest metre, is *d,* the spot on the ground?

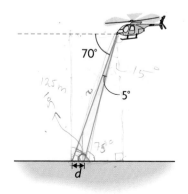

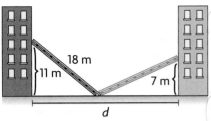

9. An 18 m long ladder is leaning against the wall of a building. The top of the ladder reaches a window 11 m above ground. If the ladder is tilted in the opposite direction, without moving its base, the top of the ladder can reach a window in another building that is 7 m above ground. How far apart, to the nearest metre, are the two buildings?

10. Kyle and Anand are standing on level ground on opposite sides of a tree. Kyle measures the angle of elevation to the treetop as 35°. Anand measures an angle of elevation of 30°. Kyle and Anand are 65 m apart. Kyle's eyes and Anand's eyes are 1.6 m above ground. How tall, to the nearest tenth of a metre, is the tree?

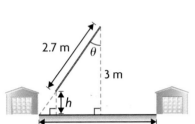

11. A tree branch 3 m above ground runs parallel to Noel's garage and his neighbour's. Noel wants to attach a rope swing on this branch. The garages are 4.2 m apart and the rope is 2.7 m long.
    a) What is the maximum angle, measured from the perpendicular, through which the rope can swing? Round your answer to the nearest degree.
    b) What is the maximum height above ground, to the nearest tenth of a metre, of the end of the rope?

12. A regular hexagon has a perimeter of 50 cm.
    a) Calculate the area of the hexagon to the nearest square centimetre.
    b) The hexagon is the base of a prism of height 100 cm. Calculate the volume, to the nearest cubic centimetre, and the surface area, to the nearest square centimetre, of the prism. Recall that volume = area of base × height. The surface area is the sum of the areas of all faces of the prism.

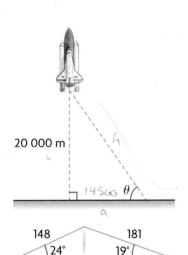

13. Lucien wants to photograph the separation of the solid rocket boosters from the space shuttle. He is standing 14 500 m from the launch pad, and the solid rocket boosters separate at 20 000 m above ground.
    a) Assume that the path of the shuttle launch is perfectly vertical up until the boosters separate. At what angle, to the nearest degree, should Lucien aim his camera?
    b) How far away is the space shuttle from Lucien at the moment the boosters separate? Round your answer to the nearest metre.

14. Sven wants to determine the unknown side of the triangle shown at the left. Even though it is not a right triangle, describe how Sven could use primary trigonometric ratios to determine x.

15. A pendulum that is 50 cm long is moved 40° from the vertical. What is the change in height from its initial position? Round your answer to the nearest centimetre.

Extending

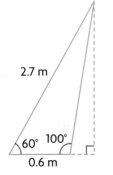

16. Calculate the area of the triangle shown at the left to the nearest tenth of a square metre.

5.3 Investigating and Applying the Sine Law in Acute Triangles

INVESTIGATE the Math

The trigonometric ratios sine, cosine, and tangent are defined only for right triangles. In an **oblique triangle**, these ratios no longer apply.

oblique triangle

a triangle (acute or obtuse) that does not contain a right angle

❓ In an oblique triangle, what is the relationship between a side and the sine of the angle opposite that side?

A. Use dynamic geometry software to construct any acute triangle.

B. Label the vertices A, B, and C. Then name the sides a, b, and c as shown.

C. Measure all three interior angles and all three sides.

D. Using one side length and the angle opposite that side, choose **Calculate ...** from the **Measure** menu to evaluate the ratio $\dfrac{\text{side length}}{\sin(\text{opposite angle})}$.

E. Repeat part D for the other sides and angles. What do you notice?

F. Drag any vertex of your triangle. What happens to the sides, angles, and ratios? Now drag the other two vertices and explain.

G. Express your findings
   • in words
   • with a mathematical relationship

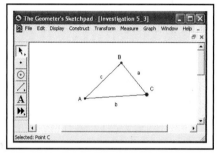

Tech | **Support**

For help using dynamic geometry software, see Technical Appendix, B-21 and B-22.

Reflecting

H. Are the reciprocals of the ratios you found equal? Explain how you know.

I. The relationship that you verified is the **sine law**. Why is this name appropriate?

J. Does the sine law apply to all types of triangles (obtuse, acute, and right)? Explain how you know.

sine law

in any acute triangle, the ratios of each side to the sine of its opposite angle are equal

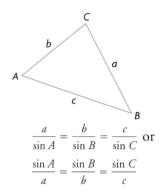

$$\frac{a}{\sin A} = \frac{b}{\sin B} = \frac{c}{\sin C} \text{ or}$$

$$\frac{\sin A}{a} = \frac{\sin B}{b} = \frac{\sin C}{c}$$

APPLY the Math

EXAMPLE 1	Using the sine law to calculate an unknown length

Determine x to the nearest tenth of a centimetre.

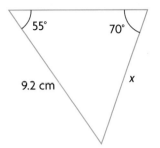

Barbara's Solution

$$\frac{x}{\sin 55°} = \frac{9.2}{\sin 70°}$$

The triangle doesn't have a 90° angle. So it isn't convenient to use the primary trigonometric ratios.

I chose the sine law because I knew two angles and the side opposite one of those angles.

I wrote the ratios with the sides in the numerators to make the calculations easier.

$$\overset{1}{\cancel{\sin 55°}} \times \frac{x}{\underset{1}{\cancel{\sin 55°}}} = \sin 55° \times \frac{9.2}{\sin 70°}$$

To solve for x, I multiplied both sides of the equation by $\sin 55°$.

$$x = \sin 55° \times \frac{9.2}{\sin 70°}$$

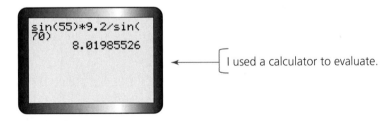

I used a calculator to evaluate.

$$x \doteq 8.0 \text{ cm}$$

The unknown side of the triangle is about 8.0 cm long.

EXAMPLE 2 | Selecting the sine law as a strategy to calculate unknown angles

Determine all the interior angles in $\triangle PQR$.
Round your answers to the nearest degree.

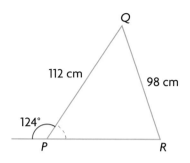

Tom's Solution

$\angle QPR = 180° - 124°$

$\quad\quad\quad = 56°$

> $\angle QPR$ and 124° add up to 180°. So, to determine $\angle QPR$, I subtracted 124° from 180°.

$\dfrac{\sin R}{112} = \dfrac{\sin 56°}{98}$

> $\triangle PQR$ doesn't have a 90° angle. So I chose the sine law, since I knew two sides and an angle opposite one of those sides.
>
> I wrote the ratios with the angles in the numerators to make it easier to solve for $\angle R$.

$\overset{1}{\cancel{112}} \times \dfrac{\sin R}{\underset{1}{\cancel{112}}} = 112 \times \dfrac{\sin 56°}{98}$

> To solve for $\angle R$, I multiplied both sides of the equation by 112.

$\sin R = 112 \times \dfrac{\sin 56°}{98}$

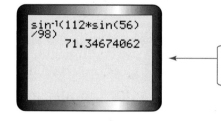

> I used the inverse sine function on a calculator to evaluate.

$\angle R \doteq 71°$

$\angle Q = 180° - \angle QPR - \angle R$

$\quad\quad = 180° - 56° - 71°$

$\quad\quad = 53°$

> All three interior angles add up to 180°. So I subtracted $\angle R$ and $\angle QPR$ from 180° to determine $\angle Q$.

The interior angles in $\triangle PQR$ are 56°, 53°, and 71°.

EXAMPLE 3 | Solving a problem by using the sine law

A wall that is 1.4 m long has started to lean and now makes an angle of 80°
with the ground. A 2.0 m board is jammed between the top of the wall and the
ground to prop the wall up. Assume that the ground is level.

a) What angle, to the nearest degree, does the
board make with the ground?

b) What angle, to the nearest degree, does the
board make with the wall?

c) How far, to the nearest tenth of a metre, is
the board from the base of the wall?

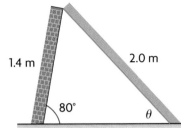

Isabelle's Solution

a)
$$\frac{\sin \theta}{1.4} = \frac{\sin 80°}{2.0}$$

> This triangle doesn't have a 90°
> angle. So I chose the sine law,
> since I knew two sides and an
> angle opposite one of those
> sides.

$$\overset{1}{\cancel{1.4}} \times \frac{\sin \theta}{\underset{1}{\cancel{1.4}}} = 1.4 \times \frac{\sin 80°}{2.0}$$

> To solve for θ, I multiplied both
> sides of the equation by 1.4.

$$\sin \theta = 1.4 \times \frac{\sin 80°}{2.0}$$

$$\theta = \sin^{-1}\left(1.4 \times \frac{\sin 80°}{2.0}\right)$$

> I used the inverse sine ratio
> on a calculator to determine
> the angle.

$$\theta \doteq 44°$$

The board makes an angle of about
44° with the ground.

b) $180° - 80° - 44° = 56°$

> The interior angles add up to
> 180°. So I subtracted the two
> known angles from 180°.

The board makes an angle of about
56° with the wall.

c)

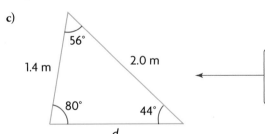

> I labelled the distance along the
> ground between the wall and
> the board as d.

$$\frac{d}{\sin 56°} = \frac{2.0}{\sin 80°}$$

Then I used the sine law.

$$\frac{\cancel{\sin 56°}^{1}}{1} \times \frac{d}{\cancel{\sin 56°}_{1}} = \sin 56° \times \frac{2.0}{\sin 80°}$$

To solve for d, I multiplied both sides of the equation by $\sin 56°$ and used a calculator to evaluate.

$$d = \sin 56° \times \frac{2.0}{\sin 80°}$$

$$d \doteq 1.7 \text{ m}$$

The board is about 1.7 m from the base of the wall.

In Summary

Key Idea

- The sine law states that, in any acute $\triangle ABC$, the ratios of each side to the sine of its opposite angle are equal.

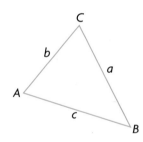

$$\frac{a}{\sin A} = \frac{b}{\sin B} = \frac{c}{\sin C}$$

or

$$\frac{\sin A}{a} = \frac{\sin B}{b} = \frac{\sin C}{c}$$

Need to Know

- The sine law can be used only when you know
  - two sides and the angle opposite a known side or

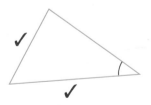

  - two angles and any side

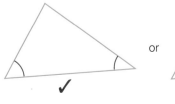   or

CHECK *Your Understanding*

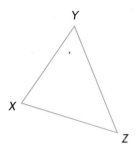

1. **a)** Given $\triangle XYZ$ at the left, label the sides with lower case letters.
   b) State the sine law for $\triangle XYZ$.

2. Solve each equation. Round x to the nearest tenth of a unit and θ to the nearest degree.

   a) $\dfrac{x}{\sin 22°} = \dfrac{11.6}{\sin 71°}$

   b) $\dfrac{13.1}{\sin \theta} = \dfrac{29.2}{\sin 65°}$

3. Use the sine law to calculate b to the nearest centimetre and $\angle D$ to the nearest degree.

   a)

   b)

   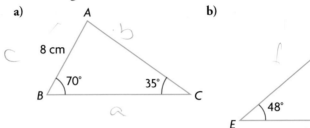

PRACTISING

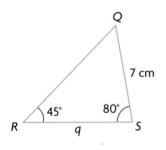

4. Given $\triangle RQS$ at the left, determine q to the nearest centimetre.
   K

5. Archimedes Avenue and Bernoulli Boulevard meet at an angle of $45°$ near a bus stop. Riemann Road intersects Archimedes Avenue at an angle of $60°$. That intersection is 112 m from the bus stop.
   a) At what angle do Riemann Road and Bernoulli Boulevard meet? Round your answer to the nearest degree.
   b) How far, to the nearest metre, is the intersection of Riemann Road and Bernoulli Boulevard from the bus stop?

   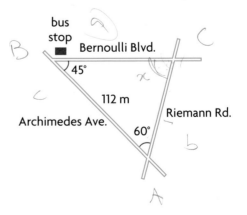

6. In △*ABC*, two sides and an angle are given. Determine the value of
   ∠*C* to the nearest degree and the length of *b* to the nearest tenth
   of a centimetre.
   a) $a = 2.4$ cm, $c = 3.2$ cm, ∠*A* = 28°
   b) $a = 9.9$ cm, $c = 11.2$ cm, ∠*A* = 58°
   c) $a = 8.6$ cm, $c = 9.4$ cm, ∠*A* = 47°
   d) $a = 5.5$ cm, $c = 10.4$ cm, ∠*A* = 30°

7. An isosceles triangle has two 5.5 cm sides and two 32° angles.
   a) Calculate the perimeter of the triangle to the nearest tenth
      of a centimetre.
   b) Calculate the area of the triangle to the nearest tenth of a
      square centimetre.

8. Solve each triangle. Round each length to the nearest centimetre and
   each angle to the nearest degree.

> **Communication | Tip**
>
> To solve a triangle, determine the measures of all unknown sides and angles.

a)

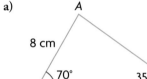

b)

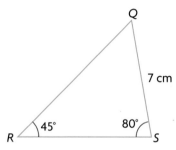

c)

d)

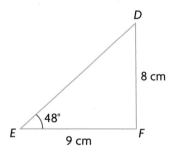

9. Solve each triangle. Round each length to the nearest tenth of a unit
   and each angle to the nearest degree.
   a) △*ABC*: $a = 10.3$, $c = 14.4$, ∠*C* = 68°
   b) △*DEF*: ∠*E* = 38°, ∠*F* = 48°, $f = 15.8$
   c) △*GHJ*: ∠*G* = 61°, $g = 5.3$, $j = 3.1$
   d) △*KMN*: $k = 12.5$, $n = 9.6$, ∠*N* = 42°
   e) △*PQR*: $p = 1.2$, $r = 1.6$, ∠*R* = 52°
   f) △*XYZ*: $z = 6.8$, ∠*X* = 42°, ∠*Y* = 77°

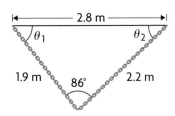

10. Toby uses chains and a winch to lift engines at his father's garage. Two hooks in the ceiling are 2.8 m apart. Each hook has a chain hanging from it. The chains are of length 1.9 m and 2.2 m. When the ends of the chains are attached, they form an angle of 86°. In this configuration, what acute angle, to the nearest degree, does each chain make with the ceiling?

11. Betsy installed cordless phones in the student centre at points *A*, *B*, and *C* as shown. Explain how you can use the given information to determine which two phones are farthest apart.

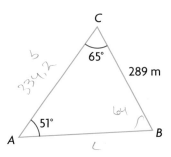

12. Tom says that he doesn't need to use the sine law because he can always determine a solution by using primary trigonometric ratios. Is Tom correct? What would you say to Tom to convince him to use the sine law to solve a problem?

Extending

13. Two angles in a triangle measure 54° and 38°. The longest side of the triangle is 24 cm longer than the shortest side. Calculate the length, to the nearest centimetre, of all three sides.

14. Use the sine law to show why the longest side of a triangle must be opposite the largest angle.

15. A triangular garden is enclosed by a fence. A dog is on a 5 m leash tethered to the fence at point *P*, 6.5 m from point *C*, as shown at the left. If ∠*ACB* = 41°, calculate the total length, to the nearest tenth of a metre, of fence that the dog can reach.

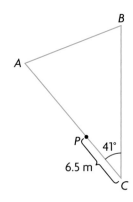

16. Tara built a sculpture in the shape of a huge equilateral triangle of side length 4.0 m. Unfortunately, the ground underneath the sculpture was not stable, and one of the vertices of the triangle sank 55 cm into the ground. Assume that Tara's sculpture was built on level ground originally.
  a) What is the length, to the nearest tenth of a metre, of the two exposed sides of Tara's triangle now?
  b) What percent of the triangle's surface area remains above ground? Round your answer to the nearest percent.

FREQUENTLY ASKED Questions

Q: **What are the primary trigonometric ratios, and how do you use them?**

A: Given $\triangle ABC$, the primary trigonometric ratios for $\angle A$ are:

Study | *Aid*

• See Lesson 5.1,
  Examples 1, 2, and 3 and
  Lesson 5.2, Examples 1 to 4.
• Try Mid-Chapter Review
  Questions 1 to 8.

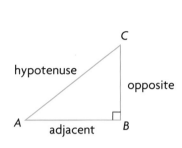

$$\sin A = \frac{\text{opposite}}{\text{hypotenuse}}$$

$$\cos A = \frac{\text{adjacent}}{\text{hypotenuse}}$$

$$\tan A = \frac{\text{opposite}}{\text{adjacent}}$$

To calculate an angle or side using a trigonometric ratio:
• Label the sides of the triangle relative to either a given angle or the one you want to determine.
• Write an equation that involves what you are trying to find by using the appropriate ratio.
• Solve your equation.

Q: **What is the sine law, and when can I use it?**

A: In any acute triangle, the ratios of a side to the sine of its opposite angle are equal.

Study | *Aid*

• See Lesson 5.3,
  Examples 1, 2, and 3.
• Try Mid-Chapter Review
  Questions 9, 10, and 11.

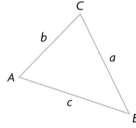

$$\frac{a}{\sin A} = \frac{b}{\sin B} = \frac{c}{\sin C}$$

or

$$\frac{\sin A}{a} = \frac{\sin B}{b} = \frac{\sin C}{c}$$

• The sine law can be used only when you know
  • two sides and the angle opposite a known side or

  • two angles and any side

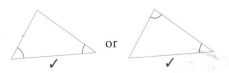

or

PRACTICE Questions

1. From a spot 25 m from the base of the Peace Tower in Ottawa, the angle of elevation to the top of the flagpole is 76°. How tall, to the nearest metre, is the Peace Tower, including the flagpole?

2. A spotlight on a 3.0 m stand shines on the surface of a swimming pool. The beam of light hits the water at a point 7.0 m away from the base of the stand.

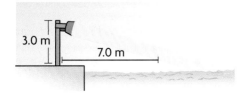

   a) Calculate the angle the beam makes with the pool surface. Round your answer to the nearest degree.

   b) The beam reflects off the pool surface and strikes a wall 4.5 m away from the reflection point. The angle the beam makes with the pool is exactly the same on either side of the reflection point. Calculate how far up the wall, to the nearest tenth of a metre, the spotlight will appear.

3. Solve each triangle. Round each length to the nearest tenth of a unit and each angle to the nearest degree.

   a)

   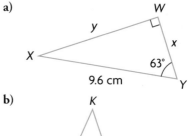

   b)

   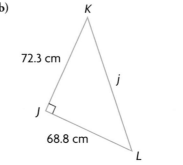

   c)

   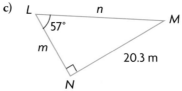

   d)

   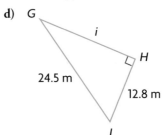

4. The sides of an isosceles triangle are 12 cm, 12 cm, and 16 cm. Use primary trigonometric ratios to determine the largest interior angle to the nearest degree. (*Hint:* Divide the triangle into two congruent right triangles.)

5. The mainsail of the small sailboat shown below measures 2.4 m across the boom and 4.4 m up the mainmast. The boom and mast meet at an angle of 74°. What is the area of the mainsail to the nearest tenth of a square metre?

6. A coast guard boat is tracking two ships using radar. At noon, the ships are 5.0 km apart and the angle between them is 90°. The closest ship is 3.1 km from the coast guard boat. How far, to the nearest tenth of a kilometre, is the other ship from the coast guard boat?

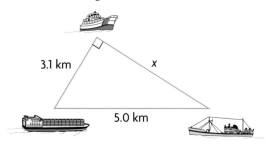

7. An overhead streetlight can illuminate a circular area of diameter 14 m. The light bulb is 6.8 m directly above a bike path. Determine the angle of elevation, to the nearest degree, from the edge of the illuminated area to the light bulb.

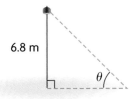

8. Martin's building is 105 m high. From the roof, he spots his car in the parking lot. He estimates that it is about 70 m from the base of the building.

a) What is the angle of depression, to the nearest degree, from Martin's eyes to the car?
b) What is the straight-line distance, to the nearest metre, from Martin to his car?

Lesson 5.3

9. Solve each triangle. Round each length to the nearest unit and each angle to the nearest degree.
   a) $\triangle DEF$: $\angle D = 67°$, $\angle F = 42°$, $e = 25$
   b) $\triangle PQR$: $\angle R = 80°$, $\angle Q = 49°$, $r = 8$
   c) $\triangle ABC$: $\angle A = 52°$, $\angle B = 70°$, $a = 20$
   d) $\triangle XYZ$: $\angle Z = 23°$, $\angle Y = 54°$, $x = 16$

10. From the bottom of a canyon, Rita stands 47 m directly below an overhead bridge. She estimates that the angle of elevation of the bridge is about 35° at the north end and about 40° at the south end. For each question, round your answer to the nearest metre.
    a) If the bridge were level, how long would it be?
    b) If the bridge were inclined 4° from north to south, how much longer would it be?

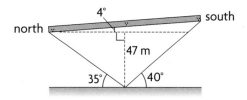

11. The manufacturer of a reclining lawn chair is planning to cut notches on the back of the chair so that you can recline at an angle of 30° as shown.
    a) What is the measure of $\angle B$, to the nearest degree, for the chair to be reclined at the proper angle?
    b) Determine the distance from A to B to the nearest centimetre.

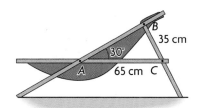

Investigating and Applying the Cosine Law in Acute Triangles

YOU WILL NEED

• dynamic geometry software

GOAL

Verify the cosine law and use it to solve real-life problems.

INVESTIGATE the Math

The sine law is useful only when you know two sides and an angle opposite one of those sides or when you know two angles and a side. But in the triangles shown, you don't know this information, so the sine law cannot be applied.

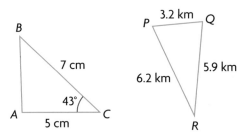

? How can the Pythagorean theorem be modified to relate the sides and angles in these types of triangles?

A. Use dynamic geometry software to construct any acute triangle.

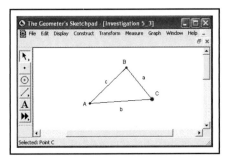

B. Label the vertices A, B, and C. Then name the sides a, b, and c as shown.

C. Measure all three interior angles and all three sides.

D. Drag vertex C until $\angle C = 90°$.

E. What is the Pythagorean relationship for this triangle? Using the measures of a and b, choose **Calculate ...** from the **Measure** menu to determine $a^2 + b^2$. Use the measure of c and repeat to determine c^2.

Tech | **Support**

For help using dynamic geometry software, see Technical Appendix, B-21 and B-22.

F. Does $a^2 + b^2 = c^2$? If not, are they close? Why might they be off by a little bit?

G. Move vertex C farther away from AB to create an acute triangle. How does the value of $a^2 + b^2$ compare with that of c^2?

H. Using the measures of $a^2 + b^2$ and c^2, choose **Calculate ...** from the **Measure** menu to determine $a^2 + b^2 - c^2$. How far off is this value from the Pythagorean theorem? Copy the table shown and record your observations.

Triangle	a	b	c	∠C	$a^2 + b^2$	c^2	$a^2 + b^2 - c^2$
1							
2							
3							
4							
5							

I. Move vertex C to four other locations and record your observations. Make sure that one of the triangles has $\angle C = 90°$, while the rest are acute.

J. Use a calculator to determine $2ab \cos C$ for each of your triangles. Add this column to your table and record these values. How do they compare with the values of $a^2 + b^2 - c^2$?

K. Based on your observations, modify the Pythagorean theorem to relate c^2 to $a^2 + b^2$.

Reflecting

L. How is the Pythagorean theorem a special case of the relationship you found? Explain.

M. Based on your observations, how did the value of $\angle C$ affect the value of $a^2 + b^2 - c^2$?

cosine law

in any acute $\triangle ABC$,
$c^2 = a^2 + b^2 - 2ab \cos C$

N. Explain how you would use the **cosine law** to relate each pair of values in any acute triangle.
  • a^2 to $b^2 + c^2$
  • b^2 to $a^2 + c^2$

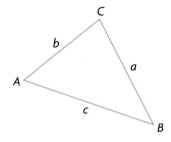

APPLY the Math

EXAMPLE 1 **Selecting the cosine law as a strategy to calculate an unknown length**

Determine the length of c to the nearest centimetre.

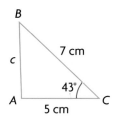

Anita's Solution

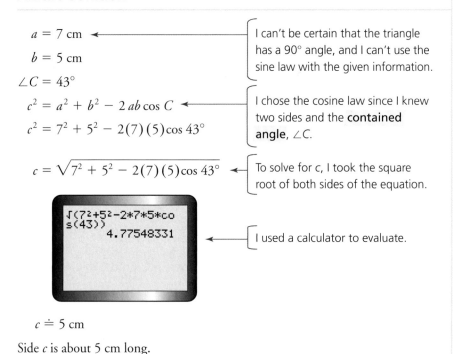

$a = 7$ cm   ← I can't be certain that the triangle has a 90° angle, and I can't use the sine law with the given information.

$b = 5$ cm

$\angle C = 43°$

$c^2 = a^2 + b^2 - 2\,ab\cos C$ ← I chose the cosine law since I knew two sides and the **contained angle**, $\angle C$.

$c^2 = 7^2 + 5^2 - 2(7)(5)\cos 43°$

contained angle

the acute angle between two known sides

$c = \sqrt{7^2 + 5^2 - 2(7)(5)\cos 43°}$ ← To solve for c, I took the square root of both sides of the equation.

```
√(7²+5²−2*7*5*co
s(43))
         4.77548331
```

I used a calculator to evaluate.

$c \doteq 5$ cm

Side c is about 5 cm long.

EXAMPLE 2   **Selecting the cosine law as a strategy to calculate an unknown angle**

Determine the measure of $\angle R$ to the nearest degree.

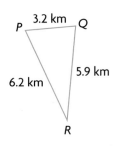

Kew's Solution

$p = 5.9$ km ← ⎤ I knew three sides and no angles.

$q = 6.2$ km

$r = 3.2$ km

$$r^2 = p^2 + q^2 - 2pq\cos R$$ ← ⎤ I chose the cosine law. I wrote the formula in terms of $\angle R$.

$$3.2^2 = 5.9^2 + 6.2^2 - 2(5.9)(6.2)\cos R$$

$$3.2^2 - 5.9^2 - 6.2^2 = -2(5.9)(6.2)\cos R$$

$$\frac{3.2^2 - 5.9^2 - 6.2^2}{-2(5.9)(6.2)} = \left(\frac{\overset{1}{\cancel{-2(5.9)(6.2)}}}{\underset{1}{\cancel{-2(5.9)(6.2)}}}\right)\cos R$$ ← ⎤ To solve for $\angle R$, I divided both sides of the equation by $-2(5.9)(6.2)$.

$$\frac{3.2^2 - 5.9^2 - 6.2^2}{-2(5.9)(6.2)} = \cos R$$

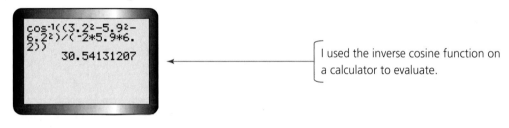

```
cos⁻¹((3.2²-5.9²-
6.2²)/(-2*5.9*6.
2))
        30.54131207
```

← ⎤ I used the inverse cosine function on a calculator to evaluate.

$$\angle R \doteq 31°$$

The measure of $\angle R$ is about $31°$.

EXAMPLE 3 | Solving a problem by using the cosine law

Ken's cell phone detects two transmission antennas, one 7 km away and the other 13 km away. From his position, the two antennas appear to be separated by an angle of $80°$. How far apart, to the nearest kilometre, are the two antennas?

Chantal's Solution

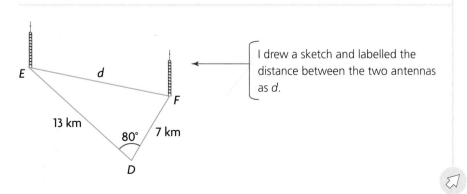

⎤ I drew a sketch and labelled the distance between the two antennas as d.

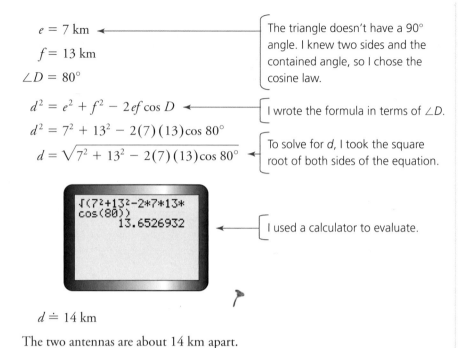

$e = 7$ km

$f = 13$ km

$\angle D = 80°$

The triangle doesn't have a 90° angle. I knew two sides and the contained angle, so I chose the cosine law.

$d^2 = e^2 + f^2 - 2ef \cos D$

I wrote the formula in terms of $\angle D$.

$d^2 = 7^2 + 13^2 - 2(7)(13)\cos 80°$

$d = \sqrt{7^2 + 13^2 - 2(7)(13)\cos 80°}$

To solve for d, I took the square root of both sides of the equation.

```
√(7²+13²-2*7*13*
cos(80))
            13.6526932
```

I used a calculator to evaluate.

$d \doteq 14$ km

The two antennas are about 14 km apart.

In Summary

Key Idea

- The cosine law states that in any $\triangle ABC$,

  $a^2 = b^2 + c^2 - 2bc \cos A$

  $b^2 = a^2 + c^2 - 2ac \cos B$

  $c^2 = a^2 + b^2 - 2ab \cos C$

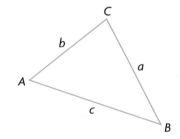

Need to Know

- The cosine law can be used only when you know
  - two sides and the contained angle or

  - all three sides

- The contained angle in a triangle is the angle between two known sides.

CHECK Your Understanding

1. i) In which triangle is it necessary to use the cosine law to calculate the third side? Justify your answer.

ii) State the formula you would use to determine the length of side *b* in △*ABC*.

a)

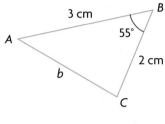

b) Sine

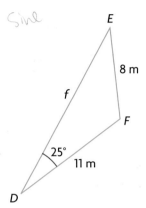

2. Determine the length of each unknown side to the nearest tenth of a centimetre.

a)

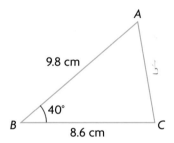

b)

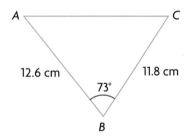

3. Determine each indicated angle to the nearest degree.

a)

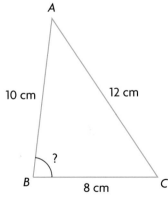

b)

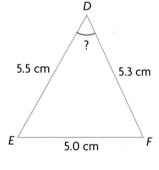

101.0.9164

PRACTISING

4. Determine each indicated length to the nearest tenth of a centimetre.

K

a)

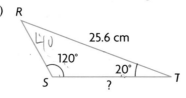

b)

c)

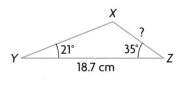

5. Solve each triangle. Round each length to the nearest tenth of a centimetre and each angle to the nearest degree.

   a) $\triangle ABC$: $\angle A = 68°$, $b = 10.1$ cm, $c = 11.1$ cm

   b) $\triangle DEF$: $\angle D = 52°$, $e = 7.2$ cm, $f = 9.6$ cm

   c) $\triangle HIF$: $\angle H = 35°$, $i = 9.3$ cm, $f = 12.5$ cm

   d) $\triangle PQR$: $p = 7.5$ cm, $q = 8.1$ cm, $r = 12.2$ cm

6. A triangle has sides that measure 5 cm, 6 cm, and 10 cm. Do any of the angles in this triangle equal 30°? Explain.

7. Two boats leave Whitby harbour at the same time. One boat heads 19 km to its destination in Lake Ontario. The second boat heads on a course 70° from the first boat and travels 11 km to its destination. How far apart, to the nearest kilometre, are the boats when they reach their destinations?

A

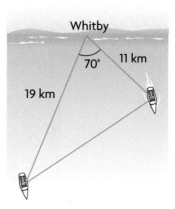

8. Louis says that he has no information to use the sine law to solve $\triangle FGH$ shown at the left and that he must use the cosine law instead. Is he correct? Describe how you would solve for each unknown side and angle.

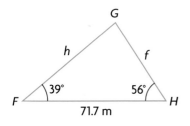

9. Driving a snowmobile across a frozen lake, Sheldon starts from the most westerly point and travels 8.0 km before he turns right at an angle of 59° and travels 6.1 km, stopping at the most easterly point of the lake. How wide, to the nearest tenth of a kilometre, is the lake?

10. What strategy would you use to calculate angle θ in $\triangle PQR$ shown at the right? Justify your choice and then use your strategy to solve for θ to the nearest degree.

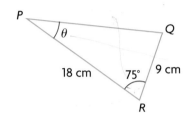

11. T A clock with a radius of 15 cm has an 11 cm minute hand and a 7 cm hour hand. How far apart, to the nearest centimetre, are the tips of the hands at each time?

  a) 3:30 p.m.   **b)** 6:38 a.m.

12. C Use the triangle shown at the right to create a problem that involves its side lengths and interior angles. Then describe how to determine d.

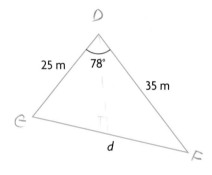

Extending

13. Two observers standing at points X and Y are 1.7 km apart. Each person measures angles of elevation to two balloons, A and B, flying overhead as shown. For each question, round your answer to the nearest tenth of a kilometre.

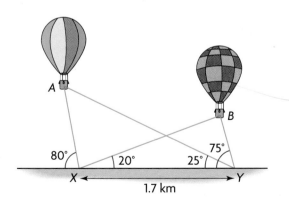

  a) How far is balloon A from point X? From point Y?
  b) How far is balloon B from point X? From point Y?
  c) How far apart are balloons A and B?

Solving Problems by Using Acute-Triangle Models

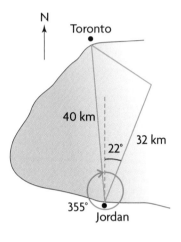

GOAL

Solve problems involving the primary trigonometric ratios and the sine and cosine laws.

LEARN ABOUT the Math

Steve leaves the marina at Jordan on a 40 km sailboat race across Lake Ontario intending to travel on a **bearing** of 355°, but an early morning fog settles. By the time it clears, Steve has travelled 32 km on a bearing of 22°.

? In which direction must Steve head to reach the finish line?

EXAMPLE 1	Solving a problem by using the sine and cosine laws

Determine the direction in which Steve should head to reach the finish line.

Liz's Solution

bearing

the direction in which you have to move in order to reach an object. A bearing is a clockwise angle from magnetic north. For example, the bearing of the lighthouse shown is 335°.

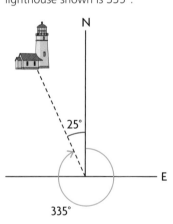

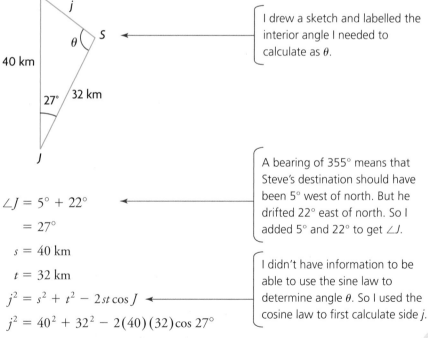

I drew a sketch and labelled the interior angle I needed to calculate as θ.

A bearing of 355° means that Steve's destination should have been 5° west of north. But he drifted 22° east of north. So I added 5° and 22° to get $\angle J$.

$\angle J = 5° + 22°$
$\quad = 27°$

$s = 40$ km

$t = 32$ km

$j^2 = s^2 + t^2 - 2st \cos J$

$j^2 = 40^2 + 32^2 - 2(40)(32)\cos 27°$

I didn't have information to be able to use the sine law to determine angle θ. So I used the cosine law to first calculate side j.

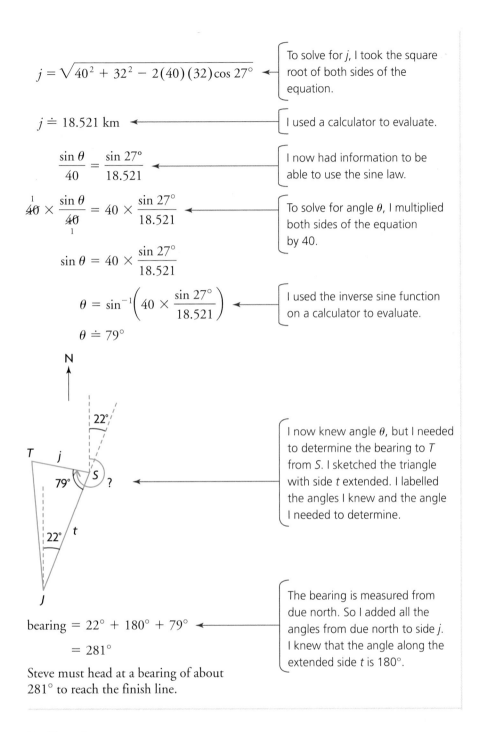

$$j = \sqrt{40^2 + 32^2 - 2(40)(32)\cos 27°}$$

To solve for j, I took the square root of both sides of the equation.

$$j \doteq 18.521 \text{ km}$$

I used a calculator to evaluate.

$$\frac{\sin \theta}{40} = \frac{\sin 27°}{18.521}$$

I now had information to be able to use the sine law.

$$\overset{1}{\cancel{40}} \times \frac{\sin \theta}{\underset{1}{\cancel{40}}} = 40 \times \frac{\sin 27°}{18.521}$$

To solve for angle θ, I multiplied both sides of the equation by 40.

$$\sin \theta = 40 \times \frac{\sin 27°}{18.521}$$

$$\theta = \sin^{-1}\left(40 \times \frac{\sin 27°}{18.521}\right)$$

I used the inverse sine function on a calculator to evaluate.

$$\theta \doteq 79°$$

I now knew angle θ, but I needed to determine the bearing to T from S. I sketched the triangle with side t extended. I labelled the angles I knew and the angle I needed to determine.

$$\text{bearing} = 22° + 180° + 79°$$

$$= 281°$$

The bearing is measured from due north. So I added all the angles from due north to side j. I knew that the angle along the extended side t is 180°.

Steve must head at a bearing of about 281° to reach the finish line.

Reflecting

A. Why didn't Liz first use the sine law in her solution?

B. How is a diagram of the situation helpful? Explain.

C. Is it possible to solve this problem *without* using the sine law or the cosine law? Justify your answer.

APPLY the Math

EXAMPLE **2** | Solving a problem by using primary trigonometric ratios

A ladder leaning against a wall makes an angle of 31° with the wall. The ladder just touches a box that is flush against the wall and the ground. The box has a height of 64 cm and a width of 27 cm. How long, to the nearest centimetre, is the ladder?

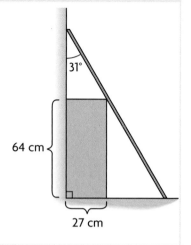

Denis's Solution

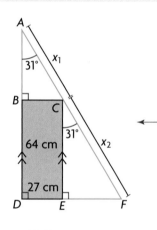

The box and ladder form two right triangles. I labelled the two parts of the ladder as x_1 and x_2. The total length of the ladder is the sum of x_1 and x_2.

Since CE is parallel to AD, $\angle ECF$ is equal to $\angle BAC$, which is 31°.

In $\triangle ABC$:

$$\sin \theta = \frac{\text{opposite}}{\text{hypotenuse}}$$

$$\sin 31° = \frac{BC}{x_1}$$

$$\sin 31° = \frac{27}{x_1}$$

$$x_1 \times \sin 31° = \overset{1}{\cancel{x_1}} \times \frac{27}{\underset{1}{\cancel{x_1}}}$$

$$x_1 \times \sin 31° = 27$$

In $\triangle CEF$:

$$\cos \theta = \frac{\text{adjacent}}{\text{hypotenuse}}$$

$$\cos 31° = \frac{CE}{x_2}$$

$$\cos 31° = \frac{64}{x_2}$$

$$x_2 \times \cos 31° = \overset{1}{\cancel{x_2}} \times \frac{64}{\underset{1}{\cancel{x_2}}}$$

$$x_2 \times \cos 31° = 64$$

In $\triangle ABC$, $BC = 27$ cm; in $\triangle CEF$, $CE = 64$ cm.

In each triangle, I knew a side and an angle and I needed to calculate the hypotenuse. So I used primary trigonometric ratios:
• sine, since BC is opposite the 31° angle
• cosine, since CE is adjacent to the 31° angle

To solve for x_1, I multiplied both sides of the equation by x_1 and divided both sides by sin 31°.

To solve for x_2, I multiplied both sides of the equation by x_2 and divided both sides by cos 31°.

$$\frac{x_1 \times \cancel{\sin 31°}^{1}}{\cancel{\sin 31°}_{1}} = \frac{27}{\sin 31°} \qquad \frac{x_2 \times \cancel{\cos 31°}^{1}}{\cancel{\cos 31°}_{1}} = \frac{64}{\cos 31°}$$

$$x_1 = \frac{27}{\sin 31°} \qquad\qquad x_2 = \frac{64}{\cos 31°}$$

$$x_1 \doteq 52.423\ 308\ 71 \qquad x_2 \doteq 74.664\ 537\ 42 \longleftarrow \text{I used a calculator to evaluate.}$$

length $= x_1 + x_2$ ⟵   I added x_1 and x_2 to determine the length of the ladder.

     $= 52.423\ 308\ 71 + 74.664\ 537\ 42$

     $\doteq 127$ cm

The ladder is about 127 cm long.

EXAMPLE 3 | ## Selecting a strategy to calculate the area of a triangle

Jim has a triangular backyard with side lengths of 27 m, 21 m, and 18 m. His bag of fertilizer covers 400 m². Does he have enough fertilizer?

Barbara's Solution

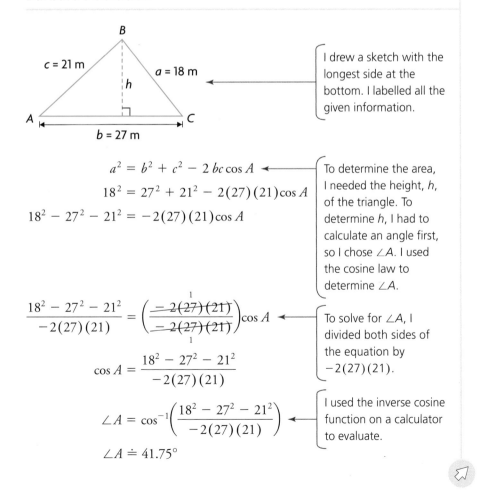

I drew a sketch with the longest side at the bottom. I labelled all the given information.

$$a^2 = b^2 + c^2 - 2\,bc\cos A$$

$$18^2 = 27^2 + 21^2 - 2(27)(21)\cos A$$

$$18^2 - 27^2 - 21^2 = -2(27)(21)\cos A$$

To determine the area, I needed the height, h, of the triangle. To determine h, I had to calculate an angle first, so I chose $\angle A$. I used the cosine law to determine $\angle A$.

$$\frac{18^2 - 27^2 - 21^2}{-2(27)(21)} = \left(\frac{\cancel{-2(27)(21)}^{1}}{\cancel{-2(27)(21)}_{1}}\right)\cos A$$

To solve for $\angle A$, I divided both sides of the equation by $-2(27)(21)$.

$$\cos A = \frac{18^2 - 27^2 - 21^2}{-2(27)(21)}$$

$$\angle A = \cos^{-1}\!\left(\frac{18^2 - 27^2 - 21^2}{-2(27)(21)}\right)$$

I used the inverse cosine function on a calculator to evaluate.

$$\angle A \doteq 41.75°$$

$$\sin A = \frac{\text{opposite}}{\text{hypotenuse}}$$

Since I knew $\angle A$, I used a primary trigonometric ratio to calculate h.

$$\sin A = \frac{h}{21}$$

$$\sin 41.75° = \frac{h}{21}$$

$$21 \times \sin 41.75° = \overset{1}{21} \times \frac{h}{\underset{1}{21}}$$

To solve for h, I multiplied both sides of the equation by 21.

$$21 \times \sin 41.75° = h$$

$$13.98 \text{ m} \doteq h$$

I used a calculator to evaluate.

$$A = \frac{1}{2} b \times h$$

I substituted the values for b and h into the formula for the area of a triangle.

$$= \left(\frac{1}{2}\right) 27 \times 13.98$$

$$\doteq 189 \text{ m}^2$$

Since 189 is less than half of 400, Jim has enough fertilizer to cover his lawn twice and still have some fertilizer left over.

EXAMPLE 4   Solving a problem to determine a perimeter

A regular octagon is inscribed in a circle of radius 15.8 cm. What is the perimeter, to the nearest tenth of a centimetre, of the octagon?

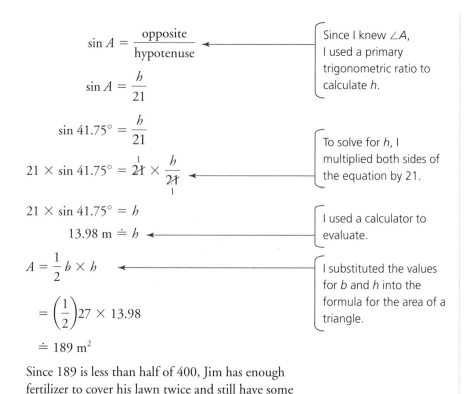

Shelley's Solution

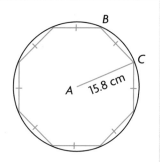

An octagon is made up of eight identical triangles, each of which is isosceles because two sides are the same length (radii of the circle).

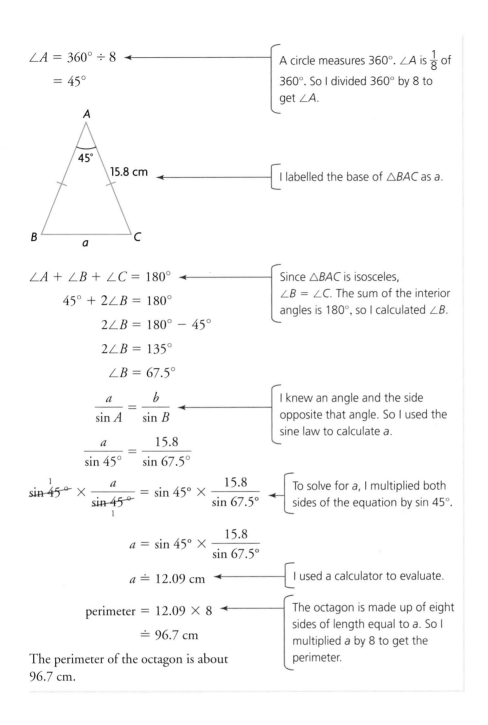

$\angle A = 360° \div 8$

 $= 45°$

A circle measures 360°. $\angle A$ is $\frac{1}{8}$ of 360°. So I divided 360° by 8 to get $\angle A$.

15.8 cm

I labelled the base of $\triangle BAC$ as a.

$\angle A + \angle B + \angle C = 180°$

$45° + 2\angle B = 180°$

$2\angle B = 180° - 45°$

$2\angle B = 135°$

$\angle B = 67.5°$

Since $\triangle BAC$ is isosceles, $\angle B = \angle C$. The sum of the interior angles is 180°, so I calculated $\angle B$.

$$\frac{a}{\sin A} = \frac{b}{\sin B}$$

I knew an angle and the side opposite that angle. So I used the sine law to calculate a.

$$\frac{a}{\sin 45°} = \frac{15.8}{\sin 67.5°}$$

$$\overset{1}{\cancel{\sin 45°}} \times \frac{a}{\underset{1}{\cancel{\sin 45°}}} = \sin 45° \times \frac{15.8}{\sin 67.5°}$$

To solve for a, I multiplied both sides of the equation by $\sin 45°$.

$$a = \sin 45° \times \frac{15.8}{\sin 67.5°}$$

$a \doteq 12.09$ cm

I used a calculator to evaluate.

perimeter $= 12.09 \times 8$

    $\doteq 96.7$ cm

The octagon is made up of eight sides of length equal to a. So I multiplied a by 8 to get the perimeter.

The perimeter of the octagon is about 96.7 cm.

In Summary

Key Idea

- The primary trigonometric ratios, the sine law, and the cosine law can be used to solve problems involving triangles. The method you use depends on the information you know about the triangle and what you want to determine.

Need to Know

- If the triangle in the problem is a right triangle, use the primary trigonometric ratios.
- If the triangle is oblique, use the sine law and/or the cosine law.

Given	Required to Find	Use
SSA	angle	sine law
ASA or AAS	side	sine law
SAS	side	cosine law
SSS	angle	cosine law

CHECK Your Understanding

1. For each triangle, describe how you would solve for the unknown side or angle.

a)

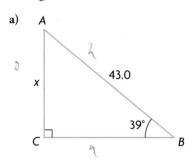

b)

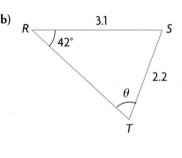

c)

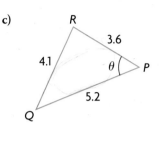

2. Complete a solution for each part of question 1. Round x to the nearest tenth of a unit and θ to the nearest degree.

PRACTISING

3. Determine the area of $\triangle ABC$, shown at the right, to the nearest
   K square centimetre.

4. The legs of a collapsible stepladder are each 2.0 m long. What is the maximum distance between the front and rear legs if the maximum angle at the top is 40°? Round your answer to the nearest tenth of a metre.

5. To get around an obstacle, a local electrical utility must lay two
   A sections of underground cable that are 371.0 m and 440.0 m long. The two sections meet at an angle of 85°. How much extra cable is necessary due to the obstacle? Round your answer to the nearest tenth of a metre.

6. A surveyor is surveying three locations (M, N, and P) for new rides in an amusement park around an artificial lake. $\angle MNP$ is measured as 57°. MN is 728.0 m and MP is 638.0 m. What is the angle at M to the nearest degree?

7. Mike's hot-air balloon is 875.0 m directly above a highway. When he is looking west, the angle of depression to Exit 81 is 11°. The exit numbers on this highway represent the number of kilometres left before the highway ends. What is the angle of depression, to the nearest degree, to Exit 74 in the east?

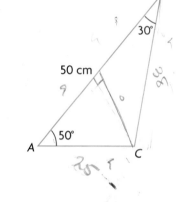

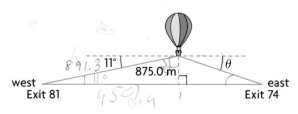

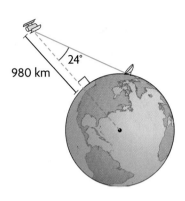

8. A satellite is orbiting Earth 980 km above Earth's surface. A receiving dish is located on Earth such that the line from the satellite to the dish and the line from the satellite to Earth's centre form an angle of 24° as shown at the left. If a signal from the satellite travels at 3×10^8 m/s, how long does it take to reach the dish? Round your answer to the nearest thousandth of a second.

9. Three circles with radii of 3 cm, 4 cm, and 5 cm are touching each other as shown. A triangle is drawn connecting the three centres. Calculate all the interior angles of the triangle. Round your answers to the nearest degree.

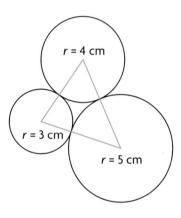

10. Given the regular pentagon shown at the left, determine its perimeter
T to the nearest tenth of a centimetre and its area to the nearest tenth of a square centimetre.

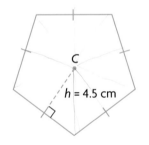

11. A 3 m high fence is on the side of a hill and tends to lean over. The hill is inclined at an angle of 20° to the horizontal. A 6.3 m brace is to be installed to prop up the fence. It will be attached to the fence at a height of 2.5 m and will be staked downhill from the base of the fence. What angle, to the nearest degree, does the brace make with the hill?

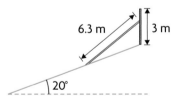

12. A surveyor wants to calculate the distance BC across a river. He selects a position, A, so that CA is 86.0 m. He measures $\angle ABC$ and $\angle BAC$ as 39° and 52°, respectively, as shown at the left. Calculate the distance BC to the nearest tenth of a metre.

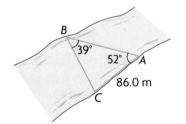

13. For best viewing, a document holder for people who work at computers should be inclined between 61° and 65° (∠ABC). A 12 cm support leg is attached to the holder 9 cm from the bottom. Calculate the minimum and maximum angle θ that the leg must make with the holder. Round your answers to the nearest degree.

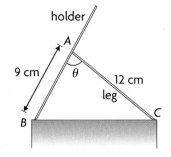

14. Match each method with a problem that can be solved by that method. Describe how each method could be used to complete a solution.

Method	Problems
Cosine law	Chris lives in a U-shaped building. From his window, he sights Bethany's window at a bearing of 328° and Josef's window at a bearing of 19°. Josef's window is 54 m from Bethany's and both windows are directly opposite each other. How far is each window from Chris's window?
Sine law	When the Sun is at an angle of elevation of 41°, Martina's treehouse casts a shadow that is 11.4 m long. Assuming that the ground is level, how tall is Martina's treehouse?
Primary trigonometric ratios	Ken walks 3.8 km west and then turns clockwise 65° before walking another 1.7 km. How far does Ken have to walk to get back to where he started?

Extending

15. Two paper strips, each 2.5 cm wide, are laying across each other at an angle of 27°, as shown at the right. What is the area of the overlapping paper? Round your answer to the nearest tenth of a square centimetre.

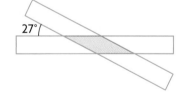

16. The diagram shows a roofing truss with AB parallel to CD. Calculate the total length of wood needed to construct the truss. Round your answer to the nearest metre.

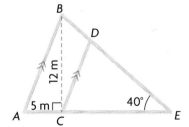

17. Lucas takes a 10.0 m rope and creates a triangle with interior angles of 30°, 70°, and 80°. How long, to the nearest tenth of a metre, is each side?

Curious | Math

Cycling Geometry

Competitive cyclists pay great attention to the geometry of their legs and of their bikes. To provide the greatest leverage and achieve optimum output, a cyclist's knees must form a right angle with the pedals at the top of a stroke and an angle of 165° at the bottom of a stroke.

To achieve these conditions, cyclists adjust the seat height to give the appropriate distances.

1. When Mark is standing, the distance from his hips to his knees is 44 cm and from his hips to the floor is 93 cm. When Mark's knees form each angle listed, how far are his hips from his heels? Round your answers to the nearest centimetre.
   a) 90°
   b) 165°

2. On Mark's bike, one pedal is 85 cm from the base of the seat. For Mark to achieve optimum output, by how much should he raise his seat? Round your answer to the nearest centimetre.

3. The table below lists the measurements of other cyclists in Mark's riding club. Would these cyclists be able to ride Mark's bike? What seat height(s) would they require? Round your answers to the nearest centimetre.

Name	Hips to Knees	Hips to Floor
Terry	38 cm	81 cm
Colleen	45 cm	96 cm
Sergio	41 cm	89 cm

FREQUENTLY ASKED *Questions*

Q: **What is the cosine law, and how do you use it to determine angles and sides in triangles?**

A: The cosine law is a relationship that is true for *all* triangles:

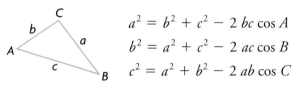

$$a^2 = b^2 + c^2 - 2\,bc \cos A$$
$$b^2 = a^2 + c^2 - 2\,ac \cos B$$
$$c^2 = a^2 + b^2 - 2\,ab \cos C$$

If you don't know a side and an angle opposite it, use the cosine law. The sine law can be used only if you know a side and the angle opposite that side. You can use the cosine law in combination with the sine law and other trigonometric ratios to solve a triangle.

Study | Aid
- See Lesson 5.4, Examples 1, 2, and 3.
- Try Chapter Review Questions 7 and 8.

Q: **How do you know which strategy to use to solve a trigonometry problem?**

Study | Aid
- See Lesson 5.5, Examples 1 to 4.
- Try Chapter Review Questions 9 and 10.

A:

Given	Use
A right triangle with any two pieces of information *(diagram of right triangle with vertices A, B, C; right angle at C)*	Primary trigonometric ratios: $\sin A = \dfrac{\text{opposite}}{\text{hypotenuse}}$ $\cos A = \dfrac{\text{adjacent}}{\text{hypotenuse}}$ $\tan A = \dfrac{\text{opposite}}{\text{adjacent}}$
A triangle with information about sides and opposite angles SSA or ASA or AAS	The sine law: $\dfrac{a}{\sin A} = \dfrac{b}{\sin B} = \dfrac{c}{\sin C}$ or $\dfrac{\sin A}{a} = \dfrac{\sin B}{b} = \dfrac{\sin C}{c}$
A triangle with no information that allows you to use the sine law SAS or SSS	The cosine law: $a^2 = b^2 + c^2 - 2\,bc \cos A$ $b^2 = a^2 + c^2 - 2\,ac \cos B$ $c^2 = a^2 + b^2 - 2\,ab \cos C$

PRACTICE Questions

Lesson 5.1

1. Determine x to the nearest unit and angle θ to the nearest degree.

a)

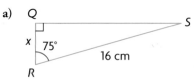

b)

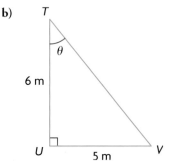

2. From Tony's seat in the classroom, his eyes are 1.0 m above ground. On the wall 4.2 m away, he can see the top of a blackboard that is 2.1 m above ground. What is the angle of elevation, to the nearest degree, to the top of the blackboard from Tony's eyes?

Lesson 5.2

3. A triangular garden has two equal sides 3.6 m long and a contained angle of $80°$.
   a) How much edging, to the nearest metre, is needed for this garden?
   b) How much area does the garden cover? Round your answer to the nearest tenth of a square metre.

4. A Bascule bridge is usually built over water and has two parts that are hinged. If each part is 64 m long and can fold up to an angle of $70°$ in the upright position, how far apart, to the nearest metre, are the two ends of the bridge when it is fully open?

Lesson 5.3

5. Use the sine law to solve each triangle. Round each length to the nearest centimetre and each angle to the nearest degree.

a)

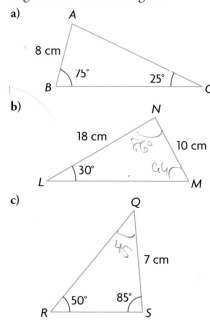

b)

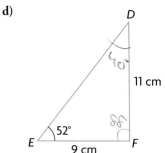

c)

d)

6. A temporary support cable for a radio antenna is 110 m long and has an angle of elevation of $30°$. Two other support cables are already attached, each at an angle of elevation of $70°$. How long, to the nearest metre, is each of the shorter cables?

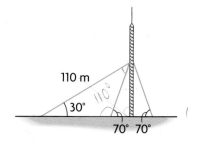

7. Use the cosine law to calculate each unknown side length to the nearest unit and each unknown angle to the nearest degree.

a)

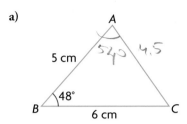

b)

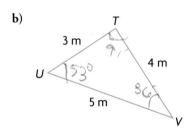

c)

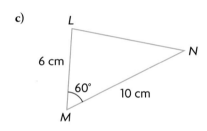

d)

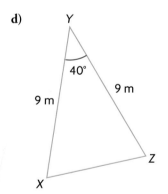

8. A security camera needs to be placed so that both the far corner of a parking lot and an entry door are visible at the same time. The entry door is 23 m from the camera, while the far corner of the parking lot is 19 m from the camera. The far corner of the parking lot is 17 m from the entry door. What angle of view for the camera, to the nearest degree, is required?

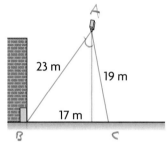

Lesson 5.5

9. Sketch and solve each triangle. Round your answers to the nearest degree and to the nearest tenth of a centimetre.

a) $\triangle ABC$: $\angle B = 90°$, $\angle C = 33°$, $b = 4.9$ cm

b) $\triangle DEF$: $\angle E = 49°$, $\angle F = 64°$, $e = 3.0$ cm

c) $\triangle GHI$: $\angle H = 43°$, $g = 7.0$ cm, $i = 6.0$ cm

d) $\triangle JKL$: $j = 17.0$ cm, $k = 18.0$ cm, $l = 21.0$ cm

10. Two sides of a parallelogram measure 7.0 cm and 9.0 cm. The longer diagonal is 12.0 cm long.

a) Calculate all the interior angles, to the nearest degree, of the parallelogram.

b) How long is the other diagonal? Round your answer to the nearest tenth of a centimetre.

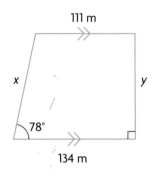

111 m

x

y

78°

134 m

1. A 3 m ladder can be used safely only at an angle of 75° with the horizontal. How high, to the nearest metre, can the ladder reach?

2. A road with an angle of elevation greater than 4.5° is steep for large vehicles. If a road rises 61 m over a horizontal distance of 540 m, is the road steep? Explain.

3. A surveyor has mapped out a property as shown at the left. Determine the length of sides *x* and *y* to the nearest metre.

4. Solve each triangle. Round each length to the nearest centimetre and each angle to the nearest degree.

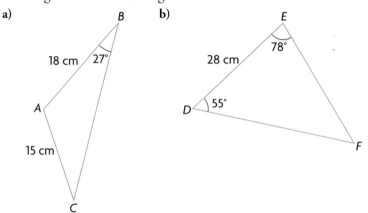

a)

B

18 cm   27°

A

15 cm

C

b)

E

78°

28 cm

D   55°

F

5. A 5.0 m tree is leaning 5° from the vertical. To prevent it from leaning any farther, a stake needs to be fastened 2 m from the top of the tree at an angle of 60° with the ground. How far from the base of the tree, to the nearest metre, must the stake be?

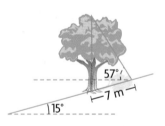

57°
7 m
15°

6. A tree is growing vertically on a hillside that is inclined at an angle of 15° to the horizontal. The tree casts a shadow uphill that extends 7 m from the base of its trunk when the angle of elevation of the Sun is 57°. How tall is the tree to the nearest metre?

7. Charmaine has planned a nature walk in the forest to visit four stations: *A*, *B*, *C*, and *D*. Use the sketch shown at the left to calculate the total length, to the nearest metre, of the nature trail, from *A* to *B*, *B* to *C*, *C* to *D*, and *D* back to *A*.

B

271 m

180 m

41°

A   17°   D

247 m   C

8. A weather balloon at a height of 117 m has an angle of elevation of 41° from one station and 62° from another. If the balloon is directly above the line joining the stations, how far apart, to the nearest metre, are the two stations?

Crime Scene Investigator

A great deal of mathematics and science is used in crime scene investigation. For example, a blood droplet is spherical when falling. If the droplet strikes a surface at 90°, it will form a circular spatter. If it strikes a surface at an angle, the spatter spreads out.

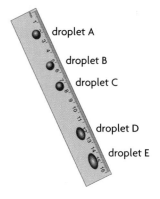

droplet A

droplet B

droplet C

droplet D

droplet E

Suppose blood droplets, each 0.8 cm wide, were found on the ruler shown at the right.

? How far above the ruler, to the nearest centimetre, did each droplet originate?

A. Use the diagram below to determine a relationship to calculate the angle of impact, θ, given a droplet of width AB.

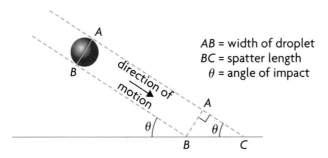

AB = width of droplet
BC = spatter length
θ = angle of impact

B. Which droplet appears most circular? What is the length of that droplet on the ruler, to the nearest tenth of a centimetre? What would be the angle of impact for that droplet, to the nearest degree? Create a table like the one shown and record your results.

Droplet	Original Width	Length	sin θ	θ
A	0.8 cm			
B	0.8 cm			

Task | Checklist

✔ Did you draw correct sketches for the problem?

✔ Did you show your work?

✔ Did you provide appropriate reasoning?

✔ Did you explain your thinking clearly?

C. Measure the length of the other four droplets. Use your relationship from part A to determine θ for each droplet.

D. If a droplet strikes the ruler head-on, the droplet originated from distance h. If the droplet strikes the ruler at some angle of impact, the length of the droplet on the ruler would be d. Use the diagram at the right to determine the distance h for droplets A to E.

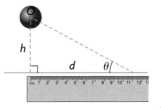

E. How far above the ruler did each droplet originate? Explain what happened with droplet B. Where appropriate, round to the nearest tenth of a centimetre.

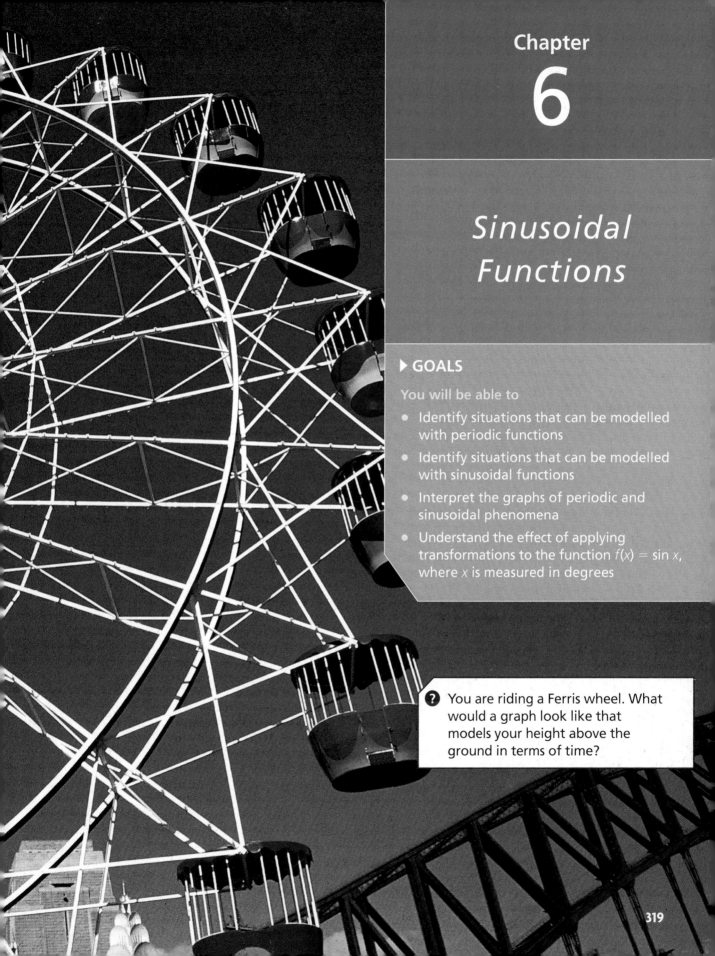

Sinusoidal Functions

▶ **GOALS**

You will be able to

- Identify situations that can be modelled with periodic functions

- Identify situations that can be modelled with sinusoidal functions

- Interpret the graphs of periodic and sinusoidal phenomena

- Understand the effect of applying transformations to the function $f(x) = \sin x$, where x is measured in degrees

? You are riding a Ferris wheel. What would a graph look like that models your height above the ground in terms of time?

WORDS *You Need to Know*

1. Match each word with the picture or example that best illustrates its definition. The blue curve is the original function $f(x) = x^2$.

a) vertical translation
b) vertical stretch

c) horizontal translation
d) vertical compression

e) minimum value
f) reflection

i)

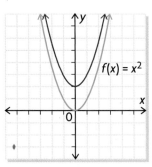

ii)

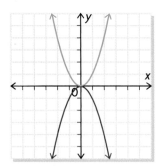

iii)

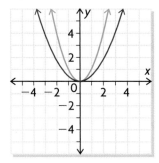

iv)

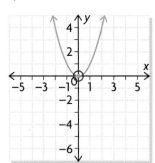

v)

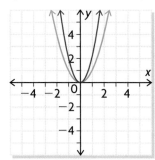

vi)

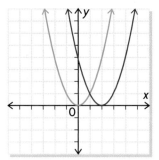

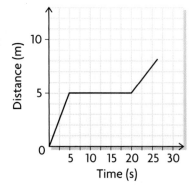

SKILLS AND CONCEPTS *You Need*

Using a Graphical Model to Predict: Interpolating and Extrapolating

Interpolate: to estimate a value that is between elements of the given data
Extrapolate: to estimate a value that is beyond the range of the given data by following a pattern

EXAMPLE

The graph represents a student using a motion sensor with a graphing calculator.
a) What is the distance of the student from the motion sensor at 15 s?
b) At what time is the student 15 m from the origin?

Solution

a) About 5 m

b) About 40 s

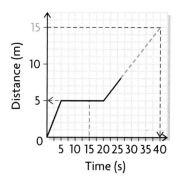

2. The graph at the bottom right represents the position of a grade 11 student participating in a 400 m race. Use the graph to determine the following.

a) What is the distance of the sprinter from the starting line at 10 s?

b) If the sprinter continues to run at a consistent rate, how long would she take to reach 500 m?

c) Describe how the sprinter's speed changed during the course of the race.

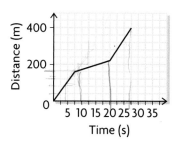

Identifying Transformations

When a quadratic function is in vertex form, $y = a(x - h)^2 + k$, several properties of the graph of the relation are obvious:

- If $k > 0$, then the graph of $y = x^2$ is translated vertically up by k units. If $k < 0$, then the graph is translated down by k units.
- If $h > 0$, then the graph of $y = x^2$ is translated horizontally h units to the right. If $h < 0$, then the graph of $y = x^2$ is translated horizontally h units to the left.
- If $-1 < a < 1$, then the graph of $y = x^2$ is compressed vertically by a factor of a. The resulting graph is wider than $y = x^2$.
- If $a > 1$ or $a < -1$, then the graph of $y = x^2$ is stretched vertically by a factor of a. The resulting graph is narrower than $y = x^2$.
- If $a > 0$, then the curve opens upward.
- If $a < 0$, then the curve is reflected through the x-axis and opens downward.
- When you use **transformations** to graph $y = a(x - h)^2 + k$ from $y = x^2$, you can apply the transformations in the following order:
  1. translation left or right
  2. vertical stretch or compression and reflection about the x-axis
  3. translation up or down

EXAMPLE

Determine the transformations you would apply to $f(x) = x^2$ to graph $g(x) = -2(x - 1)^2 + 5$.

Solution

Move to the right 1 unit, reflect in the x-axis and stretch by a factor of 2, and move up 5 units.

3. Determine the transformations you would apply to $y = x^2$ to graph each of the following.

   a) $y = 3(x - 5)^2 + 4$ **c)** $y = 0.5(x + 1)^2 - 3$

   b) $y = 2(x - 2)^2 + 1$ **d)** $y = -\frac{1}{4}(x + 2)^2 - 4$

PRACTICE

Motion of Truck

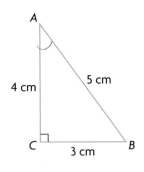

4. Earl operates a remote-control truck in front of a motion detector. He moves the truck either toward or away from the detector, never right or left. The graph shows the distance between the truck and the motion detector at regular time intervals.

   a) Determine each value, including units, and explain what they represent in this situation:

     i) the slope of the line between $t = 0$ and $t = 3$

     ii) the distance intercept

     iii) the slope of the line between $t = 3$ and $t = 5$

     iv) the time intercept

   b) State the domain and range of this function.

5. Determine the values of $\sin A$, $\cos A$, and $\tan A$ in $\triangle ABC$ at the left.

6. Use transformations of the graph of $f(x) = x^2$ to sketch the graphs of each of the following.

   a) $y = f(x) + 2$ **c)** $y = f(x - 1)$

   b) $y = 2f(x)$ **d)** $y = -f(x)$

7. For each function, determine the **maximum** or the **minimum** value, identify the **zeros**, and state the **domain** and **range**.

   a)
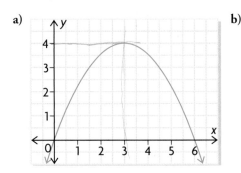

   b)
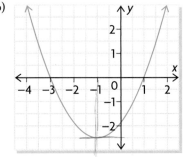

8. Complete a chart like the one below to show what you know about the term *function*.

Definition:		Ways to Test:
	Function	
Example:		Non-examples:

APPLYING *What You Know*

YOU WILL NEED
• graph paper

Maximizing Profit

ToyMart sells diecast cars.
- Its current profit per car is 45¢.
- Each week ToyMart sells 2000 cars.
- If ToyMart drops the profit per car by 5¢, it can sell 400 more cars per week.

❓ How much profit should ToyMart make on each car to maximize its weekly profit?

A. Calculate ToyMart's total profit on current sales of 2000 cars per week.

B. If n is the number of times ToyMart drops the profit by 5¢ per car and $P(n)$ is the weekly profit, explain how the function $P(n) = (2000 + 400n)(45 - 5n)$ models this relationship.

C. Complete the table and graph the ordered pairs, using n as the independent variable and $P(n)$ as the dependent variable. What does each ordered pair represent?

n	0	1	2	3	4	5	6	7	8	9
$P(n)$										

D. Calculate $P(10)$. Why should n stop at 9?

E. What is the domain of this function? What is the range?

F. What type of function is this? Explain how you know.

G. What profit is needed on each car to maximize the weekly profit?

6.1

Exploring Periodic Motion

YOU WILL NEED

- Calculator-Based Ranger (CBR)
- graphing calculator
- 20 cm × 25 cm piece of cardboard

Jordan's Graph

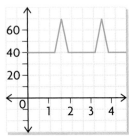

peak

the highest point(s) on the graph

trough

the lowest point(s) on the graph

period

the interval of the independent variable (often time) needed for a repeating action to complete one cycle

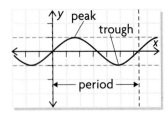

cycle

a series of events that are regularly repeated; a complete set of changes, starting from one point and returning to the same point in the same way

GOAL

Explore graphs of motion with repeating patterns.

EXPLORE the Math

Jordan produced different graphs by moving a cardboard paddle in front of a CBR in different ways. Jordan's graph is shown at the left.

? **How does the motion of the paddle influence the shape of the graph?**

A. Move the paddle to create different linear graphs.

B. Describe how you moved the paddle to control the direction and steepness of the slope of the line.

C. Describe what happens to the graph when you do the following:
- Slowly move the paddle back.
- Hold the paddle steady for a few seconds.
- Move the paddle quickly.
- Move the paddle farther away.

D. Move the paddle in different ways to create three different types of graphs that repeat in regular ways. Describe how each graph was affected by the way you moved the paddle.

E. Move the paddle to reproduce Jordan's graph. Describe how you had to move the paddle to duplicate the **peaks**, **troughs**, and **period** of the **cycle** in Jordan's graph.

Reflecting

F. How do you move the paddle to control
  a) the length of the period of a cycle?
  b) the height of the peaks and the depth of the troughs?
  c) the steepness of the line segments that form the peaks and troughs?
  d) the number of peaks or troughs in a cycle?

G. How can you predict the shape of the graph based on your motions?

In Summary

Key Idea

- When you repeat a motion in exactly the same way, the graph of that motion is periodic. By looking at the graph, you can figure out details of the motion.

Need to Know

- A linear graph results when there is no motion or the motion is performed at a constant speed in only one direction.
- If the motion is toward the motion detector, the graph falls. If the motion is away, the graph rises.
- The peaks and troughs of the graph describe the maximum and minimum distances, respectively, from the CBR.
- The steepness of the line segments describes the speed with which the object is moving. The steeper the line, the faster the motion.

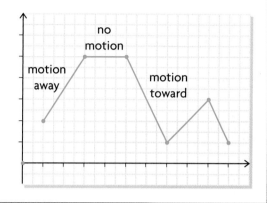

FURTHER Your Understanding

1. Sketch a graph for each motion pattern made with a paddle in front of a motion sensor.

   a)
   - Start 30 cm from the sensor.
   - Move 60 cm away from your starting point so that you end up 90 cm from the sensor.
   - Move back to the starting point.

   Repeat five times in about 10 s.

   b)
   - Start 60 cm from the sensor.
   - Move 45 cm from your starting point toward the sensor so that you end up 15 cm from the sensor.
   - Move 30 cm from this point away from the sensor so that you end up 45 cm from the sensor.
   - Move 30 cm from this point toward the sensor so that you end up 15 cm from the sensor.
   - Move 45 cm from this point away from the sensor so that you end up 60 cm from the sensor.

   Repeat three times in about 5 s.

2. Draw a graph to represent a different periodic motion of the paddle. Describe how the paddle would have to move to produce this graph.

Periodic Behaviour

GOAL

Interpret graphs that repeat at regular intervals.

LEARN ABOUT *the Math*

The Sun always shines on half the Moon. How much of the Moon we see depends on where it is in its orbit around Earth.

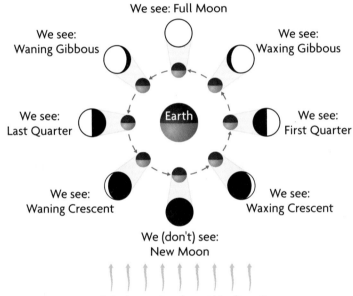

The tables show the proportion of the Moon that was visible from Southern Ontario on days 1 to 74 in the year 2006.

Day of Year	1	4	7	10	14	20	24	29	34
Proportion of Moon Visible	0.02	0.22	0.55	0.83	1.00	0.73	0.34	0.00	0.28

Day of Year	41	44	48	53	56	59	63	70	74
Proportion of Moon Visible	0.92	1.00	0.86	0.41	0.12	0.00	0.23	0.88	1.00

? What proportion of the Moon was visible on day 130?

EXAMPLE 1   Representing data as a periodic graph to make estimates

Create a graph using the Moon data, and use it to estimate what proportion of the Moon was visible on day 130.

David's Solution

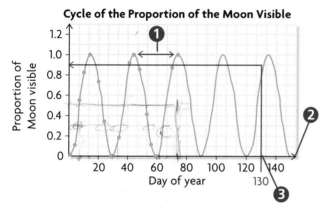

I drew a scatter plot with Day of year as the independent variable and Proportion of Moon visible as the dependent variable. Then I interpreted the pattern.

1 This graph has a repeating pattern. Its period is 30 days. I can tell because the proportion returns to 0 at 30 days, and the next part of the curve looks the same as the previous part.

2 I used the repeating pattern to extend the graph to 150 days.

3 I used the graph to estimate the proportion of the Moon visible on day 130.

At day 130, about 0.9 of the Moon is visible.

Reflecting

A. Why does it make sense to call the graph of the Moon data a periodic function?

B. How does the table help you predict the period of the graph?

C. How does knowing the period of a graph help you predict future events?

periodic function
a function whose values are repeated at equal intervals of the independent variable

APPLY the Math

If you are given a graph of a periodic function, it can be used to help you understand how the process or situation repeats.

EXAMPLE 2	Using reasoning to interpret a periodic graph

Tanya's mother works at a factory that makes rubber hoses. A chopping machine cuts each hose to 5.0 m lengths. How can Tanya interpret the graph that shows the process?

Tanya's Solution

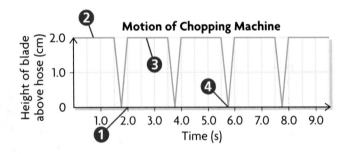

The period of this function is 2.0 s.
The maximum height of the blade is 2.0 cm.
The minimum height is 0 cm.
The blade stops for 1.5 s intervals.
The blade takes 0.5 s to go down and up.

I studied the graph and identified patterns. Then I related the patterns on the graph to the process of the chopping machine.

1 The cutting blade cuts a new section of hose every 2.0 s since the graph has a pattern that repeats every 2.0 s.

2 The height is always 2.0 cm or less, so the blade can't be higher than 2.0 cm. When the height is 0 cm, the blade is hitting the cutting surface.

3 Flat sections like the one from 2.0 to 3.5 mean that the blade stops for these intervals. The machine is probably pulling the next 5.0 m section of hose through before it's cut.

4 Parts of the graph, like from $t = 5.5$ to $t = 5.75$, show that the blade takes 0.25 s to go down. The part from $t = 5.75$ to $t = 6.0$ shows that the blade takes 0.25 s to go up.

Some graphs may appear to be periodic. They are periodic only if the pattern repeats in exactly the same way over the same interval.

EXAMPLE 3 | ## Connecting the features of a graph to the situation

The graph shows the demand for electricity in Ontario on a day in August 2002. Discuss and interpret the graph, and suggest possible reasons for its shape.

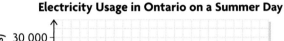
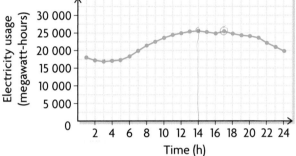

Electricity Usage in Ontario on a Summer Day

Enrique's Solution

The peak of the graph occurs at 14:00 h, or 2 p.m., and the trough of the graph occurs at 3:00 h, or 3 a.m.

> Demand for electricity is highest at 2 p.m. This could be because many factories and industries are working to capacity at this time. The demand is lowest at 3 a.m. because most people are sleeping and not using electricity.

The graph increases between 3 a.m. and 2 p.m. and again between 4 p.m. and 5 p.m.
The graph decreases between 2 p.m. and 4 p.m. and again between 5 p.m. and 3 a.m.

> As people get up to go to work, they use more electricity. Between 4 p.m. and 5 p.m., people arrive home from work so electricity usage rises. Between 2 p.m. and 4 p.m., people are travelling between home and work. Between 5 p.m. and 3 a.m., people use less electricity as they go to sleep.

I don't think the graph will be periodic. It will have this same pattern over 24 h, but I think the peaks and troughs will be different each day.

> The graph suggests that the pattern of electricity usage from day to day might repeat. But I can't be sure it repeats in exactly the same way each day. On hotter days, more people will use their air conditioners, making the demand higher than it shows for this day.

In Summary

Key Idea

- A function that produces a graph that has a regular repeating pattern over a constant interval is called a *periodic function*. It describes something that happens in a cycle, repeating in the same way over and over.

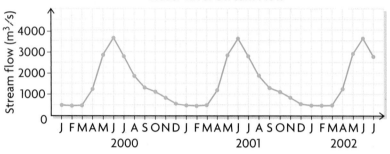

Fraser River Stream Flow

Need to Know

- A function that produces a graph that does not have a regular repeating pattern over a constant interval is called a *nonperiodic function*.
- Extending the graph of a periodic function by using the repeating pattern allows you to make reasonable predictions by extrapolating.
- The graph of a periodic function allows you to figure out information about the repeating pattern it represents.

CHECK Your Understanding

1. Which graphs represent periodic functions? Justify your decision.

a)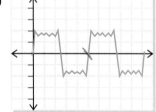

b)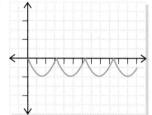

c)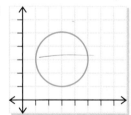

2. According to the table on page 326, there was a new Moon on day 59. When will the next two new Moons occur?

3. Use the data from the table on page 326. In the first 74 days of 2006, how many times will 50% of the Moon be visible in the clear night sky?

PRACTISING

4. A coach measures the velocity of air as a gymnast inhales and exhales
K after working out.

a) Plot the data from the table and draw a curve.

b) State the period of this function.

c) Explain why some velocities are positive and others are negative.

d) Extend the graph for the period between 9 s and 19 s.

e) Predict when the velocity of air will be 0 m/s for the period
between 9 s and 19 s.

5. The graph shows the amount of water used by an automatic
dishwasher as a function of time.

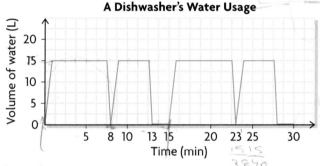

A Dishwasher's Water Usage

a) Why does the operation of the dishwasher model a periodic
function?

b) What is the period? What does one complete cycle mean?

c) Extend the graph for one more complete cycle.

d) How much water is used if the dishwasher runs through eight
complete cycles?

e) For part (d), state the domain and range of the function.

6. This is a graph of Nali's height above the ground in terms of time
A while riding a Ferris wheel.

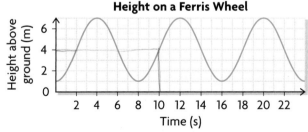

Height on a Ferris Wheel

a) What is the period of this function?

b) What does the period represent?

c) What is the diameter of the Ferris wheel? How do you know?

d) Approximately how high above the ground is Nali at 10 s?

e) At what times is Nali at the top of the wheel?

f) When is Nali 4 m above the ground?

Time (s)	Velocity of Air (L/s)
1.0	1.75
1.5	1.24
2.0	0
2.5	−1.24
3.0	−1.75
3.5	−1.24
4.0	0
4.5	1.24
5.0	1.75
5.5	1.24
6.0	0
6.5	−1.24
7.0	−1.75
7.5	−1.24
8.0	0
8.5	1.24
9.0	1.75

7. Which graphs represent periodic functions? Explain how you know.

a)
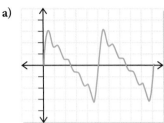

c)
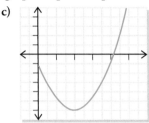

e)
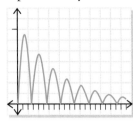

b)
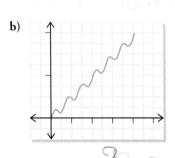

d)
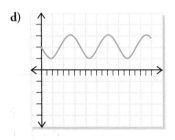

f)
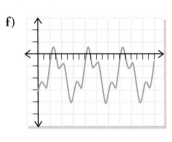

8. Which of the following situations would produce periodic graphs? Justify your decision, and draw a sketch of what you think the graph might look like.

a) An electrocardiograph monitors the electrical currents generated by the heart.
   - independent variable: time
   - dependent variable: voltage

b) Some forms of bacteria double in number every 20 min.
   - independent variable: time
   - dependent variable: number of bacteria

c) When you purchase a load of gravel, you pay for the gravel by the tonne plus a delivery fee.
   - independent variable: mass
   - dependent variable: cost

d) Alex dribbles a basketball at a steady pace and height.
   - independent variable: time
   - dependent variable: height

e) A tow truck is parked by the side of a dark road with its rotating amber light on. You are viewing it from a spot across the street.
   - independent variable: time
   - dependent variable: light intensity

f) You throw a basketball to a friend, but she is so far away that the ball bounces on the ground four times.
   - independent variable: distance
   - dependent variable: height

9. The Bay of Fundy, which is between New Brunswick and Nova Scotia, has the highest tides in the world. There can be no water on the beach at low tide, while at high tide the water covers the beach.

a) Why can you use periodic functions to model the tides?

b) What is the change in depth of the water from low to high tide?

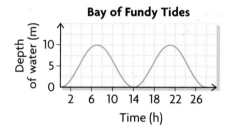

10. The graph shows the number of litres of water that a washing machine uses over several wash cycles.

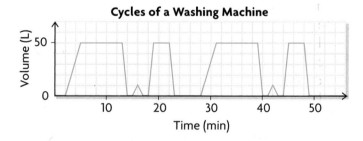

a) Explain what each part of the graph represents in terms of the cycles of the washing machine.

b) What is the period of one complete cycle?

c) What is the maximum volume of water used for each part of the cycle?

d) What is the total volume of water used for one complete cycle?

11. Describe the motion of the paddle in front of a CBR that would have produced the graph shown.

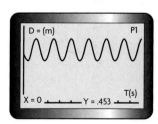

12. Denis holds a cardboard paddle 60 cm from a CBR for 3 s, and then
T within 0.5 s moves the paddle so that it is 30 cm from the detector. He
holds the paddle there for 2 s, and then within 0.5 s moves the paddle
back to the 60 cm location. Denis repeats this process three times.
Sketch the graph. Include a scale.

13. Write a definition of a periodic function. Include an example, and use
C your definition to explain why it is periodic.

Extending

14. Katrina is racing her car around the track shown. Assuming that she is
at least two laps into the race, draw a graph showing the relationship
between her speed and time as she completes two additional laps. Is
this a periodic function, or is it an approximate periodic function?
Justify your answer.

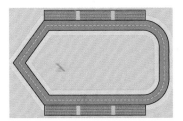

15. A traffic light at either end of a bridge under construction reduces
traffic to a single lane and changes colour over time as shown.

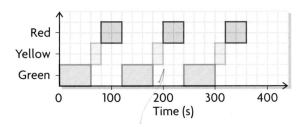

a) Describe one complete cycle.
b) What is the period?
c) A 20 s advanced green arrow is added to the beginning of the
cycle. What is the period now? Draw two full cycles of the graph.

6.3

Investigating the Sine Function

GOAL

Examine a specific type of periodic function called a sinusoidal function.

INVESTIGATE the Math

Steve uses a generator powered by a water wheel to produce his own electricity.
- Half the water wheel is below the surface of the river.
- The wheel has a radius of 1 m.
- The wheel has a nail on its circumference.
As the current flows, the wheel rotates in a counterclockwise direction to power the generator. The height of the nail, relative to the water level, as the wheel rotates is graphed in terms of the angle of rotation, *x*.

YOU WILL NEED
- cardboard
- ruler
- thumbtack
- protractor
- metre stick
- graphing calculator

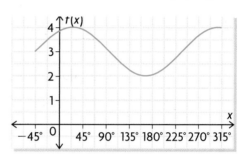

C06-F22-M11UCOSB.eps

? **How can the resulting graph be described?**

A. Construct a scale model of the water wheel by drawing a circle with a radius of 10 cm on cardboard to represent the water wheel's 1 m radius.

B. Locate the centre of the circle. Use a protractor to divide your cardboard wheel into 30° increments through the centre. Draw a dot to represent the nail on the circumference of the circle at one of the lines you drew to divide the wheel.

C. Cut out the wheel and attach it to a metre stick by pushing a thumbtack through the centre of the wheel at the 50 cm mark.

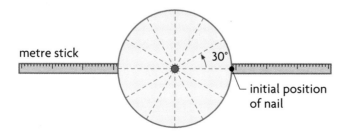

equation of the axis

the equation of the horizontal line halfway between the maximum and the minimum is determined by

$$y = \frac{(\text{maximum value} + \text{minimum value})}{2}$$

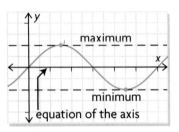

amplitude

the distance from the function's equation of the axis to either the maximum or the minimum value

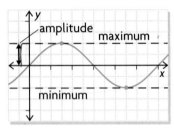

sine function

a sine function is the graph of $f(x) = \sin x$, where x is an angle measured in degrees; it is a periodic function

---

Tech | Support

For help on graphing the sine function, see Technical Appendix, B-13.

---

sinusoidal function

a type of periodic function created by transformations of $f(x) = \sin x$

D. Copy the table below. Rotate the wheel 30°, and measure the height of the "nail" (the perpendicular distance from the middle of the metre stick to the "nail"). Convert your measurement to metres and record in the table. Continue to rotate the wheel in 30° increments, and record the measurements in metres. If the "nail" goes below the surface of the centre of the metre stick, record the height as a negative value. Continue until the "nail" has rotated 720°. Use your data to graph the height versus angle of rotation.

Rotation, x (°)	0	30	60	90	120	• • •	690	720
Height of Nail, h (m)	0							

E. What are the period, the **equation of the axis**, and the **amplitude** of this function?

F. Describe how the sine ratio can be used to relate the height, h, of the nail to the angle of rotation, x.

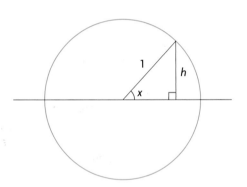

G. Use a graphing calculator to graph the **sine function**. How does this compare to your water wheel graph?

H. Describe the shape of the sine function, and determine its domain and range.

Reflecting

I. If you needed to sketch the graph $f(x) = \sin x$, what five key points would you use?

J. Which one of these equations best models this water wheel?
$x = \sin h$ $h = \sin x$ $h = \cos x$
$h = 90 x$ $x = h^2$ $h = x^2$

K. Why is the term **sinusoidal function** appropriate for this type of periodic function?

In Summary

Key Idea

- The height of a point above its starting position is a function of the angle of rotation, x, measured in degrees. The height, $f(x)$, can be modelled by the periodic function $f(x) = \sin x$, the sine function.

Need to Know

- The sine function has the following properties:
  - It has an amplitude of 1.
  - It has a period of 360°.
  - Its axis is defined by $y = 0$.
  - The domain is $\{x \in \mathbf{R}\}$, and the range is $\{y \in \mathbf{R} \mid -1 \leq y \leq 1\}$.
- The sine function passes through five key points:
  $(0°, 0)$, $(90°, 1)$, $(180°, 0)$, $(270°, -1)$, and $(360°, 0)$.

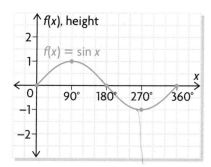

- Graphs that are periodic and have the same shape and characteristics as the sine function are called sinusoidal functions.

CHECK *Your Understanding*

1. Which graphs are sinusoidal functions? Justify your decision.

a)

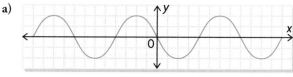

b)

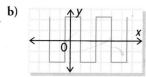

c)

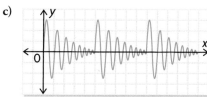

2. Nolan is jumping on a trampoline. The graph shows how high his feet are above the ground.

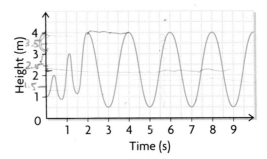

a) How long does it take for Nolan's jumping to become periodic? What is happening during these first few seconds?
b) What is the period of the curve? What does *period* mean in this context?
c) Write an equation for the axis of the periodic portion of the curve.
d) What is the amplitude of the sinusoidal portion of the curve? What does *amplitude* mean in this context?

PRACTISING

3. Alicia was swinging back and forth in front of a motion detector. Her **K** distance from the motion detector in terms of time can be modelled by the graph shown.

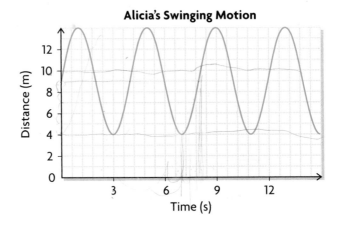

Alicia's Swinging Motion

a) What is the equation of the axis, and what does it represent in this situation?
b) What is the amplitude of this function?

c) What is the period of the function, and what does it represent in this situation?

d) How close did Alicia get to the motion detector?

e) If the motion detector was activated as soon as Alicia started to swing, how would the graph change? (You may draw a diagram or a sketch.) Would the resulting graph be sinusoidal? Why or why not?

f) At $t = 8$ s, would it be safe to run between Alicia and the motion detector? Explain your reasoning.

4. An oscilloscope hooked up to an AC (alternating current) circuit shows a sine curve on its display:

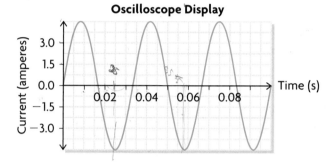

Oscilloscope Display

a) What is the period of the function?

b) What is the equation of the axis of the function?

c) What is the amplitude of the function?

d) State the units of measure for each of your answers above.

5. Using a graphing calculator in DEGREE mode, graph each function. Use the WINDOW settings shown. After you have the graph, state the period, the equation of the axis, and the amplitude for each function.

Tech | **Support**

For help on changing the window settings and graphing functions, see Technical Appendix, B-4.

```
WINDOW
 Xmin=0
 Xmax=1440
 Xscl=180
 Ymin=-2
 Ymax=6
 Yscl=1
 Xres=1
```

a) $f(x) = 2\sin x + 3$

b) $f(x) = 3\sin x + 1$

c) $f(x) = \sin(0.5x) + 2$

d) $f(x) = \sin(2x) - 1$

e) $f(x) = 2\sin(0.25x)$

f) $f(x) = 3\sin(0.5x) + 2$

6. Sketch the sinusoidal graphs that satisfy the given properties.

	Period (s)	Amplitude (m)	Equation of the Axis	Number of Cycles
a)	4	3	$y = 5$	2
b)	20	6	$y = 4$	3
c)	80	5	$y = -2$	2

7. Tides are the result of the gravitational attraction between the Sun, the Moon, and Earth. There are two high and low tides each day. The two high tides may not be the same height. The same is true of the low tides. In addition, tides differ in height on a daily basis. However, every 14.8 days, the oceans and other bodies of water that are affected by the Sun−Moon−Earth system go through a new tidal cycle.

  a) Are tides a periodic phenomenon? Why or why not?

  b) Are tides a sinusoidal phenomenon? Why or why not?

8. The table shows the hours of daylight for the city of Regina, at a latitude of 50°. The amount of daylight is calculated as the time between sunrise and sunset.

Regina, latitude of 50°

Day of Year	15	46	74	105	135	165	196	227	258	288	319	349
Hours of Daylight	8.5	10.1	11.8	13.7	16.4	17.1	15.6	14.6	12.7	10.8	9.1	8.1

  a) Graph the data, and draw a curve through the points.

  b) The hours of daylight in terms of day of the year can be modelled by a sinusoidal function. What is the period of this function?

  c) Approximate the equation of the axis. What does it represent in this situation?

  d) Approximate the amplitude. What does it represent in this situation?

9. The graph shows John's height above the ground as a function of time
T as he rides a Ferris wheel.

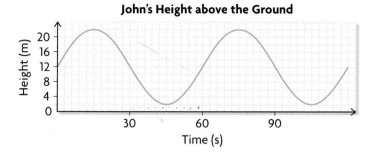

John's Height above the Ground

a) What is the diameter of the Ferris wheel?
b) What is John's initial height above the ground?
c) At what height did John board the Ferris wheel?
d) How high above the ground is the axle on the wheel?

10. a) Sketch the graph of $f(x) = \sin x$, where $-360° \le x \le 360°$.
C b) State the period, equation of the axis, and amplitude of $f(x) = \sin x$.

Extending

11. a) Create a table of values for the function defined by
$f(\theta) = \cos \theta$, where $0° \le \theta \le 360°$.

Rotation (θ)	0°	30°	60°	90°	120°	• • •	330°	360°
cos θ	1							

b) Plot these points, and draw a curve.
c) Is this a sinusoidal function? Explain.
d) Determine the period, the equation of the axis, and the amplitude of the function.
e) How does this periodic function compare with the function $f(\theta) = \sin \theta$?

12. a) Create a table of values for the function defined by
$f(\theta) = \tan \theta$, where $0° \le \theta \le 360°$.

Rotation (θ)	0°	30°	60°	90°	120°	• • •	330°	360°
tan θ	0							

b) Plot these points, and draw a curve.
c) Is this a sinusoidal function? Explain.
d) Determine the period, the equation of the axis, and the amplitude of the function.

6.4 Comparing Sinusoidal Functions

YOU WILL NEED

- graphing calculator

GOAL

Relate details of sinusoidal phenomena to their graphs.

LEARN ABOUT the Math

At an amusement park, a math teacher had different students ride two Ferris wheels. Thomas rode on Ferris wheel A, and Ryan rode on Ferris wheel B. The teacher collected data and produced two graphs.

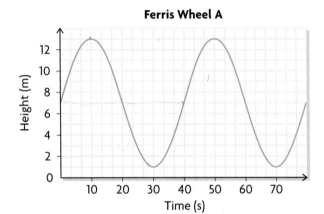

Ferris Wheel A

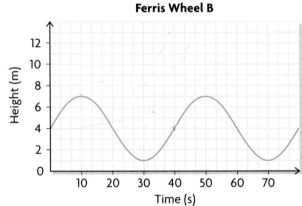

Ferris Wheel B

? What information about the Ferris wheels can you learn from the graphs of these functions?

EXAMPLE 1	Connecting the features of the graph to the situation

Nathan's Solution: Comparing Periods in Sinusoidal Functions

In both graphs, the first peak is at 10 s and the second peak is at 50 s.

$50 - 10 = 40$

> One of the easiest ways to get the period is to figure out how long it takes to go from one peak on the graph to the next peak. I read the times for the first two peaks.

The period of both Ferris wheels is 40 s.

> Both Ferris wheels take 40 s to complete one revolution.

Leila's Solution: Comparing Equations of the Axes in Sinusoidal Functions

Equation of the axis for Graph A:

$$\frac{(13 + 1)}{2} = 7$$

The equation of the axis for Graph A is $h = 7$.

Equation of the axis for Graph B:

$$\frac{(7 + 1)}{2} = 4$$

The equation of the axis for Graph B is $h = 4$.

The axle for Ferris wheel A is 7 m above the ground, and the axle for Ferris wheel B is 4 m above the ground.

> I got the equation of the axis in each case by adding the maximum and the minimum and then dividing by 2, since the answer is halfway between these values.

> In both cases, the equation of the axis represents the distance from ground level to the middle of the Ferris wheel. This distance is how high the axle is above the ground.

Dave's Solution: Comparing Amplitudes in Sinusoidal Functions

Amplitude for Graph A:

$$13 - 7 = 6$$

Amplitude for Graph B:

$$7 - 4 = 3$$

The radius of Ferris wheel A is 6 m.
The radius of Ferris wheel B is 3 m.

> I found the amplitude by getting the distance from the axis to a peak or maximum.

> I found the diameters of the wheels by subtracting the minimum from the maximum value. I divided by 2 to find the radius, which is also the amplitude of the graph.

EXAMPLE 2 | Comparing speeds

Wanda's Solution

Circumference of each wheel:

$C_A = 2\pi r_A \qquad C_B = 2\pi r_B$

$C_A = 2\pi(6) \qquad C_B = 2\pi(3)$

$C_A \doteq 37.7 \text{ m} \qquad C_B \doteq 18.8 \text{ m}$

> I want to see how fast each Ferris wheel goes. Since speed is equal to distance divided by time, I figured out how far each student travelled around the wheel in completing one revolution by finding the circumference of each wheel.

The circumference of Ferris wheel A is about 37.7 m, and the circumference of Ferris wheel B is about 18.8 m.

Speed of each rider:

$s_A = \dfrac{d_A}{t} \qquad s_B = \dfrac{d_B}{t}$

$s_A = \dfrac{37.7}{40} \qquad s_B = \dfrac{18.8}{40}$

$s_A \doteq 0.94 \text{ m/s} \qquad s_B \doteq 0.47 \text{ m/s}$

> To calculate the speed, I divided each circumference by the time taken to complete one revolution, which is 40 s.

Riders on Ferris wheel A are travelling twice as fast as those on Ferris wheel B.

> Ferris wheel A is probably scarier than Ferris wheel B because you are travelling much faster on A.

Reflecting

A. How does changing the radius of the wheel affect the sinusoidal graph?

B. How does changing the height of the axle of the wheel affect the sinusoidal graph?

C. How does changing the speed of the wheel affect the sinusoidal graph?

D. What type of information can you learn by examining the graph modelling the height of a rider in terms of time?

APPLY the Math

Comparing the differences between the maximum and minimum values of a sinusoidal function can give you insight into the periodic motion involved.

EXAMPLE 3 | Comparing peaks and troughs

As the shaft on an electric motor rotates, the shaft vibrates. The vibration causes stress to the shaft. The relationship between the stress on the shaft, y, measured in megapascals, and time, x, measured in seconds, can be modelled with a sinusoidal function. Two different equations for two different motors are given. Using a graphing calculator and the WINDOW settings shown, graph both sinusoidal functions at the same time. What can you conclude?

$$\text{Motor A: } y = \sin(9000x)^\circ + 8$$
$$\text{Motor B: } y = 3\sin(9000x)^\circ + 8$$

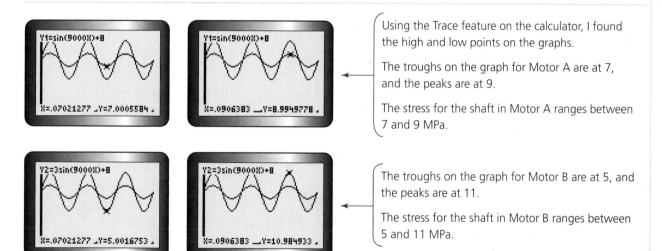

Using the Trace feature on the calculator, I found the high and low points on the graphs.

The troughs on the graph for Motor A are at 7, and the peaks are at 9.

The stress for the shaft in Motor A ranges between 7 and 9 MPa.

The troughs on the graph for Motor B are at 5, and the peaks are at 11.

The stress for the shaft in Motor B ranges between 5 and 11 MPa.

There are more extreme changes in stress to the shaft in Motor B.

In Summary

Key Idea

- Sinusoidal functions can be used as models to solve problems that involve circular or oscillating motion at a constant speed.

Need to Know

- One cycle of motion corresponds to one period of the sinusoidal function.
- The amplitude of the sinusoidal function depends on the situation being modelled.

CHECK Your Understanding

1. Tashina was on the same field trip as Thomas and Ryan. Tashina rode on Ferris wheel C. The resulting graph is shown. How does Ferris wheel C compare with Ferris wheels A and B in terms of maximum height, radius of the wheel, and speed?

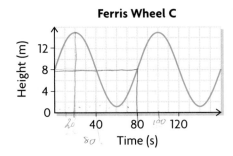

Ferris Wheel C

2. Draw two sinusoidal functions that have the same period and amplitude but have different equations of the axes.

PRACTISING

3. Sketch a height-versus-time graph of the sinusoidal function that models each situation. Assume that the first point plotted on each graph is at the lowest possible height.
   a) A Ferris wheel with a radius of 9 m, whose axle is 10 m above the ground, and that rotates once every 60 s
   b) A water wheel with a radius of 2 m, whose centre is at water level, and that rotates once every 20 s
   c) A bicycle tire with a radius of 35 cm and that rotates once every 30 s
   d) A girl lying on an air mattress in a wave pool that is 3 m deep, with waves 1 m in height that occur at 10 s intervals

4. When you breathe, the velocity of the air entering and exiting your lungs changes in terms of time. The air entering your lungs has a positive velocity, and the air exiting your lungs has a negative velocity. If a person is at rest, this relationship can be modelled using the graph shown.

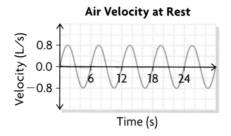

a) What is the equation of the axis, and what does it represent in this situation?
b) What is the amplitude of this function?
c) What is the period of the function, and what does it represent in this situation?
d) State the domain and range of the function.

5. When you exercise, the velocity of air entering your lungs in terms of time changes. The graph models this relationship.

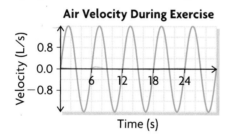

a) According to this exercise model, is the individual taking more breaths per minute or just deeper breaths than the individual in question 4? How do you know?
b) What property (period, equation of axis, or amplitude) of this graph has changed compared with the graph in question 4?
c) What is the maximum velocity of the air entering the lungs? Include the appropriate units of measure.

6. Maire is visiting a historical site that has two working water wheels.
K She notices that there is a large spike on the circumference of each wheel. She studies the motion of each spike, collects data, and constructs the following graphs:

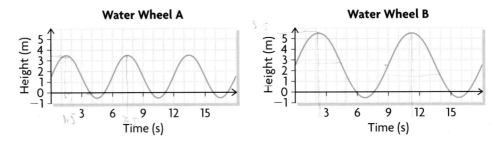

Water Wheel A

Water Wheel B

Analyze the graphs, and compare the water wheels. Refer to
- the radius of each wheel
- the height of the axle relative to the water
- the time taken to complete one revolution
- the speed of each wheel

7. Marcus has two different-sized wheels, on which he conducted four similar experiments.

Experiment 1: Marcus took the larger wheel and stuck a piece of tape on the wheel's side so that the tape would eventually touch the ground as it rolled. He rolled the wheel and recorded the height of the tape above the ground relative to the distance the wheel travelled. He graphed his data.

Experiment 2: Marcus took the same wheel and stuck the tape on the side of the wheel 10 cm in from the edge so that the tape would never touch the ground. He rolled the wheel, collected his data, and graphed the data.

Experiment 3: Marcus took the smaller wheel and stuck the tape on the side of the wheel so that the tape would eventually touch the ground as it rolled. He rolled the wheel, collected his data, and graphed the data.

Experiment 4: Using the smaller wheel, Marcus stuck the tape on the side of the wheel 10 cm in from the edge so that the tape would never touch the ground. He rolled the wheel, collected his data, and graphed the data.

a)

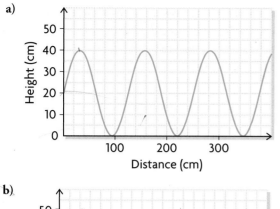

c)

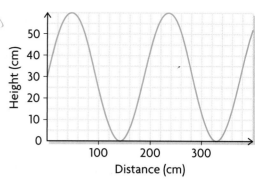

b)

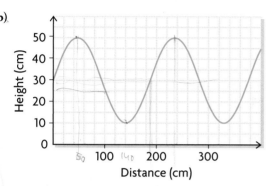

d)

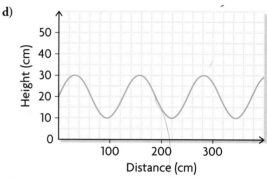

a) Match each graph to the appropriate experiment.
b) What is the equation of the axis for each graph? What does the equation of the axis represent for each graph?
c) What is the amplitude for each graph?
d) What is the period for each graph? What does the period represent in each graph?
e) What is the radius of the larger wheel?
f) What is the radius of the smaller wheel?
g) In Experiment 2, approximately how far above the ground is the tape after the wheel has travelled 100 cm from its initial position?
h) If the tape was placed on the centre of the larger wheel, what would the resulting graph look like?

8. The tables show the length of daylight for Windsor, Ontario, which is at a latitude of 40°, and for Fort Smith, Northwest Territories, which is at a latitude of 60°. The hours of daylight are calculated as the interval between sunrise and sunset.

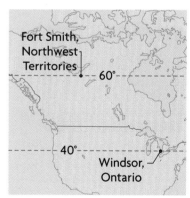

Windsor, latitude of 40°

Day of Year	15	46	74	105	135	165	196	227	258	288	319	349
Hours of Daylight	9.6	10.7	11.9	13.2	15.0	15.4	14.8	13.8	12.5	11.2	10.0	9.3

Fort Smith, latitude of 60°

Day of Year	15	46	74	105	135	165	196	227	258	288	319	349
Hours of Daylight	6.6	9.2	11.7	14.5	18.8	22.2	17.5	15.8	13.0	10.2	7.6	5.9

Tech | *Support*

For help on changing the window settings and graphing functions, see Technical Appendix, B-2 and B-4.

a) Plot the data on separate coordinate systems, and draw a curve through each set of points.

b) Compare the two curves. Refer to the periods, the equations of the axes, and the amplitudes.

c) What might you infer about the relationship between hours of daylight and the latitude at which you live?

9. In high winds, the top of a flagpole sways back and forth. The distance the tip of the pole vibrates to the left and right of its resting position can be defined by the function $d(t) = 2\sin(720t)°$, where $d(t)$ is the distance in centimetres and t is the time in seconds. If the wind speed decreases by 20 km/h, the motion of the tip of the pole can be modelled by the function $d(t) = 1.5\sin(720t)°$. Using graphing technology in DEGREE mode and the WINDOW settings shown, plot and examine the two graphs, and discuss the implications of the reduced wind speed on the flagpole. Refer to the period, axis, and amplitude of each graph.

10. The height, $h(t)$, of a basket on a water wheel at time t can be modelled by $h(t) = \sin(6t)°$, where t is in seconds and $h(t)$ is in metres.

a) Using graphing technology in DEGREE mode and the WINDOW settings shown, graph $h(t)$. Sketch this graph.

b) How long does it take for the wheel to make a complete revolution? Explain how you know.

c) What is the radius of the wheel? Explain how you know.

d) Where is the centre of the wheel located relative to the water level? Explain how you know.

11. A buoy rises and falls as it rides the waves. The function $h(t) = 1.5\sin(72t)°$ models the position of the buoy, $h(t)$, on the waves in metres at t seconds.

a) Using graphing technology in DEGREE mode and the WINDOW settings shown, graph $h(t)$. Sketch this graph.

b) How long does it take for the buoy to travel from the peak of a wave to the next peak? Explain how you know.

c) How many waves will cause the buoy to rise and fall in 1 min? Explain how you know.

d) How far does the buoy drop from its highest point to its lowest point? Explain how you know.

12. The average monthly temperature, $T(t)$, in degrees Celsius, in
T Sydney, Australia, can be modelled by the function
$T(t) = -12 \sin(30t)° + 20$, where t represents the number of
months. For $t = 0$, the month is January; for $t = 1$, the month is
February, and so on.

WINDOW
Xmin=0
Xmax=24
Xscl=1
Ymin=-25
Ymax=40
Yscl=5
Xres=1

a) Using graphing technology in DEGREE mode and the
WINDOW settings shown, graph $T(t)$. Sketch the graph.
b) What does the period represent in this situation?
c) What is the temperature range in Sydney?
d) What is the mean temperature in Sydney?

13. Explain how you would compare graphs of sinusoidal functions
C derived from similar real-world situations.

Extending

14. The diameter of a car's tire is 50 cm. While the car is being driven, the
tire picks up a nail.
a) How far does the tire travel before the nail returns to the ground
for the first time?
b) Model the height of the nail above the ground in terms of the
distance the car has travelled since the tire picked up the nail.
c) How high above the ground will the nail be after the car has
travelled 235 cm?
d) The nail reaches a height of 10 cm above the ground for the third
time. How far has the car travelled?

15. A gear of radius 1 m turns in a counterclockwise direction and drives a
larger gear of radius 3 m. Both gears have their central axes along the
horizontal.

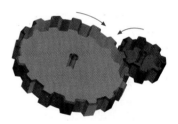

a) In which direction is the larger gear turning?
b) If the period of the smaller gear is 2 s, what is the period of the
larger gear?
c) Make a table in convenient intervals for each gear to show the
vertical displacement, d, of the point where the two gears first
touched. Begin the table at 0 s, and end it at 12 s. Graph vertical
displacement versus time.
d) What is the displacement of the point on the large gear when the
drive gear first has a displacement of –1 cm?
e) What is the displacement of the drive gear when the large gear first
has a displacement of 2 m?
f) What is the displacement of the point on the large gear at
5 min?

If It Rolls, Is It Sinusoidal?

If you roll an object, is the graph of its movement always
a sinusoidal function?

YOU WILL NEED
- ruler
- protractor
- corrugated cardboard
- graph paper
- tape
- compass
- scissors

1. Draw a circle of radius 6 cm on cardboard and
   cut it out. Using a pencil, punch a small hole in
   the cardboard circle, 5 cm from its centre. The
   hole should be big enough so that the tip of the
   pencil can fit through.

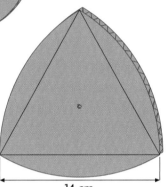

2. Draw an equilateral triangle with sides of length
   14 cm on cardboard. Using a pencil, punch a
   small hole in the centre of the triangle, large
   enough so that the tip of the pencil can fit through. Using a compass,
   draw an arc from one vertex of the triangle to another, using the third
   vertex as the centre of rotation. Continue this process until you have
   drawn three arcs. Cut out the figure.

3. Tape two sheets of graph paper together, short end to short end. Tape
   these two pieces of paper onto a wall or chalkboard so that there is a
   flat surface, such as a floor or chalk ledge, perpendicular to the paper.

4. Put a pencil through the hole in the cardboard circle, and rotate the
   circle along the flat surface. The pencil should be pressed against the
   graph paper so that it marks a curve on the paper as the circle rotates.

5. Repeat step 4 using the cardboard equilateral triangle.

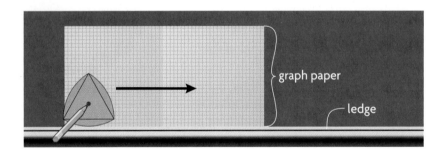

a) Does each curve represent a periodic function? Why or why not?
b) Does each curve represent a sinusoidal function? Why or why not?
c) Will rolling an object always result in a sinusoidal function?

FREQUENTLY ASKED Questions

Q: What are periodic functions?

A: Periodic functions repeat at regular intervals. As a result, their graphs have a repeating pattern.

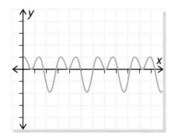

Study | *Aid*

• See Lesson 6.2,
  Examples 1 and 2.
• Try Mid-Chapter Review
  Questions 1, 2, and 3.

Q: What are the characteristics of a sinusoidal function?

A: Sinusoidal functions, like other periodic functions, repeat at regular intervals. Unlike other periodic functions, sinusoidal functions form symmetrical waves, where any portion of the wave can be horizontally translated onto another portion of the curve. The three most important characteristics of a sinusoidal function are the period, the equation of the axis, and the amplitude.

Study | *Aid*

• See Lesson 6.3,
  Examples 1 and 2.
• Try Mid-Chapter Review
  Questions 4 and 5.

Period	Equation of the Axis	Amplitude
The period is the change in *x* corresponding to one cycle. (A cycle of a sinusoidal function is a portion of the graph from one point to the point at which the graph starts to repeat.) One way to determine the period is to look at the change in *x*'s between two maximum values.	The equation of the axis is the equation of the line halfway between the maximum and minimum values on a sinusoidal function. It can be determined from the following formula: $$y = \frac{(\text{maximum value} + \text{minimum value})}{2}$$	The amplitude is the vertical distance from the function's axis to the minimum or maximum value.

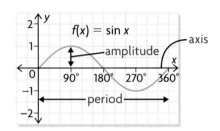

For the function $f(x) = \sin x$, the period is $360°$, the equation of the axis is $y = 0$, and the amplitude is 1.

The domain is $\{x \in \mathbf{R}\}$, and the range is $\{f(x) \in \mathbf{R} \mid -1 \leq f(x) \leq 1\}$.

Q: Why do you learn about sinusoidal functions?

A1: Many situations can be modelled using sinusoidal functions. Examples are:
- the motion of objects in a circular orbit
- the motion of a pendulum
- the motion of vibrating objects
- the number of hours of sunlight for a particular latitude
- the phase of the Moon
- the current for an AC circuit

A2: When the graph of a sinusoidal function models a repeating situation, the graph can be used to make predictions.

EXAMPLE

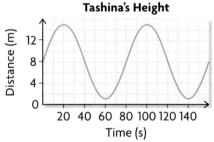

The graph represents Tashina's ride on a Ferris wheel. According to the graph,
- it takes 80 s to complete one revolution (the period)
- the axle is 7 m above the ground ($y = 7$, the equation of the axis)
- the radius of the Ferris wheel is 6.5 m (the amplitude)
- we can predict that at 200 s, Tashina's height on the Ferris wheel will be 8 m

PRACTICE Questions

Lesson 6.2

1. Which of the following situations would produce a periodic graph?
   a) Angelo is bouncing a tennis ball in the air with his racket. He strikes the ball with the same force each time such that the ball reaches the same maximum height.
      • independent variable: time
      • dependent variable: height of the ball
   b) A super ball is released from a third-storey window. The ball bounces back up to 80% of its previous height on each bounce.
      • independent variable: time
      • dependent variable: height of the ball
   c) A police cruiser is parked on the street with its siren on.
      • independent variable: time
      • dependent variable: intensity of the sound coming from the siren
   d) Alicia's investment fund doubles every eight years.
      • independent variable: time
      • dependent variable: total amount of money in the fund
   e) Lexi is driving through a parking lot that has speed bumps placed at regular intervals.
      • independent variable: the distance Lexi travels
      • dependent variable: the force exerted on the shock absorbers in her vehicle

2. Explain what each characteristic means for a periodic curve. Show each on a labelled diagram.
   a) cycle
   b) period
   c) amplitude
   d) equation of the axis
   e) maximum and minimum

3. A power nailer on an assembly line fires continuously. The compressed air that powers the nailer is contained in a large tank, and the pressure in this tank changes as the nailer is fired.

A pump maintains a certain level of pressure in the tank. The pressure in the tank in terms of time can be represented by the graph shown.

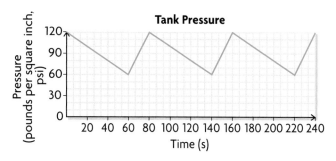

Tank Pressure

a) Is this function periodic?
b) At what pressure does the pump turn on?
c) At what pressure does the pump turn off?
d) What is the period of the function? Include the units of measure.
e) How long does the pump work at any one time?

Lesson 6.3

4. Tommy's blood pressure in terms of time can be modelled by a sinusoidal function. The graph shown represents this relationship.

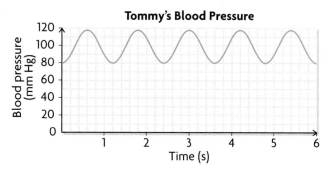

Tommy's Blood Pressure

a) What is the period of the function?
b) How many times does Tommy's heart beat each minute?
c) What is the range of the function? Explain the meaning of this range in terms of Tommy's blood pressure.

5. The pendulum on a grandfather clock swings uniformly back and forth. For a particular clock, the distance the pendulum moves to the left and right of its resting position in terms of time can be modelled by the function $d(t) = 0.25 \sin (180t)°$. The distance is measured in metres, and time is measured in seconds. Using graphing technology, in DEGREE mode, with the WINDOW settings shown, answer the following questions.

a) What is the period of the function, and what does it represent in this situation? (*Hint:* The period for this function is going to be quite short.)

b) What is the equation of the axis, and what does it represent in this situation?

c) What is the amplitude of the function, and what does it represent in this situation?

d) What will be the distance of the pendulum from its resting position at 10.2 s?

6. Sketch three cycles of a sinusoidal function that has a period of 30, an amplitude of 6, and whose equation of the axis is $y = 5$.

7. State the period, amplitude, and the equation of the axis for each function.

a)
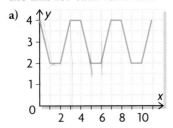

b)
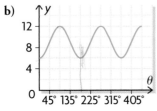

Lesson 6.4

8. Steve and Monique are swinging on separate swings beside a school. The lengths of the ropes on each swing differ. Their distances from one wall of the school in terms of time can be modelled by the graphs shown.

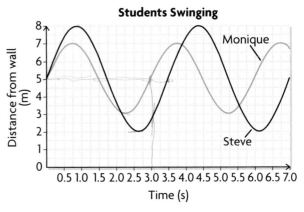

a) Compare the two curves. Refer to the periods, amplitudes, and the equations of the axes.

b) Compare Monique's motion on the swing with Steve's motion.

c) State the range of each function.

9. A Ferris wheel at the county fall fair has a radius of 12 m and rotates once every 60 s. At its lowest point, a rider is 2 m above the ground. Another Ferris wheel at an amusement park has a radius of 15 m and rotates once every 75 s. On this ride, the highest point a passenger reaches is 33 m above the ground.

a) On the same graph, sketch the height of a passenger above the ground for two complete revolutions of both wheels.

b) Compare the period, amplitude, and the equation of the axis of both graphs.

c) Which Ferris wheel is travelling faster? Explain how you know.

Transformations of the Sine Function: $f(x) = \sin(x - c)$ and $f(x) = \sin x + d$

YOU WILL NEED

• graphing calculator

GOAL

Determine how the values c and d affect the functions $f(x) = \sin x + d$ and $f(x) = \sin(x - c)$.

INVESTIGATE the Math

The table and graph show the relationship between the height of a nail on a water wheel with a radius of 1 m relative to the water level and the angle of rotation of the wheel.

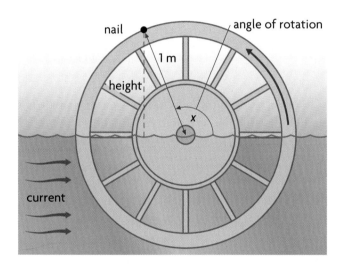

Five Key Points for $f(x) = \sin x$	
x, Angle of Rotation	$f(x)$, Height (m)
0°	0
90°	1
180°	0
270°	−1
360°	0

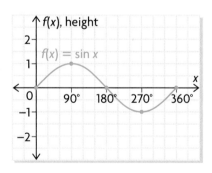

? What characteristics of the water wheel are directly related to c and d in the graphs of $f(x) = \sin(x - c)$ and $f(x) = \sin(x) + d$, and how does changing these values affect the graph of $f(x) = \sin x$?

A. Predict what will happen to the graph of $f(x) = \sin x$ when different values of d are used in functions of the form $f(x) = \sin x + d$.

Tech | **Support**

If you are graphing the functions on a graphing calculator, make sure that the calculator is in DEGREE mode and that you are using the following WINDOW settings:

```
WINDOW
  Xmin=0
  Xmax=720
  Xscl=90
  Ymin=-2
  Ymax=2
  Yscl=1
  Xres=1■
```

To change the table setting, press (2nd)(WINDOW).
To view the table, press (2nd)(GRAPH).

B. Graph $f(x) = \sin x$ in Y1 using the calculator settings in the margin. Investigate the two functions in the chart below by graphing each function in Y2, one at a time. Compare the graphs of the two functions with the graph of $f(x) = \sin x$. Compare the table of values of both graphs, using the settings given in the chart. Use this information to sketch the transformed function, and then describe how the characteristics (i.e., amplitude, range, and so on) have changed.

Equation	Table Settings	Sketch of the New Graph	Description of How the Graph Has Changed (i.e., amplitude, range, and so on)
$g(x) = \sin x + 0.5$	TblStart = 0 $\triangle$Tbl = 90		
$h(x) = \sin x - 0.5$	TblStart = 0 $\triangle$Tbl = 90		

C. Were your descriptions in the last column what you predicted? For each new function, describe the type of transformation that produced it.

D. Predict what will happen to the graph of $f(x) = \sin x$ when different values of c are used to graph functions of the form $f(x) = \sin(x - c)$.

E. Repeat part B and create a new table using the functions $g(x) = \sin(x - 90°)$ and $h(x) = \sin(x + 90°)$.

F. Were your descriptions in the last column what you predicted? For each new function, describe the type of transformation that produced it.

G. For the functions $f(x) = \sin x + 0.5$ and $f(x) = \sin(x - 90)$, how would the characteristics of the water wheel or its placement be different?

Reflecting

H. When you compare a function of the form $g(x) = \sin x + d$ with $f(x) = \sin x$, does the amplitude, period, or equation of the axis change from one function to the other?

I. When you compare a function of the form $h(x) = \sin(x - c)$ with $f(x) = \sin x$, does the amplitude, period, or equation of the axis change from one function to the other?

J. Suppose only 0.25 m of the water wheel was still exposed, due to high floodwaters. What would the graph of the sinusoidal function describing the height of the nail in terms of the rotation look like? What transformation of $f(x) = \sin x$ would you be dealing with? Sketch the situation.

K. Discuss how the domain and range change when
   a) a horizontal translation is applied
   b) a vertical translation is applied

L. Explain how you would change the equation $f(x) = \sin x$ so that the graph moves up or down the y-axis as well as left or right along the x-axis.

APPLY the Math

EXAMPLE 1   **Reasoning about a vertical translation of $f(x) = \sin x$**

a) How does the function $g(x) = \sin x - 3$ transform the function $f(x) = \sin x$?
b) What are the domain and range of the new function?
c) State the equation of the axis, amplitude, and period of the new function.

Anne's Solution

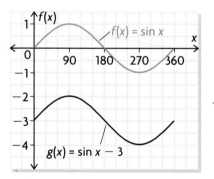

I started by graphing $f(x) = \sin x$ (in green). The function $g(x) = \sin x - 3$ has undergone a vertical translation of -3. So I slid the graph of $f(x)$ down 3 units to get the graph of $g(x) = \sin x - 3$ (in black). The equation of the axis is $y = -3$, but the period and amplitude are the same.

a) The -3 moves the graph down 3 units. The domain is $\{x \in \mathbf{R}\}$.

Anytime I add a value to or subtract a value from the sine function, the graph moves up or down.

b) The domain is $\{x \in \mathbf{R}\}$. The range is $\{y \in \mathbf{R} \,|\, -4 \le y \le -2\}$.

The graph continues to the left and the right, covering all positive and negative values of x. The largest y-value is -2, and the smallest y-value is -4.

c) The equation of the axis is $y = -3$. The amplitude is 1. The period is $360°$.

The axis is halfway between the maximum and minimum. It is also equal to the vertical translation. The amplitude is the distance from the axis to the maximum. The distance from one maximum to another gives the period.

EXAMPLE **2** | Reasoning about a horizontal translation of $f(x) = \sin x$

a) How does the function $g(x) = \sin(x - 60°)$ transform the function $f(x) = \sin x$?
b) What are the domain and range of the new function?
c) State the equation of the axis, amplitude, and period of the new function.

Teja's Solution

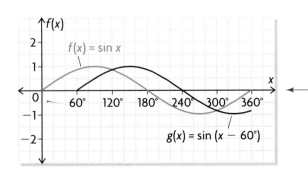

I started by graphing $f(x) = \sin x$ (in green). The function $f(x) = \sin(x - 60°)$ has undergone a horizontal translation of 60°. So I slid the graph of $g(x)$ right by 60° to get the graph of $g(x) = \sin(x - 60°)$ (in black). The equation of the axis is $y = 0$, and the period and amplitude are the same.

a) The graph moves right 60°.
b) The domain is $\{x \in \mathbf{R}\}$. The range is $\{y \in \mathbf{R} \mid -1 \le y \le 1\}$.
c) The axis is $y = 0$. The amplitude is 1. The period is 360°.

Anytime I add or subtract a number from the independent variable in the sine function, the graph moves right or left.

The graph continues to the left and the right, covering all positive and negative values of x. The largest y-value is 1, and the smallest y-value is -1.

Both a vertical and a horizontal translation can be applied to the sine function in the same manner you have seen with quadratic functions.

EXAMPLE **3** | Reasoning about translations of $f(x) = \sin x$

a) How does the function $g(x) = \sin(x + 45°) + 2$ transform the function $f(x) = \sin x$?
b) What are the domain and range of the transformed function?
c) State the equation of the axis, amplitude, and period of the new function.

David's Solution

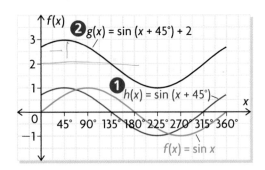

I began by graphing $f(x) = \sin x$, my green graph. The function given has two transformations, so I dealt with them one at a time.

1 I slid the graph left 45° to show the horizontal translation, my red graph, $h(x)$.

2 Then I slid this graph up 2 units to show the vertical translation, my black graph, $g(x)$.

I checked my work on my graphing calculator, and I'm confident that I did it right.

a) The graph is moved to the left 45° and up 2 units.
b) The domain is $\{x \in \mathbf{R}\}$. The range is
$\{y \in \mathbf{R} \mid 1 \le y \le 3\}$. ◄
c) The equation of the axis is $y = 2$.
The period is 360°, and the amplitude is 1.

> The graph continues to the left and the right, covering all positive and negative values of x. The largest y-value is 3, and the smallest y-value is 1.

> The vertical translation of 2 affected the axis. The horizontal translation doesn't affect the period or the amplitude.

EXAMPLE 4 | **Representing an equation from the description of a transformation of $f(x) = \sin x$**

The graph of $f(x) = \sin x$ has been translated to the right 30° and up 4 units. Write the new equation.

Ryan's Solution: Without Graphing

Horizontal translation: 30° ◄
vertical translation: +4
$f(x) = \sin(x - 30°)$

> If the graph's been translated to the right 30°, then there's been a horizontal translation of 30°. So there should be an $x - 30°$ in the equation.

$f(x) = \sin(x - 30°) + 4$ ◄

> If the graph is slid up 4 units, then there's been a vertical translation of +4. I'll add 4 to my equation.

Brianna's Solution: By Graphing

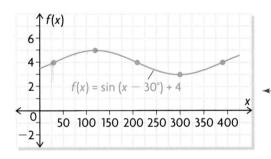

$f(x) = \sin(x - 30°) + 4$

From the graph, I determined the equation to be
$f(x) = \sin(x - 30°) + 4$.

> $f(x) = \sin x$ has five key points: (0°, 0), (90°, 1) (180°, 0), (270°, −1), and (360°, 0). These points are going to move when transformations are applied to the equation.

> Since there is a horizontal translation of 30°, all the x-values will be increased by 30. By adding 30 to the x-coordinate of each of my five key points, they change to (30°, 0), (120°, 1), (210°, 0), (300°, −1), and (390°, 0).

> Since there's a vertical translation of 4, all the y-values for my new key points will be increased by 4. By adding 4 to the y-coordinate of each of my five key points, they change to (30°, 4), (120°, 5), (210°, 4), (300°, 3), and (390°, 4).

> I plotted my new points and sketched the curve.

EXAMPLE 5 | Identifying transformations of $f(x) = \sin x$ from a graph

Explain what transformations were applied to $f(x) = \sin x$ (green curve) to create the black curve, $g(x)$.

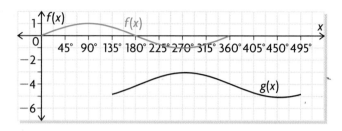

Sasha's Solution

$270° - 90 = 180°$
horizontal translation: $+180°$
$f(x) = \sin(x - 180°)$

There are two transformations; I'll start with the horizontal one and then move on to the vertical transformation. I looked at the x-coordinate of the first peak of the original graph; it was $x = 90°$. Then I looked at the x-coordinate of the first peak of the transformed graph; it was $x = 270°$. So the graph has moved to the right by $180°$. I must have a horizontal translation of $180°$.

vertical translation: -4
$f(x) = \sin(x - 180°) - 4$

I looked at the y-coordinate of the first peak of the original graph; it was $y = 1$. Then I looked at the y-coordinate of the first peak of the transformed graph; it was $y = -3$. So the graph has moved down by 4 units. I must have a vertical translation of -4.

In Summary

Key Idea

• The graph of the function $f(x) = \sin(x - c°) + d$ is congruent to the graph of $f(x) = \sin x$. The differences are only in the placement of the graph.

Need to Know

• Transformations that move a function left/right as well as up/down are called translations.
• The graph of the function $f(x) = \sin(x - c°) + d$ can be drawn from the graph of $f(x) = \sin x$ by applying the appropriate translations one at a time to the key points of $f(x) = \sin x$.
   • Move the graph of $f(x) = \sin x$ $c°$ to the right when $c > 0$.
   • Move the graph of $f(x) = \sin x$ $c°$ to the left when $c < 0$.
   • Move the graph of $f(x) = \sin x$ d units down when $d < 0$.
   • Move the graph of $f(x) = \sin x$ d units up when $d > 0$.
• A vertical translation affects the equation of the axis and the range of the function, but has no effect on the period, amplitude, or domain. A horizontal translation slides a graph to the left or right, but has no effect on the period, amplitude, equation of the axis, domain, or range.

CHECK Your Understanding

1. State the transformations for each function, and determine the domain and range.
   a) $f(x) = \sin(x + 40°)$
   b) $f(x) = \sin x + 8$
   c) $f(x) = \sin(x - 60°)$
   d) $f(x) = \sin x - 5$

2. Sketch $f(x) = \sin(x - 90°) - 5$, and verify with graphing technology.

3. a) The graph of $f(x) = \sin x$ has been translated to the left 70° and up 6 units. Write the new equation.
   b) State the amplitude, period, and equation of the axis of the new function.
   c) State the domain and range of the new function.

PRACTISING

4. State the transformations for each sinusoidal function, and then sketch its graph.
   a) $f(x) = \sin(x - 20°)$
   b) $f(x) = \sin x + 5$
   c) $f(x) = \sin(x - 150°) - 6$
   d) $f(x) = \sin(x + 40°) - 7$
   e) $f(x) = \sin(x + 30°) - 8$
   f) $f(x) = \sin(x + 120°) + 3$

5. For question 4, what feature or features of the sinusoidal functions are alike? Choose from the period, amplitude, equation of the axis, domain, and range.

6. a) The function $f(x) = \sin x$ undergoes a horizontal translation of 15° and a vertical translation of 4. Write the new equation.
   b) The function $f(x) = \sin x$ undergoes a vertical translation of -7 and a horizontal translation of 60°. Write the new equation.
   c) The graph of $f(x) = \sin x$ has been translated to the right 45° and up 3 units. Write the new equation.

7. The graph of $f(x) = \sin x$ has been translated down 5 units and to the left 30°. Write the new equation.

8. For each graph, determine whether a horizontal or vertical translation (or both) has occurred to the graph of $f(x) = \sin x$. If so, indicate how much the graph has been translated.

   a)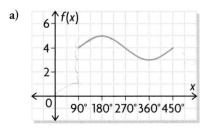

   b)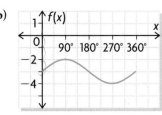

9. Sketch each sinusoidal function, and verify your answers using graphing technology.
   a) $f(x) = \sin x + 5$
   b) $f(x) = \sin(x - 120°)$
   c) $f(x) = \sin(x - 30°) + 4$
   d) $f(x) = \sin(x - 60°) + 1$
   e) $f(x) = \sin(x - 90°) - 2$
   f) $f(x) = \sin(x + 30°) - 1$

10. **A** A tire has a pebble stuck in its tread. The height of the pebble above the ground in terms of the rotation of the tire in degrees can be modelled by the graph shown.

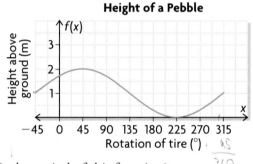

Height of a Pebble

a) What is the period of this function?
b) What is the equation of the axis? What does it represent in this situation? What transformation is characterized by the equation of the axis?
c) What is the amplitude of the function? What does it represent in this situation?
d) Besides the transformation identified in part (b), what other transformation has occurred?

11. Two tires have pebbles stuck in their treads. The heights of the pebbles above the ground, in terms of the rotation of the tires, can be modelled by the following functions, where the dependent variable is the height in metres and the independent variable is the rotation in degrees:

   Tire 1: $f(x) = \sin(x + 45°) + 1$
   Tire 2: $g(x) = \sin(x - 90°) + 1$

   a) What transformations do these functions share?
   b) What transformations are different?
   c) How are the real-world situations represented by the two functions the same? How are they different?

12. **T** Determine the equations of three sinusoidal functions that would have the range $\{y \in \mathbf{R} \mid 3 \le y \le 5\}$.

13. Create a chart that compares and contrasts horizontal and vertical translations.

14. Explain how you can determine from the equation of the function
C whether a sinusoidal function has undergone a vertical translation and a horizontal translation.

Extending

15. Write the equation for each sinusoidal function.

a)

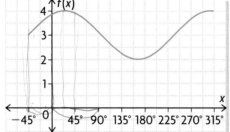

b)

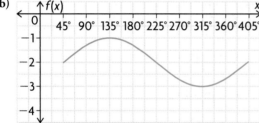

16. Three students were asked to determine the equation of the sinusoidal function shown. Whose answer is correct? Explain.

Student 1: $y = \sin(x - 90°) + 5$

Student 2: $y = \sin(x - 450°) + 5$

Student 3: $y = \sin(x - 810°) + 5$

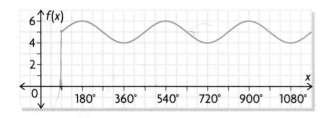

More Transformations of sin x: $f(x) = a \sin x$

GOAL

Determine how the value a affects the function $f(x) = a \sin x$.

INVESTIGATE the Math

The graph of $f(x) = \sin x$, based on the water wheel, relates the height of a point on the wheel to the water level and the angle of rotation.

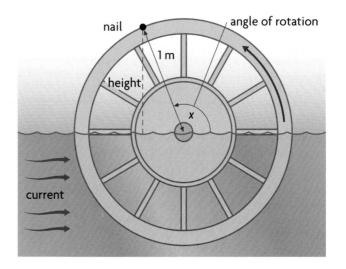

5 Key Points for $f(x) = \sin x$	
x, Angle of Rotation	f(x), Height (m)
0°	0
90°	1
180°	0
270°	−1
360°	0

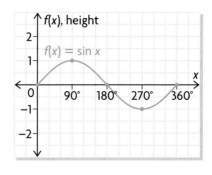

> **?** What characteristic of the water wheel is directly related to a in the graph of $f(x) = a \sin x$, and how does changing this value affect the graph of $f(x) = \sin x$?

A. Predict how the graph of $f(x) = \sin x$ will change if $\sin x$ is multiplied by
  - a positive integer
  - a positive fraction (less than 1)
  - a negative integer

Explain your predictions.

B. Graph $f(x) = \sin x$ in Y1, using the calculator settings in the margin. Investigate the three functions in the chart by graphing each function in Y2, one at a time. Compare the graph of each with the graph of $f(x) = \sin x$. Compare the table of values of both graphs using the settings given in the chart. Use this information to sketch the transformed function, and then describe how the characteristics (i.e., amplitude, range, and so on) have changed.

Tech | *Support*

If you are graphing the functions on a graphing calculator, make sure that the calculator is in DEGREE mode and that you are using the following WINDOW settings:

```
WINDOW
 Xmin=0
 Xmax=720
 Xscl=90
 Ymin=-2
 Ymax=2
 Yscl=1
 Xres=1
```

For help on using and changing the table settings, see Technical Appendix, B-6.

Equation	Table Settings and Values	Sketch of the New Graph	Description of How the Graph Has Changed (i.e., amplitude, range, and so on)
$g(x) = 2 \sin x°$	TblStart = 0 ΔTbl = 90		
$h(x) = 0.5 \sin x°$	TblStart = 0 ΔTbl = 90		
$j(x) = -\sin x°$	TblStart = 0 ΔTbl = 90		

C. Were your descriptions in the last column what you predicted?

D. For each new function, describe the type of transformation that produced it.

E. For the functions $f(x) = 2 \sin x$ and $f(x) = 0.5 \sin x$, how would the characteristics of the water wheel or its placement be different?

Reflecting

F. What characteristic (amplitude, period, or equation of the axis) of the graph is affected by multiplying or dividing the value of $\sin x$ by 2? Why might you call one transformation a stretch and the other a compression?

G. How would the function change for a water wheel of radius 3 m? What would the graph look like?

H. If $g(x) = a \sin x$, what effect does the value of a have on the function $f(x) = \sin x$ when
  a) $a > 1$?   b) $0 < a < 1$?   c) $a = -1$?

APPLY *the Math*

EXAMPLE 1	Connecting the effect of a vertical stretch on the function $f(x) = \sin x$

How does the value 5 in the function $g(x) = 5 \sin x$ change $f(x) = \sin x$?

Liz's Solution

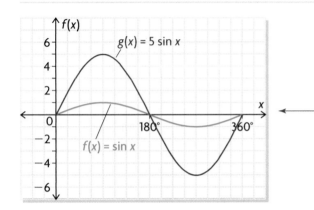

I studied the equation and looked at the parameter a.

There's a vertical stretch of a factor of 5 because $\sin x$ is multiplied by 5. Normally my amplitude is 1; now my amplitude is 5 because all the y-coordinates of the points on $f(x)$ have been multiplied by 5. The zeros of the function are not affected by the stretch.

Multiplying by 5 has changed the amplitude. The equation of the axis is still $y = 0$. The period is unchanged and is still $360°$. The domain is unchanged, $\{x \in \mathbf{R}\}$, but the range has changed to $\{y \in \mathbf{R} \mid -5 \leq y \leq 5\}$.

EXAMPLE 2	Connecting the effect of a vertical compression and reflection on the function $f(x) = \sin x$

How does the value $-\frac{1}{4}$ in the function $g(x) = -\frac{1}{4} \sin x$ change $f(x) = \sin x$?

Liz's Solution

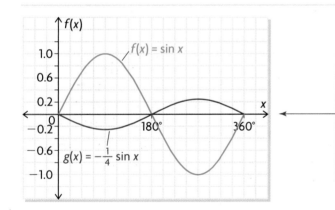

I studied the equation and looked at the parameter a.

I can tell that there's a vertical compression of a factor of $\frac{1}{4}$ and a reflection in the x-axis because $\sin x$ is multiplied by $-\frac{1}{4}$. Normally my amplitude is 1; now my amplitude is $\frac{1}{4}$ because all the y-coordinates of the points on $f(x) = \sin x$ have been divided by -4 (or multiplied by $-\frac{1}{4}$). The zeros of the function are not affected by the compression.

Multiplying by $-\frac{1}{4}$ has changed the amplitude and caused a reflection in the x-axis. The equation of the axis is still $y = 0$. The period is unchanged and is still $360°$. The domain is unchanged, $\{x \in \mathbf{R}\}$, but the range has changed to $\{y \in \mathbf{R} \mid -\frac{1}{4} \leq y \leq \frac{1}{4}\}$.

EXAMPLE **3**
Connecting the effect of a vertical stretch on the function $f(x) = \sin x$

How do the values 0.5 and 3 in the function $g(x) = 0.5 \sin x + 3$ change
$f(x) = \sin x$?

Anne's Solution

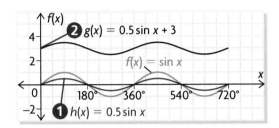

Multiplying by 0.5 has changed the amplitude, and adding
3 has changed the equation of the axis to $y = 3$. The period is
unchanged and is still $360°$. The domain is unchanged,
$\{x \in \mathbf{R}\}$, but the range has changed to $\{y \in \mathbf{R} \mid 2.5 \leq y \leq 3.5\}$.

I studied the equation and looked at each parameter, one
at a time.

❶ There's a vertical compression by a factor of 0.5
because $\sin x$ is multiplied by 0.5. Normally, my
amplitude is 1; now my amplitude is 0.5. This results
in the graph of $h(x) = 0.5 \sin x$ (in red).

❷ There's a vertical translation of 3 because 3 is added
to the height of each point, so I have to slide the
graph up 3 units. This results in the graph of
$g(x) = 0.5 \sin x + 3$ (in black).

My answer looks reasonable because the period of my
graph is still $360°$, the amplitude is 0.5, and the equation
of the axis is $y = 3$.

EXAMPLE **4**
Connecting the effect of a reflection on the function $f(x) = \sin x$

How do the values -2 and -3 in the function $g(x) = -2 \sin x - 3$ change
$f(x) = \sin x$?

Colin's Solution

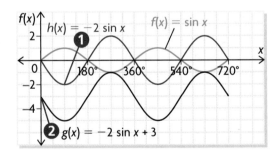

Multiplying by -2 has changed the amplitude and caused
a reflection in the x-axis. Subtracting 3 has changed the
equation of the axis to $y = -3$. The period is unchanged
and is still $360°$. The domain is unchanged, $\{x \in \mathbf{R}\}$, but
the range has changed to $\{y \in \mathbf{R} \mid -5 \leq y \leq -1\}$.

I studied the equation and looked at each parameter, one
at a time.

❶ The negative sign in front of the 2 tells me there is a
reflection in the x-axis, so my graph starts by going
down, rather than up. There is a vertical stretch of 2.
That means that my graph has an amplitude of 2.
This results in the graph of $h(x) = -2 \sin x$ (in red).

❷ There is also a vertical translation of -3, so I have to
slide the graph down 3 units. This results in the
graph of $g(x) = -2 \sin x - 3$ (in black).

I think my graph is correct. The period is $360°$, the
amplitude is 2, and the equation of the axis is $y = -3$.

EXAMPLE 5	Representing the equation of a transformed sine function

The green curve $f(x) = \sin x$ has been transformed to the black curve by means of three transformations. Identify them, and write the equation of the new function.

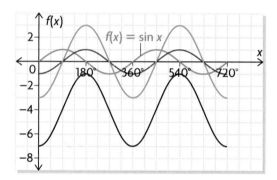

Ryan's Solution

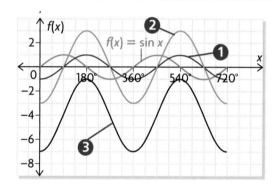

The equation for the black curve is
$f(x) = 3 \sin(x - 90°) - 4$.

I'm dealing with three transformations. There's a horizontal translation of 90° to the right (red curve), followed by a vertical stretch by a factor of 3 (blue curve), and then a vertical translation of 4 downward (black curve).

1 The equation for the red curve is
$f(x) = \sin(x - 90°)$,
since the green curve has been moved 90° to the right.

2 The equation for the blue curve is
$f(x) = 3 \sin(x - 90°)$,
since the red curve has been stretched by a factor of 3.

3 The equation for the black curve is
$f(x) = 3 \sin(x - 90°) - 4$,
since the blue curve has been shifted down 4 units.

In Summary

Key Idea

- The graph of the function $f(x) = a \sin(x - c°) + d$ looks periodic in the same way the graph of $f(x) = \sin x$ does. The differences are only in the placement of the graph and how stretched or compressed it is.

Need to Know

- If $f(x) = a \sin x$, the value of a has the following effect on the function $f(x) = \sin x$:
  - When $a > 1$, the function is stretched vertically by the factor a.
  - When $0 < a < 1$, the function is compressed vertically by the factor a.
  - When $a < -1$, the function is stretched vertically by the factor a and reflected across the x-axis.
  - When $-1 < a < 0$, the function is compressed vertically by the factor a and reflected across the x-axis.
- A function of the form $f(x) = a \sin(x - c°) + d$ results from applying transformations to the graph of $f(x) = \sin x$ in the following order:
  1. Horizontal translations: determined by the value of c
  2. Stretches/compressions: determined by the value of a;
     Reflections: necessary only when $a < 0$
  3. Vertical translations: determined by the value of d

CHECK *Your Understanding*

1. State the transformations that are applied to $f(x) = \sin x$.
   a) $f(x) = 3 \sin x$
   b) $f(x) = -2 \sin x$
   c) $f(x) = 0.1 \sin x$
   d) $f(x) = -\dfrac{1}{3} \sin x$

2. The graph of $f(x) = \sin x$ has been compressed by a factor of 5 and reflected in the x-axis. Write the new equation.

3. a) Sketch the sinusoidal function $f(x) = 4 \sin x$. Verify your answer with graphing technology.
   b) State the period, amplitude, and the equation of the axis.
   c) State the domain and range.

PRACTISING

4. State whether a vertical stretch or a compression results for each of the following.
   a) $f(x) = 2 \sin x$
   b) $f(x) = 0.5 \sin x$
   c) $f(x) = -\dfrac{1}{4} \sin x$
   d) $f(x) = -4 \sin x$
   e) $f(x) = \dfrac{1}{3} \sin x$
   f) $f(x) = 10 \sin x$

5. State the transformations for each sinusoidal function.
   a) $f(x) = 3 \sin (x + 20°)$ d) $f(x) = -2 \sin x + 6$
   b) $f(x) = -\sin x - 3$ e) $f(x) = -7 \sin (x + 10°)$
   c) $f(x) = 5 \sin(x - 50°) - 7$ f) $f(x) = -0.5 \sin (x - 30°) + 1$

6. For each of the functions in question 5:
   a) State the amplitude, period, and equation of the axis.
   b) State the domain and range.

7. For each graph, determine whether a horizontal translation, a vertical stretch and/or a reflection in the x-axis, or a vertical translation has occurred to the graph of $f(x) = \sin x$. Describe the effect of each transformation.

a)
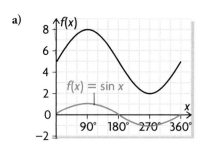

b)
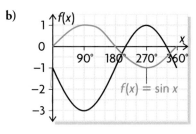

8. Complete the following statements: If a sinusoidal function is of the
   K form $f(x) = -a \sin (x - c°) + d$, then
   a) the negative sign in front of the a means that a ▒▒▒▒▒▒▒▒▒ has occurred
   b) the a means that a ▒▒▒▒▒▒▒▒ has occurred
   c) the c means that a ▒▒▒▒▒▒▒▒ has occurred
   d) the d means that a ▒▒▒▒▒▒▒▒ has occurred

9. Write the new equation for each function.
   a) The graph of $f(x) = \sin x$ has been translated to the right $135°$ and up 5 units.
   b) The graph of $f(x) = \sin x$ has been translated to the left $210°$ and down 7 units.

10. The function $f(x) = \sin x$ is stretched by a factor of 4 and then translated down 5 units.
    a) Sketch the resulting graph.
    b) What is the resulting equation of the function?

11. The function $f(x) = \sin x$ is reflected across the x-axis and translated up 4 units.
    a) Sketch the resulting graph.
    b) What is the resulting equation of the function?

12. Predict the maximum and the minimum values of $f(x)$ for each sinusoidal function. Verify your answers using graphing technology.

a) $f(x) = 3 \sin x$

b) $f(x) = -2 \sin x$

c) $f(x) = 4 \sin x + 6$

d) $f(x) = 0.5 \sin x - 3$

13. Sketch each sinusoidal function. Verify your answers using graphing technology.

a) $f(x) = 4 \sin x + 5$

b) $f(x) = -3 \sin x + 6$

c) $f(x) = -0.5 \sin x$

d) $f(x) = 2 \sin x - 3$

e) $f(x) = 3 \sin (x - 45°) + 2$

f) $f(x) = -4 \sin (x + 90°) - 5$

14. What characteristics (amplitude, period, or equation of the axis) do all the sinusoidal functions in question 13 have in common?

15. Suppose the height of a Ferris wheel is modelled by the function

A $h(\theta) = 8 \sin (\theta - 45°) + 10$, where $h(\theta)$ is the height in metres and θ is the number of degrees one has rotated from the boarding position.

a) What does the 10 mean in terms of a Ferris wheel?

b) What does the 10 mean in terms of transformations?

c) Sketch the function. Verify your answer using graphing technology.

d) Determine the range of the function.

e) What is the amplitude of the function, and what does it represent in this situation?

f) When an individual has rotated 200° from the boarding position, how high above the ground is the individual?

16. Draw a diagram of the Ferris wheel that could have formed the

T function $h(\theta) = -20 \sin (\theta + 90°) + 21$, where $h(\theta)$ is the height in metres and θ is the number of degrees the rider has rotated from the boarding position. Include all relevant measurements on your diagram, and indicate the direction the wheel is rotating.

17. a) Determine the equation of each function in the graph shown.

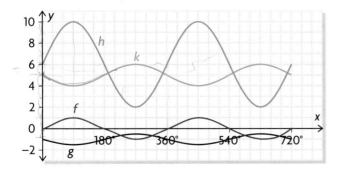

b) If the period of each function were changed from 360° to 180°, what type of transformation would you be dealing with?

18. Explain how you can graph a sinusoidal function that has undergone
C more than one transformation.

Extending

19. State the period of each function.

a) $f(x) = \sin\left(\frac{1}{2}x\right)$ c) $f(x) = \sin(2x)$

b) $f(x) = \sin\left(\frac{1}{4}x\right)$ d) $f(x) = \sin(10x)$

20. Don Quixote, a fictional character in a Spanish novel, attacked
windmills, thinking they were giants. At one point, he got snagged by
one of the blades and was hoisted into the air. The graph shows his
height above the ground in terms of time.

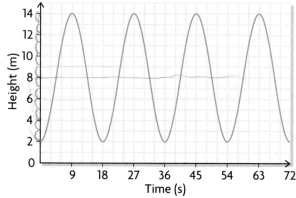

Don Quixote's Height above the Ground

a) What is the equation of the axis of the function, and what
does it represent in this situation?
b) What is the amplitude of the function, and what does it
represent in this situation?
c) What is the period of the function, and what does it
represent in this situation?
d) What transformation would generate the period for this
graph?
e) Determine the equation of the sinusoidal function.
f) If the wind speed decreased, how would that affect the graph
of the function?
g) If the axle for the windmill were 1 m higher, how would that
affect the graph of the function?

FREQUENTLY ASKED *Questions*

Q: How do you transform a sinusoidal function?

A:

Function	Type and Description of Transformations
$f(x) = \sin(x) + d$	The value d represents a vertical translation. If d is positive, then the graph shifts up the y-axis by the amount d. If d is negative, then the graph shifts down the y-axis by the amount d. This type of transformation will affect the equation of the axis and the range of the function.
$f(x) = \sin(x - c°)$	The value c represents a horizontal translation. If c is positive, then the graph shifts to the left on the x-axis by the amount c. If c is negative, then the graph shifts to the right on the x-axis by the amount c.
$f(x) = a\sin x$	The value a represents the vertical stretch/compression, which changes the amplitude of the sine function. If a is negative, it also represents a reflection of the function in the x-axis.

Study | **Aid**

• See Lesson 6.5, Examples 1, 2, and 3.
• See Lesson 6.6, Examples 1, 2, and 3.
• Try Chapter Review Questions 4, 5, and 9.

Q: How do I determine the domain and range of a sinusoidal function from its equation?

A: Sinusoidal functions cover all positive and negative values of x. Therefore, the domain of any sinusoidal function is $\{x \in \mathbf{R}\}$. To determine the range of a sinusoidal function, you must calculate the equation of the axis, based on the vertical translation, and then the amplitude, based on the vertical stretch. Determine the equation of the axis, and then go above and below that value an amount equivalent to the amplitude. For example, if the equation of the axis is $y = 7$, and the amplitude is 3, then you would go 3 units above and below 7. That would mean that the maximum value for y is 10 $(7 + 3)$, and the minimum value for y is 4 $(7 - 3)$.

Study | **Aid**

• See Lesson 6.5, Example 1.
• Try Chapter Review Questions 6, 11, and 13.

PRACTICE Questions

Lesson 6.2

1. Which of the following situations would produce periodic graphs?
   a) You are looking at a lighthouse.
      - independent variable: time
      - dependent variable: the intensity of the light from your perspective
   b) You are going to a ride a Ferris wheel.
      - independent variable: time
      - dependent variable: your height above the ground
   c) Travis has a floating dock on an ocean.
      - independent variable: time
      - dependent variable: the height of the dock relative to the ocean floor
   d) You are watching a train that is travelling at a constant speed around the track shown.
      - independent variable: time
      - dependent variable: the distance between you and the train

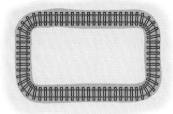

Lesson 6.3

2. A pendulum is swinging back and forth. However, the motion of the pendulum stops at regular intervals. The distance in terms of time between the pendulum and an interior wall of the device can be represented by this graph.

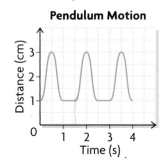

Pendulum Motion

a) Is this function periodic, sinusoidal, or both?
b) How long is the pendulum at rest at any one time?
c) What is the farthest distance the pendulum will be from the interior wall of the device?
d) What is the period of the function? Include the units of measure.
e) If the pendulum was momentarily stopped for a longer period, how would the graph change?

3. A machine used to remove asphalt from roads has a large rotating drum covered with teeth. The height, $f(t)$, in terms of time of one of these teeth relative to the ground can be modelled by the function $f(t) = 35 \sin (360t)° + 30$. The height is measured in centimetres, and time is measured in seconds. Using graphing technology with the WINDOW settings shown, answer the following questions.

a) What is the minimum height of the tooth? Does this number make sense? Explain.
b) What is the equation of the axis, and what does it represent in this situation?
c) What is the period of the function, and what does it represent in this situation? (*Hint:* The period of this function is going to be quite short.)
d) What is the amplitude of the function, and what does it represent in this situation?

4. Sketch three cycles of a sinusoidal function that has a period of $360°$, an amplitude of 3, and whose equation of the axis is $y = -1$.

5. Sketch two cycles of a sinusoidal function that has a period of $360°$ and an amplitude of 2 and whose equation of the axis is $y = 3$.

Lesson 6.4

6. Two different engines idling at different speeds are vibrating on their engine mounts. The distance, $d(t)$, in terms of time, t, that the engines vibrate to the left and right of their resting position can be modelled by the graphs shown.

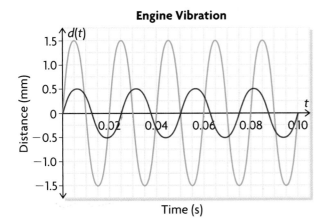

Engine Vibration

a) Determine the range of each function.
b) What is the period of each function? What does the period tell you about the engines?
c) What is the equation of the axis for each function, and what does it represent in this situation?
d) What is the amplitude of each function? What does this tell you about the engines?

Lesson 6.5

7. State the amplitude, period, equation of the axis, and maximum and minimum values of $f(x)$ for each sinusoidal function.
   a) $f(x) = \sin x + 3$
   b) $f(x) = \sin(x + 60°) - 2$

8. Sketch each sinusoidal function in question 7, and state the domain and range. Verify your answers using graphing technology.

Lesson 6.6

9. a) The function $f(x) = \sin x$ undergoes a reflection in the x-axis, a vertical stretch of 0.5, and a vertical translation of 4. What is the equation of the resulting function?
   b) Sketch the resulting graph.

10. State the amplitude, period, equation of the axis, and maximum and minimum values of $f(x)$ for each sinusoidal function. Verify your answers using graphing technology.
    a) $f(x) = 3 \sin x$
    b) $f(x) = -2 \sin x$
    c) $f(x) = 4 \sin x + 6$
    d) $f(x) = -0.25 \sin x$
    e) $f(x) = 3 \sin(x + 45°)$

11. Sketch each sinusoidal function, and state the domain and range. Verify your answers using graphing technology.
    a) $f(x) = 3 \sin x + 5$
    b) $f(x) = -\sin x - 2$
    c) $f(x) = 0.5 \sin x - 1$
    d) $f(x) = 2 \sin(x - 90°)$
    e) $f(x) = -0.5 \sin(x + 90°) + 3$

12. The height of a Ferris wheel is modelled by the function $h(\theta) = 6 \sin(\theta - 45°) + 7$, where $h(\theta)$ is in metres and θ is the number of degrees the wheel has rotated from the boarding position of a rider.
    a) Sketch the graph of the function, and verify your answer using graphing technology.
    b) Determine the range of the function.
    c) What is the amplitude of the function, and what does it represent in this situation?
    d) When the rider has rotated 400° from the boarding position, how high above the ground is the rider?

1. Graph each sinusoidal function. If you are using a graphing calculator, set it to DEGREE mode and use the WINDOW settings shown at the left. State the amplitude, period, equation of the axis, domain, and range for each function.

a) $f(x) = \sin x + 4$

b) $f(x) = \sin(x + 30°)$

c) $f(x) = 2 \sin x$

d) $f(x) = -0.5 \sin x$

2. Sketch three cycles of a sinusoidal function that has a period of 30, an amplitude of 6, and whose axis is $y = 11$.

3. The hands on a wall clock move in a predictable manner. As time passes, the distance between the tip of the minute hand and the ceiling changes. Suppose we have two different wall clocks. The two graphs model the relationship between distance and time.

a) Determine each of the following and explain what they represent in this situation: period, equation of the axis, and amplitude.

b) What was the initial position of the tip of the minute hand?

c) What is the range for each function?

d) Approximate the distance between the minute hand and the ceiling for each clock at $t = 80$ min.

e) Draw a diagram showing the position of the clocks relative to the ceiling. Include all relevant numbers on your diagram.

Distance from Ceiling

4. The function $f(x) = \sin x$ undergoes a reflection in the x-axis and a vertical stretch of 2.

a) Write the resulting equation of the function.

b) Sketch the resulting graph.

c) State the amplitude, equation of the axis, period, domain, and range.

5. Sketch the sinusoidal function $f(x) = \sin(x + 90°) + 5$. Verify your answer using graphing technology.

6. Determine the equation of each function in the graph shown.

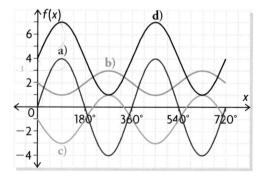

Designing a Ferris Wheel

The London Eye, the world's largest Ferris wheel, is located in London, England. Completed in March 2000, it stands 135 m high and has 32 passenger capsules. Each capsule weighs 10 tonnes, carries 25 passengers, and takes 30 min to complete one revolution.

In 2008, Shanghai, China, expects to complete an even larger wheel. The wheel will have a diameter of 170 m and sit atop a 50 m entertainment complex. Each of the 36 passenger capsules will carry 30 passengers. Like the London Eye, it will take 30 min to complete one revolution.

? **How can you design a Ferris wheel larger than the Shanghai wheel?**

A. Draw a scale diagram of your Ferris wheel beside a scale diagram of the London Eye and the Shanghai wheel.

B. On the same coordinate system, draw three graphs modelling the height of a passenger in terms of time for all three wheels.

C. Determine the speed that a passenger would travel around your wheel in metres per minute.

D. Determine the period, amplitude, and equation of the axis for each graph.

E. Determine the range for each graph.

F. Explain the process you used in designing your wheel.

Task | Checklist

✔ Do the numbers that you are using seem reasonable?

✔ Is your diagram of the Ferris wheels drawn to scale?

✔ Are the patrons riding the Ferris wheel travelling at a reasonable speed?

✔ Have you shown all necessary calculations and provided written explanations so that your teacher can understand your reasoning?

Multiple Choice

1. A T-ball player hits a baseball from a tee that is 1 m tall. The flight of the ball can be modelled by $h(t) = -5t^2 + 10t - 1$, where $h(t)$ is the height in metres and t is the time in seconds. When does the ball reach its maximum height?
   - a) 0.5 s
   - b) 1.00 s
   - c) 1.60 s
   - d) 1.5 s

2. A rock is dropped from the edge of a 180 m cliff. The function $h(t) = -5t^2 - 5t + 180$ gives the approximate height of the rock, $h(t)$, in metres t seconds after it was released. How long does it take for the rock to reach a ledge 80 m above the base of the cliff?
   - a) 5 s
   - b) 6 s
   - c) 3 s
   - d) 4 s

3. The function $f(x) = -5x^2 + 20x + 2$ in vertex form is
   - a) $f(x) = -5(x - 2)^2 + 18$
   - b) $f(x) = -5(x - 2)^2 + 22$
   - c) $f(x) = -5(x + 2)^2 - 22$
   - d) $f(x) = -5(x + 2)^2 - 18$

4. Which of the following quadratic equations has no solution?
   - a) $2x^2 - 4x = 2x - 3$
   - b) $2x^2 - 15x - 8 = 0$
   - c) $16(x + 1)^2 = 0$
   - d) $3(x + 5)^2 + 7 = 0$

5. Identify which parabola does not intersect the x-axis.
   - a) $f(x) = x^2 - 6x + 7$
   - b) $f(x) = 9 - x^2$
   - c) $f(x) = (4 + x)^2$
   - d) $f(x) = -2(x - 1)^2 - 1$

6. Two support wires are fastened to the top of a TV satellite dish tower from two points on the ground, A and B, on either side of the tower. One wire is 18 m long, and the other is 12 m long. The angle of elevation of the longer wire is 28°. How tall is the satellite dish tower?
   - a) 24.41 m
   - b) 18.2 m
   - c) 6 m
   - d) 8.45 m

7. Two airplanes leave the same airport in opposite directions. At 2:00 p.m., the angle of elevation from the airport to the first plane is 48° and to the second plane 59°. The elevation of the first plane is 5.5 km, and the elevation of the second plane is 7.2 km. Determine the air distance between the two airplanes to the nearest tenth of a kilometre.
   - a) 9.4 km
   - b) 8.5 km
   - c) 15.7 km
   - d) 7.8 km

8. Determine the length of AB to the nearest metre.

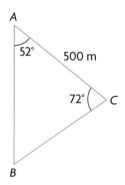

   - a) 524 m
   - b) 544 m
   - c) 574 m
   - d) 564 m

9. Use this diagram to determine the height, h, of the mountain.

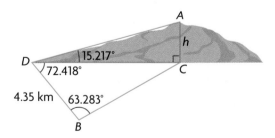

   - a) 2.5 km
   - b) 1.98 km
   - c) 1.5 km
   - d) 1.87 km

10. Which function is both periodic and sinusoidal?

a)

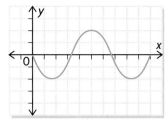

b)

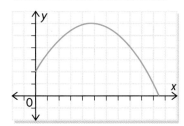

c)

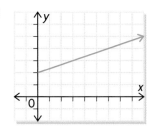

d)

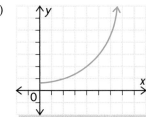

a) (b)   c) (b) and (c)

b) (a) and (d)   d) (a)

11. Identify the amplitude and equation of the axis of $f(x) = 2 \sin x + 5$.

a) amplitude: 2; equation of axis: $y = 5$

b) amplitude: 5; equation of axis: $y = 2$

c) amplitude: 2; equation of axis: $y = -5$

d) amplitude: 5; equation of axis: $y = 2$

12. What is the range of $f(x) = -4 \sin x - 2$?

a) $\{y \in \mathbf{R} \mid -6 \leq y \leq 2\}$

b) $\{y \in \mathbf{R} \mid -2 \leq y \leq 6\}$

c) $\{y \in \mathbf{R} \mid -4 \leq y \leq 4\}$

d) $\{y \in \mathbf{R} \mid -4 \leq y \leq 2\}$

13. Identify the correct amplitude and period.

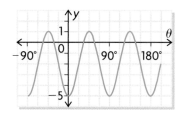

a) amplitude: 1; period: 360°

b) amplitude: 3; period: 90°

c) amplitude: −5; period: 180°

d) amplitude: 3; period: 180°

14. Identify the transformations you would apply to $f(x) = \sin x$ to graph $f(x) = 0.5 \sin(x - 30°)$.

a) vertical stretch by 0.5, shift left 30°

b) vertical compression by 0.5, shift left 30°

c) vertical stretch by 0.5, shift right 30°

d) vertical compression by 0.5, shift right 30°

15. Identify the transformations you would apply to $f(x) = \sin x$ to graph $f(x) = -\sin x + 3$.

a) reflection in the x-axis, shift down 3

b) reflection in the y-axis, shift up 3

c) reflection in the x-axis, shift up 3

d) reflection in the y-axis, shift down 3

16. The function $f(x) = -2(x - 3)^2 + 5$ in standard form is

a) $f(x) = -2x^2 + 12x - 10$

b) $f(x) = -2x^2 + 6x + 10$

c) $f(x) = -2x^2 + 12x + 18$

d) $f(x) = -2x^2 + 12x - 13$

17. For the parabola defined by $f(x) = -3(x + 1)^2 - 4$, which of the following statements is not true?

a) The vertex is $(-1, -4)$.

b) The axis of symmetry is $x = 1$.

c) The parabola opens down.

d) The domain is $\{x \in \mathbf{R}\}$.

18. Given $y = -x^2 + 12x - 16$, state the coordinates of the vertex and the maximum or minimum value of y.
    a) vertex $(6, 20)$, maximum 20
    b) vertex $(-6, 20)$, minimum -6
    c) vertex $(6, 20)$, minimum 6
    d) vertex $(6, -20)$, maximum -20

19. The profit function for a new product is given by $P(x) = -4x^2 + 28x - 40$, where x is the number sold in thousands. How many items must be sold for the company to break even?
    a) 2000 or 5000   c) 5000 or 7000
    b) 2000 or 3500   d) 3500 or 7000

20. Which of the following statements is not true for the equation of a quadratic function?
    a) In standard form, the y-intercept is clearly visible.
    b) In vertex form, the break-even points are clearly visible.
    c) In factored form, the x-intercepts are clearly visible.
    d) In vertex form, the coordinates of the vertex are clearly visible.

21. Which of the following is not a step required to complete the square for $y = 7x^2 + 21x - 2$?
    a) $7(x^2 + 3x) - 2$
    b) $7x(x + 3) - 2$
    c) $7\left(x^2 + 3x + \dfrac{9}{4}\right) - \dfrac{71}{4}$
    d) $7\left(x^2 + 3x + \dfrac{9}{4} - \dfrac{9}{4}\right) - 2$

22. Which of the following statements is not true for a given quadratic function?
    a) The y-coordinate of the vertex represents the minimum or maximum value.
    b) The axis of symmetry is given by the x-coordinate of the vertex.
    c) The axis of symmetry is given by the y-coordinate of the vertex.
    d) The midpoint between the x-intercepts is the x-coordinate of the vertex.

23. A quadratic function in standard form will have two distinct real roots when
    a) $b^2 - 4ac < 0$
    b) $a^2 - 4bc > 0$
    c) $b^2 = 4ac$
    d) $b^2 - 4ac > 0$

24. Which value of k will produce one root for $y = -2(x + 7)^2 - k$?
    a) $k = 1$ c) $k = -1$
    b) $k = 0$ d) $k = -2$

25. The period of a periodic graph is
    a) the length of one cycle
    b) the distance from the maximum to the minimum value of the relation
    c) the same as the domain
    d) the same as the range

26. The equation of the axis of a periodic graph is
    a) $y = b$
    b) $y = mx + b$
    c) $y = \dfrac{\text{maximum value} + \text{minimum value}}{2}$
    d) $y = \dfrac{\text{maximum value} - \text{minimum value}}{2}$

27. Parallelogram $ABCD$ has sides of length 35 cm and 27 cm. The contained angle is 130°. The length of the longer diagonal is
    a) 27.2 cm   c) 22.5 cm
    b) 56.3 cm   d) 20.7 cm

28. In $\triangle ABC$, $\angle A = 85°$, $c = 10$ cm, and $b = 15$ cm. The height of $\triangle ABC$ is
    a) 17.3 cm   c) 13.8 cm
    b) 8.6 cm   d) 12.5 cm

29. In $\triangle PQR$, $\angle P = 70°$, $r = 5$ cm, and $q = 8$ cm. The area of $\triangle PQR$ is
    a) 13 cm^2 c) 18.8 cm^2
    b) 20 cm^2 d) 19.2 cm^2

Investigations

30. Designing a Football Field

Have you ever walked on a football field covered with artificial turf? If so, you probably noticed that the field is not flat. The profile of the surface is arched and highest in the centre, permitting rainwater to drain away quickly.

45.75 cm

50 m

a) The diagram shows the profile of an actual field, viewed from the end of the field. Assuming that the cross-section is a parabola, determine the algebraic model that describes this shape.

b) Use your equation to determine the distance from the sidelines where the field surface is 20 cm above the base line.

31. How High Is the Tower?

A skier sees the top of a communications tower, due south of him, at an angle of elevation of 32°. He then skis on a bearing of 130° for 560 m and finds himself due east of the tower. Calculate

a) the height of the communications tower

b) the distance from the tower to the skier

c) the angle of elevation of the top of the tower from the new position

32. Amusement Rides

A popular ride at an amusement park is called the "Ring of Terror." It is like a Ferris wheel but is inside a haunted house. Riders board on a platform that is level with the centre of the "ring," and the ring moves counterclockwise. When a rider is moving above the platform, he or she meets flying creatures. When the seat descends to a level below the platform, creatures emerge from a murky, slimy pit. The radius of the ring is 6 m.

a) Graph the height of a rider with respect to the platform through three revolutions of the ride.

b) Determine the amplitude, period, equation of the axis, and the range of your graph.

c) Discuss how your graph would change if the rider got on the ride in the pit, at the bottom of the ring.

7

Exponential Functions

▶ **GOALS**

You will be able to

- Investigate the characteristics of exponential functions and their graphs
- Compare exponential functions with other familiar functions
- Work with integer and rational exponents
- Solve problems that involve exponential growth and decay

❓ Suppose you are photographing animal life in and around a coral reef. How does the available light change as you descend into the ocean?

WORDS You Need to Know

1. Match each term with its picture or example.

a) circumference   **c)** exponent   **e)** power   **g)** increasing function

b) surface area   **d)** base   **f)** rational numbers   **h)** decreasing function

i) 2^5

ii)

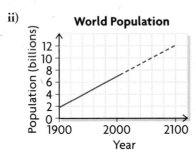

World Population

iii) 8^2

iv) $-\dfrac{3}{4}$

v) 8^2

vi)

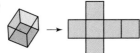

vii) Transmittance vs. Wavelength

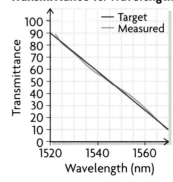

viii)

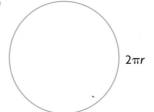

$2\pi r$

SKILLS AND CONCEPTS You Need

Square Root and Cube Root Estimates

The square root of a number is the number that multiplies by itself to give the required value. The square root symbol is $\sqrt{\ }$. The cube root of a number is the number that multiplies itself three times to give a required value. The cube root symbol is $\sqrt[3]{\ }$.

EXAMPLES

a) $\sqrt{36} = 6$ since $6 \times 6 = 36$

b) $\sqrt{0.25} = 0.5$ since $0.5 \times 0.5 = 0.25$

c) $\sqrt[3]{125} = 5$ since $5 \times 5 \times 5 = 125$

d) $\sqrt[3]{64} = 4$ since $4 \times 4 \times 4 = 64$

Estimating Square Roots and Cube Roots

EXAMPLE

Is the square root of 17 closer to 5 or to 4?

Solution: $5 \times 5 = 25$ and
$4 \times 4 = 16$

17 is closer to 16 than to 25, so $\sqrt{17}$ is closer to 4.

EXAMPLE

Is the cube root of 110 closer to 4 or 5?

Solution: $4 \times 4 \times 4 = 64$ and
$5 \times 5 \times 5 = 125$

110 is closer to 125 than to 64, so $\sqrt[3]{110}$ is closer to 5.

2. Select the best answer from those given.

   a) Which is the better estimate of $\sqrt{34}$?   5.2   5.9

   b) Which is the better estimate of $\sqrt{50}$?   7.1   7.8

   c) Which is the better estimate of $\sqrt[3]{109}$?   4.4   4.8

   d) Which is the better estimate of $\sqrt[3]{300}$?   6.4   6.7

Evaluating Powers

5^2 is called a power. 5 is the base and 2 is the exponent.
$$5^2 = 5 \times 5 \quad \text{(expanded form)}$$
$$= 25$$

Study | *Aid*
• For help, see Essential Skills
  Appendix, A-3.

Exponent Laws

Rule	Written Description	Algebraic Description	Worked Example in Standard Form
Multiplication	To multiply powers with the same base, keep the base the same and add the exponents.	$b^m \times b^n = b^{m+n}$	$3^2 \times 3^3 = 3^{2+3}$ $= 3^5$ $= 243$
Division	To divide powers with the same base, keep the base the same and subtract the exponents.	$b^m \div b^n = b^{m-n}$	$4^5 \times 4^2 = 4^{5-2}$ $= 4^3$ $= 64$
Power of a Power	To simplify a power of a power, keep the base the same and multiply the exponents.	$(b^m)^n = b^{m \times n}$	$(2^3)^2 = 2^{3 \times 2}$ $= 2^6$ $= 64$
Power of a Product	To simplify a power of a product, apply the exponent to both numbers in the product.	$(a \times b)^m = a^m \times b^m$	$(5 \times 3)^2 = 5^2 \times 3^2$ $= 25 \times 9$ $= 225$
Power of a Quotient	To simplify a power of a quotient, apply the exponent to both numbers in the quotient.	$\left(\dfrac{a}{b}\right)^m = \dfrac{a^m}{b^m}$	$\left(\dfrac{7}{3}\right)^2 = \dfrac{7^2}{3^2}$ $= \dfrac{49}{9}$

3. Write each of the following in expanded form, and then evaluate.

   a) 5^4 b) $(-4)^3$ c) -2^2 d) $(5 \times 2)^3$ e) $\left(\dfrac{3}{4}\right)^2$

PRACTICE

Evaluate questions 4 to 8 without using a calculator.

4. Evaluate.

a) 8^3

b) 11^4

c) 5^6

d) 19^2

e) 4^5

f) 2^{10}

Study **Aid**

• For help, see Essential Skills Appendix.

Question	Appendix
5, 6, 7, and 9	A-3
10	A-2
11	A-12

5. Evaluate.

a) $(-5)^2$

b) -5^2

c) $(-2)^3$

d) -2^3

e) $(-10)^4$

f) -10^4

6. Evaluate.

a) $(-3^3)^3$

b) $[(-3)^3]^3$

c) $[(-3)^4]^2$

d) $(-3^4)^2$

e) $(-3^3)^2$

f) $(-3^2)^3$

7. Evaluate.

a) $3^2 - 4^2$

b) $10^2 - 15^1 + 5^2$

c) $(1 + 7^2)^2$

d) $(6^2 - 4^2)^2$

e) $5^2 \times (-2)^3$

f) $8^2 \div (-4)^3$

8. Evaluate.

a) $\sqrt{25} + \sqrt{16}$

b) $\dfrac{\sqrt{100}}{\sqrt{25}}$

c) $\sqrt{\sqrt{81}}$

9. Determine the exponent that makes each of the following true.

a) $2^x = 16$

b) $17^m = 17$

c) $3^y = 27$

d) $4^x = 64$

e) $(-2)^n = -8$

f) $5^c = 125$

10. Evaluate.

a) $\dfrac{4}{7} - \dfrac{3}{4}$

b) $\dfrac{7}{9} \div \dfrac{4}{5}$

c) $\dfrac{2}{3}\left(\dfrac{5}{4}\right)$

d) $\dfrac{2}{3} + \left(\dfrac{5}{4}\right)$

e) $\dfrac{4}{9}\left(\dfrac{9}{5} - \dfrac{3}{10}\right)$

f) $\left(\dfrac{9}{10}\right)\dfrac{3}{7} \div \dfrac{3}{14}$

11. Determine the first and second finite differences for each set of data. State whether each set represents a linear or a quadratic relationship.

a)

x	y
−3	14
−2	10
−1	6
0	2
1	−2
2	−6
3	−10

b)

x	y
−3	11.5
−2	6.5
−1	3.5
0	2.5
1	3.5
2	6.5
3	11.5

c)

x	y
−6	15
−4	−3
−2	−13
0	−15
2	−9
4	5
6	27

APPLYING *What You Know*

Comparing Soccer Ball Sizes

Soccer is an old game played worldwide. Early soccer balls were made from pig bladders that varied in size.

Today, the Fédération Internationale de Football Association, or FIFA, determines the "qualities and measurements" of the ball.

According to FIFA, the ball must
• be spherical
• be made of leather or other suitable material
• have a circumference that is 68 cm – 70 cm
• have a mass of 410 g – 450 g
• have a pressure of 600 g/cm² – 1100 g/cm²

? **How do changes in the circumference of the ball affect its volume and surface area?**

A. Use the minimum circumference of 68 cm. Determine the radius of the ball to the nearest tenth of a centimetre where $\pi \doteq 3.14$.

B. Repeat part A for the maximum circumference of 70 cm.

C. Use each radius from parts A and B to calculate the volume of each sphere to the nearest tenth of a cubic centimetre.

D. Calculate the difference in volume of the two spheres.

E. Use each radius from parts A and B to calculate the surface area of each sphere to the nearest tenth of a square centimetre.

F. Calculate the difference in surface area of the two spheres.

G. Is the difference in surface area greater than the difference in volume? Explain how you made your comparison.

> Communication | *Tip*
> $C(r) = 2\pi r$
> $V(r) = \frac{4}{3}\pi r^3$
> $A(r) = 4\pi r^2$

Collecting Exponential Data

YOU WILL NEED

- some water and a way to heat it (a microwave oven or a kettle)
- thermometer (a temperature probe if linked to a graphing calculator)
- stopwatch or clock with a second hand
- graph paper (if you are not using a graphing calculator)
- Styrofoam or ceramic cup

GOAL

Collect data that lead to an exponential relationship and study its characteristics.

INVESTIGATE the Math

Hot objects left to cool have temperatures that drop quickly over time. A hot cup of tea is left on a teacher's desk as he is called away unexpectedly. He returns to find that his tea is lukewarm.

? How long does it take for a hot liquid to cool to room temperature?

A. Predict how long it will take for the liquid to cool. Sketch a graph of what you think the cooling curve of temperature versus time might look like.

B. Fill a Styrofoam or ceramic cup with a hot liquid.

C. Record the room temperature with a thermometer (or temperature probe).

D. Place the thermometer (or temperature probe) into the cup of hot liquid and wait until it gives a steady reading. Record the temperature in Trial 0 in a table like the one shown. Then take about 20 readings, one every minute, and record the temperatures in your table.

Trial	Time after Initial Reading (min)	Temperature (°C)
0	0	
1	1	
2	2	
3	3	

E. Using time as the independent variable, draw a scatter plot of the data. Draw a smooth curve that you think fits the data.

F. Describe the shape of the curve from left to right.

G. On your graph, draw a horizontal line that represents the room temperature you recorded. Use your curve to estimate the time it will take for the liquid to cool to room temperature.

Reflecting

H. Did the temperature fall at a steady rate? Describe how it fell.

I. Is this relation a function? Explain.

J. State the domain and range of the relation. Explain how they relate to the changes in time and temperature.

In Summary

Key Ideas

- A scatter plot of data that appears to have a rapidly increasing or decreasing nonlinear pattern can be modelled by an exponential function.
- The range of an exponential function with a positive base always has a lower limit.
- The domain of the exponential model may need to be restricted for the situation you are dealing with.

Need to Know

- Situations that show increases in the value of a function, where the increases grow larger with time in a predictable way, can be modelled by exponential functions. This is called *exponential growth*.

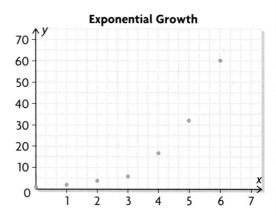

- Situations that show decreases in the value of a function, where the decreases grow smaller with time in a predictable way, can be modelled by exponential functions. This is called *exponential decay*.

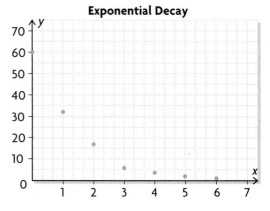

FURTHER *Your Understanding*

1. What do you think would happen to the curve if you took many more measurements beyond the 20 min? Explain your reasoning.

2. Describe how you think the shape of the curve would change if
   - the initial water temperature were higher
   - the room temperature were much lower

   Sketch both situations on the same set of axes, along with the curve of your estimated data, to illustrate your thinking.

3. Marnie's grandmother won the "Really Big Lottery." Her prize is an initial payment of $5000 and a payment on her birthday for the next 25 years. Each year the payment doubles in size.

Year	Annual Payment ($1000)
1	5
2	10
3	20
4	
5	
⋮	
10	

a) Complete the table of values for the first 10 years.
b) Create a scatter plot of your data and draw a curve of best fit.
c) Compare your payment curve with the cooling curve. Discuss the similarities and the differences.

7.2 The Laws of Exponents

GOAL

Investigate the rules for simplifying numerical expressions involving products, quotients, and powers of powers.

YOU WILL NEED
- graphing calculator

INVESTIGATE the Math

The Senior Girls hockey team is due back from a tournament, but a snowstorm delays their flight. The principal calls the parents of two girls to inform them of the delay. The parents call two other households to tell them the news. Each of these households call two other households, and the pattern continues.

The number of people who are contacted in each round of calls is summarized in the table.

Household C

Household A

Household D

Principal

Household E

Household B

Household F

Round of Calls	Number of People Contacted	Power of Two
0	1	
1	2	
2	4	2^2
3	8	2^3
4	16	
5	32	
6	64	

? What are some of the relationships between powers of two in the table?

A. Copy and complete the table.

B. Select any two numbers (other than the last two) from the second column of the table. Find their product. Write the two numbers and their product as powers of two.

C. Repeat part B for two other numbers.

D. What is the relationship between the exponents of the powers that you multiplied and the exponent of the resulting power?

E. Select any two numbers (other than the first two) from the second column. Divide the greater value by the lesser one. Write the two numbers and their quotient as powers of two.

F. Repeat part E using two other numbers.

G. What is the relationship between the exponents of the powers that you divided and the exponent of the quotient?

Reflecting

H. Suggest a rule for multiplying powers with the *same* base. Create an example that shows this rule.

I. Suggest a rule for dividing powers with the *same* base. Create an example that shows this rule.

APPLY the Math

EXAMPLE 1	Connecting a power of a power to multiplication

Evaluate $(5^2)^4$.

Dylan's Solution

$(5^2)^4 = (5^2)(5^2)(5^2)(5^2)$ ← $(5^2)^4$ means four "5^2"s multiplied together.

$= 5^{2+2+2+2}$ ← Since the four powers of 5 are multiplied together and all have the same base, I added the exponents.

$= 5^8$

$(5^2)^4 = 5^{2 \times 4} = 5^8$ ← This leads to the answer. I noticed that I could get the same result if I had multiplied the exponents in the original question.

If a numerical expression contains several operations, evaluate by following the order of operations.

| EXAMPLE **2** | Selecting a strategy to evaluate an expression involving powers |

Evaluate $3^4(3^8) \div 3^7$.

Lesley's Solution: Applying Exponent Rules

$3^4(3^8) \div 3^7 = 3^{4+8} \div 3^7$

$\qquad = 3^{12} \div 3^7$

$\qquad = 3^{12-7}$

$\qquad = 3^5$

$\qquad = 243$

Since all of the powers in the expression have the same base, I followed the order of operations. I used exponent rules to multiply the first two powers and then divide the result.

I added the exponents for the first two powers and then subtracted the third power's exponent from the result.

Graphing calculators perform calculations following the order of operations. You can enter the expression as it appears and the calculator will determine the result.

Meredith's Solution: Using a Calculator

Evaluate $3^4(3^8) \div 3^7$.

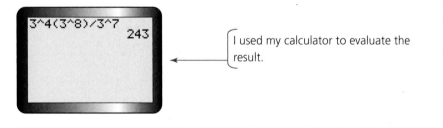

I used my calculator to evaluate the result.

Tech | *Support*

Most scientific calculators have an exponent key that can be used to evaluate powers, such as x^y. Check your calculator's manual to help you identify this key. For help using a graphing calculator to evaluate powers, see Technical Appendix, B-14.

Evaluate $\dfrac{6(6^7)}{(6^3)^2}$.

Tom's Solution: Applying the Exponent Rules

$$\frac{6(6^7)}{(6^3)^2} = \frac{6^{1+7}}{6^{3\times 2}}$$

I evaluated the numerator and denominator separately by using the exponent rule for multiplication in the numerator and the power rule in the denominator.

$$= \frac{6^8}{6^6}$$

$$= 6^{8-6}$$

Then I divided the numerator by the denominator. Since the bases are the same, I subtracted the exponents.

$$= 6^2$$

$$= 36$$

Bronwyn's Solution: Using a Calculator

Evaluate $\dfrac{6(6^7)}{(6^3)^2}$.

```
(6(6^7))/((6²)^3
)
              36
```

I entered the expression with a set of brackets around the numerator and a different set around the denominator. I did this to make sure that the calculator evaluated the numerator and denominator separately before dividing.

In Summary

Key Ideas

- To multiply powers with the *same* base, add the exponents.

$$b^n \times b^m = b^{n+m}$$

- To divide powers with the *same* base, subtract the exponents.

$$b^n \div b^m = b^{n-m}$$

- To simplify a power of a power, multiply the exponents.

$$(b^n)^m = b^{n \times m}$$

Need to Know

- Use the order of operations to simplify expressions involving powers. If there are powers of powers in an expression, simplify them first. Then simplify the multiplications and divisions in order from left to right. For example,

$$
\begin{aligned}
(2^3)^2 \times 2^5 \div 2^7 &= 2^{3 \times 2} \times 2^5 \div 2^7 \\
&= 2^6 \times 2^5 \div 2^7 \\
&= 2^{6+5-7} \\
&= 2^4 \\
&= 16
\end{aligned}
$$

- In a rational expression, simplify the numerator and denominator first. Then divide the numerator by the denominator.
- Remember, these rules apply only to powers that have the same base (e.g., $2^3 \times 2^5 = 2^8$).
- When evaluating powers, it is common practice to write answers as rational numbers.

CHECK *Your Understanding*

1. Write each expression as a single power.
   a) $4^3 \times 4^5$
   b) $13^3 \times 13^{11}$
   c) $(7^2)(7)$
   d) $(12^3)(12^3)(12^3)$

2. Write each expression as a single power.
   a) $7^4 \div 7^3$
   b) $6^{16} \div 6^{11}$
   c) $\dfrac{9^3}{9}$
   d) $\dfrac{5^6}{5^3}$

3. Write each expression as a single power.
   a) $(2^3)^4$
   b) $(12^3)^3$
   c) $(10^7)^4$
   d) $((3^2)^3)^4$

PRACTISING

4. Create three examples to help a classmate learn about the following
C relationships:
- **a)** the result of multiplying powers with the same base
- **b)** the result of dividing powers with the same base
- **c)** the result of raising a power to an exponent

5. Simplify. Write as a single power.
K

a) $3(3^5) \div 3^3$

b) $10^9 \div (10^3 \times 10^2)$

c) $(7^8 \div 7^5)(7^2)$

d) $\dfrac{(6^2)(6^{11})}{6^8}$

e) $\dfrac{9^{12}}{9(9^{10})}$

f) $\dfrac{(8^7)(8^3)}{8^6(8^2)}$

6. Simplify. Write as a single power.

a) $(2^5)^3 \times 2^3$

b) $5^9 \div (5^3)^2$

c) $(7^8)(7^5)^2$

d) $\dfrac{(8^2)^5}{8^8}$

e) $\dfrac{10(10^9)}{(10^2)^3}$

f) $\dfrac{(4^7)^3}{4^9(4^{11})}$

7. Simplify. Write as a single power.

a) $10(10^5)(10^3) \div (10^3)^2$

b) $\dfrac{(8^8)(8^3)^3}{8^3(8^{11})}$

c) $\left(\dfrac{13(13^{12})}{13^7}\right)^2$

d) $\dfrac{(5^4)^2(5^5)^2}{5^2(5^{13})}$

8. Simplify, then evaluate without using a calculator.

a) $\left(\dfrac{4}{3}\right)\left(\dfrac{4}{3}\right)^2$

b) $\left(\dfrac{1}{9}\right)^4 \div \left(\dfrac{1}{9}\right)^2$

c) $\left(\left(\dfrac{2}{5}\right)^2\right)^2$

d) $\left(\dfrac{5}{4}\right)^5\left(\dfrac{5}{4}\right)^3 \div \left(\dfrac{5}{4}\right)^6$

9. Simplify.

a) $x^4(x^2)^2$

b) $\dfrac{(m^5)^2}{m^8}$

c) $(y(y^6))^3$

d) $((a^2)^2)^2$

e) $a^2\, a^2\, a^2$

f) $\dfrac{b(b^5)b^4}{b^5}$

10. Write each power in simplified form.

a) 4^3 as a base 2 power

b) 9^5 as a base 3 power

c) 27^5 as a base 3 power

d) $(-8)^4$ as a base -2 power

e) $\left(\dfrac{1}{4}\right)^3$ as a base $\dfrac{1}{2}$ power

f) $\left(\dfrac{1}{25}\right)^4$ as a base $\dfrac{1}{5}$ power

11. Simplify, then evaluate without using a calculator.

a) $(2^2)^3 \div 4^2$

b) $\dfrac{9^2}{(3^2)^2}$

c) $\dfrac{(10^2)^5}{100^3}$

d) $5^2\left(\dfrac{(5^4)^3}{5^{10}}\right)$

12. Clare was asked to simplify the expression $3^2(2^2)^2$. This is her solution:

$$3^2(2^2)^2 = 3^2(2^4)$$
$$= 6^6$$
$$= 46\ 656$$

Her solution is incorrect.

a) Identify her error.

b) Determine the correct answer. Show your steps.

13. a) Explain the steps involved in simplifying the expression $\dfrac{x^4(y(y^5))}{xy^4}$.

A b) Simplify the expression.

c) If $x = -2$ and $y = 3$, evaluate the expression.

14. a) What can you conclude about the numbers a and b if

T $3^a \times 3^b = 3^n$ and n is an even number?

b) If 5^m is a perfect square, what can you conclude about m?

Extending

15. a) Evaluate $[(-5)^2]^3$ and $(-5^2)^3$. Do these expressions have the same value? Justify your answer.

b) Evaluate $(-5)^2$ and $(-5)^3$. Make a conjecture about the sign of a base and how the exponent may affect the value of the power.

16. Simplify, then determine the number that makes each statement true.

a) $(n^2)^5 \div n^5 = 243$ b) $\left(\dfrac{m^7}{m^6}\right)^2 = 196$

17. Simplify.

a) $(4x^3)^2$

b) $(5x^3y^4)^3$

c) $\left(\dfrac{3x^3}{2y^4}\right)^2\left(\dfrac{2y^2}{3x^4}\right)^3$

18. If $A = \dfrac{(2x^2)^3}{(x^3)(4x)}$, determine a possible value for x where

a) $0 < A < 1$ b) $A = 1$ c) $A > 1$ d) $A < 0$

Working with Integer Exponents

Evaluate numerical expressions involving integer exponents.

INVESTIGATE the Math

A yeast culture grows by doubling its number of cells every 20 min. Bill is working with someone else's lab notes. The notes include a table of the number of yeast cells counted, in thousands, every 20 min since noon. However, the first three readings are missing.

Time	Number of Cells (thousands)	Power of Two
1:00 p.m.		
1:20		
1:40		
2:00 p.m.	8	2^3
2:10	16	2^4
2:20	32	2^5

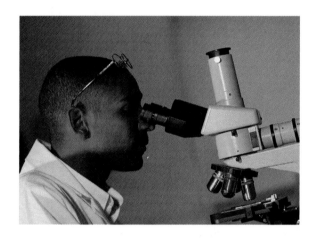

❓ **How can Bill determine the number of cells that were present at noon?**

A. Copy the table. Fill in the missing times in the first column.

B. What is the pattern of the numbers in the second column? Work backward to fill in the second column to 1:00 p.m.

C. What is the pattern of the powers of two in the third column? Work backward to fill in the third column to 1:00 p.m.

D. What does the pattern suggest about the value of 2^0?

E. Continue to work backward to complete each column.

F. What are the signs of the exponents of the first three entries in the third column?

G. Write the values in the second column as fractions in lowest terms. Write their denominators as powers of two.

H. Compare the fractions you wrote in part G with the powers of two in the first three entries.

I. How many cells did this culture have at noon?

Reflecting

J. How does the pattern in the table suggest that the value of 2^0 must be 1?

K. How can you write a power with a negative exponent so that the exponent is positive?

L. Do you think that the rules for multiplying and dividing powers with the same base still apply if the exponents are zero or negative? Create some examples and test your conjecture with and without a calculator.

APPLY the Math

EXAMPLE **1**	Representing the value of a number with a zero exponent

Explain why a number with a zero exponent has a value of 1.

Alifiyah's Solution: Starting with a Value of 1

$1 = \dfrac{7^4}{7^4}$ ⟵

$\quad = 7^{4-4}$

$\quad = 7^0$

Therefore, $7^0 = 7^{4-4} = 1$

I know that any number divided by itself is 1. I chose the power 7^4. Using this power, I wrote 1 as a rational number. I used the rule for dividing powers of the same base and subtracted the exponents. The result is a power that has an exponent of zero.

This shows that a number with a zero exponent is equal to 1.

Matt's Solution: Starting with the Exponent 0

$6^0 = 6^{3-3}$ ← I began by looking at 6^0. I can write this as 6^{3-3}. (Any number other than 3 will work as well.)

$\quad = \dfrac{6^3}{6^3}$ Subtracting exponents means that I am dividing the two powers. This leads to the answer 1.

$\quad = 1$

In an expression with a negative exponent, the sign of the exponent changes when you take the reciprocal of the original base.

EXAMPLE 2	**Representing a number raised to a negative exponent**

Evaluate $(-3)^{-4}$.

Liz's Solution

$(-3)^{-4} = \left(-\dfrac{1}{3}\right)^4$ ← I know that a number raised to a negative exponent is equal to a power of the reciprocal of the original base, raised to the positive exponent. I changed -3 to its reciprocal and changed the sign of the exponent.

$\quad = \left(-\dfrac{1}{3}\right)\left(-\dfrac{1}{3}\right)\left(-\dfrac{1}{3}\right)\left(-\dfrac{1}{3}\right)$

$\quad = \dfrac{1}{81}$ ← I multiplied.

EXAMPLE **3**	Representing a number with a rational base and a negative exponent

Evaluate $\left(\dfrac{5}{6}\right)^{-2}$.

Jill's Solution: Using Division

$\left(\dfrac{5}{6}\right)^{-2} = \dfrac{1}{\left(\dfrac{5}{6}\right)^{2}}$ ←   I wrote a 1 over the entire expression and changed -2 to 2.

$= \dfrac{1}{\dfrac{25}{36}}$ ←   I evaluated the power.

$= 1 \times \dfrac{36}{25}$ ←   I divided 1 by the fraction by taking its reciprocal and multiplying.

$= \dfrac{36}{25}$

Martin's Solution: Using the Reciprocal

$\left(\dfrac{5}{6}\right)^{-2} = \left(\dfrac{6}{5}\right)^{2}$ ←   I evaluated this expression by taking the reciprocal of the base. When I did this, I changed the sign of the exponent (from negative to positive).

$= \dfrac{6^{2}}{5^{2}}$ ←   The rational expression is squared. So I squared the numerator and squared the denominator.

$= \dfrac{36}{25}$

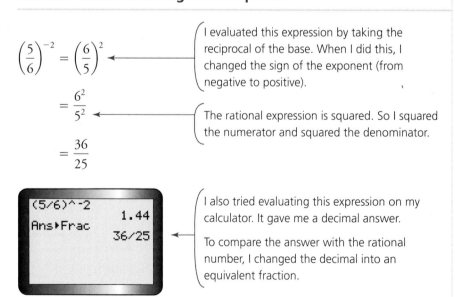

I also tried evaluating this expression on my calculator. It gave me a decimal answer.

To compare the answer with the rational number, I changed the decimal into an equivalent fraction.

Tech | *Support*

You can use the "Frac" command, which is found in the MATH menu on the graphing calculator, to change most decimals into equivalent fractions.

EXAMPLE **4**

Evaluate $\dfrac{5^{-4}(5^8)}{(5^{-3})^2}$.

Lesley's Solution: Applying the Exponent Laws

$$\frac{5^{-4}(5^8)}{(5^{-2})^2} = \frac{5^4}{5^{-4}}$$

$$= 5^{4-(-4)}$$

$$= 5^8$$

$$= 390\ 625$$

Since this expression is written as a fraction, I evaluated the numerator and denominator separately. The numerator has two powers of 5 multiplied, so I added their exponents. The denominator is a power of a power, so I multiplied these exponents.

To divide the numerator by the denominator I subtracted the exponents (since they have the same base).

Aneesh's Solution: Using a Calculator

$$\frac{5^{-4}(5^8)}{(5^{-2})^2}$$

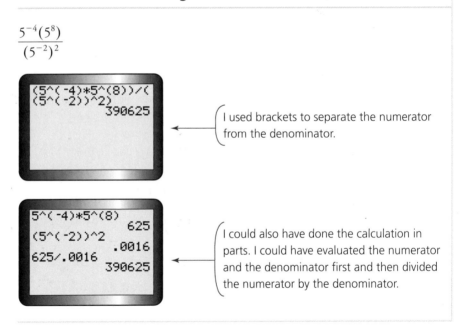

```
(5^(-4)*5^(8))/(
(5^(-2))^2)
             390625
```

I used brackets to separate the numerator from the denominator.

```
5^(-4)*5^(8)
             625
(5^(-2))^2
             .0016
625/.0016
             390625
```

I could also have done the calculation in parts. I could have evaluated the numerator and the denominator first and then divided the numerator by the denominator.

In Summary

Key Ideas

- A power with a negative exponent and an integer base is equivalent to the reciprocal of that base to the opposite exponent.

$$b^{-n} = \frac{1}{b^n} \text{, where } b \neq 0$$

- A power with a negative exponent and a fractional base is equivalent to the reciprocal of that base to the opposite exponent.

$$\left(\frac{a}{b}\right)^{-n} = \left(\frac{b}{a}\right)^{n} \text{, where } a \neq 0, b \neq 0$$

- Any number (or expression) to the exponent 0 is equal to 1.

$$b^0 = 1 \text{, where } b \neq 0$$

Need to Know

- When simplifying numerical expressions involving powers, present the answer as an integer, a fraction, or a decimal.
- When simplifying algebraic expressions involving exponents, it is common practice to present the answer with positive exponents.

CHECK Your Understanding

1. Rewrite each expression as an equivalent expression with a positive exponent.

   a) 4^{-6}
   b) $\left(\frac{7}{3}\right)^{-5}$
   c) $\frac{1}{8^{-2}}$
   d) $(-3)^{-2}$

2. Evaluate each expression without using a calculator.

   a) $(8)^0$
   b) 5^{-3}
   c) $\left(\frac{3}{2}\right)^{-3}$
   d) $(-2)^{-4}$

3. Use your calculator to evaluate each expression.

   a) 8^{-2}
   b) 4^{-3}
   c) $\left(\frac{5}{2}\right)^{-3}$
   d) $\left(-\frac{1}{2}\right)^{-3}$

PRACTISING

4. Evaluate.

a) 10^{-2}

c) $\left(\dfrac{1}{2}\right)^{-5}$

e) $\dfrac{1}{(-9)^2}$

b) $(-4)^{-2}$

d) $\left(\dfrac{1}{7}\right)^{-3}$

f) $(-5)^0$

5. Simplify. Write each expression as a single power with a positive exponent.

a) $9^7 \times 9^{-3}$

c) $8^6 \div 8^{-5}$

e) $(-3)^{-8} \times (-3)^9$

b) $6^{-3} \div 6^{-5}$

d) $17^{-4} \div 17^{-6}$

f) $(-4)^{-5} \times (-4)^5$

6. Simplify. Write each expression as a single power with a positive exponent.

a) $2^4(2^2) \div 2^{-6}$

c) $\dfrac{(-12^3)^{-1}}{(-12)^7}$

e) $\dfrac{9^4(9^3)}{9^{12}}$

b) $-5 \times (-5^4)^{-3}$

d) $\left(\dfrac{3^4}{3^6}\right)^{-1}$

f) $((7^2)^{-3})^{-4}$

7. Simplify. Write each expression as a single power with a positive exponent.

a) $\dfrac{11^{-2}(11^3)}{(11^{-2})^4}$

c) $\left(\dfrac{4^{-3}}{4^{-2}}\right)^{-3}$

e) $\dfrac{(-8^{-1})(-8^{-5})}{(-8^{-2})^3}$

b) $\left(\dfrac{9^{-2}}{(9^2)^2}\right)^2$

d) $\left(\dfrac{10}{10^{-3}}\right)^2\left(\dfrac{10^5}{10^7}\right)$

f) $\left(\dfrac{(5^3)^2}{5(5^6)}\right)^{-1}$

8. Simplify, then evaluate each expression. Leave answers as fractions or integers.

a) $13^3 \times 13^{-4}$

c) $\left(\dfrac{10^{-3}}{10^{-5}}\right)^2$

e) $\dfrac{-2(-2^{-3})}{(-2)^4}$

b) $\dfrac{3^{-2}}{3^{-6}}$

d) $6^{-2}(6^{-2})^{-1}$

f) $\left(\dfrac{5^{-2}}{5}\right)^{-1}$

9. Evaluate. Leave answers as fractions or integers.

a) $3^{-2} - 9^{-1}$

c) $8^{-2} + (4^{-1})^2$

e) $12(4^0 - 3^{-2})$

b) $4^{-2} + 3^0 - 2^{-3}$

d) $\left(\dfrac{1}{2}\right)^{-1} + \left(\dfrac{1}{3}\right)^{-1}$

f) $\dfrac{4^2}{2^5}$

10. Scientific notation can be used to represent very large and very small numbers. The diameter of Earth is about 1.276×10^7 m, while the diameter of a plant cell is about 1.276×10^{-5} m. Explain why negative exponents are used in scientific notation to represent very small numbers.

11. Simplify. Write each expression as a single power with a positive exponent.

a) $x^4(x^{-2})^2$

c) $\left(\dfrac{w^2}{w^4}\right)^{-3}$

e) $a^{-2} \times a^3 \times a^4$ ✓

b) $\dfrac{(m^5)^2}{m^{-8}}$

d) $((a^{-2})^2)^{-3}$

f) $\dfrac{b(b^{-5})b^{-2}}{b(b^8)}$

12. Determine the value of the variable that makes each of the following true.
T

a) $12^x = 1$

c) $10^n = 0.0001$

e) $2^b = \dfrac{1}{32}$

b) $10^m = 100\ 000$

d) $2^k = \dfrac{1}{2}$

f) $2^{2a} = 64$

13. Francesca is helping her friends Sasha and Vanessa study for a quiz.
C They are working on simplifying $2^{-2} \times 2$. Francesca notices errors in each of her friends' solutions, shown here:

Sasha's solution

$2^{-2} \times 2$
$= 4^{-1}$
$= -\dfrac{1}{4^1}$
$= -\dfrac{1}{4}$

Vanessa's solution

$2^{-2} \times 2$
$= 2^{-2}$
$= \dfrac{1}{2^2}$
$= \dfrac{1}{4}$

a) Explain where each student went wrong.
b) Write the correct solution.

Extending

14. Consider the following powers: 2^{12}, 4^6, 8^4, 16^3.
a) Use your calculator to show that the powers above are equivalent.
b) Can you think of a way to explain why each power above is equivalent to the preceding power without referring to your calculator?
c) Create a similar list using 3 as the base.

15. If $x = -2$ and $y = 3$, write the following three expressions in order from least to greatest.

$$\dfrac{y^{-4}(x^2)^{-3}y^{-3}}{x^{-5}(y^{-4})^2},\quad \dfrac{x^{-3}(y^{-1})^{-2}}{(x^{-5})(y^4)},\quad (y-5)(x^5) - 2(y^2)(x-3) - 4$$

Working with Rational Exponents

GOAL

Determine the meaning of a power with a rational exponent, and evaluate expressions containing such powers.

INVESTIGATE the Math

On August 16, 1896, gold was discovered near Dawson, in the Yukon region of Canada.

The population of Dawson City experienced rapid growth during this time. The population was approximately 1000 in April and 3000 in July (and grew to 30 000 at one point). The population is given in the table and is also shown on the graph.

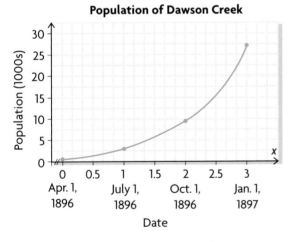

Population of Dawson Creek

Date	x	Population (1000s)
Apr. 1, 1896	0	1, or 3^0
July 1, 1896	1	3, or 3^1
Oct. 1, 1896	2	9, or 3^2
Jan. 1, 1897	3	27, or 3^3

? About how many people were in Dawson City in mid-May and mid-August of 1896?

A. What *x*-value can be used to represent the middle of May? Explain how you know.

B. Use the graph to estimate the population in the middle of May.

C. What power can be used to represent the population in the middle of May? Multiply this power by itself. What do you notice?

D. Use the exponent button on your calculator to calculate $3^{\frac{1}{2}}$. How does this value relate to your estimate in part B?

E. Use the graph to estimate the population at the beginning of May (when $x = \frac{1}{3}$). Calculate $3^{\frac{1}{3}}$ and compare it with your estimate.

F. Calculate $3^{\frac{1}{3}} \times 3^{\frac{1}{3}} \times 3^{\frac{1}{3}}$. What do you notice?

G. What is another way to express $3^{\frac{1}{3}}$ using a radical?

H. Write a power of 3 that would estimate the population in the middle of August. Evaluate the power with your calculator and check your answer by using the graph shown.

radical

the indicated root of a quantity; for example, $\sqrt[3]{8} = 2$ since $2 \times 2 \times 2 = 2^3 = 8$

Reflecting

I. Explain why $x^{\frac{1}{2}}$ is equivalent to $\sqrt{x}$.

J. Make a conjecture about powers with the exponent $\frac{1}{n}$.

K. Do the rules for multiplying powers with the same base still apply if the exponents are rational? Use the numbers in your table to investigate.

APPLY the Math

EXAMPLE 1	Representing a power with a positive rational exponent

Evaluate $65^{\frac{1}{3}}$.

Elaine's Solution

$65^{\frac{1}{3}} = \sqrt[3]{65}$ ◄——— I know that the exponent $\frac{1}{3}$ is a way of representing the 3rd root of the base. I wrote it in radical notation.

$\sqrt[3]{65} \doteq 4$

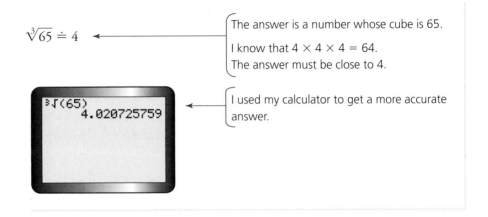

The answer is a number whose cube is 65.
I know that $4 \times 4 \times 4 = 64$.
The answer must be close to 4.

I used my calculator to get a more accurate answer.

When an exponent is written as a decimal, it is often easier to evaluate the power if the exponent is written as its equivalent fraction.

| EXAMPLE 2 | Representing a power with a decimal rational exponent |

Evaluate $32^{0.2}$.

Tosh's Solution

$32^{0.2} = 32^{\frac{1}{5}}$

I rewrote the power changing the exponent from 0.2 to its equivalent fraction.
$$0.2 = \frac{2}{10} = \frac{1}{5}$$

$= \sqrt[5]{32}$

I know that an exponent $\frac{1}{5}$ is a way of representing the 5th root of the number. I wrote it in radical notation.

$= 2$

Since $2^5 = 32$, the 5th root of 32 must be 2.

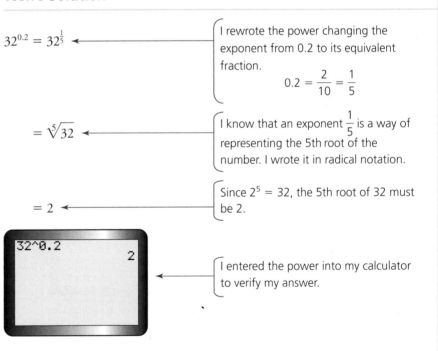

I entered the power into my calculator to verify my answer.

If a power involves a negative rational exponent, then you can write the exponent as the product of a fraction and an integer.

example 3 **Representing a power with a negative rational exponent**

Evaluate $(-27)^{-\frac{2}{3}}$.

George's Solution: Using $\dfrac{1}{3}$ and -2 as Exponents

$$(-27)^{-\frac{2}{3}} = \left((-27)^{\frac{1}{3}}\right)^{-2}$$

I separated the exponent into two parts, since the exponent $-\dfrac{2}{3}$ can be written as a product: $\dfrac{1}{3} \times (-2)$.

$$= \frac{1}{\left((-27)^{\frac{1}{3}}\right)^{2}}$$

I expressed $\left((-27)^{\frac{1}{3}}\right)^{-2}$ as a rational number, using 1 as the numerator and $\left((-27)^{\frac{1}{3}}\right)^{2}$ as the denominator.

$$= \frac{1}{\left((-27)^{\frac{1}{3}} \times (-27)^{\frac{1}{3}}\right)}$$

$$= \frac{1}{\left(\sqrt[3]{-27} \times \sqrt[3]{-27}\right)}$$

I determined the cube root of -27. I know that $(-3) \times (-3) \times (-3) = -27$.

$$= \frac{1}{(-3 \times -3)}$$

$$= \frac{1}{9}$$

Nadia's Solution: Using $-\dfrac{1}{3}$ and 2 as Exponents

$$(-27)^{-\frac{2}{3}} = \left((-27)^{-\frac{1}{3}}\right)^{2}$$

I separated the exponent into two parts. The exponent $-\dfrac{2}{3}$ can be written as a product: $-\dfrac{1}{3} \times 2$.

$$= \left(\frac{1}{(-27)^{\frac{1}{3}}}\right)^{2}$$

$$= \left(\frac{1}{\sqrt[3]{-27}}\right)^{2}$$

I expressed $(-27)^{\frac{1}{3}}$ as a radical.

$$= \left(\frac{1}{-3}\right)^{2}$$

I evaluated the root and squared the result.

$$= \frac{1}{9}$$

Anjali's Solution: Using a Calculator

$(-27)^{-\frac{2}{3}}$

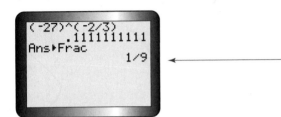

I used my calculator to determine the answer and then changed it from a decimal to a fraction.

The strategies you have used to evaluate numerical expressions involving integer exponents also apply when exponents are rational.

EXAMPLE 4	Selecting a strategy to evaluate an expression involving rational exponents

Simplify, then evaluate $\dfrac{\left(8^{\frac{1}{6}}\right)^7}{8^{\frac{1}{2}}8^{\frac{1}{3}}}$.

Aaron's Solution

$$\frac{\left(8^{\frac{1}{6}}\right)^7}{8^{\frac{1}{2}}8^{\frac{1}{3}}} = \frac{8^{\frac{7}{6}}}{8^{\frac{5}{6}}}$$

The numerator is a power of a power, so I multiplied the exponents: $\dfrac{1}{6} \times 7 = \dfrac{7}{6}$

The denominator is a product of powers, so I added the exponents: $\dfrac{1}{2} + \dfrac{1}{3} = \dfrac{3}{6} + \dfrac{2}{6} = \dfrac{5}{6}$

$$= 8^{\frac{7}{6}-\frac{5}{6}}$$

To divide powers (numerator by denominator), I subtracted the exponents.

$$= 8^{\frac{2}{6}}$$

$$= 8^{\frac{1}{3}}$$

I wrote the fraction in lowest terms.

$$= \sqrt[3]{8}$$

$$= 2$$

I found the cube root of 8.

In Summary

Key Ideas

- A power with a rational exponent is equivalent to a radical. The rational exponent $\frac{1}{n}$ indicates the nth root of the base. If $n > 1$ and $n \in \mathbf{N}$, then $b^{\frac{1}{n}} = \sqrt[n]{b}$, where $b \neq 0$.

- If $m \neq 1$ and if m and n are both positive integers, then $b^{\frac{m}{n}} = \left(\sqrt[n]{b}\right)^m = \sqrt[n]{b^m}$, where $b \neq 0$.

Need to Know

- The exponent laws that apply to powers with integer exponents also apply to powers with rational exponents.

- The power button on a scientific calculator can be used to evaluate rational exponents.

- Some roots of negative numbers cannot be determined. For example, -16 does not have a real-number square root, since $(-4)^2 = (-4) \times (-4) = +16$. Odd roots can have negative bases, but even ones cannot.

- Since radicals can be written as powers with rational exponents:
  - Their products are equivalent to the products of powers. This means that $\sqrt{a} \times \sqrt{b} \times \sqrt{c} = \sqrt{a \times b \times c}$, because $a^{\frac{1}{2}} \times b^{\frac{1}{2}} \times c^{\frac{1}{2}} = (abc)^{\frac{1}{2}}$, where a, b, and $c > 0$.
  - Their quotients are equivalent to the quotient of powers. This means that $\frac{\sqrt{a}}{\sqrt{b}} = \sqrt{\frac{a}{b}}$, because $\frac{a^{\frac{1}{2}}}{b^{\frac{1}{2}}} = \left(\frac{a}{b}\right)^{\frac{1}{2}}$, where a, b, and $c > 0$.

CHECK Your Understanding

1. Write in radical form. Then evaluate without using a calculator.
   a) $49^{\frac{1}{2}}$
   b) $(-125)^{\frac{1}{3}}$
   c) $81^{\frac{1}{4}}$
   d) $100^{\frac{1}{2}}$
   e) $16^{0.25}$
   f) $-(144)^{0.5}$

2. Write in exponent form. Then evaluate.
   a) $\sqrt[10]{1024}$
   b) $\sqrt[5]{1024}$
   c) $\sqrt[3]{27^4}$
   d) $\left(\sqrt[3]{-216}\right)^5$
   e) $\sqrt[4]{16}$
   f) $(\sqrt{25})^{-1}$

3. Use your calculator to evaluate each expression to the nearest hundredth.
   a) $6^{\frac{2}{5}}$
   b) $0.0625^{\frac{1}{4}}$
   c) $\sqrt[15]{4421}$
   d) $144^{0.25}$
   e) $10^{\frac{-2}{3}}$
   f) $200^{-0.4}$

PRACTICE

4. Evaluate each expression. Use the fact that $2^5 = 32$ and $3^4 = 81$.

K

a) $-32^{\frac{1}{5}}$

c) $\sqrt[10]{(-32)^2}$

e) $-81^{-0.25}$

b) $\sqrt[5]{-32}$

d) $(-2)^5$

f) $\sqrt[4]{81}$

5. The volume of a cube is 0.027cm^3. What does $\sqrt[3]{0.027}$ represent in
A the situation?

6. Write each expression as a single power with a positive exponent.

a) $7^{\frac{1}{2}} \times 7^2$

c) $\left(16^{\frac{2}{3}}\right)^6$

e) $\left(10^{\frac{5}{8}}\right)^{-2}$

b) $3^4 \div 3^{\frac{1}{2}}$

d) $\dfrac{12^{\frac{2}{3}}}{12^{\frac{1}{6}}}$

f) $2^{0.5} \times 2^2 \times 2^{2.5}$

7. Write each expression as a single power with a positive exponent.

a) $3^{-\frac{1}{2}} \div 3^{\frac{3}{2}}$

c) $8^{\frac{5}{2}} \div 8^{-\frac{5}{2}}$

e) $9^{0.5} \times 3^0$

b) $\left(5^{-\frac{2}{5}}\right)^{10}$

d) $4^{0.3} \div 4^{0.8} \times 4^{-0.7}$

f) $2^3 \times 4^{-2} \div 8$

8. Given that $10^{0.5} \doteq 3.16$, determine the value of $10^{1.5}$ and $10^{-0.5}$.
Explain your reasoning.

9. Write each expression using powers, then simplify. Evaluate your
simplified expression.

a) $\dfrac{\sqrt{200}}{\sqrt{2}}$

c) $\dfrac{\sqrt{98}}{\sqrt{2}}$

e) $\dfrac{\sqrt{15}\sqrt{10}}{\sqrt{6}}$

b) $\sqrt{3}\sqrt{6}\sqrt{2}$

d) $\dfrac{\sqrt{12}}{\sqrt{8}\sqrt{6}}$

f) $3\sqrt{12}\sqrt{3}$

10. Write each expression as a single power with positive exponents.

a) $\dfrac{6^{\frac{3}{2}} \times 6^5}{6}$

c) $\dfrac{12^{-\frac{3}{4}}}{12^{\frac{-1}{2}}}$

e) $4^{\frac{2}{3}} \div 4^{\frac{-1}{2}} \times 4^{\frac{5}{6}}$

b) $\dfrac{\left(8^2\right)^{\frac{1}{2}}}{8\left(8^2\right)}$

d) $\dfrac{\left(11^{-\frac{3}{4}}\right)\left(11^{\frac{5}{8}}\right)}{11^{\frac{3}{2}}}$

f) $\left(\left(16^{-\frac{1}{2}}\right)^2\right)^{-\frac{1}{4}}$

11. Simplify. Express final answers in radical form.

a) $4^{-\frac{3}{8}}\left(4^2\right)$

c) $10^{\frac{9}{4}}\left(10^{-2}\right)$

e) $5^{\frac{1}{2}} \times 5^{-1}$

b) $\dfrac{9^{\frac{4}{3}}}{9^{\frac{7}{10}}}$

d) $\left(8^{\frac{1}{5}}\right)\left(8^{-\frac{2}{15}}\right)$

f) $4^3 \div 4^{\frac{3}{4}}$

12. Rewrite each of the following expressions using a rational exponent.
Then evaluate using your calculator. Express answers to the nearest
thousandth.

a) $\sqrt[3]{120}$

c) $\sqrt[5]{13^{-2}}$

e) $\left(\sqrt[3]{216}\right)^{-2}$

b) $25^{0.75}$

d) $10^{-0.8}$

f) $\left(\sqrt[3]{-15}\right)^{-2}$

13. Evaluate $-(8)^{\frac{1}{3}}$ and $-(4)^{\frac{1}{2}}$ using your calculator. Compare the results.

14. Simplify. Write each answer with positive exponents.

a) $m\left(m^{\frac{2}{3}}\right)$

c) $\left(c^3\right)^{\frac{5}{6}}$

e) $\dfrac{s\left(s^{0.25}\right)}{\left(s^{1.5}\right)^{0.3}}$

b) $\dfrac{x}{x^{-\frac{4}{3}}}$

d) $\left(b^{\frac{8}{9}}\right)^{\frac{9}{4}}$

f) $m\left(m^{\frac{2}{3}}\right)m^{-\frac{5}{3}}$

15. Simplify. Write each answer with positive exponents.

a) $\dfrac{t\left(t^{-\frac{8}{5}}\right)}{t^{\frac{3}{5}}}$

c) $\dfrac{\left(y^5\right)^{-\frac{9}{5}}}{\left(y^{-\frac{3}{2}}\right)^4}$

e) $\left(x^{\frac{1}{3}} \div x^{\frac{2}{3}}\right)^{-3}$

b) $\dfrac{\left(x^{\frac{7}{4}}\right)^{\frac{1}{2}}}{x^{-\frac{5}{6}}}$

d) $\left(\dfrac{a^{\frac{1}{2}}a^{\frac{3}{2}}}{\left(a^{-2}\right)^{\frac{1}{2}}}\right)$

f) $\left(\left(b^{-8}\right)^{\frac{-1}{2}}\right)^{\frac{-3}{4}}$

16. Evaluate $64^{-\frac{5}{3}}$ without a calculator. Explain each of the steps in your evaluation.

17. Determine the value of the variable that makes each of the following true. Express each answer to the nearest hundredth.

a) $1.05 = \sqrt[3]{M}$

c) $N^{\frac{1}{5}} - 3 = 0$

e) $x^{\frac{2}{3}} = 4$

b) $2.5 = \sqrt[4]{T}$

d) $\dfrac{x^5}{x^2} = 125$

f) $y^{-0.25} = \dfrac{1}{3}$

18. Write in exponential form. Use the exponent laws to simplify and then evaluate.

a) $\sqrt{1000} \times \sqrt[3]{1000} \div \sqrt[6]{1000}$

b) $\dfrac{\left(\sqrt{64}\right)^2}{\sqrt[3]{64}}$

19. Use your knowledge of exponents to express $32^{\frac{4}{5}}$ in two other ways. Which one is easier to evaluate if you do not use a calculator?

Extending

20. Simplify.

a) $\dfrac{4 + 4^{-1}}{4 - 4^{-1}}$

b) $\dfrac{5^{-2} - 5^{-1}}{5^{-2} + 5^{-1}}$

c) $\dfrac{\sqrt{4^3}\left(\sqrt[5]{4^4}\right)}{\sqrt{2^{10}}}$

21. Write as a radical: $\left(21^6\right)^{-\frac{1}{4}}$

22. If $a = 2$ and $b = -1$, which expression has the greater value?

a) $\dfrac{a^{-2b}a^{-b+2}}{\left(a^{-2}\right)^b}$

b) $\dfrac{\left(a^b\right)^{-3}a^{-1(-2b)}}{\left(a^{-b}\right)^3}$

FREQUENTLY ASKED Questions

Study | **Aid**

• See Lesson 7.3, Example 1.

Q: **What is the value of an experssion with a zero exponent?**

A: When there is a zero exponent, the value is 1.

EXAMPLE

$$12^0 = 1 \qquad \left(\frac{7}{11}\right)^0 = 1$$

$$(-6)^0 = 1 \qquad \left(-3(8) - \frac{2}{5}\left(\frac{13}{20}\right)\right)^0 = 1$$

Study | **Aid**

• See Lesson 7.3, Examples 2, 3, and 4.
• Try Mid-Chapter Review Questions 2 and 3.

Q: **What does it mean when a power has a negative exponent and how do you evaluate this kind of power?**

A: A power with a negative exponent is equivalent to a power whose base is the reciprocal of the original base, and whose exponent is the opposite of the original exponent.

To evaluate such a power, take the reciprocal of the base and change the sign of the exponent. Then, multiply the base by itself the number of times indicated by the exponent.

EXAMPLE

$$5^{-3} = \left(\frac{1}{5}\right)^3 \qquad\qquad \left(\frac{3}{4}\right)^{-2} = \left(\frac{4}{3}\right)^2$$

$$= \frac{1^3}{5^3} \qquad\qquad\qquad = \frac{4^2}{3^2}$$

$$= \frac{1}{125} \qquad\qquad\qquad = \frac{16}{9}$$

Study | **Aid**

• See Lesson 7.4, Examples 1 to 4.
• Try Mid-Chapter Review Questions 4 to 8.

Q: **What does it mean when a power has a rational exponent, and how do you evaluate this kind of power?**

A: The denominator of a rational exponent indicates the required root of the base. The numerator has the same meaning as an integer exponent. Using the fact that powers with rational exponents can be expressed as powers of powers, this can be evaluated in two steps, in two different ways.

EXAMPLE

$$27^{\frac{2}{3}} = \left(27^{\frac{1}{3}}\right)^2 \qquad\qquad 27^{\frac{2}{3}} = \left(27^2\right)^{\frac{1}{3}}$$

$$= \left(\sqrt[3]{27}\right)^2 \qquad\qquad\quad = (243)^{\frac{1}{3}}$$

$$= (3)^2 \qquad\qquad\qquad = \sqrt[3]{243}$$

$$= 9 \qquad\qquad\qquad\quad = 9$$

PRACTICE Questions

Lesson 7.2

1. Write as a single power. Express your answers with positive exponents.

a) $5(5^4)$

b) $\dfrac{8^8}{8^6}$

c) $(16^2)^5$

d) $\dfrac{(-4)^6(-4)^3}{((-4)^9)^2}$

e) $\left(\dfrac{1}{10}\right)^6\left(\dfrac{1}{10}\right)^{-4}$

f) $\left(\dfrac{(7)^2}{(7)^4}\right)^{-5}$

Lesson 7.3

2. Write each power with only positive exponents.

a) x^{-2}

b) $(m^{-4})^2$

c) $b^{-3} \times b^{-2}$

d) $\left(\dfrac{1}{y}\right)^{-2}$

e) $(n^{-7})^{-2}$

f) $y^{-3} \div y$

3. Evaluate without using a calculator.

a) $5^{-2} + 10^{-1}$

b) $4^0 + 8^{-2} - 2^{-2}$

c) $9^{-1} - (3^{-1})^2$

d) $(6^{-2})^{-1} + \left(\dfrac{1}{3}\right)^{-2}$

e) $\left(-\dfrac{1}{2}\right)^3 + 4^{-3}$

f) $25^{-1} + \left(-\dfrac{5}{2}\right)^{-2}$

Lesson 7.4

4. Evaluate without using a calculator.

a) $\left(\dfrac{2}{3}\right)^{-1}$

b) $\left(-\dfrac{2}{5}\right)^{-3}$

c) $\left(\dfrac{81}{16}\right)^{\frac{1}{2}}$

d) $\left(64^{\frac{1}{3}}\right)^4$

e) $\left(\dfrac{16}{81}\right)^{\frac{1}{4}}$

f) $[(2^2)(4^2)]^{-1}$

5. Simplify. Write each expression with only positive exponents.

a) $a^{\frac{1}{5}} \times a^{\frac{2}{3}}$

b) $\dfrac{b^2}{b^{\frac{3}{2}}}$

c) $\dfrac{c^{-3}}{c^2}$

d) $\dfrac{d^{-3}}{d^{-5}}$

e) $e(e^{-5})^{-2}$

f) $\left(f^{-\frac{2}{3}}\right)^{\frac{5}{8}}$

6. Copy and complete the table.

	Exponential Form	Radical Form	Evaluation of Expression
a)	$100^{\frac{1}{2}}$		
b)	$16^{0.25}$		
c)		$\sqrt{121}$	
d)	$(-27)^{\frac{5}{3}}$		
e)	$49^{2.5}$		
f)		$\sqrt[10]{1024}$	
g)		$\sqrt[3]{\left(\dfrac{1}{2}\right)^9}$	

7. Evaluate. Express each answer to three decimals.

a) $\sqrt[6]{2400}$

b) $120^{0.8}$

c) $9^{\frac{-6}{5}}$

d) $0.5^{-0.5}$

e) $\sqrt[9]{-1024}$

f) 0.2^{-2}

8. Use trial and error to determine the value of the variable that makes each of the following true.

a) $\sqrt[3]{x} = 125$

b) $m^{\frac{3}{2}} = 64$

c) $p^{-3} = \dfrac{1}{27}$

d) $\sqrt{x^3} = 8$

7.5 Exploring the Properties of Exponential Functions

YOU WILL NEED

- graphing calculator
- graph paper

GOAL

Examine the features of the graphs of exponential functions and compare them with graphs of linear and quadratic functions.

EXPLORE the Math

The temperature of a cup of hot liquid as it cools with time is modelled by an exponential function.

? What are the characteristics of the graph of the exponential function $f(x) = b^x$, and how does this graph compare with the graphs of quadratic and linear functions?

A. Copy and complete the tables of values for the functions $g(x) = x$, $h(x) = x^2$, and $k(x) = 2^x$.

x	−3	−2	−1	0	1	2	3	4	5
g(x) = x									

x	−3	−2	−1	0	1	2	3	4	5
h(x) = x²									

x	−3	−2	−1	0	1	2	3	4	5
k(x) = 2ˣ									

B. Add two rows to each table and calculate the first and second differences. Discuss the difference patterns for each type of function.

C. Graph each function on graph paper, and draw a smooth curve or line through each set of points. Label each with the appropriate equation.

D. State the domain and range of each function.

E. For each function, describe how the y-values change as the x-values increase and decrease.

F. Use a graphing calculator to graph the functions $y = 2^x$, $y = 5^x$, and $y = 10^x$. Use the WINDOW settings shown to graph these functions on the same axes.

G. For each function, state
  • the domain and range
  • the intercepts
  • the equations of any **asymptotes**

H. Use the trace key to examine the y-values as x increases and as x decreases. Which curve increases faster as you trace to the right? Which one decreases faster as you trace to the left?

I. Delete $y = 5^x$ and $y = 10^x$, and replace them with $y = \left(\frac{1}{2}\right)^x$ and $y = \left(\frac{1}{10}\right)^x$. Graph these functions on the same axes.

J. For each new function, state
  • the domain and range
  • the intercepts
  • the equations of any asymptotes

K. Describe how the graphs of $y = \left(\frac{1}{2}\right)^x$ and $y = \left(\frac{1}{10}\right)^x$ differ from the graph of $y = 2^x$.

L. What happens when the base of an exponential function is negative. Try $y = (-2)^x$. Discuss your findings.

Reflecting

M. Describe how the graph of an exponential function differs from the graph of a linear and a quadratic function.

N. How do the first and second differences of exponential functions differ from those of linear and quadratic functions? How can you tell that a function is exponential from its finite differences?

O. What type of function is $f(x) = b^x$ when $b = 1$?

P. Investigate the graphs of the exponential function $f(x) = b^x$ for various values of b, listing all similarities and differences in their features (such as the domain, the range, and any intercepts and asymptotes). Generalize their features for the cases $b > 1$ and $0 < b < 1$.

asymptote
a line that a curve approaches, but never reaches on some part of its domain

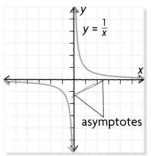

Tech | *Support*

For help tracing functions on the graphing calculator, see Technical Appendix, B-2.

In Summary

Key Ideas

- Linear, quadratic, and exponential functions have unique difference patterns that allow them to be recognized.

Linear	Quadratic	Exponential
Linear functions have constant first differences.	Quadratic functions have first differences that are related by an addition pattern. As a result, their second differences are constant.	Exponential functions have first differences that are related by a multiplication pattern. As a result, their second differences are not constant.

Linear graph: $f(x) = x$, with $\triangle y = 1$ repeated, $\triangle x = 1$

Quadratic graph: $f(x) = x^2$, with $\triangle y = 7 = 5 + 2$, $\triangle y = 5 = 3 + 2$, $\triangle y = 3 = 1 + 2$, $\triangle y = 1$, $\triangle x = 1$

Exponential graph: $f(x) = 2^x$, with $\triangle y = 8 = 4 \times 2$, $\triangle y = 4 = 2 \times 2$, $\triangle y = 2 = 1 \times 2$, $\triangle y = 1$, $\triangle x = 1$

- The exponential function $f(x) = b^x$ is
  - an increasing function representing rapid growth when $b > 1$
  - a decreasing function representing rapid decay when $0 < b < 1$

Need to Know

- The exponential function $f(x) = b^x$ has the following characteristics:
  - The function is exponential only if $b > 0$ and $b \neq 1$; its domain is the set of real numbers, and its range is the set of all positive real numbers.
  - If $b > 1$, the greater the value, the faster the growth.
  - If $0 < b < 1$, the lesser the value, the faster the decay.
  - The function has a horizontal *asymptote*, which is the x-axis.
  - The function has a y-intercept of 1.

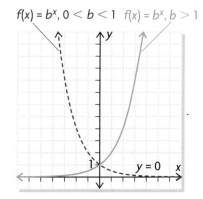

$f(x) = b^x, 0 < b < 1$ $f(x) = b^x, b > 1$

FURTHER *Your Understanding*

1. Use each table of values to identify each function as linear, quadratic, or exponential. Justify your answer.

a)

x	y
−2	0.3
−1	0.6
0	1.2
1	2.4
2	4.8
3	9.6
4	19.2

b)

x	y
0.2	2
0.6	5
1.0	8
1.4	11
1.8	14
2.2	17
2.6	20

c)

x	y
−6	29
−1	20
4	13
9	8
14	5
19	4
24	5

2. For each exponential function,
   i) determine the *y*-intercept
   ii) sketch the function
   iii) state the domain and range
   iv) state the equation of the horizontal asymptote

   a) $y = 3^x$
   b) $y = 0.25^x$
   c) $y = -(2^x)$
   d) $y = 2(0.3)^x$
   e) $y = (2^x) - 3$
   f) $y = 4(0.5)^x + 5$

3. You are given the functions $f(x) = 2x + 5$, $g(x) = (x - 2)^2 + 3$, and $h(x) = 4^x$.
   a) State which function is exponential. Explain how you know.
   b) Use a difference table to justify your answer to part (a).
   c) Which function is linear and which is quadratic? Explain how you know.
   d) Use difference tables to justify your answer to part (c).

4. Select the function that matches each graph.

a) $y = 0.5x + 1$ c) $y = 0.5x^2$ e) $y = 2^x$

b) $y = 0.5^x$ d) $y = 8^x$ f) $y = 2(0.5)^x$

i)
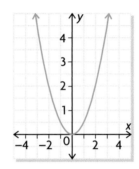

iv)
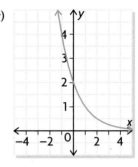

ii)
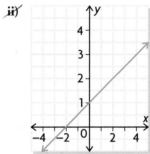

v)
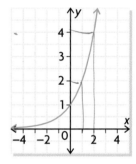

iii)
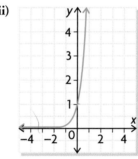

vi)
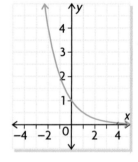

7.6 Solving Problems Involving Exponential Growth

GOAL

Use exponential functions to model and solve problems involving exponential growth.

YOU WILL NEED

- graphing calculator
- graph paper

LEARN ABOUT the Math

In 1978, researchers placed a group of 20 deer on a large island in the middle of a lake. The deer had no natural predators on the island. Researchers collected population data and created a graphical model.

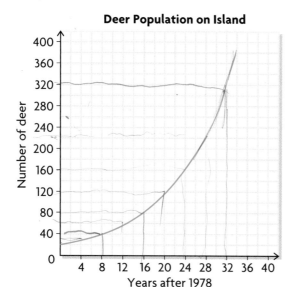

Deer Population on Island

Number of deer (vertical axis: 0, 40, 80, 120, 160, 200, 240, 280, 320, 360, 400)

Years after 1978 (horizontal axis: 4, 8, 12, 16, 20, 24, 28, 32, 36, 40)

The researchers estimate that the island has enough resources to support a population of 320 deer.

? In what year will the deer population be too large for the island?

EXAMPLE **1**   Making a prediction from data

Use the graph to estimate when the deer population will be greater than 320.

Liu's Solution: Using a Graph

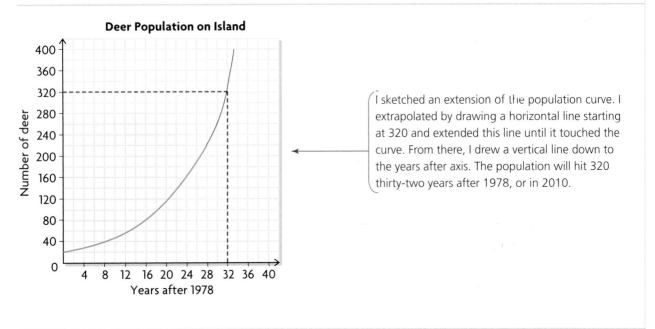

Deer Population on Island

I sketched an extension of the population curve. I extrapolated by drawing a horizontal line starting at 320 and extended this line until it touched the curve. From there, I drew a vertical line down to the years after axis. The population will hit 320 thirty-two years after 1978, or in 2010.

Petra's Solution: Using a Table of Values

Year	Number of Deer
0	20
4	30
8	40
12	60
16	40 × 2 = 80
20	60 × 2 = 120
24	80 × 2 = 160
28	120 × 2 = 240
32	160 × 2 = 320

I looked at some points in the table and noticed that the population doubled from 0 to 8 years and also from 4 to 12 years. It seems that every 8 years the population doubles.

I assumed that this doubling would continue and used this pattern to extend my table.

If the trend continues, then the population should reach 320 about 32 years after 1978. The year would be 2010.

Reflecting

A. What feature(s) of the graph indicate that this set of data might represent an exponential relationship?

B. Describe two ways that you can use data from a graph to determine whether or not there is an exponential relationship.

C. What are some of the advantages and disadvantages of making predictions from a graph?

APPLY the Math

EXAMPLE **2**	Making a prediction from an algebraic model

There are 5000 yeast cells in a culture. The number of cells grows at a rate of 25% per day. The function that models the growth of the yeast cells is $N(d) = N_0(1 + r)^d$, where N is the number of yeast cells d days after the culture is started, N_0 is the initial population, and r is the growth rate. How many cells will there be one week later?

Joelle's Solution

Initial population, $N_0 = 5000$ ←   I made a list of all of the given information.

Growth rate, $r = 25\% = 0.25$

Number of days, $d = 7$

$N(d) = N_0(1 + r)^d$

$N(7) = 5000(1 + 0.25)^7$ ←   I substituted the given values into the function and solved for N.

$\quad = 5000(1.25)^7$

$\quad \doteq 5000(4.768\ 371\ 582)$ I realized that the answer would be approximate since 1.25^7 is a decimal with many digits.

$\quad \doteq 23\ 841$

One week later, there are approximately 23 841 yeast cells.

EXAMPLE 3	Selecting a problem-solving strategy to make a prediction

According to the U.S. Census Bureau's World Population Clock, the number of
people in the world in 1950 was 2.5 billion. The population has grown at a rate of
approximately 1.7% per year since then. Create an algebraic model to predict the
population in the future.

Daniel's Solution

Growth each year is 1.7%
(or $1.7 \div 100 = 0.017$).

> To determine the population in the next year, I calculated
> 1.7% of 2.5 billion, and added it to the original population.

In 1950, $P_0 = 2.5$ billion;
In 1951, growth $= 0.017 \times P_0 = 0.0425$

> To make the notation easier, I used *subscripts*. For
> example, the population in year zero (1950) can be
> represented by P_0, and for year 1 (1951) it is P_1.

$P_1 = P_0 + 0.0425$
$P_1 = 2.5 + 0.0425 = 2.5425$ billion people

or

$P_1 = 2.5 + 0.017(2.5)$
$\quad = 2.5(1 + 0.017)$
$\quad = 2.5(1.017)$
$\quad = 2.5425$

> After a few calculations, I realized that I could do this in
> one step. I multiplied the population by 1 + 1.7% (or
> 1.017) to find the population in the next year.

So, $P_1 = P_0(1.017)$

In 1952, $P_2 = P_1(1.017)$

$P_2 = [P_0(1.017)](1.017)$

$\quad = P_0(1.017)^2$

Similarly,
In 1953, $P_3 = P_0(1.017)^3$

> If 1.017 is the growth rate each year, then I can find the
> population for any year if I know the population for the
> previous year. So I can calculate P_1 from P_0, P_2 from P_1, and
> so on. I can also calculate P_2 from P_0 just by substituting
> the formula for P_1.
>
> The number of the year is the same as the exponent on
> 1.017.

$P_n = P_0(1.017)^n$

where P_n is the population n years after 1950,
and $P_0 = 2.5$ billion

> If I continue this pattern, I can calculate the population for
> any year after 1950 (provided the growth rate stays the
> same). I tried the year 2020 as an example.

In 2020, $n = 70$

$P_{70} = 2.5(1.017)^{70}$

$\quad \doteq 8.1$ billion

> Since 2020 is 70 years after 1950 and the population was
> 2.5 billion in 1950, I calculated that the population in
> 2020 should be approximately 8.1 billion people.

In Summary

Key Ideas

- A graph can be used to estimate answers to problems involving exponential growth by interpolating and extrapolating where necessary.
- If you are given a function that models exponential growth, it can be used to make accurate predictions.
- If you know the initial amount and growth rate, the exponential function $P(n) = P_0(1 + r)^n$ can be used as a model to solve problems involving exponential growth where
  - $P(n)$ is the final amount or number
  - P_0 is the initial amount or number
  - r is the rate of growth
  - n is the number of growth periods

Need to Know

- The growth rate needs to be expressed as a fraction or decimal, *then* added to 1 in the exponential-growth equation.
- The units for the growth rate and for the number of growth periods must be compatible. For example, if a population growth rate is given as "per hour," then the number of growth periods in the equation is measured in hours.

CHECK *Your Understanding*

1. A brochure for a financial services company has a graph showing the value of a $4000 savings bond since 1996. Suppose the bond continued to increase in value at the same rate.
   a) How much will the bond be worth in 2008?
   b) How much would it be worth in 2025?
   c) Is it possible to determine the length of time needed for the savings bond to double its value from the graph? Explain.

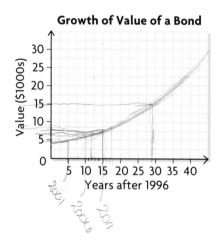

Growth of Value of a Bond

2. Algae in a pond grow at a rate of 10% per week. Currently the algae cover one-quarter of the pond. An algebraic model of this situation is $C(w) = 0.25(1 + 0.1)^w$, where $C(w)$ is the percent covered after w weeks.
   a) Explain what 0.25 represents in the equation.
   b) Explain what 0.1 represents in the equation.
   c) Determine how much of the pond (to the nearest percent) will be covered 10 weeks from now.

3. According to the 1991 census, the region of Niagara, Ontario, had 364 552 residents. For the next few years, the population grew at a rate of 2.2% per year. For planning purposes, the regional government needs to determine the population of the region in 2010. The algebraic model for this case is $P(n) = P_0(1 + r)^n$.
   a) What is the initial population, P_0?
   b) What is the growth rate, r?
   c) How many growth periods, n, are there?
   d) Write the algebraic model for this situation.
   e) Use the model to determine the population in 2010.

PRACTISING

4. a) In each of the algebraic models that follow, identify
   • the initial amount
   • the growth rate
   • the number of growth periods

   i) $M(n) = (18)(1.05)^{22}$
   ii) $P(n) = 64\,000(1.1)^{12}$
   iii) $N(n) = 750(2)^4$

   b) Evaluate each equation in part (a) to two decimal places.

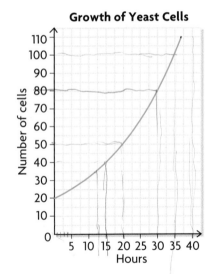

Growth of Yeast Cells

5. The size of a yeast culture is measured each hour, and the results are displayed on the graph shown.
   a) From the graph, estimate the initial number of yeast cells.
   b) Estimate the number of cells after 30 h.
   c) Use the initial amount to estimate the length of time required for the number of cells to double.
   d) Use an amount (other than the initial amount) to determine the length of time required for the number of cells present at that time to double.
   e) Describe what you noticed about the two doubling times. Check what you found with a different amount from the graph.

6. The number of guppies in an aquarium is modelled by the function
   K $N(t) = 12(1 + 0.04)^t$, where t is measured in weeks.
   a) Describe what each part of the equation represents.
   b) Determine the number of guppies in the aquarium after 10 weeks.
   c) Will this equation *always* model the population in the aquarium?

7. In each case, write an equation that models the situation described.
   a) An antique is purchased for $5000 in 1990. It appreciates in value by 3.25% each year.
   b) A town had 2500 residents in 1990. It grew at a rate of 0.5% per year for t years.
   c) A single bacterium of a particular type takes one day to double. The population is P after t days.

8. Mari invests $2000 in a bond that pays 6% per year.
   a) Write an equation that models the growth of her investment.
   b) How much money does she have if she cashes the bond at the end of the 4th year?
   c) How much will the bond be worth at the end of the 5th year? How can you determine the amount earned during the 4th year?
   d) Determine the amounts Mari will earn at the end of the 20th and 21st years to find the amount earned during the 20th year.
   e) Compare the money earned in the 4th and 20th years. What does this tell you about exponential growth?

9. Five hundred yeast cells in a bowl of warm water doubled in number every 40 min.
   A
   a) Create a graph of the number of yeast cells versus time.
   b) Use the graph to determine how long it would take for the total number of cells to triple (to the nearest minute).
   c) Describe how you can adapt your graph to determine the number of cells for the time *before* they were monitored.

10. An ant colony triples in number every month. Currently, there are 24 000 in the nest.
    a) What is the monthly growth rate of the colony? What is the initial population?
    b) Write an equation that models the number of ants in the colony, given the number of months.
    c) Use your equation to predict the size of the colony in three months.
    d) Use your equation to predict the size of the colony five months ago.

11. The number of franchises of a popular café has been growing
    C exponentially since the first store opened in 1971. Since then, the number of stores has grown at a rate of 33% per year.
    a) Explain how you could create an algebraic model that gives the number of stores in any year after 1971. Discuss how the information in the problem relates to your algebraic model.
    b) Use your model to predict the number of stores in 2010.

12. The population of Upper Canada between 1784 and 1830 grew at a
T rate of about 8% each year.

Year	Year Since 1784	Population (1000s)
1784	0	6
1791	7	10
1811	27	77
1824	40	150
1827	43	177
1830	46	213

a) Use a graphing calculator and enter the years since 1784 in L1 and population in L2, and create a scatter plot in a suitable window.

b) Consider an exponential equation that would fit this data to the form $P(n) = P_0(1 + r)^n$. Estimate the values of P_0 and r, and write the equation. Enter your equation into the equation editor and graph it.

c) Does your equation fit the data?

d) The population in 1842 was 487 000. Does your equation "predict" this number? Explain.

Extending

13. The height, $h(t)$, of a tree with respect to its age, t, in years is given by the formula

$$h(t) = \frac{20}{1 + 200(10)^{-0.3t}}$$

a) Determine the height of a 6-year-old tree.

b) Use technology to graph the function, h. Explain why the shape of the curve is appropriate for this situation.

14. The city of Mississauga has experienced rapid growth in recent years. It had a population of 234 975 in 1975 and 610 700 in 2000. Determine the annual growth rate of the population over the 1975–2005 period.

15. An old riddle says: Water lilies in a pond double each day. It takes 30 days for the lilies to completely cover the pond. On what day was the pond half full of water lilies?

7.7 Problems Involving Exponential Decay

GOAL

Use exponential functions to model and solve problems involving exponential decay.

YOU WILL NEED
- graphing calculator

LEARN ABOUT *the Math*

Scuba divers who photograph shipwrecks or coral reefs need to adjust to decreasing amounts of light as they descend into the ocean.

Frank knows that light intensity reduces by 2% for each metre below the water surface. His underwater camera requires a minimum of 60% of the light at the surface to operate without an additional light source. He has seen coral that he would like to photograph 23 m below the surface. His user manual gives the function $I(n) = I_0(0.98)^n$ for determining light intensity, where I_0 is the initial amount of light at the surface and n is the depth below the surface, in metres.

? Is Frank able to take pictures at a depth of 23 m without an additional light source?

EXAMPLE 1   Reasoning to solve an exponential decay problem

Determine whether Frank's dive to 23 m will require him to use an additional light source to take pictures.

Kira's Solution: Using an Algebraic Model

$I(n) = I_0 (0.98)^n$

> Since light intensity *decreases* with increasing depth, the decay factor must be less than 1. Since the light intensity diminishes by 2% per metre, I can represent the amount of light 1 m deep as 98% of the light intensity at the surface (100% − 2%). For every metre below the surface, I multiply the surface light intensity by 0.98.

$I(23) = 100(0.98)^{23}$

> I substituted 23 m for n and 100 for I_0 and evaluated.

$I(23) \doteq 100(0.628\ 347\ 28)$

$I(23) \doteq 62.8\%$

> $I(23)$ is above 60%.

Frank's camera will not require an additional light source.

David's Solution: Using a Calculator to Create a Graphical Model

Tech | *Support*

For help determining the point(s) of intersection between two functions, see Technical Appendix, B-11.

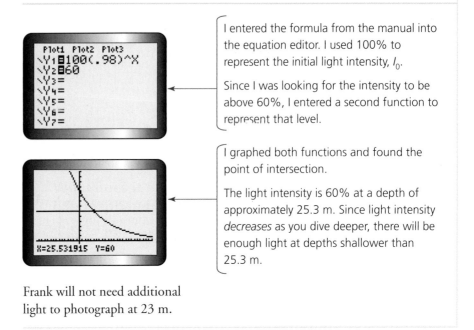

I entered the formula from the manual into the equation editor. I used 100% to represent the initial light intensity, I_0.

Since I was looking for the intensity to be above 60%, I entered a second function to represent that level.

I graphed both functions and found the point of intersection.

The light intensity is 60% at a depth of approximately 25.3 m. Since light intensity *decreases* as you dive deeper, there will be enough light at depths shallower than 25.3 m.

Frank will not need additional light to photograph at 23 m.

Reflecting

A. What part of the exponential function $I(n) = 100(0.98)^n$ indicates that it models exponential decay?

B. Compare the exponential-decay graph with the exponential-growth graph. Explain the differences.

C. Explain the differences in the approaches that Kira and David took to answer the question.

APPLY *the Math*

EXAMPLE **2** | Connecting graphs to half-life models

Archaeologists use carbon-14 to estimate the age of the artifacts they discover. Carbon-14 has a half-life of 5730 years. This means that the amount of carbon-14 in an artifact will be reduced by half every 5730 years.

An ancient animal bone was found near a construction site. Tests were conducted, and the bone was found to contain 20% of the amount of carbon-14 in a present day bone. Use the decay curve shown to estimate the age of the bone.

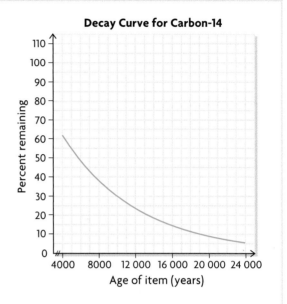

Eric's Solution

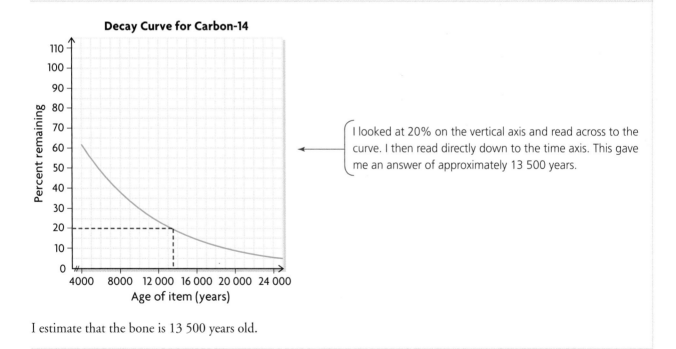

I looked at 20% on the vertical axis and read across to the curve. I then read directly down to the time axis. This gave me an answer of approximately 13 500 years.

I estimate that the bone is 13 500 years old.

EXAMPLE 3   Connecting functions to half-life models

A hospital uses cobalt-60 in its radiotherapy treatment for cancer patients. Cobalt-60 has a half-life of 5.2 years. This means that every 5.2 years, 50% of the original sample of cobalt-60 has decayed. The hospital has 80 g of cobalt-60. It requires at least 64 mg for the therapy machine to provide an effective amount of radiation.

a) How often must the staff replace the cobalt in their machine?
b) How much of the orignal sample will there be after 1 year?

Liz's Solution

a) $P(n) = P_0(1 - r)^n$

$P(n) = 80(0.5)^n$

> I created a function to model this situation. Since the rate of decay is $\frac{1}{2}$, I used the decimal 0.5. I used 80 for the initial amount, since that is the starting mass. $P(n)$ represents the amount of cobalt-60 remaining after n periods of 5.2 years.

n	M
1	$80(0.5)^1 = 40$
0.5	56.5685
0.4	60.6287
0.3	64.98
0.32	64.1

> I need to determine the exponent to get an answer of about 64 mg. I know that one half-life ($n = 1$) gives an answer of 40 mg. So I need a number less than 1 to start. I made a table to record my guesses. $n = 0.3$ is quite close, but too high, and $n = 0.4$ is too low. So I tried $n = 0.32$, which gave me a good approximation.

$n \doteq 0.32$

$\text{time} = 0.32 \times 5.2$

Therefore, $t \doteq 1.664$ years, or 20 months.

> To determine the number of years the cobalt-60 will last, I converted the number of half-lives, 0.32, to years by multiplying by 5.2, the length of one half-life.

The hospital should replenish its supply of cobalt-60 every 20 months to maintain an effective amount of radiation.

b) $P(n) = 80(0.5)^n$

> I used the function I created above.

$n = 1 \div 5.2 = 0.19$

> I need to determine the number of half-lives, n, that 1 year is equivalent to. I divided 1 by 5.2, since each half-life is 5.2 years.

$P(0.19) = 80(0.5)^{0.19}$

> I substituted 0.19 for n and evaluated.

$= 70.13$

About 70.13 g of cobalt-60 will remain after 1 year.

CHECK Your Understanding

1. **a)** In each of the algebraic models that follow, identify
   - the initial amount
   - the decay rate
   - the number of decay periods
     - **i)** $M(n) = 100(1 - 0.25)^{28}$
     - **ii)** $P(n) = 32\,000(1 - 0.44)^{12}$
     - **iii)** $N(n) = 500(1 - 0.025)^{20}$

   b) Evaluate each expression to the nearest hundredth of a unit.

2. The radioactive decay curve for strontium-90 is shown.
   a) Estimate the percent of strontium-90 remaining after 20 years.
   b) How much time is required for a sample to decay to 25% of its original mass?

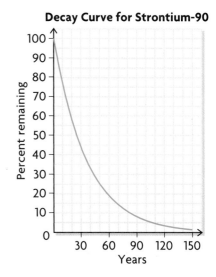

Decay Curve for Strontium-90

3. A new car costs $24 000. It loses 18% of its value each year after it is purchased. This type of loss is called *depreciation*. The value of the car is given by $V(n) = V_0(1 - 0.18)^n$, where V_0 is the original value of the car, and n is the number of years after the car was purchased.

   a) Use the formula to determine how much of the car's initial value is lost after 5 years.

   b) Use the formula to determine the value of the car after 30 months.

PRACTISING

4. Solve each equation. Express each answer to the nearest hundredth.

   a) $y = 52(0.87)^7$

   b) $N(t) = 100(0.99)^6$

   c) $P(t) = 512\left(\dfrac{1}{2}\right)^9$

5. After being filled, a basketball loses 3.2% of its air every day. An equation that models this situation is $V(d) = V_0(1 - 0.032)^d$.

   a) Describe what each part of the equation represents.

   b) The initial volume of air in the ball was 840 cm³. Determine the volume after 4 days.

   c) Will this model be valid after several weeks? Explain why or why not.

6. A lab has 200 grams of an unknown radioactive substance. The scientists in the lab measure the mass of the substance each minute and plot the curve shown.

Decay Curve for Unknown Substance

(graph: y-axis "Mass remaining (g)" with values 0, 40, 80, 120, 160, 200; x-axis "Time (min)" with values 5, 10, 15, 20, 25, 30)

   a) How many grams of the substance remain after 18 min?

   b) Use the graph to determine the half-life of the substance.

7. Gels used to change the colour of spotlights each reduce the intensity of the light by 4%. The algebraic model for this situation is $I = 100(0.96)^n$.

   a) Describe what each part of the equation represents.

   b) Determine the intensity of the spotlight if three gels are used.

   c) How many gels would reduce the intensity by more than 75%?

8. The value of a car after it is purchased depreciates according to the
A formula $V(n) = 25\,000(0.85)^n$, where $V(n)$ is the car's value in the
nth year after it was purchased.

a) What is the purchase price of the car?
b) What is the annual rate of depreciation?
c) What is the car's value at the end of 3 years?
d) How much value does the car lose in its first year?
e) How much value does it lose in its fifth year?
f) After how many years will the value of the car be half of the
   original purchase price?

9. A hot cup of coffee cools according to the equation

$$T(t) = 68\left(\frac{1}{2}\right)^{\frac{t}{22}} + 18$$

where $T(t)$ is the temperature, in degrees Celsius, and t is the time in
minutes.

a) Which part of the equation indicates that it models exponential
   decay?
b) What value of t makes the exponent in the equation equal to 1?
c) What is the significance of this value?
d) What was the initial temperature of the coffee?
e) Determine the temperature of the coffee after 40 min, to the
   nearest degree.

10. Old Aboriginal wooden tools were found at an archaeological dig
in the Brantford area. Carbon-14 dating was used to determine
the age of the tools. Carbon-14 has a half-life of 5730 years. The
general equation that models radioactive decay is

$$A(t) = 100\left(\frac{1}{2}\right)^{\frac{t}{H}}$$

where the initial radioactivity of the tools is 100%, t is time in
years, $A(t)$ is the radioactivity of the tools today, and H is the
half-life of the radioactive substance.

a) Archaeologists guessed that the tools were about 6000 years old.
   What percent of the present-day radioactivity would the tools
   emit if that were the case? Express your answer to the nearest tenth
   of a percent.
b) The tools' actual radioactivity was 56% of the radioactivity of the
   same type of present-day material. Determine the age of the wood
   to the nearest hundred years.

11. Write an equation to model each situation. In each case, describe how you can tell that it is an example of exponential decay.

    a) The value of my car was \$25 000 when I purchased it. The car depreciates at a rate of 25% each year.

    b) The radioactive isotope U_{238} has a half-life of 4.5×10^9 years.

    c) Light intensity in a pond is reduced by 12% per metre of depth, relative to the light intensity at the surface.

12. The population of a small mining town was 13 700 in 2000. Each year, the population decreases by an average of 5%.

    a) Explain why the situation described is an example of exponential decay.

    b) Write an exponential function that models this situation. Explain each part of the equation.

    c) Use your equation to estimate when the population will drop below 5000.

13. A rubber ball is dropped from a height of 5 m. It bounces to a height that is 80% of its previous maximum height after each bounce.

    a) Complete the table to determine the height of the ball after the 4th bounce.

    b) The equation that models the maximum height of the ball after each bounce is $H(n) = 5(0.80)^n$. Use the equation to verify the height of the ball after 4 bounces

    c) Determine the height of the ball after 7 bounces.

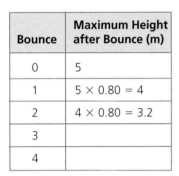

Bounce	Maximum Height after Bounce (m)
0	5
1	$5 \times 0.80 = 4$
2	$4 \times 0.80 = 3.2$
3	
4	

Extending

14. A blue shirt loses 0.5% of its colour every time it is washed. Once the shirt has lost 15% of its colour, it is no longer desirable to wear.

    a) Describe two methods for determining how many times the shirt can be washed before it must be discarded.

    b) Determine how many times the shirt can be washed before it must be discarded.

15. Frozen carbon dioxide changes directly from a solid to a gas in a process known as *sublimation*. For this reason, it is sometimes called "dry ice." Suppose a 50 kg block of dry ice lost 10 kg in 24 h due to sublimation.

    a) Determine the percent lost in 24 h.

    b) Write an equation that models the mass, M, of dry ice left after d days.

    c) Use your equation to determine the amount of dry ice left after 54 h.

Disappearing Coins

You drop 100 pennies onto a table and remove all of those that turn up "tails." You repeat this over and over for the coins that remain. What function models this situation?

1. Drop your pennies onto a table. Remove all the coins that turned up "tails."

2. Record your data in a table like the one shown.

Trial #	Number of Coins	Number of Tails Removed	Number of Coins that Remain
0	100		
1			
2			

3. Repeat parts 1 and 2 with the remaining coins until you have no coins left.

4. Create a scatter plot of number of coins versus trial number. What type of function does this appear to be?

5. Compare your results with other classmates. How many trials are needed until all the coins are gone?

6. Determine the equation of a function that shows the relationship between the number of coins remaining and the trial number.

7. If you were to use thumbtacks instead of coins and took away the tacks that landed tip up, do you think the scatter plot and function would be the same? Explain.

FREQUENTLY ASKED Questions

Q: How can you identify an exponential function from
- its equation?
- its graph?
- a table of values?

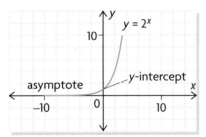

A: The basic exponential function has the form $f(x) = b^x$, where $b > 0$.

The shape of its graph depends upon the parameter, b.

If $b > 1$, the curve increases as x increases.
If $0 < b < 1$, the curve decreases as x increases.

Each of the functions has the x-axis as its horizontal *asymptote*.

If you create a difference table for an exponential function, you will notice that the differences are never constant, as they are for linear and quadratic functions. Instead, the differences are related by a multiplication pattern.

x	$f(x) = 2^x$	First Differences	Second Differences
0	1		
		1	
1	2		1
		2	
2	4		2
		4	
3	8		4
		8	
4	16		8
		16	
5	32		16
		32	
6	64		

Study | Aid

- See Lesson 7.5, Key Ideas.
- Try Chapter Review Question 7.

Q: How can exponential functions model growth and decay? How can you use them to solve problems?

A: Exponential functions can be used to model situations that show repeated multiplication by the same factor. Here are some examples:

Growth	Decay
Population $P(n) = P_0(1 + r)^n$	Depreciation of assets $V(n) = V_0 (1 - r)^n$
Cell division (bacteria, yeast cells, etc.) $P(n) = P_0 (1 + r)^n$	Radioactivity or half-life $Q(n) = 100\left(\dfrac{1}{2}\right)^n$
Compound interest $A(n) = P(1 + i)^n$	Light intensity in water $V(n) = 100(1 - r)^n$

If you are given a graph, you can estimate growth and decay by interpolating and extrapolating as needed.

If you are given a function, you can make accurate predictions by substituting the given information into the function and solving for the unknown quantity.

You can develop a function based on the general exponential function and use it to make accurate predictions if you know the initial amount, P_0, and the rate of growth or decay, r.

Study Aid

- See Lessons 7.6 and 7.7, Examples 1, 2, and 3 in each.
- Try Chapter Review Questions 8 to 13.

Exponential Growth	Exponential Decay
$P(n) = P_0(1 + r)^n$	$P(n) = P_0(1 - r)^n$

n is the number of growth/decay periods and $P(n)$ is the amount in the future.

PRACTICE Questions

Lesson 7.2

1. Write as a single power. Express answers with positive exponents.

a) $3^4 \times 3^8 \times 3$

b) $\dfrac{(-5)^6}{(-5)^4}$

c) $\dfrac{11^5}{11^9}$

d) $\left((-9)^2\right)^5$

e) $\dfrac{4^7 4^5}{4^{12}}$

f) $\dfrac{6^{10}}{(6^6)^2}$

Lesson 7.3

2. Write as a single power. Express answers with positive exponents.

a) $12^{-6} \times 12^8 \times 12^0$

b) $\dfrac{\left((-8)^6\right)^{-2}}{\left((-8)^{-4}\right)^3}$

c) $\dfrac{(20^{-1})^8}{20^2 20^6}$

d) $\left(10(10^3)^{-1}\right)^{-2}$

3. Write as a single power. Express answers with positive exponents.

a) $\dfrac{a^5}{a^3}$

b) $(b)(b^4)(b^2)$

c) $\dfrac{c^3}{c^9}$

d) $(d^6)^3$

e) $\dfrac{e(e^5)}{e^7}$

f) $(f^{-3})^{-2}$

Lesson 7.4

4. Copy and complete the table.

	Exponential Form	Radical Form	Evaluation of Expression
a)	$36^{\frac{1}{2}}$		
b)	$16^{\frac{5}{4}}$		
c)		$\sqrt[5]{1024}$	
d)	$16\,807^{0.2}$		
e)		$\sqrt[3]{-216^4}$	

5. Use your calculator to evaluate each expression. Express answers to two decimals.

a) $125^{0.33}$

b) $\sqrt[3]{-1953.125}$

c) $\sqrt[7]{-180}$

d) $16^{\frac{2}{3}}$

e) $10^{-\frac{3}{2}}$

f) $\sqrt[12]{1.9}$

6. Evaluate each expression without using a calculator.

a) $-125^{\frac{1}{3}}$

b) $81^{0.25}$

c) $\sqrt[3]{27}$

d) $16^{\frac{3}{2}}$

e) $256^{-\frac{5}{4}}$

f) $\sqrt[5]{-32}$

Lesson 7.5

7. Calculate the finite differences for each table of values. Then use the finite differences to classify each function as linear, quadratic, or exponential.

a)

x	y
1	−1
2	3
3	9
4	17
5	27
6	39

b)

x	y
−2	3
−1	8
0	13
1	18
2	23
3	28

c)

x	y
−1	0.25
1	0.5
3	1
5	2
7	4
9	8

d)

x	y
−2	10
−1	30
0	90
1	270
2	810
3	2430

Lesson 7.6

8. A city has a population increase of 3% per year from 1990 to 2007. In 1990 the population was 110 000.

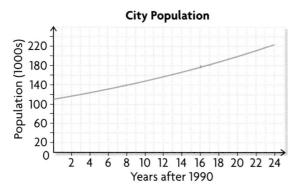

City Population

a) Use the graph to estimate the population in 2006.
b) How long did it take for the population to increase by 30 000 from its 1990 value?
c) Estimate the *doubling period* of the population if it continues to grow at this rate. Explain your steps.

9. A biologist measures 1000 yeast cells in a culture at 12:00 noon. She predicts that, under current conditions, the culture should double every 50 min.
a) Sketch a graph of the population of this culture for 200 min after noon.
b) Use your graph to determine the population of the culture in 2.5 h.
c) How can you adapt the graph to determine what the population of cells was 3 h *before* noon?

10. In each case, write a function that models the situation described.
a) A rare coin is purchased for $2500 in 2000. It appreciates in value by 5% each year for t years.
b) A school has 750 students in 2003. It grew at a rate of 2% per year for t years.
c) Evaluate the functions you found in parts (a) and (b) when $t = 10$. Explain what these numbers represent.

11. Simon has a Wayne Gretzky rookie hockey card. He bought it in 2000 for $500 on e-Bay. He estimates that it will appreciate in value by 7% each year. The dealer told him the function $V(t) = 500(1 + 0.07)^t$ can be used to estimate the card's future value.
a) Explain how the numbers in the equation are related to the situation.
b) How much will the card be worth in 2020?
c) How long will it take for the card to double its value?

Lesson 7.7

12. A police diver is searching a harbour for stolen goods. The equation that models the intensity of light per metre of depth is $I(n) = 100(0.94)^n$.
a) Describe each part of the equation.
b) Determine the amount of sunlight the diver will have at a depth of 16 m, relative to the intensity at the surface.

13. Copy and complete the table.

Function	Exponential Growth or Decay	Initial Value (*y*-intercept)	Growth/ Decay Rate
$V(t) = 125(0.78)^t$			
$P(t) = 0.12(1.05)^t$			
$A(x) = (2)^x$			
$Q(x) = 0.85\left(\dfrac{1}{3}\right)^x$			

14. Jerry invests $500 in a bond that pays 5% per year. He will need the money for college in 3 years.
a) Write an equation that models the growth of the money.
b) Use the equation to determine how much Jerry will have at the end of 3 years.
c) How much money did his $500 earn in the 3 years?
d) Jerry thinks that if he keeps his money invested for twice as long (6 years), he will earn twice as much. Is this true? Explain your reasoning.

1. Evaluate without using a calculator.

   a) 5^{-3}

   c) $8^{\frac{1}{3}}$

   e) -7^0

   b) $\left(\dfrac{3}{4}\right)^{-2}$

   d) $16^{-0.75}$

   f) $100^{\frac{-3}{2}}$

2. Write as a single power. Express answers with a positive exponent.

   a) $(6)^{-\frac{1}{3}} \times (6)^{\frac{5}{6}}$

   c) $\dfrac{10}{10^{-4}}$

   e) $a^7(a^6)^{-2}$

   b) $4\left(\dfrac{1}{4}\right)^{-4}$

   d) $\dfrac{7^8}{(7^2)^3}$

   f) $\dfrac{b^3(b^{-2})}{b^4}$

3. Write $\sqrt[6]{4^3}$ in exponent form, then evaluate.

4. Sketch the graph of each function. If applicable, label the x- and y-intercepts and asymptotes.

   a) $y = 2^x$

   b) $y = 0.5^x$

5. The values of two different automobiles over time are shown in the graph.

   a) Compare the initial value of each car with its value through the first 6 years of ownership.

   b) Which car has the higher depreciation rate? Explain your reasoning.

Depreciation of Car

6. An archaeologist discovers an ancient settlement. To determine the age of the settlement, she measures the radioactivity of a fragment of bone recovered at the site. Carbon-14 has a half-life of 5730 years. The algebraic model for the radioactivity of carbon-14 is

   $$A(t) = 100\left(\dfrac{1}{2}\right)^{\frac{t}{5730}}$$

   Determine the radioactivity of the bone, to the nearest percent, if it is 12 000 years old.

7. The population of a small town has increased at a rate of 1.5% per year since 1980. The town had a population of 1600 that year.

   a) Write the equation that models the growth in population of the town. Describe each part of your equation.

   b) Use your equation to determine the population of the town in 2008.

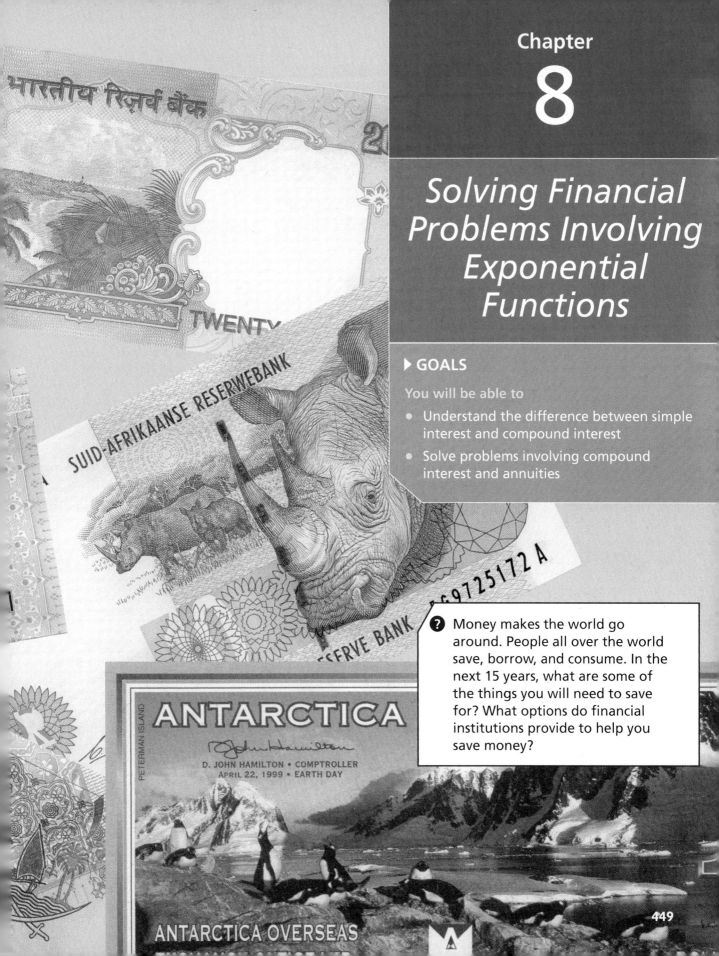

Solving Financial Problems Involving Exponential Functions

▶ **GOALS**

You will be able to

- Understand the difference between simple interest and compound interest
- Solve problems involving compound interest and annuities

❓ Money makes the world go around. People all over the world save, borrow, and consume. In the next 15 years, what are some of the things you will need to save for? What options do financial institutions provide to help you save money?

WORDS *You Need to Know*

1. Match each term with the example that illustrates it.
   - **a)** interest
   - **b)** interest rate
   - **c)** loan
   - **d)** investment

   i) 5% per year
   ii) Last month I earned $12.50 in my savings account.
   iii) Each month I deposit $50 into a Registered Retirement Savings Plan.
   iv) I bought a new TV and financed most of the cost.

SKILLS AND CONCEPTS *You Need*

Expressing Percent as a Decimal

A percent is a way of expressing a number as a fraction of 100 (*per cent* means "per hundred").

EXAMPLE

Convert each percent to a decimal.
a) 27%   **b)** 104%   **c)** 0.5%

Solution

$$27\% = \frac{27}{100}$$
$$= 0.27$$

$$104\% = \frac{104}{100}$$
$$= 1.04$$

$$0.5\% = \frac{0.5}{100}$$
$$= 0.005$$

2. Convert each percent to a decimal.
   - **a)** 35%
   - **b)** 67%
   - **c)** 8.5%
   - **d)** 2.75%

Calculating the Percent of a Number

To determine the percent of a number, change the percent into a decimal by dividing by 100 and multiply by the number you are finding the percent of.

EXAMPLE

Determine each amount.
a) 5% of 60   **b)** 3.5% of 220

Solution

$$5\% = 0.05 \times 60$$
$$= 3$$

$$3.5\% = 0.035 \times 220$$
$$= 7.7$$

3. Determine each amount.
   - **a)** 15% of 75
   - **b)** 75% of 68
   - **c)** 3.5% of 60
   - **d)** 7.25% of 2000

Evaluating Exponential Functions

When evaluating exponential functions, substitute the given value for x, then calculate the value of the function. Remember BEDMAS!

EXAMPLE

If $f(x) = 4(3)^{2x}$, determine $f(-2)$.

Solution

$$f(-2) = 4(3)^{2(-2)}$$
$$= 4(3)^{-4}$$
$$= 4\left(\frac{1}{3^4}\right)$$
$$= 4\left(\frac{1}{81}\right)$$
$$= \frac{4}{81}$$

4. a) If $f(x) = 3(4)^{2x}$, determine $f(3)$ and $f(-3)$.
   b) If $f(x) = 5(2)^{3x}$, determine $f(2)$ and $f(-2)$.
   c) If $f(x) = 6(1.5)^x$, determine $f(1)$ and $f(-1)$.
   d) If $f(x) = 20\ 000(1.02)^x$, determine $f(10)$ and $f(20)$.

Relating Units of Time

One year is the length of time it takes for Earth to complete its orbit around the Sun. One day is the length of time for Earth to rotate once, $360°$, on its axis.

1 year = 365 days or 52 weeks or 12 months

EXAMPLE

Express each of the following as a fraction of a year.
a) 4 weeks   **b)** 3 months   **c)** 200 days

Solution

a) 4 weeks
$= \frac{4}{52}$ of a year
$= \frac{1}{13}$ of a year

b) 3 months
$= \frac{3}{12}$ of a year
$= \frac{1}{4}$ of a year

c) 200 days
$= \frac{200}{365}$
$= \frac{40}{73}$ of a year

5. Express each of the following as a fraction of a year.
  a) 30 days
  b) 26 weeks
  c) 8 months
  d) 400 days
  e) 100 weeks
  f) 18 months

6. a) Determine the number of months in 3.5 years.
  b) Determine the number of weeks in 2.25 years.
  c) Determine the number of days in 5 years.
  d) Determine the number of months in 0.25 years.

PRACTICE

Study | **Aid**

• For help, see Essential Skills Appendix.

Question	Appendix
10	A-7

7. Write each percent as a decimal.
  a) 35%
  b) 5%
  c) 14.6 %
  d) 115%
  e) $2\frac{3}{4}\%$
  f) $14\frac{3}{4}\%$

8. Evaluate.
  a) 3.25% of $150
  b) 14% of $28
  c) 110% of $225
  d) 1.5% of $2000
  e) 4% of $75
  f) 400% of $500

9. Evaluate to two decimal places.
  a) $(1.10)^{12}$
  b) $(1.03)^{-24}$
  c) $1000(1.10)^4$
  d) $5000(1.07)^{-12}$
  e) $100 \times \dfrac{(1.12)^{10} - 1}{0.12}$
  f) $575 \times \dfrac{1 - (1.08)^{48}}{0.08}$

10. Sketch the graph of each function.
  a) $f(x) = 4x - 5$
  b) $f(x) = -\dfrac{2}{3}x + 6$
  c) $f(x) = 1.5x + 3$
  d) $f(x) = -2.25x - 10$

11. Sketch the graph of each function.
  a) $f(x) = 2^x$
  b) $f(x) = \left(\dfrac{1}{3}\right)^x$
  c) $f(x) = 5(3)^x$
  d) $f(x) = 10\left(\dfrac{1}{2}\right)^x$

12. An antique postage stamp appreciates by 9% of its value each year. The stamp was worth $0.34 in 1969. What will its value be in 2012?

13. A new car bought for $25 600 loses 12% of its value each year. Determine its value at the end of 7 years.

APPLYING *What You Know*

Saving Money

Suppose you deposit $40 into a bank account at the beginning of January and continue to put $40 into the account at the beginning of every month for the rest of the year. The bank account earns 6% interest per year and the interest is paid into the account at the end of every month.

❓ **How much money will be in the account at the end of the year?**

A. Why is the monthly interest rate 0.5%? Why can you write it as 0.005?

B. The table shows the interest earned in January and the end-of-month balance. At the beginning of February, another $40 deposit is made. Copy the table and then enter the new balance at the start of February.

Month	$40, 6% Annual Interest			
	Deposit Made on First Day of Month ($)	**Start-of-Month Balance ($)**	**Interest Earned During the Month ($)**	**End-of-Month Balance ($)**
January	40.00	40.00	0.005 × 40 = 0.20	40 + 0.20 = 40.20
February	40.00			
March	40.00			
April	40.00			
May	40.00			
June	40.00			
July	40.00			
August	40.00			
September	40.00			
October	40.00			
November	40.00			
December	40.00			

C. Calculate the interest earned in February and the new balance at the end of the month. Enter these amounts into the table.

D. Repeat parts B and C for each of the remaining months in the table.

E. Determine the total amount of interest earned for the year.

F. Determine the amount in the account at the end of the year.

8.1

Investigating Interest and Rates of Change

YOU WILL NEED

- graphing calculator (optional)
- graph paper

simple interest

interest earned or paid only on the original sum of money invested or borrowed

compound interest

interest calculated at regular periods and added to the principal for the next period

principal

a sum of money that is borrowed or invested

Communication | **Tip**

The duration of an investment, or the time required to pay off a loan, is called the **term**.

GOAL

Identify the difference between simple interest and compound interest for a given principal.

INVESTIGATE *the Math*

The Canada Savings Bond (CSB) was created in 1946 by the government to help people reach their savings and investment goals. Regular-interest CSBs earn **simple interest**. The interest earned each year is deposited directly into the investor's bank account. **Compound-interest** CSBs also earn interest annually. However, the interest earned each year is reinvested, so that each year interest is earned not only on the **principal**, but also on the interest earned in previous years.

Sonia buys a $1000 regular-interest CSB. It has a 10-year term and earns 5% interest annually.

Zuhal invests $1000 in a compound-interest CSB. It has a 10-year term and earns 5% interest annually.

? How does the amount of interest earned on Canada Savings Bonds differ for simple interest and compound interest?

A. Copy and complete the savings tables for Sonia and Zuhal.

Year	Principal for Year ($)	Interest Earned ($)	Accumulated Interest ($)	Amount at End of Year ($)
1	1000	0.05 × 1000 = 50	50	1000 + 50 = 1050
2	1000	0.05 × 1000 = 50	50 + 50 = 100	1050 + 50 = 1100
3	1000	0.05 × 1000 = 50	50 + 100 = 150	1100 + 50 = 1150
4	1000			
5	1000			
6	1000			
7	1000			
8	1000			
9	1000			
10	1000			

Zuhal's Table

Year	Principal for Year ($)	Interest Earned ($)	Accumulated Interest ($)	Amount at End of Year ($)
1	1000	0.05 × 1000 = 50	50	1000 + 50 = 1050
2	1050	0.05 × 1050 = 52.50	50 + 52.50 = 102.50	1050 + 52.50 = 1102.50
3	1102.50	0.05 × 1102.50 = 53.13	102.50 + 53.13 = 157.63	1102.50 + 53.13 = 1157.63
4	1157.63			
5				
6				
7				
8				
9				
10				

Sonia's Table

B. On a single graph, plot the amount at the end of year versus time for both of the investments. Label the plots as simple interest and compound interest. How are the graphs the same? How are they different?

C. Create a table of first differences for the amounts at the end of the year for both types of savings bonds. Describe any patterns you see.

D. For each type of CSB, what kind of relationship exists between the year and the amount at the end of the year? Explain how you know.

E. If both investments were kept for another two years,
   a) how much simple interest would be earned in year 11 and year 12? Explain.
   b) how much compound interest would be earned in year 11 and year 12? Explain.

F. Which type of savings bond earns more interest by the end of year 10? Why?

Reflecting

G. Which type of Canada Savings Bond will double its value faster? Explain how you know.

H. Examine Sonia's table. How can you determine the amount of simple interest earned on a principal of P dollars invested at an interest rate of r% for t years?

APPLY the Math

EXAMPLE 1	Using a formula to determine the value of an investment earning simple interest

Kevin invested \$4500 in a two-year regular-interest CSB that earns $7\frac{1}{2}\%$ annually.

a) How much interest did he earn?

b) How much will his investment be worth at the end of the term?

Patricia's Solution

a) $I = Prt$

$P = \$4500$

$r = 7\frac{1}{2}\%$

$ = \dfrac{7.5}{100}$

$ = 0.075$

$t = 2$

$I = 4500 \times 0.075 \times 2$

$ = 675$

Kevin earned \$675 in interest over two years.

> The total interest earned over more than one year is the product of the yearly interest, Pr, and the number of years, t.
>
> I used the values for P, r, and t for Kevin's situation. To describe the interest rate as a decimal, I divided the percent by 100. Then I substituted into the formula.

amount

the sum of the original principal and the interest; given by $A = P + I$, where A is the amount, P is the principal, and I is the interest

b) $A = P + I$

$ = 4500 + 675$

$ = 5175$

The savings bond is worth \$5175 at the end of two years.

> I calculated the final **amount** of the investment by adding the principal to the interest he earned.

EXAMPLE **2**

Selecting a strategy to determine the amount of simple interest earned for part of a year

Keila had a credit card balance of $550 that was 31 days overdue. The annual interest rate on the card is 23.9%.

a) How much interest did Keila have to pay?

b) Explain why paying interest on an outstanding credit card balance is sometimes referred to as "the cost of borrowing money."

Eusebio's Solution

a) $P = \$550$

$r = 23.9\% = \dfrac{23.9}{100} = 0.239$

> I knew I could use the formula $I = Prt$ to calculate the interest over t years. So I needed to identify P, r, and t.

$t = \dfrac{31}{365}$

$I = Prt$

> Since the interest rate is expressed in terms of a year, the time must also be written in terms of years. There are 365 days in a year.

$= 550 \times 0.239 \times \dfrac{31}{365}$

> I substituted the values for P, r, and t, and multiplied.

$\doteq 11.16$

Therefore, Keila paid $11.16 in interest for the 31 days.

b) $11.16 was the interest she owed for the late payment. It was like borrowing the money for the extra 31 days. The interest is what borrowing the money cost her.

> If Keila had paid the credit card balance when it was due, the amount paid would have been $550. Because she waited 31 days, she had to pay $561.16.

EXAMPLE **3**

Selecting a strategy to determine the principal of a simple-interest loan

Five years ago, Jason lent Matt money. Matt repaid Jason a total of $2100, which included simple interest charged at 10%. How much did Jason originally lend Matt?

Kendra's Solution

$I = Prt$

$A = P + I$

$= P + Prt$

$= P(1 + rt)$

> I wrote the formula for determining simple interest. I added the interest to the principal to determine the amount. I then factored out the common factor P.

$A = \$2100$

$r = 10\% = 0.1$

$t = 5$ years

> I substituted these values to calculate the principal, P.

$2100 = P(1 + 0.1 \times 5)$

$2100 = 1.5P$

$P = \dfrac{2100}{1.5}$

$P = 1400.00$

Five years ago, Jason loaned $1400 to Matt.

EXAMPLE 4 | **Selecting a strategy to calculate compound interest**

Mohsin bought a $500 Guaranteed Investment Certificate (GIC). It has a 3-year term and earns 3.25% compounded annually. How much interest will the GIC have earned at maturity?

Edwin's Solution

Year	Principal (Amount at Start of Year) ($)	Interest Earned ($)	Amount at the End of the Year ($)
1	500	$0.0325 \times 500 = 16.25$	$500 + 16.25 = 516.25$
2	516.25	$0.0325 \times 516.25 = 16.78$	$516.25 + 16.78 = 533.03$
3	533.03	$0.0325 \times 533.03 = 17.32$	$533.03 + 17.32 = 550.35$

> The original principal is $P = \$500$.
>
> The interest rate is 3.25%, or 0.0325 as a decimal.
>
> The time is 3 years.
>
> Since the interest is compounded annually, the interest earned will be added to the principal at the beginning of the next year. I set up a table to help me keep track of the principal, interest, and year-end amounts.

$550.35 - 500 = 50.35$

The interest earned in 3 years is $50.35.

> Mohsin started with $500.00 and ended with $550.35. The interest he earned is the difference.

Check:

$16.25 + 16.78 + 17.32 = 50.35$

> I checked my calculation. First I added the interest earned each year. Then I added the total interest earned to the original investment. The result was the same.

In Summary

Key Ideas

- Simple interest is calculated by applying the interest rate only to the original principal amount, resulting in linear growth.
- Compound interest is calculated by applying the interest rate to the original principal and any accumulated interest, resulting in exponential growth.

Need to Know

- The interest rate is converted to decimal form prior to calculating interest earned.
- Simple interest can be calculated with the formula $I = Prt$, where
  - I is the interest, earned in dollars
  - P is the principal invested or borrowed, in dollars
  - r is the annual interest rate, expressed as a decimal
  - t is the time, in years
- The amount that a simple-interest investment or loan is worth can be calculated with the formula $A = P + Prt$ or its factored form, $A = P(1 + rt)$, where
  - A is the final amount of the investment or loan, in dollars
  - P is the principal, in dollars
  - r is the annual interest rate, expressed as a decimal
  - t is the time, in years
- Tables are useful as tools for organizing calculations involving compound interest.

CHECK Your Understanding

1. Calculate the simple interest earned in 1 year.
   a) $100 invested at 3%   **b)** $100 invested at 4.5%

2. Each simple-interest investment matures in 2 years. Calculate the interest and the final amount.
   a) $675 invested at 7.25%   **b)** $4261 invested at 13.75%

3. Each investment matures in 3 years. The interest compounds annually. Calculate the interest and the final amount.
   a) $600 invested at 5%   **b)** $750 invested at $4\frac{3}{4}$%

PRACTISING

4. Calculate the simple interest earned or due and the amount at the end of each term.
  a) $500 invested at 6% for 4 years
  b) $2000 invested at 4.8% for 5 years
  c) $1250 borrowed at 3% for 18 months
  d) $1000 borrowed at 10% for 12 weeks
  e) $5000 borrowed at $5\frac{1}{2}$% for 40 days

5. Copy and complete the tables for an investment of $500.
K

	Regular-Interest CSB, $11\frac{1}{4}$%, 5 years		
Year	Interest Earned ($)	Accumulated Interest ($)	Amount at End of Year ($)
1			
2			
3			
4			
5			

	Compound-Interest CSB, $11\frac{1}{4}$%, 5 years		
Year	Interest Earned ($)	Accumulated Interest ($)	Amount at End of Year ($)
1			
2			
3			
4			
5			

6. Calculate the missing information in the table.

	Principal, P ($)	Interest Rate, r (%)	Time, t	Simple Interest, I ($)
a)	735.00	$5\frac{1}{2}$	27 days	
b)		8.25	240 days	138.25
c)	182.65	6.75		23.28
d)	260.00		2 months	16.50

7. An investment of $1500 earned $35.20 at a simple-interest rate of 5.5% per year. How long was the investment held?

8. Jennifer's investment account balance grew from $400 to $432.76 in 5 months. What annual rate of simple interest does her account pay?

9. Gordon has a Canada Savings Bond that pays $150 in simple interest
A each year. The annual interest rate is 4.75%. What principal did Gordon invest in the bond?

10. Natasha borrows $3600 to help pay for university tuition. The annual interest rate is 8%. She will start to repay the loan in 4 years.

  a) How much interest will have been added to the original amount if the interest is simple?

  b) How much interest will have been added to the original amount if the interest is compound?

11. Afzal can buy a $2000 GIC from a bank that earns 6.5% compounded annually for 5 years. At another bank, he can buy a $2000 GIC that earns 7% simple interest for 5 years.

  a) Which GIC earns more interest?

  b) How much more interest does it earn?

12. Explain the difference between simple interest and compound interest. Use examples to support your explanation.

Extending

13. Tricia invested in a 90-day term deposit that earned 5.75% simple interest annually. When it matured, she received $760.63, which she reinvested in a 270-day term deposit so that it would earn 6.25% annually.

  a) How much was originally invested?

  b) How much will she receive when the second term deposit matures?

> **Communication | Tip**
>
> An investment purchased from a bank, trust company, or credit union for a fixed period or term is called a **term deposit**.

14. Barry deposited $9000 in an account that pays 10% each year, where interest is compounded 4 times a year at the end of each 3-month period. How much will be in the account at the end of 3 years?

15. Azif has deposited $4800 into a savings account that pays 2.8% simple interest per year. When his balance passes $5000, the interest rate increases to 3.5% simple interest per year, with the amount of his investment so far becoming the new principal. If Azif leaves the money in the account for 5 years and makes no other deposits, how much will he have?

Compound Interest: Determining Future Value

Solve problems that involve calculating the amount of an investment for a variety of compounding periods.

INVESTIGATE the Math

(Photo courtesy Art Gallery of Ontario/Sean Weaver)

This new frontage and hosting centre at Toronto's Art Gallery of Ontario is designed by Canadian-born Frank Gehry, one of the world's leading architects. It has been funded by governments, corporate sponsors, and many individuals.

The gallery's president, Charles Baillie, and his wife, Marilyn, personally pledged $2 million, and later increased their donation to $5 million. The Art Gallery received this money in early 2005, but construction was not set to be completed until late 2008, so they deposited it into a bank account. The new hosting centre has been named Baillie Court in their honour.

❓ How can you develop an expression to calculate what the Baillie's $5 million new donation will be worth in 2008, if it earns interest at 10%/a compounded annually?

A. Copy and complete the investment table showing the growth of the donation to determine its future value.

Time Invested (years)	Calculation of Amount, Using Formula $A = P(1 + rt)$	Amount at End of Year, or Future Value ($)
1 (2005)	$A = 5\ 000\ 000(1 + 0.10(1))$ $= 5\ 000\ 000(1.1)$	5 500 000
2 (2005–2006)	$A = 5\ 500\ 000(1 + 0.10(1))$ $= 5\ 000\ 000(1.10)(1.10)$ $= 5\ 000\ 000(1.10)^2$	6 050 000
3 (2005–2007)		
4 (2005–2008)		

future value

the final amount (principal plus interest) of an investment or loan when it matures at the end of the investment or loan period

B. Examine the calculations for the amount at the end of each year. By what factor does the amount increase by each year? What expression could you use to represent this factor if i represents the interest rate, expressed as a decimal?

C. Consider how the investment would grow if the Art Gallery of Ontario decided to leave the money invested until 2014. Predict the amount of the investment at the end of 2014.

D. Check your prediction by extending the table you created in part A to show 6 more years of calculations.

E. Plot the amount at the end of each year versus time for the 10-year investment. What kind of growth does the graph suggest? Explain.

F. Examine the calculations for determining the future value at the end of each year. What is the expression that represents the future value at the end of 2008?

G. Calculate the future value at the end of 2008 from part F. How much of this amount is earned interest?

Reflecting

H. Write expressions that will represent the amount of the investment at the end of the 5th year, 10th year, and nth year.

I. Write a formula that will represent the amount of an investment earned on a principal invested for several years. Let
- A represent the amount or future value of the investment in dollars
- P represent the principal in dollars
- i represent the interest rate expressed as a decimal
- n represent the number of periods over which the investment is compounded

APPLY *the Math*

EXAMPLE 1 **Selecting a strategy to determine the amount for different compounding periods**

compounding period

each period over which compound interest is earned or charged in an investment or loan

Jayesh has $2000 to invest in a compound-interest account in which he would like to leave the money for 3 years. He considers three different **compounding periods:**

Option A: 8%/a compounded annually

Option B: 8%/a compounded semi-annually

Option C: 8%/a compounded quarterly

Communication | Tip

The expression 8%/a (read "8 percent per annum") is shorthand for "8% per year." So, for example, the quarterly interest rate is 8% divided by 4.

---

a) For each option, what is the amount at the end of 3 years? What is the interest earned?

b) How does changing the compounding period affect the amount of the investment? Why?

c) For each option, create a timeline to show the annual value of Jayesh's investment.

Soo-Lin's Solution

a) Option A:

$$A = P(1 + i)^n$$

$P = \$2000$

$i = 8\% = 0.08$

$n = 3 \times 1 = 3$

I wrote the formula for calculating the amount when a principal, P, earns compound interest at a rate of $i\%$ for each of n compounding periods.

The compounding period is 1 year.

I substituted these values into the formula.

$A = 2000(1 + 0.08)^3$

$ = 2000(1.08)^3$

$ \doteq 2519.42$

The amount after 3 years using Option A will be $2519.42.

Interest $= 2519.42 - 2000.00 = 519.42$

I subtracted the principal from the amount to get the interest.

The interest earned in 3 years is $519.42.

Option B:

$P = \$2000$

$i = \dfrac{0.08}{2} = 0.04$

Interest is compounded semi-annually. I had to divide the interest rate by 2 to get the rate for each compounding period.

$n = 3 \times 2 = 6$

I multiplied the number of years by 2 to calculate the number of compounding periods.

$A = 2000(1.04)^6$

$\quad \doteq 2530.64$

The amount after 3 years using Option B will be $2530.64.

Interest $= 2530.64 - 2000.00 = 530.64$

The interest earned in 3 years is $530.64.

Option C:

$P = \$2000$

$i = \dfrac{0.08}{4} = 0.02$ ⟵——————— Interest is compounded quarterly. I had to divide the interest rate by 4 to get the rate for each compounding period.

$n = 3 \times 4 = 12$ ⟵——————— I multiplied the number of years by 4 to calculate the number of compounding periods.

$A = 2000(1.02)^{12}$

$\quad \doteq 2536.48$

The amount after 3 years using Option C will be $2536.48.

Interest $= 2536.48 - 2000.00 = 536.48$

The interest earned in 3 years is $536.48.

b) The amount of an investment increases as the number of compounding periods increases. Since changing the compounding period changes both the interest rate and the number of periods, the amount, A, of the investment is also changed.

⟵ Changing the compounding period changes the interest, i, used in the formula. This is because the annual interest rate has to be divided by the fraction of the year the compounding period is. The compounding period also affects the number of compounding periods over the 3 years.

c)

Option	Amount after Year t	Timeline	Interest Earned ($)
A	$2000(1 + 0.08)^t$	0 1 2 3 ⟶ 2000 2160.00 2332.80 2519.42	519.42
B	$2000\left(1 + \dfrac{0.08}{2}\right)^{2t}$	0 1 2 3 ⟶ 2000 2163.20 2339.72 2530.64	530.64
C	$2000\left(1 + \dfrac{0.08}{4}\right)^{4t}$	0 1 2 3 ⟶ 2000 2164.86 2343.32 2536.48	536.48

EXAMPLE **2**

Using a graph to represent and compare the amount of an investment over time

Alwynn invests $500 in an account that earns 6%/a compounded monthly.

Peter invests $500 at the same time, but in an account that earns 6%/a at simple interest.

a)   Determine the difference between their investments at the end of the 5th year.

b)   Using graphing technology, compare the balances in the accounts at the end of each month for 10 years.

c)   Discuss how the two investments are different.

Ella's Solution

a)   Alwynn's investment:

$$A = P(1 + i)^n$$

I used the formula for future value with compound interest.

$$P = 500$$

$$i = \frac{0.06}{12} = 0.005$$

I divided the annual rate by 12 to calculate the monthly rate.

$$n = 5 \times 12 = 60$$

I multiplied the number of years by 12 to calculate the number of compounding periods.

$$A = 500(1.005)^{60}$$

$$\doteq 674.43$$

Alwynn's investment is worth $674.43 after 5 years.

Peter's investment:

$$I = Prt$$

I used the formula for simple interest.

$$P = 500$$

$$r = 6\% = 0.06$$

$$t = 5$$

$$I = 500 \times 0.06 \times 5 = 150$$

$$A = P + I$$

$$= 500 + 150 = 650$$

I added the interest to the principal to determine the amount.

Peter's investment is worth $650 after 5 years.

The difference is $674.43 − $650 = $24.43.

b)

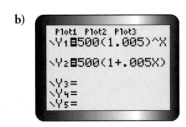
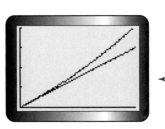

To compare the balances, I used the monthly interest rate of 0.005 for both investments.

For Alwynn, I graphed $A = 500(1.005)^n$ in Y_1.
For Peter, I graphed $A = 500(1 + 0.005n)$ in Y_2.

In both cases, I graphed values of n from 1 to 120 (10 years = 120 months).

c) From the graphs, Alwynn's investment is growing much faster than Peter's.

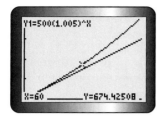
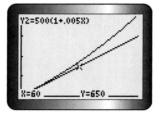

I used the value operation to compare the amounts of each investment after 5 years.

As time passes, the difference between the monthly balances will increase. Peter's monthly balance grows at a constant rate, but Alwynn's grows by more and more each month.

Tech | **Support**

For help graphing functions and determining a value, see Technical Appendix, B-2 and B-3.

In Summary

Key Idea

- The formula for calculating the amount of an investment earning compound interest is $A = P(1 + i)^n$, where
  - A is the amount or future value, in dollars
  - P is the principal, in dollars
  - i is the interest rate per compounding period
  - n is the number of compounding periods

Need to Know

- The compounding frequency determines the number of compounding periods per year. The compounding period changes the total number of periods, n, over which the interest is compounded during the term of the investment. Changing the compounding period changes the interest, i, because the annual interest rate must be adjusted to the rate that would be used for each compounding period.

Annually	once per year	i = annual interest rate	n = number of years
Semi-annually	2 times per year	i = annual interest rate $\div$ 2	n = number of years $\times$ 2
Quarterly	4 times per year	i = annual interest rate $\div$ 4	n = number of years $\times$ 4
Monthly	12 times per year	i = annual interest rate $\div$ 12	n = number of years $\times$ 12
Daily	365 times per year	i = annual interest rate $\div$ 365	n = number of years $\times$ 365

- The amount, or future value, of an investment increases with the number of compounding periods. However, the amount of the increase is not usually significant when interest is compounded more often than monthly.
- A timeline is useful for organizing and visualizing the information required to solve a compound-interest problem.

CHECK Your Understanding

1. An investment earns 9%/a. Calculate i and n when the interest is compounded
   a) annually for 4 years
   c) quarterly for 2 years
   b) semi-annually for 6 years
   d) monthly for 3 years

2. Complete the table.

	Principal ($)	Annual Interest Rate (%)	Time (years)	Compounding Frequency	Rate for the Compounding Period, i (%)	Number of Compounding Periods, n	Amount ($)	Interest Earned ($)
a)	400.00	5	15	annually				
b)	750.00	13	5	semi-annually				
c)	350.00	2.45	8	monthly				
d)	150.00	7.6	3	quarterly				
e)	1000.00	4.75	4	daily				

PRACTISING

3. An investment earns 5.75%/a. Calculate i and n when the interest is compounded.
   a) annually for 3 years
   c) quarterly for 3 years
   b) semi-annually for 5 years
   d) monthly for 2 years

4. Complete the table.

	Principal ($)	Interest Rate (%)	Years	Compounding Frequency	i	n	Amount ($)	Interest Earned ($)
a)	800	8	10	annually				
b)	1500	9.6	3	semi-annually				
c)	700	$3\frac{1}{2}$	5	monthly				
d)	300	7.25	2	quarterly				
e)	2000	$4\frac{1}{4}$	$\frac{1}{2}$	daily				

5. Calculate the amount you would end up with if you invested $5000 at
   K 14.6%/a compounded annually for 10 years.

6. Mario neglected to pay a credit card bill of $1550 at 17%/a, compounded daily, for 2 weeks after it was due. What is the amount he must pay to settle the bill at the end of the 2 weeks?

7. Use a timeline to show the growth in value of a $300 bond at 9%/a for 1 year, compounded monthly. What is the interest earned?

8. If $350 grows to $500 in 3 years, what is the annual interest rate assuming that interest is compounded annually?

9. Wasantha deposits $750 into a savings account that pays compound interest annually. The table at the right shows his annual balance for this investment. What interest rate did the bank give Wasantha?

Year	Final Balance ($)
1	795.00
2	842.70
3	893.26
4	946.86
5	1003.67

10. In about how many years will $600 grow to $1000 if it is invested at 8%/a compounded annually?

11. A donor gives $50 000 to the high school he graduated from. The amount must be invested for 3 years, and the accumulated interest will be used to buy books for the school library. If the money earns 7.75%/a compounded monthly, how much will be available for the books?

12. Josephine is purchasing a used car. Her bank has offered a loan of **A** $5000 at 5%/a compounded monthly. The used-car dealer has offered a loan of 5.25%/a compounded semi-annually.
 a) What is the amount owing on each loan after 1 year?
 b) Which loan should Josephine take and why?

13. Create a financial problem whose solution could be represented by the function $f(x) = 500(1.01)^x$.

14. a) $1000 is invested for 1 year at 10%/a. Copy and complete the **T** table, showing the amounts that $1000 invested at 10%/a would grow to in 1 year for different compounding frequencies.

Compounding Frequency	Number of Compounding Periods per Year	Formula	Amount ($)
annually	1	1000(1 + 0.10)	
semi-annually	2	1000(1 + 0.05)²	
quarterly			
monthly			
weekly			
daily			
hourly			

 b) How do the amounts change as the number of compounding periods per year increases?
 c) Is there a maximum amount that can be earned in 1 year? Explain.
 d) Why don't banks offer hourly compounding frequencies?

15. For each of the following expressions, identify the principal, compounding frequency, interest rate per compounding period, number of compounding periods, annual interest rate, and number of years. Evaluate the expression to determine the amount; then determine the interest.

	Formula	P ($)	Compounding Frequency	i (%)	n	Annual Interest Rate (%)	Number of Years	A ($)	I ($)
a)	$A = 145(1 + 0.0475)^{12}$								
b)	$A = 850(1 + 0.195)^{5}$								
c)	$A = 4500\left(1 + \dfrac{0.0525}{365}\right)^{1095}$								
d)	$A = 4500\left(1 + \dfrac{0.15}{12}\right)^{78}$								
e)	$A = 4500\left(1 + \dfrac{0.03}{4}\right)^{20}$								

16. Banks offer a variety of terms (for example, 90 days, 3 months, 1 year, [C] 3 years, 10 years, where the interest rate is greater for larger terms) and a variety of compounding periods. What advantages does the variety of terms and compounding periods provide for bank customers? Use different scenarios to support your explanation.

Extending

17. In the 1980s, financial experts used to talk about doubling your investment in 7 years. This was related to the 1980 Canada Savings Bond, which had an interest rate that would double the investment in 7 years. Determine that interest rate.

18. Mustafa wants his investment to be worth $10 000 in 5 years. The bank will give him 6%/a interest compounded annually. How much does Mustafa have to invest now?

19. If $500 was invested at 8%/a, compounded annually, in one account, and $600 was invested at 6%/a, compounded annually, in another account, when would the amounts in both accounts be equal?

Compound Interest: Determining Present Value

Solve problems that involve calculating the principal that must be invested today to obtain a given amount in the future.

LEARN ABOUT the Math

When Hua was born, her parents decided to invest some money so that she could have a gift of $20 000 on her 16th birthday. They decided on a compound-interest government bond that paid 10% interest per year, compounded monthly. After the initial amount was invested, there would be no further transactions until the bond reached maturity.

? How much money must be invested today to guarantee Hua's future amount of $20 000?

EXAMPLE 1 **Selecting a strategy to determine the principal needed to grow to a given amount**

a) What is the **present value** that Hua's parents must invest today to reach their savings goal of $20 000 by her 16th birthday?

b) If Hua's parents decide to wait until she is 13 and then invest a lump sum to save for the gift of $20 000, what is the present value if the investment earns the same rate of interest?

present value

the principal that must be invested today to obtain a given amount in the future

Martha's Solution: Using a Timeline

a)
$$A = P(1 + i)^n$$
$$\frac{A}{(1 + i)^n} = \frac{P(1 + i)^n}{(1 + i)^n}$$
$$\frac{A}{(1 + i)^n} = P$$

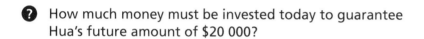

> I knew I could use the formula $A = P(1 + i)^n$ to find the amount of the principal. What I needed to find was the principal that must be invested now to get an amount of $20 000. I saw that if I divided the future value of an investment by $(1 + i)^n$, I could work my way back to the present value, or principal, to be invested.
>
> I rearranged the formula to solve for P, giving me an expression for the present value.

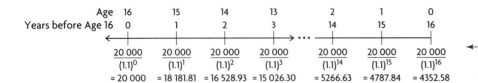

| Age | 16 | 15 | 14 | 13 | 2 | 1 | 0 |
| Years before Age 16 | 0 | 1 | 2 | 3 | 14 | 15 | 16 |

| $\frac{20\,000}{(1.1)^0}$ | $\frac{20\,000}{(1.1)^1}$ | $\frac{20\,000}{(1.1)^2}$ | $\frac{20\,000}{(1.1)^3}$ | $\frac{20\,000}{(1.1)^{14}}$ | $\frac{20\,000}{(1.1)^{15}}$ | $\frac{20\,000}{(1.1)^{16}}$ |
| $= 20\,000$ | $= 18\,181.81$ | $= 16\,528.93$ | $= 15\,026.30$ | $= 5266.63$ | $= 4787.84$ | $= 4352.58$ |

I drew a timeline showing how the present value needed to reach $20 000 decreases as the time before Hua's 16th birthday increases. It makes sense that if they invest it longer, they don't have to invest as much, since there would be more time to earn extra interest.

$$A = P(1 + i)^n$$

$$20\,000 = P(1 + 0.1)^{16}$$

$$P = \frac{20\,000}{(1 + 0.10)^{16}}$$

$$= \frac{20\,000}{(1.1)^{16}}$$

$$= 4352.58$$

I divided each amount by $1 + 0.10 = 1.1$ to get the amount in the next column.

The future amount of money is $A = \$20\,000$.

The annual interest rate is $i = 10\%$, or 0.1.

The interest is compounded annually for $n = 16$ years.

Hua's parents must invest $4352.58 now.

I checked my answer by creating a table of values showing the year-end amount of investment, $A = 4352.58\,(1 + 0.1)^n$, for each year.

At the end of year 16, the initial investment of $4352.58 is worth $20 000.

This means that if the parents invest $4352.58 today, their investment will be worth $20 000 when Hua turns 16.

b) $A = P(1 + i)^n$

$$P = \frac{A}{(1 + i)^n}$$

$$A = \$20\,000$$

$$i = 10\% \text{ or } 0.1$$

$$n = 3$$

$$P = \frac{20\,000}{(1 + 0.10)^3}$$

$$= 15\,026.30$$

Hua's parents must invest $15 026.30 when she turns 13.

To find the principal Hua's parents must invest, I rearranged the formula to solve for P.

I knew the future amount of money (A), the annual interest rate (i), and the number of years of compounding (n).

Communication | Tip

$A = P(1 + i)^n$ is sometimes written as $FV = PV(1 + i)^n$, where FV is future value and PV is present value.

Reflecting

A. What is the total interest earned over the 16 years of the investment in part (a)? What is the total interest over the 3 years of the investment in part (b)?

B. What are the advantages and disadvantages of investing earlier rather than later?

C. The formula $A = P(1 + i)^n$ can be used to calculate both present value and future value. State what you need to know and how you would use the formula to calculate the following.
  i) future value   **ii)** present value

APPLY the Math

EXAMPLE **2**	Determining present value with a compounding period of less than one year

An investment earns $7\frac{3}{4}$%/a compounded semi-annually. Determine the present value if the investment is worth $800 five years from now.

Luc's Solution

$$P = \frac{A}{(1 + i)^n}$$

$$= A(1 + i)^{-n}$$

Since I was dividing by $(1 + i)^n$, I multiplied by the reciprocal, using a negative exponent.

$$A = 800$$

The amount of the investment is $800.

$$i = \frac{0.0775}{2} = 0.038\,75$$

The annual interest rate is $7\frac{3}{4}\% = 0.0775$.

The semi-annual interest rate is $\frac{1}{2}$ of the annual rate.

$$n = 5 \times 2 = 10$$

I multiplied the number of years by 2 to calculate the number of compounding periods.

$$P = 800(1.038\,75)^{-10}$$

$$= 546.99$$

I substituted the values of A, i, and n, and evaluated P.

The present value of the investment is $546.99.

EXAMPLE **3** | Solving a problem involving present value

Tony has $3000 in his savings account. He intends to buy a laptop computer and printer and invest the remainder for 2 years, compounding monthly at an annual interest rate of 3%. He wants to have $2000 in his account 2 years from now. How much can he spend on the laptop and printer?

Martha's Solution

$$P = \frac{A}{(1 + i)^n}$$

The future value of the investment is $2000.

$A = 2000$

The annual interest rate is 3% = 0.03.

$$i = \frac{0.03}{12} = 0.0025$$

The monthly rate of interest is $\frac{1}{12}$ of the annual rate.

$n = 12 \times 2 = 24$

I multiplied the number of years by 12 to calculate the number of compounding periods.

$$P = \frac{2000}{(1 + 0.0025)^{24}}$$

Then I substituted into the formula.

$$= \frac{2000}{(1.0025)^{24}}$$

$$= 1883.67$$

The present value is $1883.67.

Amount that can be spent

I needed to determine the amount of Tony's savings that he needs to keep invested to reach $2000 in 2 years. Whatever he has left after this amount is set aside is what he can spend.

= Amount in savings account −
Present value

= $3000.00 − $1883.67

= $1116.33

Tony can spend $1116.33 on a laptop and printer.

I subtracted the present value from $3000 to determine how much Tony can spend.

In Summary

Key Idea

- The formula for calculating future value can be rearranged to give the present value of an investment earning compound interest. The rearranged formula is

$$P = \frac{A}{(1 + i)^n} \quad \text{or} \quad P = A(1 + i)^{-n}$$

where
  - A is the amount, or future value, in dollars
  - P is the principal, or present value, in dollars
  - i is the interest rate per compounding period
  - n is the number of compounding periods

Need to Know

- The amount, P, that must be invested now in order to grow to a specific amount later on can be calculated from the future value by dividing by $(1 + i)^n$, where
  - i is the interest rate per compounding period
  - n is the number of compounding periods
- Drawing a timeline can help you decide whether you need to determine the future value (or amount) or the present value (or principal) of an investment or loan.

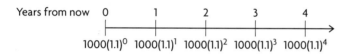

Years from now   0   1   2   3   4

$1000(1.1)^0 \quad 1000(1.1)^1 \quad 1000(1.1)^2 \quad 1000(1.1)^3 \quad 1000(1.1)^4$

The future value of $1000 invested at 10%/a compounded annually for 4 years.

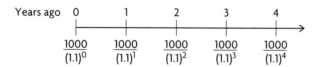

Years ago   0   1   2   3   4

$\dfrac{1000}{(1.1)^0} \quad \dfrac{1000}{(1.1)^1} \quad \dfrac{1000}{(1.1)^2} \quad \dfrac{1000}{(1.1)^3} \quad \dfrac{1000}{(1.1)^4}$

The present value of $1000 invested at 10%/a compounded annually for 4 years.

- Interest earned can be calculated by subtracting the present value (principal) from the future value (amount):

$$I = FV - PV, \quad \text{or} \quad I = A - P$$

CHECK Your Understanding

1. Solve for the principal, P.
   a) $100 = P(1.05)^3$ b) $500 = P(1 + 0.00375)^{48}$

2. Copy and complete the table.

	Future Value ($)	Annual Interest Rate (%)	Time Invested (years)	Compounding Frequency	i (%)	n	Present Value ($)	Interest Earned ($)
a)	4 000	5	15	annually				
b)	3 500	2.45	8	monthly				
c)	10 000	4.75	4	daily				

3. The first timeline that follows visually represents the future value of $100 invested at 5%/a compounded annually for 4 years. Copy and complete the second timeline to show the calculations of present value in each year for an investment whose future value is $150.

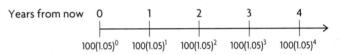

The future value of $100 invested at 5%/a compounded annually for 4 years.

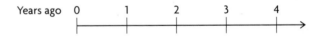

The present value of $150 invested at 5%/a compounded annually for 4 years.

PRACTISING

4. Copy and complete the table.

	Future Value ($)	Annual Interest Rate (%)	Time Invested (years)	Compounding Frequency	i (%)	n	Present Value ($)	Interest Earned ($)
a)	8000	10	7	annually				
b)	7500	13	5	semi-annually				
c)	1500	7.6	3	quarterly				

5. Use a timeline to illustrate the present value of an investment worth $5750 in 3 years at 12%/a compounded semi-annually.

6. **K** How much should Jethro invest now to have $10 000 in 3 years' time? The money will be invested at 5%/a compounded monthly.

7. Tim has arranged to pay $2000 toward a debt now and $3000 two years from now. What amount of money would settle the entire debt today if the interest is 10.5%/a compounded semi-annually?

8. On Abby's 21st birthday, she receives a gift of $10 000, the accumulated amount of an investment her grandparents made for her when she was born. Determine the amount of their investment and the interest earned if the interest rate was 8.75%/a compounded
   a) annually   b) semi-annually

9. Daveed has a savings account that pays interest at 4.25%/a compounded monthly. She has not made any deposits or withdrawals for the past 6 months. There is $3542.16 in the account today. How much interest has the account earned in the past 6 months?

10. Jason borrowed money that he will pay back in 3 years' time. The interest rate was 5.25%/a compounded monthly. He will repay $3350 after 3 years. How much money did Jason borrow?

11. **A** Betty plans to send her parents on a $15 000 vacation for their 30th wedding anniversary 10 years from now. She would like to invest the money today in a GIC term deposit earning 6%/a compounded semi-annually and split the cost of its purchase with her sister and brother. How much will each person contribute toward the purchase of the GIC?

12. **T** Clem inherits $250 000. He wants to save $150 000 for college or university costs in 4 years.
   a) How much should Clem invest in a GIC earning 10.5%/a compounded monthly to ensure that he has $150 000 in savings 4 years from now?
   b) How much of Clem's inheritance remains after his investment?
   c) How much interest would Clem's inheritance earn in 4 years if he invested the entire amount in the GIC now?

13. For each situation, determine
   i) the present value   ii) the interest earned
   a) A loan of $21 500 is due in in 6 years. The interest rate is 8%/a, compounded quarterly.
   b) A loan of $100 000 is due in 5 years. The interest rate is 5%/a, compounded semi-annually.

14. Copy the table and fill in the missing entries.

	Future-Value Formula	A ($)	Compounding Frequency	i (%)	n	Annual Interest Rate (%)	Number of Years	Present Value ($)
a)	$280\,000 = P(1 + 0.0575)^{24}$		semi-annually					
b)	$16\,000 = P(1 + 0.20)^{5}$		annually					
c)	$10\,000 = P\left(1 + \dfrac{0.0425}{365}\right)^{1460}$							
d)	$9500 = P\left(1 + \dfrac{0.15}{12}\right)^{50}$							
e)	$1500 = P\left(1 + \dfrac{0.03}{4}\right)^{24}$							

15. Marshall wants to have $5000 in 4 years. He has two options for investment: A savings account will pay 3.5%/a compounded monthly; a GIC will pay 3.4%/a compounded semi-annually. Write an explanation of which investment Marshall should pick and why.

Extending

16. A loan at 12%/a compounded semi-annually must be repaid in one single payment of $2837.04 in 3 years. What is the principal borrowed?

17. What equal deposits, one made now and another made one year from now, will accumulate to $2000 two years from now at 6.25%/a compounded semi-annually?

18. Gina agrees to pay $25 000 now and $75 000 in 4 years for a studio condominium. If she can invest at 10.5%/a compounded annually, what sum of money does she need now to buy the condominium?

Compound Interest: Solving Financial Problems

YOU WILL NEED
- graphing calculator with TVM Solver program

LEARN ABOUT the Math

There are a variety of technological tools for calculating financial information involving compound interest. These include spreadsheets, calculators on websites of financial institutions, and graphing calculator programs.

Some graphing calculators include financial programs such as the Time Value of Money (TVM) Solver. This program can be used to quickly investigate and solve many compound-interest problems.

? How can the TVM Solver be used to solve problems involving compound interest, and how does it compare with using a formula?

EXAMPLE 1	Selecting a strategy to determine the amount of an investment

Peggy's employer has loaned her $5000 to pay for university course tuition and textbooks. The interest rate of the loan is 2.5%/a compounded monthly, and the loan is to be paid back in one payment at the end of 2 years. How much will Peggy have to pay back?

Jeremy's Solution: Using a Formula

$A = P(1 + i)^n$ ← The principal is $P = \$5000$.

$P = 5000$

$i = \dfrac{0.025}{12}$

$n = 2 \times 12 = 24$

$A = 5000\left(1 + \dfrac{0.025}{12}\right)^{24}$

$A = 5256.09$

Peggy will pay $5256.08 at the end of 2 years.

The annual interest rate is 2.5%, or 0.025. Since it is compounded monthly, I divided it by 12.

I multiplied the number of years by 12 to determine the number of compounding periods.

Mei-Mei's Solution: Using the TVM Solver

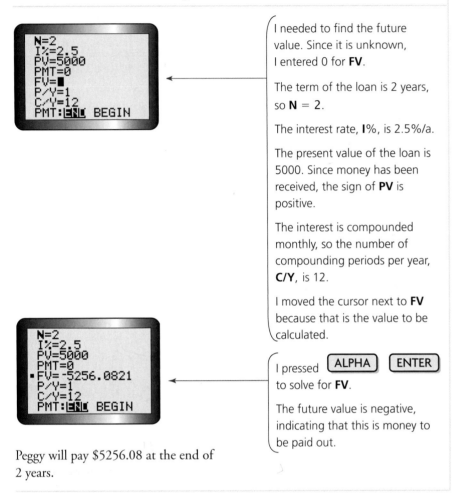

I needed to find the future value. Since it is unknown, I entered 0 for **FV**.

The term of the loan is 2 years, so **N** = 2.

The interest rate, **I**%, is 2.5%/a.

The present value of the loan is 5000. Since money has been received, the sign of **PV** is positive.

The interest is compounded monthly, so the number of compounding periods per year, **C/Y**, is 12.

I moved the cursor next to **FV** because that is the value to be calculated.

I pressed [ALPHA] [ENTER] to solve for **FV**.

The future value is negative, indicating that this is money to be paid out.

Peggy will pay $5256.08 at the end of 2 years.

EXAMPLE **2** Selecting a strategy to determine the present value of an investment

How much was invested at 4%/a compounded semi-annually for 3 years if the final amount was $7500?

Martin's Solution: Using a Formula

$A = 7500$

$i = \dfrac{0.04}{2} = 0.02$

The future value of the amount of the investment is A = $7500.

The annual interest rate is 4%, or 0.04. Since it is compounded semi-annually, I divided it by 2.

The Half-Life of Caffeine

Caffeine is a stimulant found in many products, such as coffee, tea, pop, and chocolate. When you drink a cup of tea or eat a chocolate bar, you ingest caffeine. Your body breaks down caffeine slowly. As with other drugs, caffeine has a half-life in the body. The half-life of caffeine in a typical nonsmoking adult is 5.5 h.

The caffeine content of many popular foods and drinks is listed in the table at the right. You can use the information from the table to write an equation to model the amount of caffeine in your system after drinking or eating foods that contain caffeine.

A. A cup of brewed coffee has approximately 130 mg of caffeine in it. Suppose you drink a cup of drip coffee at 9 a.m. How much caffeine is left in your body at

    i) 12:00 noon?   **ii)** 8 p.m.?   **iii)** midnight?

B. Suppose that in addition to your 9 a.m. cup of coffee at noon, you ate a chocolate bar and drank a cup of green tea. How can you calculate the additonal amount of caffeine in your system? Determine the total amount of caffeine in your body at noon.

C. If you had nothing else to eat or drink that contained caffeine from noon onward, predict the amount of caffeine in your body at 8 p.m.

D. Make a list of the food or drinks with caffeine you have had today. Write the time you had each food or drink. Calculate the amount of caffeine that will be left in your body at 10 p.m. tonight. Use a table like the one below to record your data.

Time Ingested	Food or Drink	Caffeine Content (mg)	Number of Hours until 10 p.m.

Caffeine Content	(mg)
250 mL of Coffee:	
Drip	165
Brewed	130
Instant	95
Decaffeinated	4
250 mL of Tea:	
Brewed	45
Instant	35
Green tea	30
Soft drinks (mean)	43
Chocolate bar	20
Energy drinks	80
Espresso, double (2 oz.)	70

Task | *Checklist*

✔ Did you show the calculations you used to determine the caffeine levels for part A?

✔ Did you remember to include the caffeine from the 9 a.m. cup of coffee in your calculations for part B?

✔ Did you remember to write an equation that models this situation?

✔ Did you include the table you created for part D?

$n = 3 \times 2 = 6$

$P = A(1 + i)^{-n}$

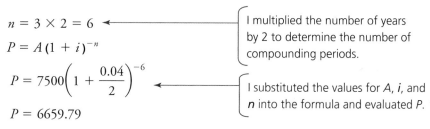

I multiplied the number of years by 2 to determine the number of compounding periods.

$P = 7500\left(1 + \dfrac{0.04}{2}\right)^{-6}$

I substituted the values for A, i, and n into the formula and evaluated P.

$P = 6659.79$

The original present value was $ 6659.79.

Rebecca's Solution: Using the TVM Solver

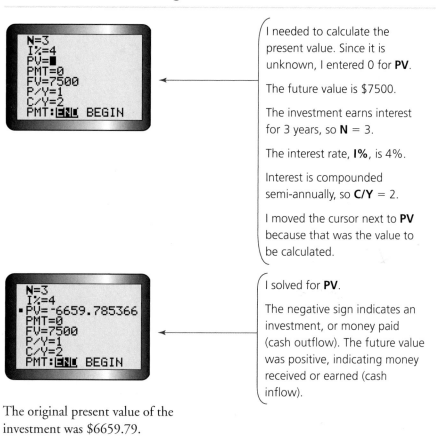

```
N=3
I%=4
PV=■
PMT=0
FV=7500
P/Y=1
C/Y=2
PMT:END BEGIN
```

I needed to calculate the present value. Since it is unknown, I entered 0 for **PV**.

The future value is $7500.

The investment earns interest for 3 years, so **N** = 3.

The interest rate, **I%**, is 4%.

Interest is compounded semi-annually, so **C/Y** = 2.

I moved the cursor next to **PV** because that was the value to be calculated.

```
N=3
I%=4
•PV=-6659.785366
PMT=0
FV=7500
P/Y=1
C/Y=2
PMT:END BEGIN
```

I solved for **PV**.

The negative sign indicates an investment, or money paid (cash outflow). The future value was positive, indicating money received or earned (cash inflow).

The original present value of the investment was $6659.79.

Reflecting

A. If you are using the TVM Solver, when is the present value entered as positive and when is it entered as negative? Explain, using examples.

B. How is using the TVM Solver to solve compound-interest problems similar to using the formula $A = P(1 + i)^n$? How is it different?

C. Which method do you prefer? Explain why.

APPLY the Math

EXAMPLE 3 | **Selecting a strategy to determine the annual interest rate**

What annual interest rate was charged if an $800 credit card bill grew to $920.99 in 6 months and interest was compounded monthly?

Delacey's Solution: Using a Formula

$A = \$920.99$ ◄──────────────────────

$P = \$800$

The number of compounding periods is $n = 6$.

> I listed the values I knew and substituted into the formula for A.
>
> I needed to solve for i.

$$A = P(1 + i)^n$$

$$(1 + i)^n = \frac{A}{P}$$

$$(1 + i)^6 = \frac{920.99}{800}$$

$$\left((1 + i)^6\right)^{\frac{1}{6}} = \left(\frac{920.99}{800}\right)^{\frac{1}{6}}$$ ◄──────

> To solve for i, I raised each side of the equation to the power of $\frac{1}{6}$.

$$1 + i = \sqrt[6]{\frac{920.99}{800}}$$ ◄──────

> By using the power-of-a-power rule, I was able to get the exponent on $1 + i$ to be 1. To do this, I had to calculate the 6th root of the number on the right side.

$$i = \sqrt[6]{\frac{920.99}{800}} - 1$$ ◄──────

> I solved for i by subtracting 1 from both sides.

$$i = 0.023\,75$$

$$i \times 12 = 0.023\,75 \times 12$$ ◄──────

> Since i is the monthly interest rate, I multiplied it by 12 to determine the annual interest rate.

$$= 0.285$$

$$= 28.5\%$$

The annual interest rate is 28.5%.

Kara's Solution: Using the TVM Solver

I needed to determine the interest rate. I entered 0.5 for **N**, since the investment earns interest for 0.5 years. I entered 0 for **I%**, since the interest rate is unknown. The present value, **PV**, is 800. The future value, **FV**, is −920.99 because the money will eventually be paid out. The number of compounding periods per year, **C/Y**, is 12, because interest is compounded monthly.

I moved the cursor next to **I%** because that was the value to be calculated.

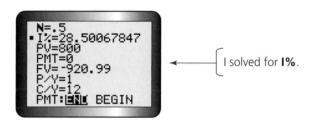

I solved for **I%**.

The interest rate is 28.5%/a.

$$A = P(1 + i)^n$$

I checked the answer with the formula $A = P(1 + i)^n$.

$$= 800\left(1 + \frac{0.285}{12}\right)^6$$

$$= 920.99$$

EXAMPLE 4

Selecting a strategy to determine the number of years required to double an investment

Approximately how long would it take for a $15 000 investment to double if it earns 10%/a interest compounded semi-annually?

Marita's Solution: Using a Formula with Guess-and-Check

$$A = P(1 + i)^n$$

$$(1 + i)^n = \frac{A}{P}$$

> The present value is $P = \$15\ 000$.
> The future value is $A = \$30\ 000$.
>
> The annual interest rate is 10%.
> The semi-annual interest rate is $i = 0.05$.

$$1.05^n = \frac{30\ 000}{15\ 000}$$

$$1.05^n = 2$$

> I substituted these values to get an equation involving n, the number of compounding periods.

$$1.05^2 = 1.1025$$

$$1.05^6 = 1.3400$$

$$1.05^{14} = 1.98$$

$$1.05^{15} = 2.08$$

> Since n is an exponent, I tried different values of n to solve the equation.
>
> I started with 2, then 6, but I got values that were too low. 14 was really close and 15 was too high.

The number of compounding periods is approximately 14.

> n must be a number between 14 and 15, but closer to 14.

The number of years is $\frac{14}{2} = 7$.

> I divided 14 by 2 because each year has 2 compounding periods.

It will take approximately 7 years for the investment to double in value.

Samir's Solution: Using the TVM Solver

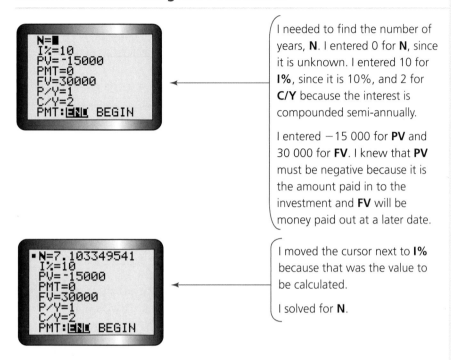

I needed to find the number of years, **N**. I entered 0 for **N**, since it is unknown. I entered 10 for **I%**, since it is 10%, and 2 for **C/Y** because the interest is compounded semi-annually.

I entered −15 000 for **PV** and 30 000 for **FV**. I knew that **PV** must be negative because it is the amount paid in to the investment and **FV** will be money paid out at a later date.

I moved the cursor next to **I%** because that was the value to be calculated.

I solved for **N**.

It will take about 7 years for the investment to double in value.

In Summary

Key Ideas

- The TVM Solver is a program on some graphing calculators. It can be used to investigate and solve financial problems involving compound interest.
- For compound-interest problems, you can use the two forms of the compound-interest formula

$$A = P(1 + i)^n \quad \text{and} \quad P = A(1 + i)^{-n}$$

either as an alternative to using the TVM Solver or as a check.

(continued)

CHECK Your Understanding

1. Copy the table that follows. For each of problems (a) through (d), record the values you would enter for the known TVM Solver variables. Record 0 for the unknown. Then solve the problem and indicate the solution by marking it with *.

  a) Determine the amount of an investment if $600 is invested at 4.5%/a interest for 8 years, compounded quarterly.

  b) How long would it take $6000 to grow to $8000 if it is invested at 2.5%/a compounded semi-annually?

  c) What interest rate is needed for $20 000 to double in 5 years if interest is compounded quarterly?

  d) What amount needs to be invested at 6%/a interest compounded weekly if you want to have $900 after 1 year?

	N	I%	PV	PMT	FV	P/Y	C/Y
a)							
b)							
c)							
d)							

PRACTISING

2. Guo is a civic employee. His last contract negotiated a 2.75% increase each year for the next 4 years. Guo's current salary is $48 500 per year. What will his salary be in 4 years?

3. Beverley plans to invest $675 in a GIC for 2 years. She has researched two plans: Plan A offers 5.9%/a interest compounded semi-annually. Plan B offers 5.75%/a interest compounded monthly. In which plan should Beverley invest to earn the most?

4. Determine the future value of an investment of $10 000 compounded annually at 5%/a for
   a) 10 years   b) 20 years   c) 30 years

5. For each situation, determine both the present value and the earned interest.
   a) An investment that will be worth $5000 in 3 years. The interest rate is 4%/a compounded annually.
   b) An investment that will be worth $13 500 in 4 years. The interest rate is 6%/a compounded monthly.
   c) A loan repayment of $11 200 paid after 5 years, with interest of 4.4%/a compounded monthly.
   d) An investment that will be worth $128 500 in 8 years. The interest rate is 6.5%/a compounded semi-annually.
   e) A loan repayment of $850 paid after 400 days, with interest of 5.84%/a compounded daily.
   f) An investment that will be worth $6225 in 100 weeks. The interest rate is 13%/a compounded weekly.

6. At what interest rate will an investment compounded annually for 12 years double in value?

7. How long does it take for an investment to triple in value at 10%/a interest compounded monthly?

8. When Ron was born, a $5000 deposit was made into an account that pays interest compounded quarterly. The money was left until Ron's 21st birthday, when he was presented with a cheque for $12 148.79. What was the annual interest rate?

9. Shirley redeemed a $2000 GIC and received $2220. The GIC paid interest at 5.25%/a compounded quarterly. For how long was the money invested?

10. A $3000 GIC pays 5%/a interest compounded annually for a 3-year term. At maturity, the accumulated amount is reinvested in another GIC at 6.5%/a compounded annually for 5 years. What is the final amount when the second investment matures?

11. Today Sigrid has $7424.83 in her bank account. For the last 2 years, her account has paid 6%/a compounded monthly. Before then, her account paid 6%/a compounded semi-annually for 4 years. If she made only one deposit 6 years ago, determine the original principal.

12. On June 1, 2001, Anna invested $2000 in a money market fund that paid 6%/a compounded monthly. After 5 years, her financial advisor moved the accumulated amount to a new account that paid 8%/a compounded quarterly. Determine the balance in her account on January 1, 2013.

13. On the day Sarah was born, her grandparents deposited $500 in a
T savings account that earns 4.8%/a compounded monthly. They deposited the same amount on her 5th, 10th, and 15th birthdays. Determine the balance in the account on Sarah's 18th birthday.

14. Tresha paid for household purchases with her credit card. The credit
K card company charges 18%/a compounded monthly. Tresha forgot to pay the monthly bill of $465 for 3 months after it was due to be paid.
  a) How much does Tresha owe at the end of each of the 3 months?
  b) How much of each amount in part (a) is interest?

15. Do an Internet search of the phrase "compound interest calculators."
C Try out two different online calculators. In what ways are they similar to the TVM Solver? In what ways are they different?

Extending

16. Asif bought an oil painting at a yard sale for $15 in 2001. Five years later, he took the painting to have it appraised. To his surprise, it was worth between $20 000 and $30 000. What annual interest rate corresponds to the growth in the value of his purchase?

17. A used car costs $32 000. The dealer offers a finance plan at 2.4%/a compounded monthly for 5 years with monthly payments. If you pay cash for the car, its cost is $29 000. The bank will loan you cash for 5.4%/a compounded monthly, also with monthly repayments for 5 years. Should you finance the purchase of the car through the dealer or through the bank? Explain.

18. Barry bought a boat 2 years ago, paying $10 000 toward the cost. Today he must pay the $7500 he still owes, which includes the interest charge on the balance due. Barry financed the purchase at 6.2%/a compounded semi-annually. Determine the purchase price of the boat.

FREQUENTLY ASKED Questions

Q: **What is the difference between simple interest and compound interest?**

Study | *Aid*
- See Lesson 8.1, Examples 1 to 4.
- Try Mid-Chapter Review Questions 1 to 3.

A: Simple interest is calculated only on the original principal. The formulas used are $I = Prt$ and $A = P(1 + rt)$, where
- I is the interest earned, in dollars
- P is the principal invested, in dollars
- r is the annual interest rate, expressed as a decimal
- t is the time, in years
- A is the final amount earned, in dollars

Simple-interest investments grow at a constant, or linear, rate over time.

Compound interest is calculated at regular periods, and the interest is added to the principal for the next period. The formula used is $A = P(1 + i)^n$, where
- A is the amount, or future value, of the investment, in dollars
- P is the principal, in dollars
- i is the interest rate, expressed as a decimal
- n is number of compounding periods

Compound-interest investments grow exponentially, as a function of the number of compounding periods.

Q: **What is the difference between future value and present value, and how are they calculated for situations involving compound interest?**

Study | *Aid*
- See Lesson 8.2, Examples 1 and 2, and Lesson 8.3, Examples 1 to 3.
- Try Mid-Chapter Review Questions 4 to 10.

A: When an investment matures, both the principal and the interest are paid to the investor. This total amount is called the amount of the investment, or the future value of the investment.

The formula $A = P(1 + i)^n$ can be used to determine the future value. This formula is sometimes written as $FV = PV(1 + i)^n$, where
- FV is the future value, or amount of the investment, in dollars
- PV is the present value, or principal, in dollars

To see how this formula comes about, consider this example: Marina invests \$5000 in a savings account that pays 5.25%/a, compounded annually. In this case, $r = 5.25\%$ or 0.0525 and $P = 5000$. The amount at the end of the first year is

$$A = P + I$$
$$= P + Prt$$
$$= P(1 + rt)$$

$$A = 5000(1 + 0.0525 \times 1)$$
$$= 5000(1.0525)^1$$
$$= 5262.50$$

This amount becomes the new principal at the beginning of the second year and earns interest at the same rate. The amount at the end of the second year is

$$A = 5262.50(1.0525)$$
$$= 5000(1.0525)^2$$
$$= 5538.78$$

The amount at the end of the third year is

$$A = 5538.78(1.0525)$$
$$= 5000(1.0525)^3$$
$$= 5829.57$$

The general term is

$$A = 5538.78(1.0525)^n$$

The terms are $5000(1.0525)^1$, $5000(1.0525)^2$, $5000(1.0525)^3$, ...

Therefore, the sequence of year-end amounts is \$5262.50, \$5538.78, \$5829.57, ...

Present value is the amount of money that must be invested today at a given rate and compounding frequency in order to provide for a given amount in the future. The formula $A = P(1 + i)^n$ can be rearranged to $P = A(1 + i)^{-n}$ or $PV = FV(1 + i)^{-n}$ to determine present value.

Study **Aid**

• See Lesson 8.4, Examples 1 to 4.
• Try Mid-Chapter Review Questions 11 to 14.

Q: **What are the advantages and disadvantages of using the TVM Solver to solve compound-interest problems?**

A: Advantages: The TVM Solver is useful for quickly calculating values of unknown financial variables, such as present value, future value, interest rate, and number of payments. These equations are handled easily on the TVM Solver.

Disadvantages: You need to decide whether present value or future value should be entered as a negative or positive number. If you make a data entry error, it is not always easy to identify the specific error. You need to remember what each of the variables in the TVM Solver means.

PRACTICE Questions

Lesson 8.1

1. Copy and complete the table.

	Principal ($)	Annual Interest Rate (%)	Time	Simple Interest Paid ($)	Amount ($)
a)	250	2	3 years		
b)		2.5	200 weeks	38.46	
c)	1000	3.1	18 months	46.50	
d)	5000	5	30 weeks		
e)		4.2	5 years	157.50	
f)		3	54 months	202.50	

2. Tami earned $20.64 in simple interest by investing a principal of $400 in a Treasury bill. If the interest rate was 1.72%/a, for how many years did she have her investment?

3. You invest $500 for 10 years at 10%/a simple interest. Your friend invests $500 for 10 years at 10%/a interest compounded annually. Copy and complete the tables to compare the investments.

	Your Investment (10% Simple Interest)	
Year	Interest Earned ($)	Accumulated Interest ($)
1		
2		
3		
⋮		
10		

Friend's Investment (10% Compound Interest)	
Interest Earned ($)	Accumulated Interest ($)
⋮	

Lesson 8.2

4. Copy and complete the table.

	Principal ($)	Annual Interest Rate (%)	Years Invested	Compounding Frequency	Amount ($)	Interest Earned ($)
a)	400	5	15	annually		
b)	350	2.45	8	monthly		
c)		3.5	5	quarterly	500	
d)	120		7	semi-annually	150	
e)	2 500	7.6		monthly		350
f)	10 000	7.5	3	quarterly		

5. Sam invests $800 at 6%/a compounded annually for 5 years. What is the total interest earned?

6. $1000 is invested at 8%/a compounded daily for 10 years. What is the total interest earned?

7. You have inherited $30 000 and want to invest it for 20 years. You have two options: a Treasury bill that earns 3.48%/a interest compounded monthly and a GIC that earns 3.5%/a compounded semi-annually. Determine which investment is the better choice.

Lesson 8.3

8. Anthony wants to have $10 000 in 5 years. His bank will pay him 6%/a interest compounded monthly. How much does he have to invest now?

9. How much should be invested at $12\frac{1}{2}$%/a compounded semi-annually to amount to $1150 in $3\frac{1}{2}$ years?

10. Determine the present value of $3500 if the original deposit can earn 8%/a compounded quarterly.

Lesson 8.4

11. Steve's investment doubled from $1000 to $2000 over 8 years. He knows that the interest was compounded quarterly. What annual rate did he get on his investment?

12. How many years will it take to see a $500 investment grow to $937.70 if the annual interest rate is 4.5%/a compounded monthly? Round to the nearest year.

13. How long will it take an amount of money invested at 5%/a compounded annually to grow to five times as large as the principal? Round your answer to the nearest year.

14. Sarah has saved $500 from babysitting. She would like to put this money into a savings account. Bank A offers an account that pays 6 %/a compounded monthly while Bank B offers an account that pays 7% compounded semi-annually. Which bank will provide her with the most interest over a 2-year period?

8.5 Regular Annuities: Determining Future Value

GOAL

GOAL

Solve future-value problems involving regular payments or deposits.

YOU WILL NEED

- graphing calculator with TVM Solver
- spreadsheet software

INVESTIGATE *the Math*

When Jessica turned 13 years old, her grandmother gave her $500. Jessica deposited the money into a savings account that paid 4.5%/a compounded annually. Since then, Jessica's grandmother has made an automatic deposit of $500 into Jessica's account on her birthday each year. Each payment earns a different amount of interest because it compounds for a different amount of time.

❓ **How much money is in Jessica's bank account on her 18th birthday, and how much interest has it earned?**

A. The timeline shows the deposits of the **annuity** made into Jessica's account. For how many compounding periods has the $500 deposit on her 17th birthday earned interest? For how many compounding periods has each of the 16th, 15th, 14th, and 13th birthday deposits earned interest?

annuity

a series of equal deposits or payments made at regular intervals; a **simple** annuity is an annuity in which the payments coincide with the compounding period, or *conversion* period; an **ordinary** annuity is an annuity in which the payments are made at the end of each interval; unless otherwise stated, each annuity in this chapter is a simple, ordinary annuity

Future value of Jessica's annuity

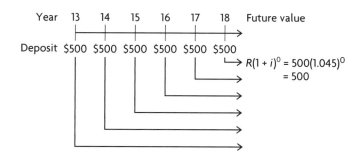

B. Complete the timeline by calculating the amount, or future value, of each $500 birthday deposit. The deposit made on Jessica's 18th birthday has not earned any interest. (In the diagram, R is like the variable P you have been using in earlier sections.)

C. What is the total amount of the annuity on Jessica's 18th birthday?

D. How much of the annuity is interest earned?

Reflecting

E. In what ways is calculating the amount, or future value, of an annuity the same as calculating the amount, or future value, of a single deposit? In what ways is it different?

F. Why can't the formula $A = 5[500(1 + 0.03)^5]$ be used to calculate the amount of Jessica's annuity?

G. Create an expression that represents the sum of the amounts, or future value, of the six deposits. Use expressions of the form $R(1 + i)^n$, where
- R is the regular deposit or payment, in dollars
- i is the interest rate per compounding period, expressed as a decimal
- for each deposit, n is the number of compounding periods

APPLY the Math

| EXAMPLE **1** | Selecting a strategy to calculate the amount of an annuity |

Steve makes deposits of $300 semi-annually into an account that pays 4%/a interest compounded semi-annually.

a) How much money will be in the account after 5 years?

b) How much interest will Steve have earned over the 5-year term?

Marta's Solution: Using a Timeline

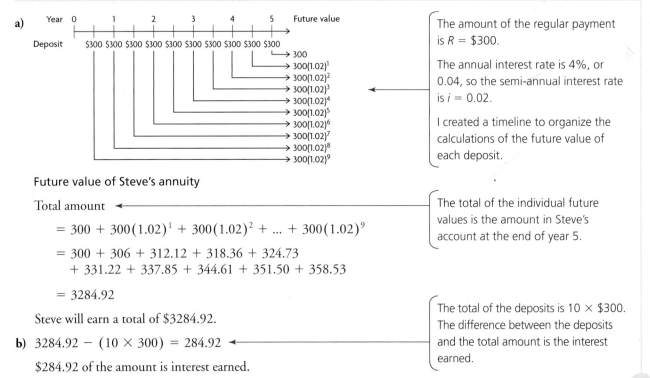

a)

The amount of the regular payment is $R = \$300$.

The annual interest rate is 4%, or 0.04, so the semi-annual interest rate is $i = 0.02$.

I created a timeline to organize the calculations of the future value of each deposit.

Future value of Steve's annuity

Total amount

$= 300 + 300(1.02)^1 + 300(1.02)^2 + ... + 300(1.02)^9$

$= 300 + 306 + 312.12 + 318.36 + 324.73$
$\quad + 331.22 + 337.85 + 344.61 + 351.50 + 358.53$

$= 3284.92$

Steve will earn a total of $3284.92.

The total of the individual future values is the amount in Steve's account at the end of year 5.

b) $3284.92 - (10 \times 300) = 284.92$

$284.92 of the amount is interest earned.

The total of the deposits is $10 \times \$300$. The difference between the deposits and the total amount is the interest earned.

Joseph's Solution: Using a Spreadsheet

Formulas (as entered)

	A	B
1	Deposit	Future Value
2	0	=300*(1.02)^A2
3	=1+A2	=300*(1.02)^A3
	⋮	
12	Sum	=SUM(B2:B11)
13	Interest	=B12 − (10*300)

Values (as displayed)

	A	B
1	Deposit	Future Value
2	0	300.00
3	1	306.00
4	2	312.12
	⋮	
11	9	358.53
12	Sum	3284.92
13	Interest	284.92

I entered the deposit numbers into the first column. In the second column, I used the formula $300(1.02)^n$ to calculate the future value of each deposit.

I used the Fill Down command to complete and display the future values.

To determine the future value of the annuity, I calculated the total of all these amounts. I then subtracted the amount of the actual deposits to calculate the interest.

a) Steve will earn a total of $3284.92.

b) $284.92 of the amount is interest earned.

Sergei's Solution: Use the TVM Solver

a)

I entered 10 beside **N** for the 10 compounding periods (not the number of years) and 4 beside **I** for 4% annual interest. **PV** is not required, so it is 0. I entered − 300 beside **PMT** because this is money Asif must pay. **P/Y** and **C/Y** are both 2 because the payments are made and compounded semi-annually.

I wanted to solve for the future value, so I entered 0 for **FV**. I placed the cursor beside **FV** because I was solving for it.

I solved for **FV**. The future value is positive because it is money that Asif will receive.

Steve will earn a total of $3284.92.

b) $3284.92 − (10 \times 300) = 284.92$

$284.92 of the amount is interest earned.

I calculated the interest earned by subtracting the total of the payments, $10 \times \$300$, from the future value.

If technology is not available to help you calculate the future value of an annuity, you can do so using a formula.

EXAMPLE 2 **Using a formula to determine the future value of an annuity**

Jay deposits $1500 every 3 months for 2 years into a savings account that earns 10%/a compounded quarterly. How much money will have accumulated at the end of 2 years?

Measha's Solution

$$A = \frac{R[(1 + i)^n - 1]}{i}$$

I wrote the formula for calculating the future value of an annuity, where

- A is the amount, or future value, in dollars
- R is the regular deposit, or payment, in dollars
- i is the interest rate per compounding period, expressed as a decimal
- n is the total number of deposits

$R = \$1500$

$i = 10\% \div 4$

$\quad = 2.5\% = 0.025$

$n = 2 \times 4 = 8$

$$A = \frac{1500[(1 + 0.025)^8 - 1]}{0.025}$$

$\quad = 13\ 104.17$

The interest rate is 10% compounded quarterly. To determine i, I divided it by 4.

I multiplied the number of years by 4 to determine the number of compounding periods.

Jay will have $13 107.17 at the end of 2 years.

EXAMPLE 3 **Selecting a strategy to calculate the regular payment**

An investor wants to retire in 25 years with $1 000 000 in savings. Her current investments are earning, on average, 12%/a compounded annually.

a) What regular annual deposit must she make to have the required amount at retirement?

b) How much of the $1 000 000 is interest earned?

Andrea's Solution: Using a Formula

a) $$A = \frac{R[(1 + i)^n - 1]}{i}$$

I wrote the formula for calculating the future value of an annuity.

$$1\ 000\ 000 = \frac{R[(1 + 0.12)^{25} - 1]}{0.12}$$

The future value is $A = \$1\ 000\ 000$.

The annual interest rate is $i = 0.12$.

Regular deposits are made every year, so $n = 25$.

$$120\ 000 = R[(1.12)^{25} - 1]$$

I needed to solve for *R*, so I multiplied each side by 0.12.

$$R = \frac{120\ 000}{[(1.12)^{25} - 1]}$$

I isolated *R* by dividing both sides by the expression in square brackets.

$$= 7499.97$$

The investor needs to deposit $7499.97 annually.

I solved for *R* by simplifying the right side.

Pradesh's Solution: Using the TVM Solver

a)

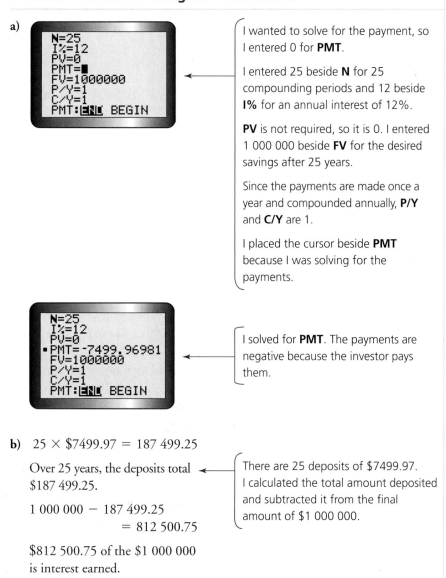

I wanted to solve for the payment, so I entered 0 for **PMT**.

I entered 25 beside **N** for 25 compounding periods and 12 beside **I%** for an annual interest of 12%.

PV is not required, so it is 0. I entered 1 000 000 beside **FV** for the desired savings after 25 years.

Since the payments are made once a year and compounded annually, **P/Y** and **C/Y** are 1.

I placed the cursor beside **PMT** because I was solving for the payments.

I solved for **PMT**. The payments are negative because the investor pays them.

b) $25 \times \$7499.97 = 187\ 499.25$

Over 25 years, the deposits total $187 499.25.

$$1\ 000\ 000 - 187\ 499.25$$
$$= 812\ 500.75$$

$812 500.75 of the $1 000 000 is interest earned.

There are 25 deposits of $7499.97. I calculated the total amount deposited and subtracted it from the final amount of $1 000 000.

Chapter 8 Solving Financial Problems Involving Exponential Functions   **497**

In Summary

Key Ideas

- Since an annuity is a series of equal deposits made at regular intervals, the amount, or future value, can be found by determining the sum of all the future values for each regular payment.
- The amount, or future value, of an annuity is the sum of all deposits and the accumulated interest and can be found with the formula

$$A = \frac{R[(1 + i)^n - 1]}{i}$$

where
  - A is the amount, or future value, in dollars
  - R is the regular deposit, or payment, in dollars
  - i is the interest rate per compounding period, expressed as a decimal
  - n is the total number of deposits

Need to Know

- Problems involving annuities can be solved with a formula, spreadsheet software, or financial software such as the TVM Solver.
- When the TVM Solver is used to solve problems that involve regular payments, the present value, **PV**, and future value, **FV**, are set to 0.

CHECK Your Understanding

1. Draw a timeline representing an annuity of semi-annual payments of $450 for 3 years at 12%/a compounded semi-annually. Use the timeline to show how the future value of each payment contributes toward the future value of the annuity.

2. Geoff and Marilynn are each investing in a 3-year Registered Retirement Savings Plan (RRSP) fund at 6%/a compounded quarterly. Geoff will make one deposit of $3600 at the beginning of the first year. Marilynn will make a $300 deposit at the end of March and will continue to contribute $300 every quarter until the end of the third year. Determine the difference in the future values.

PRACTISING

3. Determine the amount of each annuity.
   a) Regular deposits of $500 every 6 months for 4 years at 8%/a compounded semi-annually
   b) Regular deposits of $200 every month for 8 years at 10%/a compounded monthly

4. For each situation, identify R, i, and n. Then use the formula

$$A = \frac{R\left[(1 + i)^n - 1\right]}{i}$$

to determine the amount of the annuity.

	Payment ($)	Interest Rate	Compounding Period	Term of Annuity	Amount ($)
a)	1000	8%/a	annually	3 years	
b)	500	$7\frac{1}{2}$%/a	quarterly	8.5 years	
c)	200	3.25%/a	monthly	5 years	

5. Calculate the regular deposit made twice a year for 5 years at 6%/a compounded semi-annually to accumulate an amount of $4000.

6. Carollynne has found her dream home in Pictou, Nova Scotia. It is selling for $500 000. When she retires 2 years from now, she plans to sell her present house for $450 000 and move. She decides to set aside $900 every two weeks until she retires in a fund earning 10.5%/a, compounded every second week. What is the difference between the future value of Carollynne's investment and the extra $50 000 she needs for her dream home?

7. Yanmei has contributed $250 to an RRSP at the end of each 3-month period for the past 35 years. During this time, the RRSP has earned an average of 11.5%/a compounded quarterly.
  a) How much will the RRSP be worth at maturity?
  b) How much of the investment will be interest earned over the 35 years?

8. Miguel wants to buy an entertainment system as a gift for his sister's wedding. He estimates that when she marries 1 year from now, the system will cost $2499, plus GST (government sales tax) at 6% and PST (provincial sales tax) at 8%. He knows he can deposit $225 a month into an account earning 3.5%/a compounded monthly. Will he have enough money to buy the gift? Explain.

9. At the end of every 6 months, Marcia deposits $100 in a savings account
[T] that pays 4%/a compounded semi-annually. She made the first deposit when her son was 6 months old, and she made the last deposit on her son's 21st birthday. The money remained in the account until her son turned 25, when Marcia gave it to him. How much did he receive?

10. Marcel would like to take a vacation to Mexico during March break, 6 months from today. The trip will cost $3600. Marcel deposits $195 into an account at the end of each month for the next 8 months at 9%/a compounded monthly. Will he have enough money to pay for his trip? Explain.

11. Mario deposits $25 at the end of each month for 4 years into an account that pays 9.6%/a compounded monthly. He then makes no further deposits and no withdrawals. Determine the balance 10 years after his last deposit.

12. Darcey would like to accumulate $80 000 in savings before she retires
A 20 years from now. She intends to make the same deposit at the end of each month in an RRSP that pays 6.3%/a compounded monthly.
   a) Draw a timeline to represent the annuity.
   b) What regular payment will let Darcey reach her goal?
   c) Suppose Darcey decides to wait 5 years before starting her deposits. What regular payment would she have to make to reach the same goal?

13. Describe the payments, interest rates, and type of compounding necessary for a 15-year annuity with a future value between $10 000 and $12 000. Use two different compounding periods, each at a different interest rate, to modify the amounts shown.

14. Which annuity will earn the greater amount at the end of 2 years?
K Justify your answer.
   a) $50 at the end of every week at 5%/a compounded weekly
   b) $2600 at the end of every year at 5%/a compounded annually

15. Explain why the formula for calculating the accumulated value of a
C simple annuity would not work if the interest-compounding period did not coincide with the payment interval.

Extending

16. Byron has just bought a car for $25 000. He plans to replace it with a similar car in 3 years. At that time, his current car will be worth about one-third of its current value, and he will trade it for the new car. He will start saving for the rest of the cost by investing every month into an account paying 4.5%/a compounded monthly. How much should each payment be so that he can pay cash for the new car?

17. Nastassia borrowed $4831 at 7.5%/a compounded monthly. She has made 30 monthly payments of $130 each. She is now in a position to pay off the balance. What is that balance?

Regular Annuities: Determining Present Value

GOAL

Find the present value when payments or deposits are made at regular intervals.

LEARN ABOUT the Math

Harry has money in an account that pays 9%/a compounded annually. One year from now he will go to college. While Harry attends college, the annuity must provide him with 4 equal annual payments of $5000 for tuition.

? How much must be in Harry's account now if the first payment starts in a year?

EXAMPLE 1 Selecting a strategy to determine the present value of an annuity

Determine the present value of Harry's annuity.

Rahiv's Solution: Using a Timeline

Present Value of Annuity at 9%/a Compounded Annually

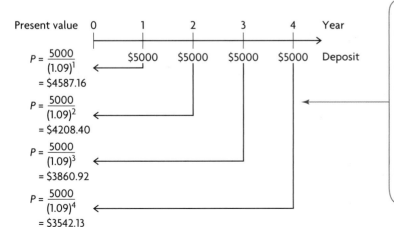

I used a timeline to organize the solution.

The annual interest rate is 9%, or 0.09. At the end of 1 year, the first payment's present value will have earned interest for 1 year: $n = 1$.

The second payment's present value will have earned interest for 2 years: $n = 2$.

The third payment's present value will have earned interest for 3 years: $n = 3$.

The fourth payment's present value will have earned interest for 4 years: $n = 4$.

$P = \dfrac{5000}{(1.09)^1}$
$= \$4587.16$

$P = \dfrac{5000}{(1.09)^2}$
$= \$4208.40$

$P = \dfrac{5000}{(1.09)^3}$
$= \$3860.92$

$P = \dfrac{5000}{(1.09)^4}$
$= \$3542.13$

$$PV = \frac{5000}{(1.09)^1} + \frac{5000}{(1.09)^2} + \frac{5000}{(1.09)^3} + \frac{5000}{(1.09)^4}$$

The present value is the total of the present values of all the payments.

$$= 4587.16 + 4208.40 + 3860.92 + 3542.13$$

$$= 16\ 198.61$$

To have four equal annual payments of $5000 starting 1 year from now, Harry needs $16 198.61 in his account now.

Tamika's Solution: Use a Spreadsheet

Formulas (as entered)

	A	B
1	End of Year	Present Value
2	1	$= \dfrac{5000}{(1+0.09)^{\wedge}A2}$
3	= A2 + 1	$= \dfrac{5000}{(1+0.09)^{\wedge}A3}$
4	= A3 + 1	$= \dfrac{5000}{(1+0.09)^{\wedge}A4}$
5	= A4 + 1	$= \dfrac{5000}{(1+0.09)^{\wedge}A5}$
6	Total PV	= SUM(B2:B5)

Values (as displayed)

	A	B
1	End of Year	Present Value
2	1	4587.16
3	2	4208.40
4	3	3860.92
5	4	3542.13
6	Total PV	16 198.60

The annual interest rate is 9%, or 0.09.

The formulas I used in the spreadsheet are in columns A and B. I knew that the amount paid at the end of each year had earned interest from the beginning of the annuity. So I used the year number, located in column A, in the formula to calculate the present value of each payment.

I read the total present value of the annuity from cell B6.

Reflecting

A. In calculating the present value of an annuity, the amount of each payment is divided by a factor of $(1 + i)$ for each additional compounding period. Why does this factor increase the distance in the future the payment is made?

B. Compare the methods used in the two solutions. What are the advantages and disadvantages of each?

C. If Harry were to receive payments every month instead of every year, which method would you use? Explain.

APPLY the Math

If technology is not available to help you calculate the present value of an annuity, then you can also use a formula.

EXAMPLE 2   Using a formula to determine the present value of an annuity

Roshan has set up an annuity to help his son pay living expenses over the next 5 years. The annuity will pay $50 a month. The first payment will be made 1 month from now. The annuity earns 7.75%/a compounded monthly.
a) How much money did Roshan put in the annuity?
b) How much interest will the annuity earn over its term?
c) Verify your results using the TVM solver.

Tim's Solution

a) $PV = \dfrac{R[1 - (1 + i)^{-n}}{i}$

    $R = 50$

> I wrote the formula for calculating the present value of an annuity, where
> - PV is the present value, in dollars
> - R is the regular payment, in dollars
> - i is the interest rate per compounding period, expressed as a decimal
> - n is the total number of payments

    $i = \dfrac{0.0775}{12} = 0.006\ 458\ 333\ 3$

    $n = 5 \times 12 = 60$

> The annual interest rate is $7\frac{3}{4}\%$. The monthly interest rate is $\frac{1}{12}$ of the annual rate, so I divided it by 12. I multiplied the number of years by 12 to determine the number of compounding periods.

    $PV = \dfrac{50[1 - (1 + 0.006\ 458\ 333\ 3)^{-60}]}{0.006\ 458\ 333\ 3}$

> I substituted the values for R, i, and n, and calculated PV.

    $PV = 2480.53$

Roshan put $2480.53 in the annuity.

b) $60 \times 50 = 3000$

   $3000 - 2480.53 = 519.47$

> I determined the total interest earned over the term of the annuity by subtracting the present value of the annuity from the total value of payments.

The annuity earned $519.47 in interest over its term.

c)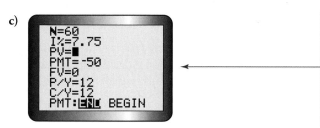

> I entered 60 beside **N** for the 60 compounding periods. I entered 7.75 beside **I%** for 7.75% annual interest. The regular payment of $50 means that Roshan has paid this amount out to his son, so I entered −50 beside **PMT**. **FV** is not required, so I entered a value of 0.
>
> I entered 12 for both **P/Y** and **C/Y** because the payments are made and compounded monthly.

```
N=60
I%=7.75
■PV=2480.532809
PMT=-50
FV=0
P/Y=12
C/Y=12
PMT:END BEGIN
```

I placed the cursor beside **PV** because I was solving for the present value.

I solved for **PV**. The present value is positive because it is money that is deposited into the bank at the beginning of the annuity to provide for the monthly withdrawls.

I could have also used **PMT** = +50, and the **PV** would have been − 2480.532 809.

Roshan put $2480.53 in the annuity.

```
ΣInt(1,60)
        -519.4671906
```

I used the TVM Solver's interest function and entered 1 for the starting payment number and 60 for the ending payment number.

Tech | Support

Before using the ΣInt function, make sure that values have been entered into the TVM Solver for **N, I%, PV, PMT, P/Y, C/Y**, and **PMT:END**. For more help with using the TVM Solver to solve problems involving compound interest, see Technical Appendix, B-15.

The annuity earned $519.47 in interest.

EXAMPLE 3 **Selecting a strategy that uses present value to calculate the payment of an annuity**

Robin bought a bicycle for $1500. She arranged to make a payment to the store at the end of every month for 1 year. The store is charging 11%/a interest compounded monthly.

a) How much is each monthly payment?
b) How much interest is Robin paying?

Leshawn's Solution: Using a Formula

a) $PV = \dfrac{R[1 - (1 + i)^{-n}]}{i}$

I used the formula for present value so that I could solve for the payment.

$PV = 1500$

$i = \dfrac{11}{12}\% = 0.009\ 166\ 666\ 7$

$n = 12$

The present value is $PV = \$1500$. The annual interest rate is 11%, so the monthly interest rate is $\frac{1}{12}$ as much.

She makes 12 payments in a year.

$$1500 = \frac{R[1 - (1 + 0.009\ 17)^{-12}]}{0.009\ 17}$$

$$\frac{0.009\ 17(1500)}{1 - (1 + 0.009\ 17)^{-12}} = R$$

> I multiplied both sides of the equation by 0.009 17, then I divided both sides by $1 - (1 + 0.009\ 17)^{-12}$ to solve for R.

$$132.57 = R$$

Robin makes monthly payments of $132.57.

b) $\qquad 132.57 \times 12 = 1590.84$

> To calculate the interest paid, I found the total of the payments and subtracted the cost of the bicycle. The difference was the interest paid to the store.

$$1590.84 - 1500.00 = 90.84$$

The interest paid to the store is $90.84.

Henrique's Solution: Using the TVM Solver

a)

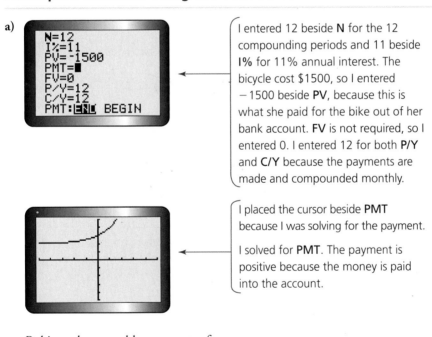

> I entered 12 beside **N** for the 12 compounding periods and 11 beside **I%** for 11% annual interest. The bicycle cost $1500, so I entered −1500 beside **PV**, because this is what she paid for the bike out of her bank account. **FV** is not required, so I entered 0. I entered 12 for both **P/Y** and **C/Y** because the payments are made and compounded monthly.

> I placed the cursor beside **PMT** because I was solving for the payment.

> I solved for **PMT**. The payment is positive because the money is paid into the account.

Robin makes monthly payments of $132.57.

b)

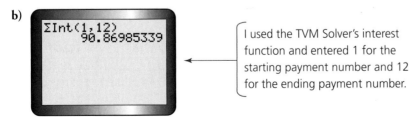

> I used the TVM Solver's interest function and entered 1 for the starting payment number and 12 for the ending payment number.

The interest paid to the store is $90.87.

In Summary

Key Ideas

- The present value of an annuity is (1) the amount that must be invested now to provide payments of a specific amount at regular intervals over a certain term or (2) the amount borrowed or financed now that must be paid for by deposits of a specific amount at regular intervals over a certain term.
- The present value of an annuity is the sum of the present values of all of the regular payments.
- The formula for calculating the present value of an annuity is

$$PV = \frac{R[1 - (1 + i)^{-n}]}{i}$$

where
- PV is the present value, in dollars
- R is the regular payment, in dollars
- i is the interest rate per compounding period, expressed as a decimal
- n is the number of payments or withdrawals

Need to Know

- Problems involving annuities can be solved with a formula, spreadsheet software, or financial software such as the TVM Solver.

CHECK Your Understanding

1. Draw a timeline to represent an annuity of semi-annual payments of $300 for 3 years at 8%/a compounded semi-annually. Use the timeline to organize a solution that shows how the present value of each payment contributes toward the present value of the annuity.

2. For each situation, identify R, i, and n in the formula

$$PV = \frac{R[1 - (1 + i)^{-n}]}{i}.$$

Then determine the present value of the annuity.

	Withdrawal ($)	Annual Interest Rate (%)	Compounding Period	Term of Annuity
a)	750	8	annual	3 years
b)	450	$7\frac{1}{2}$	quarterly	8.5 years
c)	225	3.25	monthly	5 years

3. Solve for each unknown.

a) $PV = \dfrac{450(1 - 1.055^{-8})}{0.055}$

b) $40\,000 = \dfrac{R(1 - 1.006^{-12})}{0.006}$

PRACTISING

4. Kevin came into an inheritance of $36 000, to be paid in equal monthly instalments for the next 5 years, starting 1 month from now. The money earns 7.5%/a compounded monthly. Determine the monthly instalment.

5. Mary needs $750 a year for 3 years to buy textbooks. She will start university in 1 year. Her savings account pays 4%/a compounded annually. How much needs to be in her account now to pay for the books?

6. Your grandmother has set up an annuity of $4000 in an account that pays 5.2%/a compounded monthly. What equal monthly payments will the annuity provide for in the next 4 years?

7. May Sum has saved $125 000 in an investment account. She will use it to buy an annuity that pays 6.5%/a compounded quarterly. She will receive quarterly payments for the next 25 years. The first payment will be made 3 months from now.
   a) What is the quarterly payment she will receive?
   b) What is the interest earned over the duration of the annuity?

8. Claire buys a snowboard for $150 down and pays $35 at the end of each month for 1.5 years. If the finance charge is 16%/a compounded monthly, determine the selling price of the snowboard.

9. Felix's family has decided to deposit $350 into an annuity every 3 months for 4 years. The account will earn 3.75%/a compounded quarterly. Starting 3 months after the last deposit, Felix will withdraw the money every 3 months in equal payments for 2 years. What is the amount of each withdrawal?

10. **K** Nick plans to buy a used car today. He can afford to make payments of $250 each month for a maximum of 3 years. The best interest rate he can find is 9.8%/a compounded monthly. What is the most he can spend?

11. **A** Shimon wants to buy a speedboat that sells for $22 000, including all taxes. The dealer offers either a $2000 discount, if Shimon pays the total amount in cash, or a finance rate of 2.4%/a compounded monthly, if Shimon makes equal monthly payments for 5 years.
   a) Determine the monthly payment that Shimon must make if he chooses the second offer.
   b) What is the total cost of the dealer's finance plan for the speedboat?
   c) To pay for the boat with cash now, Shimon can borrow the money from the bank at 6%/a, compounded monthly, over the same 5-year period. Which offer should Shimon choose, the bank's or the dealer's? Justify your answer.

12. René buys a computer system for $80 down and 18 monthly payments of $55 each. The first payment is due next month.
    a) The interest rate is 15%/a compounded monthly. What is the selling price of the computer system?
    b) What is the finance charge?

13. Betty is retiring. She has $100 000 in savings. She is concerned that she will not have enough money to live on. She would like to know how much an annuity, compounded monthly, will pay her each month for a variety of interest rates. She needs to know the monthly payments over the next 10 years, starting next month. Use a spreadsheet and different annual interest rates to prepare three different schedules of payments for Betty.

14. The present value of the last payment of an annuity is $2500(1.05)^{-36}$.
    T a) Describe two annuities, with different compounding periods, that can be represented by the present value of the last payment.
    b) Calculate the present values of the total payments for each annuity in part (a).

15. The screens shown were obtained from the TVM Solver. Write a
    C problem that corresponds to the information from each screen.
    a)

    b)

Extending

16. Do the following situations double the amount of an annuity at maturity?
    a) Double the duration of the annuity.
    b) Double each payment made.
    Use examples to support your explanation.

17. Rudi deposited $100 at the end of each month into an annuity that paid 7.5%/a compounded monthly. At the end of 6 years, the interest rate increased to 8.5%/a . The deposits were continued for another 5 years.
    a) What is the amount of the annuity on the date of the last deposit?
    b) What is the interest rate earned on the annuity over the 11 years?

18. Kyla must repay her $17 000 student loan. She can afford to make monthly payments of $325. The bank's interest rate is 7.2%/a compounded monthly. Determine how long it will take Kyla to repay her loan.

Saving Plans and Loans: Creating Amortization Tables

YOU WILL NEED
- graphing calculator with TVM Solver
- spreadsheet software

GOAL

Examine how changing the conditions of an annuity affects the interest earned or paid.

INVESTIGATE the Math

One day you may wish to buy a home or a car. You might not be able to pay for it all at once. However, you might be able to make smaller payments toward its purchase over time. You can take out a loan or mortgage from a financial institution. You sign a contract agreeing to repay, or amortize, the value of the loan, plus interest, by making regular equal payments for the term, or **amortization** period, of the loan. A portion of each payment is interest. The rest of the payment is applied to reduce the principal, or amount borrowed.

amortization

the process of gradually reducing a debt through instalment payments of principal and interest

Tyler bought a used car for $10 000. He made a **down payment** of $3500. The car dealer offered to finance the rest of the purchase with a 1-year loan at 8%/a compounded monthly.

down payment

the partial amount of a purchase paid at the time of purchase

❓ What conditions affect the monthly payment and the amount of interest paid to the car dealer?

A. Study amortization table A, on the next page, which shows the repayment schedule of the $6500 loan over the period of 1 year. Determine
   - the payment made each month
   - total payments made to the car dealership, including down payment
   - the total interest paid to the car dealership

B. Tyler thinks he can negotiate a 6%/a interest rate. Using the amortization spreadsheet, he replaces the annual rate of 0.08 with 0.06. The spreadsheet automatically recalculates information. Study amortization table B, on the next page. What information is recalculated? How does the lower interest rate affect the interest paid to the car dealer and the total cost of the car?

A.

Cost of Car	$10 000	Annual Rate	0.08
Down Payment	$3500	Monthly Rate	0.0067
Loan	$6500	Number of	
Payment	$565.42	Payments	12

Payment Number	Payment ($)	Interest Paid ($)	Principal Paid ($)	Outstanding Balance ($)
0				6500.00
1	565.42	43.33	522.09	5977.91
2	565.42	39.85	525.57	5452.34
3	565.42	36.35	529.08	4923.26
4	565.42	32.82	532.60	4390.66
5	565.42	29.27	536.15	3854.50
6	565.42	25.70	539.73	3314.78
7	565.42	22.10	543.33	2771.45
8	565.42	18.48	546.95	2224.50
9	565.42	14.83	550.59	1673.91
10	565.42	11.16	554.27	1119.64
11	565.42	7.46	557.96	561.68
12	565.42	3.74	561.68	0.00
Totals	6785.04	285.04	6500.00	

B.

Cost of Car	$10 000	Annual Rate	0.06
Down Payment	$3500	Monthly Rate	0.0050
Loan	$6500	Number of	
Payment	$559.43	Payments	12

Payment Number	Payment ($)	Interest Paid ($)	Principal Paid ($)	Outstanding Balance ($)
0				6500.00
1	559.43	32.50	526.93	5973.07
2	559.43	29.87	529.57	5443.50
3	559.43	27.22	532.21	4911.29
4	559.43	24.56	534.88	4376.41
⋮				⋮
10	559.43	8.31	551.12	1110.53
11	559.43	5.55	553.88	556.65
12	559.43	2.78	556.65	0.00
Totals	6713.16	213.16	6500.00	

C. To change the payments so that the loan is amortized over 2 years, Tyler enters 24 beside **Number of Payments**. What information is recalculated? How does the extended term affect the interest paid to the car dealer and the total cost of the car?

Cost of Car	$10 000	Annual Rate	0.06
Down Payment	$3500	Monthly Rate	0.005
Loan	$6500	Number of	
Payment	$288.08	Payments	24

Payment Number	Payment ($)	Interest Paid ($)	Principal Paid ($)	Outstanding Balance ($)
0				6500.00
1	288.08	32.50	255.58	6244.42
2	288.08	31.22	256.86	5987.55
3	288.08	29.94	258.15	5729.41
4	288.08	28.65	259.44	5469.97
⋮				⋮
22	288.08	4.28	283.81	571.88
23	288.08	2.86	285.22	286.65
24	288.08	1.43	286.65	0.00
Totals	6913.92	413.92	6500.00	

D. Tyler changes the spreadsheet by increasing the down payment and changing the interest and payment periods back to 1 year at 8%/a compounded monthly. How does increasing the down payment affect the interest paid and total cost of the car as determined in part A?

Cost of Car	$10 000	Annual Rate	0.08
Down Payment	$5000	Monthly Rate	0.0067
Loan	$5000	Number of	
Payment	$434.94	Payments	12

Payment Number	Payment ($)	Interest Paid ($)	Principal Paid ($)	Outstanding Balance ($)
0				5000.00
1	434.94	33.33	401.61	4598.39
2	434.94	30.66	404.29	4194.10
⋮				⋮
11	434.94	5.74	429.20	432.06
12	434.94	2.88	432.06	0.00
Totals	5219.28	219.28	5000.00	

Tech | *Support*

For help creating amortization schedules using a spreadsheet, see Technical Appendix, B-18.

Reflecting

E. What are the advantages and disadvantages of using an amortization table that has been created with spreadsheet software?

F. The formulas that are placed in the cells of the spreadsheet amortization table are shown here. For each quantity, locate the formula and describe its calculation.
   a) present value
   b) interest rate for compounding period
   c) payment
   d) interest paid

	A	B	C	D	E
1	Cost of Car		10 000	Annual Rate	0.08
2	Down Payment		3500	Monthly Rate	=E1/E3
3	Loan		=C1−C2	Number of Payments	12
4	Payment		=E2*C3/(1−(1+E2)^(−E3))		
5	Payment	Payment	Interest	Principal	Outstanding
6	Number		Paid	Paid	Balance
7	0				=C3
8	=A7+1	=$C$4	=E7*E$2	=B8−C8	=E7−D8
9	=A8+1	=$C$4	=E8*E$2	=B9−C9	=E8−D9
⋮					⋮
18	=A17+1	=$C$4	=E17*E$2	=B18−C18	=E17−D18
19	=A18+1	=$C$4	=E18*E$2	=B19−C19	=E18−D19
20	Totals	=SUM(B8:B19)	=SUM(C8:C19)	=SUM(D8:D19)	

APPLY the Math

EXAMPLE **1**	Selecting a strategy to compare the growth of annuities

Compare the amounts at age 65 that an RRSP at 6%/a compounded annually would earn under each option.

Option 1: making an annual deposit of $1000 starting at age 20
Option 2: making an annual deposit of $3000 starting at age 50

What is the total of the deposits in each situation?

Raina's Solution: Using Amortization Tables

Option 1:

Annual Rate	0.0600	Start Year	20
Rate per Period	0.0600	End Year	65
Compounding Periods/Year	1	Final Year End	45
Contribution	$1000		

End of Year	Interest ($)	Payment Made ($1000/year + interest) ($)	New Balance ($)
1		1000.00	1000.00
2	60.00	1060.00	2060.00
⋮			⋮
44	11 250.45	12 250.45	199 758.03
45	11 985.48	12 985.48	212 743.51

From this amortization table, after 45 years of investing $1000, the balance of the RRSP is $212 743.51. The interest earned over 45 years is $167 743.51, on a total deposit of $45 000.

Option 2:

Annual Rate	0.0600	Start Year	50
Rate per Period	0.0600	End Year	65
Compounding Periods/Year	1	Final Year End	15
Contribution	$3000		

End of Year	Interest ($)	Payment Made ($3000/year + interest) ($)	New Balance ($)
1		3000.00	3000.00
2	180.00	3180.00	6180.00
⋮			⋮
14	3398.78	6398.78	63 045.20
15	3782.71	6782.71	69 827.91

If the start year is changed to 50 and the contribution is changed to $3000, the balance of the RRSP is $69 827.91 and the interest earned over 15 years is $24 827.91, on a total deposit of $45 000.

Derek's Solution: Using the TVM Solver

Option 1:

The future value of the RRSP deposits is $212 743.51.

There are 45 annual deposits from age 20 to age 65, so I entered 45 for **N**.

I entered 6 beside **I%**.

I entered 0 for **PV** because the first deposit does not go in until the end of the first year.

The annual RRSP deposit is $1000, so I entered −1000 for **PMT**. I entered 1 for both **P/Y** and **C/Y**.

I solved for **FV**.

Using the **ΣInt** function of the Finance Program, I calculated the interest over the given period.

The interest earned over 45 years is $167 743.51.

Option 2:

The future value of the RRSP deposits is $69 827.91.

There are 15 annual deposits from age 50 to age 65, so I changed **N** to 15.

The annual RRSP deposit is $3000, so I changed **PMT** to −3000. I solved for **FV**.

Using the **ΣInt** function of the Finance Program, I calculated the interest.

The interest earned over 15 years is $24 827.91.

EXAMPLE **2** Changing the payment frequency

Show how changing the payment frequency from semi-annual to weekly affects the amount of interest paid and the length of time needed to repay a loan of $5000.00 at 11%/a. For example, Joe makes semi-annual payments of $520 and interest is charged semi-annually, while Sarit makes weekly payments of $20 and interest is charged weekly.

Carmen's Solution

Compounding Period	TVM Solver Screen	Number of Payments	Time

Joe: semi-annual payments of $520	`•N=14.05602953`
	`I%=11`
	`PV=5000`
	`PMT=-520`
	`FV=0`
	`P/Y=2`
	`C/Y=2`
	`PMT:END BEGIN`
		14.06	$\dfrac{14.06}{2} = 7.03$ years

	`ΣInt(1,14)` `        -2308.319852`	The interest paid was $2308.32.	

Sarit: weekly payments of $20	`•N=356.1368005`
	`I%=11`
	`PV=5000`
	`PMT=-20`
	`FV=0`
	`P/Y=52`
	`C/Y=52`
	`PMT:END BEGIN`
		356	$\dfrac{356}{52} = 6.85$ years

	`ΣInt(1,356)` `        -2122.732726`	The interest paid was $2122.74	

Both Joe and Sarit are repaying the same amount each year.
$520 × 2 = $1040
$20 × 52 = $1040
I used the TVM Solver to compare the two options.

$2308.32 - $2122.74 = $185.58

The time needed to repay the loan decreases from 7.03 to 6.85 years under the weekly option, resulting in a savings of $185.58 in interest paid out.

EXAMPLE 3 | Changing the payment

David has a credit card balance of $10 000 that is charged 23%/a interest, compounded monthly. He decides not to use his credit card again and to make monthly payments to pay off his debt.
a) How long does it take to reduce his credit card balance of $10 000 to 0 if he pays $1000 a month?
b) If he increases his monthly payment to $1600, how much sooner is the debt paid off?
c) How much will he save in finance charges if his payment is $1600 rather than $1000?

Mandy's Solution

a)

 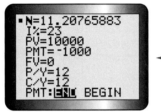

I entered 23 beside **I%** and 10 000 beside **PV**.

I entered −1000 for **PMT** because it is money David pays out.

The future value of the debt will be 0 when it is paid off, so I entered 0 for **FV**.

I entered 12 for **P/Y** and **C/Y**.

I placed the cursor beside **N** because I was solving for the number of monthly payments.

I solved for the number of compounding periods. I rounded up to 12 because David would have to make the monthly payment 11 times and a smaller payment the last time.

The credit card balance will be paid off after 12 months.

b)

I changed **PMT** to −1600 and placed the cursor beside **N**. I solved for the number of compounding periods, rounding up to 7.

The credit card balance will be paid off after 7 months.

$12 - 7 = 5$

I calculated the difference of the two numbers of monthly payments.

The debt will be paid off about 5 months earlier if David pays $1600 a month.

c) $(11.207\ 658\ 83 \times 1000) -$
   $(6.720\ 813\ 35 \times 1600) = 454.36$

I could have used the TVM ΣInt function to compare the interests paid, but I decided to calculate the total amount paid for payments of $1000 and subtracted the total amount paid for payments of $1600.

The amount saved if David pays $1600 a month is $454.36.

Key Ideas

- Amortization tables show the amount of each loan payment or savings deposit, the interest portion, the principal portion, and the new balance after each payment. An amortization table can be created with spreadsheet software.
- Once an amortization table has been set up on a spreadsheet, it is easy to change the conditions of the situation and examine the impact the change has on quantities such as the duration of a loan, the amount of interest paid, and the total amount of money paid out.

Need to Know

- The TVM Solver can be used to investigate the effects of changing the conditions when borrowing or investing.
- **ΣInt** is a financial function that may be used after entering information in the TVM Solver. **ΣInt** is used to calculate the total interest paid from a starting payment number to an ending payment number.

CHECK Your Understanding

1. The formulas in the amortization table used to solve Example 1 (Option 2) are shown.

	A	B	C	D
1	Annual Rate	0.06	Start Year	50
2	Rate per Period	=B1/B3	End Year	65
3	Compounding Periods per Year	1	Final Year End	=D2−D1
4	Contribution	3000		
5	End of Year	Interest	Payment Made	New Balance
6	1			=B4
7	=A6+1	=D6*$B$2	=$B$4+B7	=D6+C7
⋮	⋮	⋮	⋮	⋮
14	=A13+1	=D13*$B$2	=$B$4+B14	=D13+C14
15	=A14+1	=D14*$B$2	=$B$4+B15	=D14+C15

a) How is the interest calculated?

b) **Payment Made** is the sum of two numbers. What do these numbers represent?

2. An $18\,000 car loan is charged 4%/a interest compounded quarterly.
   a) Determine the quarterly payments needed to pay the loan off in 5 years.
   b) How much faster would the loan be paid off with the same payments if the interest rate was lowered to 2.9%/a?
   c) What would be the total cost of the car, including interest, in parts (a) and (b)?

3. For each situation, show what would be entered into each TVM Solver variable. Do not solve the problem.
   a) A $10\,000 loan is repaid with monthly payments of $350 for 13 years. Determine the annual interest rate, with monthly compounding.
   b) Martin took out a student loan for $15\,000 at 7.5%/a compounded monthly. He is working now and wants to pay it off. When will it be paid off if he makes monthly payments of $500?

PRACTISING

4. Susan's parents would like to save $12\,000 over the next four years to pay for her first year at McGill University in Montréal.

   a) How much should they deposit at the end of each month into an account that pays 7.25%/a, compounded monthly, to attain their goal?
   b) How much should they deposit at the end of every three months into an account that pays 7.25%/a, compounded quarterly, to attain their goal?
   c) Why is the payment in part (b) slightly over three times the payment in part (a)?

5. Joel and Katerina are each paying off loans of $5000. Joel makes monthly payments of $75 and interest is charged at 9%/a compounded monthly. Katerina pays the loan off in the same amount of time, but her monthly payments are only $65. Determine the annual interest rate that Katerina is charged if her interest is also compounded monthly.

6. Bernice will repay a $30\,000 loan with monthly payments. The term of the loan is 5 years. The interest rate is 7.25%/a compounded monthly.
   a) What is the monthly payment for this loan?
   b) What is the outstanding balance on the loan after each of the first 5 years?
   c) What is the interest and principal that she has paid at the end of the 5-year term?

7. If $1000 is deposited at the end of each year in an account that pays 13.5%/a compounded annually, about how many years will it take to accumulate to $20 000?

8. Jack's life savings total $320 000. He wants to use the money to buy an annuity earning interest at 10%/a compounded semi-annually so that he will receive equal semi-annual payments for 20 years. How much is each payment if the first is 6 months from the date of purchase?

9. An account pays 9.2%/a compounded annually. What deposit on January 1 of this year will allow you to make 10 annual withdrawals of $5000, beginning January two years from now?

10. Michael has a student loan of $15 000 at 8.5%/a compounded monthly. He will pay off the loan over the next 5 years. The first payment will be made 1 month from now.

  a) What is his monthly loan payment?
  b) What is the interest that Michael will pay over the term of the loan?
  c) At the end of one year, Michael decides to pay off the rest of the loan. How much interest did he save by repaying the loan in 1 year?

11. Describe three different ways to save money when taking out a loan. Why do they work?

Extending

12. The repayment schedule for a loan lists each payment, and shows how much of each payment is interest and how much goes to reduce the principal. It also shows the outstanding balance after each payment.
  a) Identify an item that you would like to buy for which you might need a loan.
  b) The prime interest rate is the interest rate that banks charge their preferred customers. Determine the current prime lending rate of a major bank.
  c) Create a repayment schedule for the loan, showing the amount of the loan, annual interest rate, issue date, number of payments, and monthly payment. The loan is repaid with equal monthly payments over 2 years.

13. Sumiko buys a car with a $22 000 loan at 9.25%/a compounded monthly. She will repay the loan with 60 equal monthly payments.
  a) Determine the monthly payment.
  b) After a year, Sumiko decides to increase her payments by $150 a month. How many more payments are required to pay off her loan?
  c) How much interest does she save by making the greater monthly payment?

Buy Now, No Payments for a Year: Is It Always a Good Deal?

A local home-furnishing store offers "no payments, no interest" for a year if you charge your purchase to the store's credit card. However, the store will charge a $35 administration fee. This fee, with 6% GST and 8% PST, must be paid within 30 days. If the balance of the purchase is not paid at the end of the year, interest on the full amount is added to the bill. The interest rate is 24%/a compounded annually. A customer makes a purchase of $3495, not including tax.

1. What is the cost of borrowing if the bill is paid at the end of the year? Explain.

2. What is the cost of borrowing if the bill is not paid at the end of the year? Explain.

3. If the customer is unable to pay after 1 year, the store usually arranges to have the customer pay the bill with 12 equal monthly payments. As the payments are made, interest is charged on any unpaid balance at 24%/a compounded monthly. Calculate the monthly payment and the total paid in cash by the customer.

4. Suppose that, after 1 year, the customer is able to transfer the unpaid balance to another credit card that charges 9%/a compounded monthly. The customer intends to make the same monthly payment. How much sooner will the debt be paid off? How much interest will the customer save by transferring the debt to the credit card company offering the lesser interest rate?

FREQUENTLY ASKED *Questions*

Q: **How can a timeline help you better visualize how an annuity works?**

A: An annuity is a series of payments or deposits made at regular intervals. Annuities earn interest on each regular deposit from the time the money is deposited to the end of the term of the annuity. A timeline is a tool to help you visualize how the factor $(1 + i)$ affects the present value of withdrawals or future value of deposits. The future values of deposits are multiplied by $(1 + i)$ for each compounding period. The present values of withdrawals are divided by $(1 + i)$ for each compounding period.

> **Study | *Aid***
> * See Lessons 8.5 and 8.6, Example 1.
> * Try Chapter Review Question 7.

Q: **How can you calculate the future value or the present value of an annuity?**

A: You can use
* the formula $FV = \dfrac{R[(1 + i)^n - 1]}{i}$ or $PV = \dfrac{R[1 - (1 + i)^{-n}]}{i}$

    where
    - FV is the future value (amount), in dollars
    - PV is the present value (principal), in dollars
    - R is the regular payment, in dollars
    - i is the interest rate per compounding period, expressed as a decimal
    - n is the total number of payments
* the TVM Solver
* a spreadsheet

> **Study | *Aid***
> * See Lessons 8.5 and 8.6, Examples 2 and 3.
> * Try Chapter Review Questions 8 to 15.

Q: **What is an amortization table?**

A: For each payment made during the term of the amortization, an amortization table shows the payment number, the payment amount, how much of each payment is interest, how much of each payment goes to reduce the principal, and the outstanding principal. An amortization table can be created with spreadsheet software. It is designed to allow you to change the parameters of an annuity problem and analyze the effects of the changes.

> **Study | *Aid***
> * See Lesson 8.7, Examples 1, 2, and 3.
> * Try Chapter Review Question 16.

PRACTICE Questions

Lesson 8.1

1. Copy and complete the table.

	Principal ($)	Annual Interest Rate (%)	Years	Compounding Frequency	Amount ($)	Interest Earned ($)
a)	400.00	5%	15	semi-annually		
b)	450.00	4.5%	10	monthly		
c)		3.4%	10	weekly	875.00	
d)	508.75		3	semi-annually	568.24	
e)	10 000.00	2.34%		quarterly		1000.00

2. Suppose you were to graph the accumulated simple interest on an investment and the accumulated compound interest on the same investment on the same graph. What would be the similarities and differences between the two graphs? Be as detailed as possible.

Lesson 8.2

3. Determine the future value of a $5000 compound-interest Canada Savings Bond at $8\frac{1}{2}$%/a compounded annually after each amount of time.

   a) 4 years   **b)** 8 years

Lesson 8.3

4. Your mother wants to give you $25 000 in 15 years' time. How much should she invest now at 8%/a interest compounded monthly to meet this goal?

Lesson 8.4

5. An investment of $1500 grows to $3312.06 in 10 years. What is the interest rate of the investment if interest is compounded quarterly?

6. Kadie invested $3000 at 6%/a compounded quarterly. How long will it take for the investment to be worth $8500?

Lesson 8.5

7. For 10 years, Sheila deposits $750 at the end of every 3 months in a savings account that pays 8%/a compounded quarterly.

   a) Draw a partial timeline to represent the first 3 months and the last 3 months of the annuity.

   b) Calculate the amount of the annuity and the total interest earned.

8. At the end of every 6 months, Parvati deposited $200 into a savings account that paid 3.5%/a compounded semi-annually. She made the first deposit when her son was 6 months old and the last deposit on his 18th birthday. The money remained in the account until he turned 21. How much did Parvati's son receive?

9. David is 8 years old when his parents start an education fund. They deposit $450 at the end of every 3 months in a fund that pays 8%/a compounded quarterly.

   a) How old is David when the fund is worth $20 000?

   b) How much less time would it take to build the fund to $20 000 if the regular deposit were $550?

10. The Huang family borrowed $30 000 at 9%/a compounded monthly to buy a motor home. The Huangs will make payments at the end of each month. They have two choices for the term: 5 years or 8 years.
    a) Determine the monthly payment for each term.
    b) How much interest would they save by selecting the shorter term?

11. Raymond has $53 400 in his savings account, and he withdraws $250 at the end of every 3 months. If the account earns 5%/a compounded quarterly, what will his bank balance be at the end of 4 years?

Lesson 8.6

12. Adrianna wants to buy a used car. She can afford payments of $300 each month and wants to pay off the debt in 3 years. The bank offers a rate of 9.8%/a compounded monthly. What is the most Adrianna can spend on a vehicle?

13. For each situation described, indicate the values you would enter beside each variable for the TVM Solver.

    a) Cecilia deposits $1500 at the end of each year in a savings account that pays 4.5%/a compounded annually. What is her balance after 5 years?
    b) Farouk would like to have $200 000 in his account in 15 years. How much should he deposit at the end of each month in an account that pays 3.75%/a compounded monthly?

c) A $10 000 loan is repaid with monthly payments of $334.54 for 3 years. Determine the annual interest rate, compounded monthly.
d) Determine the total amount of interest earned on an annuity consisting of quarterly deposits of $2000 for 8 years if the annuity earns 9%/a interest compounded quarterly.

14. How much must be in a fund paying 6%/a compounded semi-annually if you wish to withdraw $1000 every 6 months, starting 6 months from now, for the next 5 years?

15. Karsten is preparing his will. He wants to leave the same amount of money to his two daughters. His elder daughter is careful with money, but the younger daughter spends it carelessly, so he decides to give them the money in different ways. How much must his estate pay his younger daughter each month over 20 years so that the accumulated present value will be equal to the $50 000 cash his elder daughter will receive upon his death? Assume that the younger daughter's inheritance earns 6%/a compounded monthly over the 20 years.

Lesson 8.7

16. Create an amortization table showing the amortization of a loan that will be repaid with equal monthly payments over 5 years. The loan of $10 000 has an interest rate of 8%/a compounded monthly.

17. Jerzy borrowed $4831 at 7.5%/a compounded monthly. He has made 30 monthly payments of $130 each. He is now in a position to pay off the balance. What is his remaining balance?

1. Explain why more money accumulates if the interest is compounded than if the interest is simple.

2. Determine the amount and interest earned when $10 500 is invested for 4 years at 4.8%/a compounded monthly.

3. Determine the present value and the interest earned on a loan of $21 500 due in 6 years. The interest rate is 8%/a compounded quarterly.

4. Carter deposits $12 000 in an account that pays 6%/a compounded semi-annually. After 5 years, the interest rate changes to 6%/a compounded monthly. Calculate the value of the money 8 months after the change in the interest rate.

5. Janet and her fiancé plan to buy a new house 3 years from now. They intend to make a down payment of $20 000. They can invest money in an account offering 9.25%/a compounded monthly. How much money must they invest today to reach their goal?

6. A principal of $500 grew to $620 in 11 years. Determine the annual interest rate, compounded quarterly.

7. Raj invested $1000 in a GIC that paid 4%/a compounded weekly. He received $1350 at the end of the term. For how long was the money invested?

8. Draw a timeline showing the future value of $350 deposited semi-annually for 1.5 years at 3.75%/a compounded semi-annually.

9. Kay Chung wants to travel to China in 20 months. The trip will cost $3200. How much should she deposit at the end of each month in an account that pays 9%/a compounded monthly to save $3200?

10. Since the birth of their daughter, the Tranters have deposited $450 every 3 months in an education savings plan. The interest rate is 7.5%/a compounded quarterly. What is the plan's value when their daughter turns 17?

11. What are the components of an amortization table and what is its purpose?

Investigating RRSP Investments

Teresa began to contribute to her RRSP at age 20. She made monthly contributions of $50, starting 1 month after her birthday. Her RRSP earned interest at an average rate of 7.5%/a compounded monthly, until her 60th birthday, when she retired. One month later, she started to withdraw a monthly amount.

A. Determine the amount of her RRSP on her 60th birthday.

B. Suppose Teresa transferred her RRSP into a Registered Retirement Income Fund (RRIF)—an annuity that uses the RRSP savings to provide the holder with a regular retirement income over a term of 25 years. What monthly pension can Teresa withdraw for the next 25 years?

C. Choose different contribution amounts, times for investments, and interest rates. Determine what effects they have on monthly retirement pensions. Prepare a report of your findings that compares these different scenarios to Teresa's situation.

Task | *Checklist*

In your report,

✔ Did you show the necessary calculations to support your answers for parts A and B?

✔ Did you vary
  – the contribution amount?
  – the length of time?
  – the interest rates?

✔ Did you discuss how these changes affect the amount of monthly retirement payment?

✔ Did you compare your finding to Teresa's situation?

Multiple Choice

1. Which expression has a value of 64?
   a) $16^{\frac{3}{2}}$ **c)** $16^{\frac{1}{2}}$
   b) $-4^{\frac{3}{2}}$ **d)** $\sqrt[3]{4}$

2. If the value of the variable is 3, which of the following is true?
   a) $(p \times p^3)^3 = 3^9$ **c)** $(n^2)^3 \div n = 243$
   b) $(t^2)^2 \times t^0 = 10\,000$ **d)** $c^3 = 512$

3. Which number is equivalent to $16^{\frac{3}{4}}$?
   a) 4   **c)** -4
   b) 8   **d)** $\dfrac{1}{2}$

4. Identify the expression that is false.
   a) $(9^{\frac{1}{2}})(4^{\frac{1}{2}}) = (9 \times 4)^{\frac{1}{2}}$
   b) $\left(\dfrac{1}{9} \times \dfrac{1}{4}\right)^{-1} = (9)(4)$
   c) $9^{\frac{1}{2}} + 4^{\frac{1}{2}} = (9 + 4)^{\frac{1}{2}}$
   d) $[(9^{\frac{1}{3}})(4^{\frac{1}{3}})]^6 = 9^2 4^2$

5. Which expression does not have a value of 9 when $a = 1$, $b = 3$, and $c = 2$?
   a) $(-a \div b)^{-c}$ **c)** $(ab)^{-c}$
   b) $a^c b^c$ **d)** $(a^b b^a)^c$

6. Identify the exponential function whose equation of the asymptote is $y = 2$.
   a)
   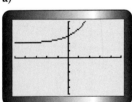
   c)
   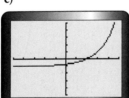
   b)
   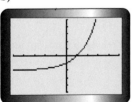
   d)
   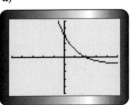

7. A bacteria culture doubles in size every 15 min. Given the formula $P(n) = 20(2)^{\frac{n}{15}}$, how long will it take for a culture of 20 bacteria to grow to a population of 163 840?
   a) 2048 min   **c)** 65 min
   b) 12 h   **d)** 195 min

8. Thorium-227 has a half-life of 18.4 days. Given the formula $M(t) = 50(\frac{1}{2})^{\frac{t}{18.4}}$, how many days, t, will a 50 mg sample take to decompose to 10 mg?
   a) 73.6   **c)** 42.72
   b) 21.09   **d)** 7.36

9. Four years ago, Sam invested a sum of money at 5%/a compounded semi-annually. Today there is $921.35 in Sam's account. How much did she invest?
   a) $756.19   **c)** $920.00
   b) $46.06   **d)** 875.29

10. How much will $7500 be worth if it is invested now for 10 years at 6%/a compounded annually?
   a) $12 000   **c)** $13 431.36
   b) $16 637.84   **d)** $4500

11. Phong wants to purchase a motorcycle. He can borrow $6500 at 10%/a compounded quarterly, if he agrees to repay the loan by making equal quarterly payments for 4 years. Determine a reasonable quarterly payment.
   a) $500   **c)** $650
   b) $300   **d)** $65

12. In order to repay a loan in less time, you could
   a) increase the periodic payment and increase the interest rate
   b) increase the periodic payment and decrease the interest rate
   c) decrease the periodic payment and increase the interest rate
   d) none of the above

Investigations

13. Ball Bounce

Marisa drops a small rubber ball from a height of 6 m onto a hard surface. After each bounce, the ball rebounds to 60% of the maximum height of the previous bounce.

a) Create a table to show the height of the ball after each bounce for the first 5 bounces.

b) Graph the height versus bounce number.

c) Create a function that models the height of the ball as a function of bounce number.

d) Estimate the height of the ball after 12 bounces from your graph. Verify this result with your function.

e) Use your graph to estimate when the ball's maximum height will be 28 cm. Verify this result using your function.

14. Retirement Plans

Sara and Ritu have just finished school and have started their first full-time jobs. Sara had decided that she is going to start putting away $100/month in a Retirement Plan. Ritu thinks that Sara is crazy because Sara is only 22 and should enjoy her money now and worry about saving for retirement later. Ritu is not going to worry about saving money until she is 45.

Suppose both young women retire at age 60 and that they can invest their money at 9%/a compound monthly.

How much money will Ritu have to contribute monthly to retire with the same savings as Sara?

15. Buying a Car

Matt is going to buy his first car for $17 500. He will make a down payment of $3500 and finance the rest at 9%/a compounded monthly. He will make regular monthly payments for 4 years. He estimates that his car will depreciate in value at a rate of 18% per year. Provide a complete analysis of this situation by determining the amount of Matt's monthly payments, the total interest that he will be paying, and the value of his car when he has completed his payments. Include graphs, charts, and tables with your analysis.

Review of Essential Skills and Knowledge

A–1 Operations with Integers

Set of integers $\mathbf{I} = \{..., -3, -2, -1, 0, 1, 2, 3, ...\}$

Addition

To add two integers,
- if the signs are the same, then the sum has the same sign as well:
  $(-12) + (-5) = -17$
- if the signs are different, then the sum takes the sign of the larger number:
  $18 + (-5) = 13$

Subtraction

Add the opposite:
$$-15 - (-8) = -15 + 8$$
$$= -7$$

Multiplication and Division

To multiply or divide two integers,
- if the two integers have the same sign, then the answer is positive:
  $6 \times 8 = 48, (-36) \div (-9) = 4$
- if the two integers have different signs, then the answer is negative:
  $(-5) \times 9 = -45, 54 \div (-6) = -9$

More Than One Operation

Follow the order of operations.

B	Brackets	
E	Exponents	
D	Division	} from left to right
M	Multiplication	
A	Addition	} from left to right
S	Subtraction	

EXAMPLE

Evaluate.

 a) $-10 + (-12)$

 b) $(-12) + 7$

 c) $(-11) + (-4) + 12 + (-7) + 18$

 d) $(-6) \times 9 \div 3$

 e) $\dfrac{20 + (-12) \div (-3)}{(-4 + 12) \div (-2)}$

Solution

a) $-10 + (-12) = -22$

b) $(-12) + 7 = -5$

c) $(-11) + (-4) + 12 + (-7) + 18$
$= (-22) + 30$
$= 8$

d) $(-6) \times 9 \div 3$
$= -54 \div 3$
$= -18$

e) $\dfrac{20 + (-12) \div (-3)}{(-4 + 12) \div (-2)}$

$= \dfrac{20 + 4}{8 \div (-2)}$

$= \dfrac{24}{-4}$

$= -6$

Practising

1. Evaluate.
  a) $6 + (-3)$
  b) $12 - (-13)$
  c) $-17 - 7$
  d) $(-23) + 9 - (-4)$
  e) $24 - 36 - (-6)$
  f) $32 + (-10) + (-12) - 18 - (-14)$

2. Which choice would make each statement true: $>$, $<$, or $=$?
  a) $-5 - 4 - 3 + 3 \ \blacksquare\ -4 - 3 - 1 - (-2)$
  b) $4 - 6 + 6 - 8 \ \blacksquare\ -3 - 5 - (-7) - 4$
  c) $8 - 6 - (-4) - 5 \ \blacksquare\ 5 - 13 - 7 - (-8)$
  d) $5 - 13 + 7 - 2 \ \blacksquare\ 4 - 5 - (-3) - 5$

3. Evaluate.
  a) $(-11) \times (-5)$ **d)** $(-72) \div (-9)$
  b) $(-3)(5)(-4)$ **e)** $(5)(-9) \div (-3)(7)$
  c) $35 \div (-5)$ **f)** $56 \div [(8)(7)] \div 49$

4. Evaluate.
  a) $(-3)^2 - (-2)^2$
  b) $(-5)^2 - (-7) + (-12)$
  c) $-4 + 20 \div (-4)$
  d) $-3(-4) + 8^2$
  e) $(-16) - [(-8) \div 2]$
  f) $8 \div (-4) + 4 \div (-2)^2$

5. Evaluate.
  a) $\dfrac{-12 - 3}{-3 - 2}$

  b) $\dfrac{-18 + 6}{(-3)(-4)}$

  c) $\dfrac{(-16 + 4) \div 2}{8 \div (-8) + 4}$

  d) $\dfrac{-5 + (-3)(-6)}{(-2)^2 + (-3)^2}$

A–2 Operations with Rational Numbers

Set of rational numbers $\mathbf{Q} = \left\{ \dfrac{a}{b} \middle| a, b \in I, b \neq 0 \right\}$

Addition and Subtraction

To add or subtract rational numbers, you need to find a common denominator.

Division

To divide by a rational number, multiply by the reciprocal.

$$\frac{a}{b} \div \frac{c}{d} = \frac{a}{b} \times \frac{d}{c}$$

$$= \frac{ad}{bc}$$

Multiplication

$\dfrac{a}{b} \times \dfrac{c}{d} = \dfrac{ac}{bd}$, but first reduce to lowest terms where possible.

More Than One Operation

Follow the order of operations.

EXAMPLE 1

Simplify $\dfrac{-2}{5} + \dfrac{3}{-2} - \dfrac{3}{10}$.

Solution

$$\frac{-2}{5} + \frac{3}{-2} - \frac{3}{10} = \frac{-4}{10} + \frac{-15}{10} - \frac{3}{10}$$

$$= \frac{-4 - 15 - 3}{10}$$

$$= \frac{-22}{10}$$

$$= -\frac{11}{5} \text{ or } -2\frac{1}{5}$$

EXAMPLE 2

Simplify $\dfrac{3}{4} \times \dfrac{-4}{5} \div \dfrac{-3}{7}$.

Solution

$$\frac{3}{4} \times \frac{-4}{5} \div \frac{-3}{7} = \frac{3}{4} \times \frac{-4}{5} \times \frac{7}{-3}$$

$$= \frac{\overset{1}{\cancel{3}}}{\underset{1}{\cancel{4}}} \times \frac{\overset{-1}{\cancel{-4}}}{5} \times \frac{7}{\underset{-1}{\cancel{-3}}}$$

$$= \frac{7}{5} \text{ or } 1\frac{2}{5}$$

Practising

1. Evaluate.

a) $\dfrac{1}{4} + \dfrac{-3}{4}$

b) $\dfrac{1}{2} - \dfrac{-2}{3}$

c) $\dfrac{-1}{4} - 1\dfrac{1}{3}$

d) $-8\dfrac{1}{4} - \dfrac{-1}{-3}$

e) $\dfrac{-3}{5} + \dfrac{-3}{4} - \dfrac{7}{10}$

f) $\dfrac{2}{3} - \dfrac{-1}{2} - \dfrac{1}{6}$

2. Evaluate.

a) $\dfrac{4}{5} \times \dfrac{-20}{25}$

b) $\dfrac{3}{-2} \times \dfrac{6}{5}$

c) $\left(\dfrac{-1}{3}\right)\left(\dfrac{2}{-5}\right)$

d) $\left(\dfrac{9}{4}\right)\left(\dfrac{-2}{-3}\right)$

e) $\left(-1\dfrac{1}{10}\right)\left(3\dfrac{1}{11}\right)$

f) $-4\dfrac{1}{6} \times \left(-7\dfrac{3}{4}\right)$

3. Evaluate.

a) $\dfrac{-4}{3} \div \dfrac{2}{-3}$

b) $-7\dfrac{1}{8} \div \dfrac{3}{2}$

c) $\dfrac{-2}{3} \div \dfrac{-3}{8}$

d) $\dfrac{-3}{-2} \div \left(\dfrac{-1}{3}\right)$

e) $-6 \div \left(\dfrac{-4}{5}\right)$

f) $\left(-2\dfrac{1}{3}\right) \div \left(-3\dfrac{1}{2}\right)$

4. Simplify.

a) $\dfrac{-2}{5} - \left(\dfrac{-1}{10} + \dfrac{1}{-2}\right)$

b) $\dfrac{-3}{5}\left(\dfrac{-3}{4} - \dfrac{-1}{4}\right)$

c) $\left(\dfrac{3}{5}\right)\left(\dfrac{1}{-6}\right)\left(\dfrac{-2}{3}\right)$

d) $\left(\dfrac{-2}{3}\right)^2\left(\dfrac{1}{-2}\right)^3$

e) $\left(\dfrac{-2}{5} + \dfrac{1}{-2}\right) \div \left(\dfrac{5}{-8} - \dfrac{-1}{2}\right)$

f) $\dfrac{\dfrac{-4}{5} - \dfrac{-3}{5}}{\dfrac{1}{3} - \dfrac{-1}{5}}$

A–3 Exponent Laws

3^4 and a^n are called powers

exponent → 3^4 ← base

4 factors of 3
$3^4 = (3)(3)(3)(3)$

n factors of a
$a^n = (a)(a)(a)...(a)$

Operations with powers follow a set of procedures or rules.

Rule	Description	Algebraic Expression	Example
Multiplication	When the bases are the same, keep the base the same and add exponents.	$(a^m)(a^n) = a^{m+n}$	$(5^4)(5^{-3}) = 5^{4+(-3)}$ $= 5^{4-3}$ $= 5^1$ $= 5$
Division	When the bases are the same, keep the base the same and subtract exponents.	$\dfrac{a^m}{a^n} = a^{m-n}$	$\dfrac{4^6}{4^{-2}} = 4^{6-(-2)}$ $= 4^{6+2}$ $= 4^8$
Power of a Power	Keep the base and multiply the exponents.	$(a^m)^n = a^{mn}$	$(3^2)^4 = 3^{(2)(4)}$ $= 3^8$

EXAMPLE

Simplify and evaluate.
$3(3^7) \div (3^3)^2$

Solution

$$3(3^7) \div (3^3)^2 = 3^{1+7} \div 3^{3 \times 2}$$
$$= 3^8 \div 3^6$$
$$= 3^{8-6}$$
$$= 3^2$$
$$= 9$$

Practising

1. Evaluate to three decimal places where necessary.
   a) 4^2
   b) 5^0
   c) 3^2
   d) -3^2
   e) $(-5)^3$
   f) $\left(\dfrac{1}{2}\right)^3$

2. Evaluate.
   a) $3^0 + 5^0$
   b) $2^2 + 3^3$
   c) $5^2 - 4^2$
   d) $\left(\dfrac{1}{2}\right)^3 \left(\dfrac{2}{3}\right)^2$
   e) $-2^5 + 2^4$
   f) $\left(\dfrac{1}{2}\right)^2 + \left(\dfrac{1}{3}\right)^2$

3. Evaluate to an exact answer.
   a) $\dfrac{9^8}{9^7}$
   b) $\dfrac{2(5^5)}{5^3}$
   c) $(4^5)(4^2)^3$
   d) $\dfrac{(3^2)(3^3)}{(3^4)^2}$

4. Simplify.
   a) $(x)^5(x)^3$
   b) $(m)^2(m)^4(m)^3$
   c) $(y)^5(y)^2$
   d) $(a^b)^c$
   e) $\dfrac{(x^5)(x^3)}{x^2}$
   f) $\left(\dfrac{x^4}{y^3}\right)^3$

5. Simplify.
   a) $(x^2y^4)(x^3y^2)$
   b) $(-2m^3)^2(3m^2)^3$
   c) $\dfrac{(5x^2)^2}{(5x^2)^0}$
   d) $(4u^3v^2)^2 \div (-2u^2v^3)^2$

A–4   The Pythagorean Theorem

The three sides of a right triangle are related to each other in a unique way. Every right triangle has a longest side, called the **hypotenuse**, which is always opposite the right angle. One of the important relationships in mathematics is known as the **Pythagorean theorem**. It states that the area of the square of the hypotenuse is equal to the sum of the areas of the squares of the other two sides.

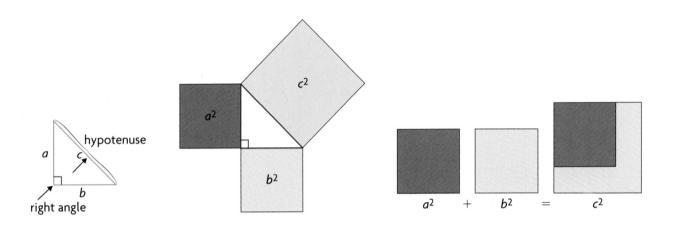

Practising

1. For each right triangle, write the equation for the Pythagorean theorem.

a)

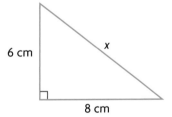

6 cm, x, 8 cm

b)

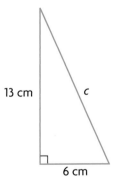

13 cm, c, 6 cm

c)

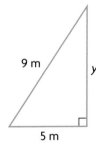

9 m, y, 5 m

d)

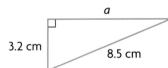

a, 3.2 cm, 8.5 cm

2. Calculate the length of the unknown side of each triangle in question 1. Round all answers to one decimal place.

3. Find the value of each unknown measure to the nearest hundredth.
   a) $a^2 = 5^2 + 13^2$
   b) $10^2 = 8^2 + m^2$
   c) $26^2 = b^2 + 12^2$
   d) $2.3^2 + 4.7^2 = c^2$

4. Determine the length of the diagonals of each rectangle to the nearest tenth.
   a)

5 m
10 m

   b)

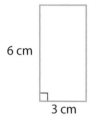
6 cm
3 cm

c)

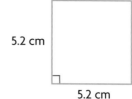
5.2 cm
5.2 cm

d)   1.2 m
4.8 m

5. An isosceles triangle has a hypotenuse 15 cm long. Determine the length of the two equal sides.

6. An apartment building casts a shadow. From the tip of the shadow to the top of the building is 100 m. The tip of the shadow is 72 m from the base of the building. How tall is the building?

A–5 Evaluating Algebraic Expressions and Formulas

To evaluate algebraic expressions and formulas, substitute the given numbers for the variables. Then follow the order of operations to calculate the answer.

EXAMPLE 1

Find the value of $2x^2 - y$ if $x = -2$ and $y = 3$.

Solution

$$2x^2 - y = 2(-2)^2 - 3$$
$$= 2(4) - 3$$
$$= 8 - 3$$
$$= 5$$

EXAMPLE 2

The formula for finding the volume of a cylinder is $V = \pi r^2 h$. Find the volume of a cylinder with a radius of 2.5 cm and a height of 7.5 cm.

Solution

$$V = \pi r^2 h$$
$$\doteq (3.14)(2.5)^2(7.5)$$
$$= (3.14)(6.25)(7.5)$$
$$\doteq 147 \text{ cm}^3$$

Practising

1. Find the value of each expression for $x = -5$ and $y = -4$.

a) $-4x - 2y$
b) $-3x - 2y^2$
c) $(3x - 4y)^2$
d) $\left(\dfrac{x}{y}\right) - \left(\dfrac{y}{x}\right)$

2. If $x = -\dfrac{1}{2}$ and $y = \dfrac{2}{3}$, find the value of each expression.

a) $x + y$
b) $x + 2y$
c) $3x - 2y$
d) $\dfrac{1}{2}x - \dfrac{1}{2}y$

3. a) The formula for the area of a triangle is $A = \dfrac{1}{2}bh$. Find the area of a triangle when $b = 13.5$ cm and $h = 12.2$ cm.

b) The area of a circle is found using the formula $A = \pi r^2$. Find the area of a circle with a radius of 4.3 m.

c) The hypotenuse of a right triangle, c, is found using the formula $c = \sqrt{a^2 + b^2}$. Find the length of the hypotenuse when $a = 6$ m and $b = 8$ m.

d) A sphere's volume is calculated using the formula $V = \dfrac{4}{3}\pi r^3$. Determine the volume of a sphere with a radius of 10.5 cm.

A–6 Finding Intercepts of Linear Relations

A linear relation of the general form $Ax + By + C = 0$ has an x-intercept and a y-intercept—the points where the line $Ax + By + C = 0$ crosses the x-axis and y-axis, respectively.

EXAMPLE 1 $Ax + By + C = 0$ FORM

Determine the x- and y-intercepts of the linear relation $2x + y - 6 = 0$.

Solution

The x-intercept is where the relation crosses the x-axis. The x-axis has equation $y = 0$, so substitute $y = 0$ into $2x + y - 6 = 0$:

$2x + y - 6 = 0$

$2x + 0 - 6 = 0$

$\qquad 2x = 6$

$\qquad\quad x = 3$

To find the y-intercept, substitute $x = 0$ into $2x + y - 6 = 0$:

$\quad 2x + y - 6 = 0$

$2(0) + y - 6 = 0$

$\qquad\qquad y = 6$

The x-intercept is at $(3, 0)$ and the y-intercept is at $(0, 6)$.

EXAMPLE 2   ANY FORM

Determine the x- and y-intercepts of the linear relation $3y = 18 - 2x$.

Solution

To find the x- and y-intercepts, substitute $y = 0$ and $x = 0$, respectively.

$3y = 18 - 2x$	$3y = 18 - 2x$
$3(0) = 18 - 2x$	$3y = 18 - 2(0)$
$2x = 18$	$3y = 18$
$x = 9$	$y = 6$

The x-intercept is at $(9, 0)$ and the y-intercept is at $(0, 6)$.

A special case is when the linear relation is a horizontal or vertical line.

EXAMPLE 3

Find either the x- or the y-intercept of each linear relation.

a) $2x = -14$ 　　　　　　　　b) $3y + 48 = 0$

Solution

a) $2x = -14$ is a vertical line, so it has no y-intercept. To find the x-intercept, solve for x.

$$2x = -14$$
$$x = -7$$

The x-intercept is at $(-7, 0)$.

b) $3y + 48 = 0$ is a horizontal line, so it has no x-intercept. To find the y-intercept, solve for y.

$$3y + 48 = 0$$
$$3y = -48$$
$$y = -16$$

The y-intercept is at $(0, -16)$.

Practising

1. Determine the x- and y-intercepts of each linear relation.
   a) $x + 3y - 3 = 0$
   b) $2x - y + 14 = 0$
   c) $-x + 2y + 6 = 0$
   d) $5x + 3y - 15 = 0$
   e) $10x - 10y + 100 = 0$
   f) $-2x + 5y - 15 = 0$

2. Determine the x- and y-intercepts of each linear relation.
   a) $x = y + 7$
   b) $3y = 2x + 6$
   c) $y = 4x + 12$
   d) $3x = 5y - 30$
   e) $2y - x = 7$
   f) $12 = 6x - 5y$

3. Find either the x- or the y-intercept of each linear relation.
   a) $x = 13$
   b) $y = -6$
   c) $2x = 14$
   d) $3x + 30 = 0$
   e) $4y = -6$
   f) $24 - 3y = 0$

4. A ladder resting against a wall is modelled by the linear relation $2y + 9x = 13.5$. The x-axis represents the ground and the y-axis represents the wall.
   a) Determine the intercepts of the relation.
   b) Using the intercept points, graph the relation.
   c) What can you conclude about the foot and the top of the ladder?

A–7   Graphing Linear Relationships

The graph of a linear relationship ($Ax + By + C = 0$) is a straight line. The graph can be drawn if at least two ordered pairs of the relationship are known. This information can be determined in several different ways.

EXAMPLE 1   TABLE OF VALUES

Sketch the graph of $2y = 4x - 2$.

Solution

A table of values can be created. Express the equation in the form $y = mx + b$.

$$\frac{2y}{2} = \frac{4x - 2}{2}$$

$$y = 2x - 1$$

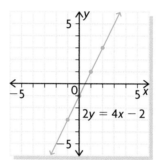

x	y
−1	2(−1) − 1 = −3
0	2(0) − 1 = −1
1	2(1) − 1 = 1
2	2(2) − 1 = 3

EXAMPLE 2   USING INTERCEPTS

Sketch the graph of $2x + 4y = 8$.

Solution

The intercepts of the line can be found. For the x-intercept, let $y = 0$.

$$2x + 4(0) = 8$$
$$2x = 8$$
$$x = 4$$

x	y
4	0
0	2

For the y-intercept, let $x = 0$.

$$2(0) + 4y = 8$$
$$4y = 8$$
$$y = 2$$

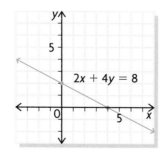

EXAMPLE 3 USING THE SLOPE AND Y-INTERCEPT

Sketch the graph of $y = 3x + 4$.

Solution

When the equation is in the form $y = mx + b$, m is the slope and b is the y-intercept.

For $y = 3x + 4$, the line has a slope of 3 and a y-intercept of 4.

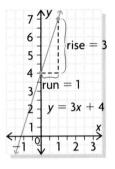

Practising

1. Express each equation in the form $y = mx + b$.
   a) $3y = 6x + 9$
   b) $2x - 4y = 8$
   c) $3x + 6y - 12 = 0$
   d) $5x = y - 9$

2. Graph each equation, using a table of values where $x \in \{-2, -1, 0, 1, 2\}$.
   a) $y = 3x - 1$
   b) $y = \frac{1}{2}x + 4$
   c) $2x + 3y = 6$
   d) $y = 4$

3. Determine the x- and y-intercepts of each equation.
   a) $x + y = 10$
   b) $2x + 4y = 16$
   c) $50 - 10x - y = 0$
   d) $\frac{x}{2} + \frac{y}{4} = 1$

4. Graph each equation by determining the intercepts.
   a) $x + y = 4$
   b) $x - y = 3$
   c) $2x + 5y = 10$
   d) $3x - 4y = 12$

5. Graph each equation, using the slope and y-intercept.
   a) $y = 2x + 3$
   b) $y = \frac{2}{3}x + 1$
   c) $y = -\frac{3}{4}x - 2$
   d) $2y = x + 6$

6. Graph each equation. Use the most suitable method.
   a) $y = 5x + 2$
   b) $3x - y = 6$
   c) $y = -\frac{2}{3}x + 4$
   d) $4x = 20 - 5y$

A–8 Graphing Quadratic Relations

The graph of a quadratic relation is called a **parabola**. All parabolas have a vertex (the lowest or highest point of the curve), an axis of symmetry (the vertical line through the vertex), and a y-intercept. However, a parabola may have 0, 1, or 2 x-intercepts, or **zeros**. To graph a quadratic relation, begin by creating a table of values.

EXAMPLE 1

Graph each quadratic relation. Use your graph to determine
i)   the vertex of the parabola
ii)   the axis of symmetry
iii) the y-intercept
iv)   the x-intercept(s), if any

    a) $y = x^2 - 4$ **b)** $y = -2x^2 + 4x - 2$

Solution

  a) Create a table of values.   Use these values to plot the graph.

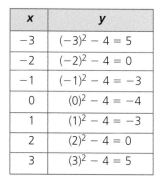
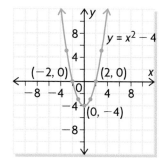

x	y
-3	$(-3)^2 - 4 = 5$
-2	$(-2)^2 - 4 = 0$
-1	$(-1)^2 - 4 = -3$
0	$(0)^2 - 4 = -4$
1	$(1)^2 - 4 = -3$
2	$(2)^2 - 4 = 0$
3	$(3)^2 - 4 = 5$

i)   From the graph, the vertex is the point $(0, -4)$ (a minimum for this relation).
ii) The axis of symmetry (the y-axis, in this case) passes through $(0, -4)$, so its equation is $x = 0$.
iii) The y-intercept is at $(0, -4)$.
iv) There are two x-intercepts, at $(-2, 0)$ and $(2, 0)$.

b) Create a table of values.   Use these values to plot the graph.

x	y
-2	$-2(-2)^2 + 4(-2) - 2 = -18$
-1	$-2(-1)^2 + 4(-1) - 2 = -8$
0	$-2(0)^2 + 4(0) - 2 = -2$
1	$-2(1)^2 + 4(1) - 2 = 0$
2	$-2(2)^2 + 4(2) - 2 = -2$
3	$-2(3)^2 + 4(3) - 2 = -8$
4	$-2(4)^2 + 4(4) - 2 = -18$

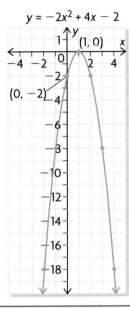

$y = -2x^2 + 4x - 2$

i) The vertex is the point $(1, 0)$ (a maximum for this relation).

ii) The axis of symmetry passes through $(1, 0)$, so its equation is $x = 1$.

iii) The y-intercept is at $(0, -2)$.

iv) There is one x-intercept, at $(1, 0)$.

Practising

1. Graph each quadratic relation. Use your graph to determine
   i) the vertex of the parabola
   ii) the axis of symmetry
   iii) the y-intercept
   iv) the x-intercept(s), if any

   a) $y = x^2 - 1$
   b) $y = 3x^2$
   c) $y = 2x^2 + 2$
   d) $y = x^2 + 4x + 4$
   e) $y = 8 - 2x^2$
   f) $y = (x - 2)^2 + 3$
   g) $y = (x + 3)(x - 2)$
   h) $y = 4 - 2(x + 1)^2$

2. A stone is thrown from a cliff. Its motion is described by $y = -5x^2 + 10x + 20$. In this quadratic relation, y is the stone's height above the sea (in metres) and x is the horizontal distance (in metres) the stone has travelled.
   a) Graph the quadratic relation.
   b) Locate the vertex. What is the stone's maximum height, and how far is it horizontally from the edge of the cliff at this point?
   c) What information does the y-intercept give you about the cliff?
   d) Locate the x-intercepts. Is the intercept with x negative meaningful? Why or why not? What is the meaning of the other x-intercept?

A–9 Expanding and Simplifying Algebraic Expressions

Type	Description	Example
Collecting Like Terms $2a + 3a = 5a$	Add or subtract the coefficients of the terms that have the same variables and exponents.	$3a - 2b - 5a + b$ $= 3a - 5a - 2b + b$ $= -2a - b$
Distributive Property $a(b + c) = ab + ac$	Multiply each term of the binomial by the monomial.	$-4a(2a - 3b)$ $= -8a^2 + 12ab$
Product of Two Binomials $(a + b)(c + d)$ $= ac + ad + bc + bd$	Multiply the first term of the first binomial by the second binomial, and then multiply the second term of the first binomial by the second binomial. Collect like terms if possible.	$(2x^2 - 3)(5x^2 + 2)$ $= 10x^4 + 4x^2 - 15x^2 - 6$ $= 10x^4 - 11x^2 - 6$

Practising

1. Simplify.
  a) $3x + 2y - 5x - 7y$
  b) $5x^2 - 4x^3 + 6x^2$
  c) $(4x - 5y) - (6x + 3y) - (7x + 2y)$
  d) $m^2n + p - (2p - 3m^2n)$

2. Expand.
  a) $3(2x + 5y - 2)$
  b) $5x(x^2 - x + y)$
  c) $m^2(3m^2 - 2n)$
  d) $x^5y^3(4x^2y^4 - 2xy^5)$

3. Expand and simplify.
  a) $3x(x + 2) + 5x(x - 2)$
  b) $-7h(2h + 5) - 4h(5h - 3)$
  c) $2m^2n(m^3 - n) - 5m^2n(3m^3 + 4n)$
  d) $-3xy^3(5x + 2y + 1) + 2xy^3(-3y - 2 + 7x)$

4. Expand and simplify.
  a) $(3x - 2)(4x + 5)$
  b) $(7 - 3y)(2 + 4y)$
  c) $(5x - 7y)(4x + y)$
  d) $(3x^3 - 4y^2)(5x^3 + 2y^2)$

A–10 Factoring Algebraic Expressions

Common Factors

Sometimes, the terms in an algebraic expression have a common factor.

Factoring Trinomials

$ax^2 + bx + c$, when $a = 1$:
Write as the product of two binomials. Determine two numbers whose sum is b and whose product is c.

EXAMPLE 1

Factor the expression $3x + 12xy$.

Solution

Divide out a common factor from each term.
$3x + 12xy$
$= 3x(1) + 3x(4y)$
$= 3x(1 + 4y)$

EXAMPLE 2

Factor the expression $x^2 - 5x + 6$.

Solution

Find two numbers whose product is 6 and whose sum is -5.
$6 = (-2)(-3)$
and $-5 = (-2) + (-3)$
$x^2 - 5x + 6$
$= (x - 2)(x - 3)$

A special form of algebraic expression is called the difference of two squares; for example, the expression $x^2 - 49$.

EXAMPLE 3

Factor each expression.
a) $x^2 - 49$ **b)** $9y^2 - 16x^2$

Solution

a) $x^2 - 49$
$= x^2 - 7^2$
$= (x + 7)(x - 7)$
Check:
$(x + 7)(x - 7)$
$= x^2 + 7x - 7x + 49$ Eliminate middle terms.
$= x^2 - 49$

b) $9y^2 - 16x^2$
$= (3y)^2 - (4x)^2$
$= (3y + 4x)(3y - 4x)$

Sometimes you may need to use several strategies to factor an expression.

EXAMPLE 4

Factor $2x^2 - 10x - 48$.

Solution

$2x^2 - 10x - 48$ Divide out the common factor of 2.
$= 2(x^2 - 5x - 24)$ Factor the trinomial.
$= 2(x - 8)(x + 3)$

Practising

1. Factor each expression.
   a) $ab + 2a$
   b) $4x + 6$
   c) $3y - 9xy$
   d) $7a + a^2$
   e) $77b^3 + 55b^2$
   f) $21a + 6ab - 15a^2$

2. Factor each trinomial.
   a) $x^2 - 2x + 1$
   b) $a^2 + 3a + 2$
   c) $x^2 - 3x - 28$
   d) $z^2 - 2z - 8$
   e) $3a^2 + 6a + 3$
   f) $5x^2 - 10x - 15$

3. Which expressions are differences of two squares?
   a) $x^2 - 9$
   b) $2x^2 - 27$
   c) $9a^2 - 25b^2$
   d) $25 - 3x^2$
   e) $z^2 - 441$
   f) $16xy - z^2$

4. Factor these expressions, if possible.
   a) $x^2 - 81$
   b) $4 - 18z^2$
   c) $4a^2 - 1$
   d) $16x^2 - 16$
   e) $369 - 4x^2$
   f) $400 - 16xy$

A–11 Solving Linear Equations Algebraically

To solve a linear equation, first eliminate any fractions by multiplying each term in the equation by the lowest common denominator. Eliminate any brackets by using the distributive property, and then isolate the variable. A linear equation has only one solution.

EXAMPLE 1

Solve $-3(x + 2) - 3x = 4(2 - 5x)$.

Solution

$$-3(x + 2) - 3x = 4(2 - 5x)$$
$$-3x - 6 - 3x = 8 - 20x$$
$$-3x - 3x + 20x = 8 + 6$$
$$14x = 14$$
$$x = \frac{14}{14}$$
$$x = 1$$

EXAMPLE 2

Solve $\dfrac{y - 7}{3} = \dfrac{y - 2}{4}$.

Solution

$$\frac{y - 7}{3} = \frac{y - 2}{4}$$
$$12\left(\frac{y - 7}{3}\right) = 12\left(\frac{y - 2}{4}\right)$$
$$4(y - 7) = 3(y - 2)$$
$$4y - 28 = 3y - 6$$
$$4y - 3y = -6 + 28$$
$$y = 22$$

Practising

1. Solve.
a) $6x - 8 = 4x + 10$
b) $2x + 7.8 = 9.4$
c) $13 = 5m - 2$
d) $13.5 - 2m = 5m + 41.5$
e) $8(y - 1) = 4(y + 4)$
f) $4(5 - r) = 3(2r - 1)$

2. Determine the root of each equation.
a) $\dfrac{x}{5} = 20$
b) $\dfrac{2}{5}x = 8$
c) $4 = \dfrac{3}{2}m + 3$
d) $\dfrac{5}{7}y = 3 + 12$
e) $3y - \dfrac{1}{2} = \dfrac{2}{3}$
f) $4 - \dfrac{m}{3} = 5 + \dfrac{m}{2}$

3. a) What is the height of a triangle with an area of 15 cm^2 and a base of 5 cm?
   b) A rectangular lot has a perimeter of 58 m and is 13 m wide. How long is the lot?

4. At the December concert, 209 tickets were sold. There were 23 more student tickets sold than twice the number of adult tickets. How many of each were sold?

A–12 Pattern Recognition and Difference Tables

When the independent variable in a relation changes by a steady amount, the dependent variable often changes according to a pattern.

Type of Pattern	Description	Example
Linear	The first differences between dependent variables are constant.	**Number of Books (× 1000):** 1, 2, 3, 4, 5 **Cost of Books ($) (× 1000):** 60, 70, 80, 90, 100 **First Differences (× 1000):** 10, 10, 10, 10 In a linear relation, each first difference is always the same. Here, each first difference is $10 000.
Quadratic	The first differences between dependent variables change, but the second differences are constant.	**Time (s):** 0, 1, 2, 3, 4, 5 **Height of Ball (m):** 5, 30.1, 45.4, 50.9, 46.6, 32.5 **First Differences:** 25.1, 15.3, 5.5, −4.3, −14 **Second Differences:** −9.8, −9.8, −9.8, −9.8 In this case, the first differences are not the same, so the relation is nonlinear. In a quadratic relation, each second difference is always the same. Here, each second difference is −9.8.

Practising

1. Examine each pattern.

a)

b)

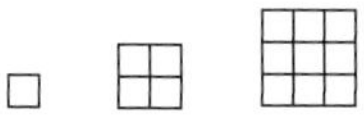

c)

d)

e)

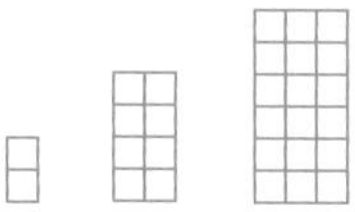

f)

i) Draw the next diagram.

ii) Write the four numbers that represent the four diagrams, in terms of the number of squares or triangles.

iii) Determine the fifth number.

2. Suppose each diagram in question 1 were made from toothpicks.

i) What numbers represent each pattern of four diagrams, in terms of the number of toothpicks?

ii) How many toothpicks will be in the fifth diagram?

3. Examine the data in each table.

  i) Is the pattern of the dependent variable linear or quadratic?

  ii) State the next value of y.

a)

x	3	4	5	6
y	9	11	13	15

b)

x	2	4	6	8
y	−13	−49	−109	−193

c)

x	0	1	2	3
y	1	0.5	0.1	−0.2

d)

x	0	2	4	6
y	1	142	283	424

4. Neville has 49 m of fencing to build a dog run. Here are some dimensions he considered to maximize the area. Describe the type of pattern displayed by **(a)** the width and **(b)** the area.

Length (m)	1	2	3	4	5
Width (m)	23.5	22.5	21.5	20.5	19.5
Area (m²)	23.5	45.0	64.5	82.0	97.5

5. What pattern does the change in the cost of renting a pickup truck show?

Distance (km)	0	100	200	300	400
Rental Cost ($)		35.00	41.00	47.00	53.00	59.00

6. Predict the next number in each pattern.

  a) $-3, 5, -7, 9, -11,$ ▨

  b) $-4, -16, -36, -64,$ ▨

  c) $\dfrac{1}{8}, \dfrac{11}{24}, \dfrac{19}{24}, \dfrac{27}{24}, \dfrac{35}{24},$ ▨

  d) $0.3, 0.6, 1.2, 2.4,$ ▨

A–13 Creating Scatter Plots and Lines or Curves of Good Fit

A **scatter plot** is a graph that shows the relationship between two sets of numeric data. The points in a scatter plot often show a general pattern, or **trend**. A line that approximates a trend for the data in a scatter plot is called a **line of best fit**. A line of best fit passes through as many points as possible, with the remaining points grouped equally above and below the line.

Data that have a **positive correlation** have a pattern that slopes up and to the right. Data that have a **negative correlation** have a pattern that slopes down and to the right. If the points nearly form a line, then the correlation is strong. If the points are dispersed, but still form some linear pattern, then the correlation is weak.

EXAMPLE 1

a) Make a scatter plot of the data and describe the kind of correlation the scatter plot shows.

b) Draw the line of best fit.

Long-Term Trends in Average Number of Cigarettes Smoked per Day by Smokers Aged 15–19

Year	1981	1983	1985	1986	1989	1990	1991	1994	1995	1996
Number per Day	16.0	16.6	15.1	15.4	12.9	13.5	14.8	12.6	11.4	12.2

Solution

a)

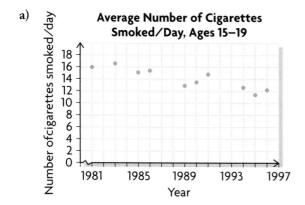

The scatter plot shows a negative correlation.

b)

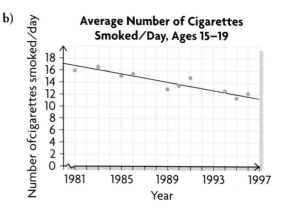

EXAMPLE 2

A professional golfer is taking part in a scientific investigation. Each time she drives the ball from the tee, a motion sensor records the initial speed of the ball. The final horizontal distance of the ball from the tee is also recorded. Here are the results:

Speed (m/s)	10	16	19	22	38	43	50	54
Distance (m)	10	25	47	43	142	182	244	280

Draw the line or curve of good fit.

Solution

The scatter plot shows that a line of best fit does not fit the data as well as an upward-sloping curve does. Therefore, sketch a curve of good fit.

Horizontal Distance of a Golf Ball

Practising

1. For each set of data,
   i) create a scatter plot and draw the line of best fit
   ii) describe the type of correlation the trend in the data displays

   a) **Population of the Hamilton–Wentworth, Ontario, Region**

Year	1966	1976	1986	1996	1998
Population	449 116	529 371	557 029	624 360	618 658

   b) **Percent of Canadians with Less than Grade 9 Education**

Year	1976	1981	1986	1991	1996
Percent of the Population	25.4	20.7	17.7	14.3	12.4

2. In an experiment for a physics project, marbles are rolled up a ramp. A motion sensor detects the speed of the marble at the start of the ramp, and the final height of the marble is recorded. However, the motion sensor may not be measuring accurately. Here are the data:

Speed (m/s)	1.2	2.1	2.8	3.3	4.0	4.5	5.1	5.6
Final Height (m)	0.07	0.21	0.38	0.49	0.86	1.02	1.36	1.51

   a) Draw a curve of good fit for the data.
   b) How consistent are the motion sensor's measurements? Explain.

A–14 Interpolating and Extrapolating

A graph can be used to make predictions about values not actually recorded and plotted. When the prediction involves a point within the range of the values of the independent variable, this is called **interpolating**. When the value of the independent variable falls outside the range of recorded data, it is called **extrapolating**. With a scatter plot, estimates are more reliable if the data show a strong positive or negative correlation.

EXAMPLE **1**

The Summer Olympics were cancelled in 1940 and 1944 because of World War II. Estimate what the 100 m run winning times might have been in these years if the Olympics had been held as scheduled.

Winning Times of 100 m Run

Year	Name (Country)	Time (s)
1928	Williams (Canada)	10.80
1932	Tolan (U.S.)	10.30
1936	Owens (U.S.)	10.30
1948	Dillard (U.S.)	10.30
1952	Remigino (U.S.)	10.40
1956	Morrow (U.S.)	10.50
1960	Hary (Germany)	10.20
1964	Hayes (U.S.)	10.00
1968	Hines (U.S.)	9.95
1972	Borzov (U.S.S.R.)	10.14
1976	Crawford (Trinidad)	10.06
1980	Wells (Great Britain)	10.25
1984	Lewis (U.S.)	9.99
1988	Lewis (U.S.)	9.92
1992	Christie (Great Britain)	9.96
1996	Bailey (Canada)	9.84

Solution

Draw a scatter plot and find the line of best fit.

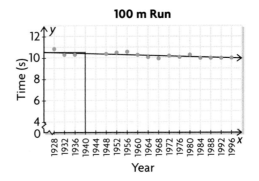

100 m Run

Find 1940 on the *x*-axis. Follow the vertical line for 1940 up until it meets the line of best fit. This occurs at about 10.8 s. For 1944, a reasonable estimate would be about 10.7 s.

EXAMPLE 2

Use the graph from Example 1 to predict what the winning time might be in 2044.

Solution

Extend the *x*-axis to 2044. Then extend the line of best fit to the vertical line through 2044.

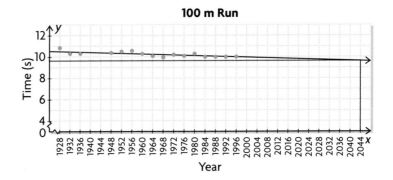

100 m Run

The vertical line for 2044 crosses the line of best fit at about 9.5, so the winning time in 2044 might be about 9.5 s.

It would be difficult to forecast much further into the future, since the winning times cannot continue to decline indefinitely. For example, a runner would likely never be able to run 100 m in 1 s and would certainly never run it in less than 0 s.

Practising

1. The scatter plot shows the gold medal throws in the discus competition in the Summer Olympics for 1908 to 1992.

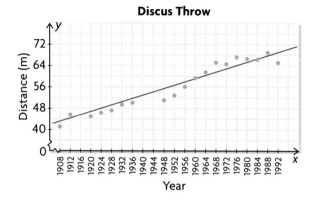

Discus Throw

a) Estimate the winning distance for 1940 and 1944.
b) Estimate the winning distance for 1996 and 2000.

2. As an object falls freely toward the ground, it accelerates at a steady rate due to gravity. The data show the speed, or velocity, an object would reach at one-second intervals during its fall.

Time from Start (s)	Velocity (m/s)
0	0
1	9.8
2	19.6
3	29.4
4	39.2
5	49.0

a) Graph the data.
b) Determine the object's velocity at 2.5 s, 3.5 s, and 4.75 s.
c) Find the object's velocity at 6 s, 9 s, and 10 s.

3. Explain why values you find by extrapolation are less reliable than those found by interpolation.

4. A school principal wants to know if there is a relationship between attendance and marks. You have been hired to collect data and analyze the results. You start by taking a sample of 12 students.

Days Absent	0	3	4	2	0	6	4	1	3	7	8	4
Average (%)	93	79	81	87	87	75	77	90	77	72	61	80

a) Create a scatter plot. Draw the line of best fit.
b) What appears to be the average decrease in marks for 1 day's absence?
c) Predict the average of a student who is absent for 6 days.
d) About how many days could a student likely miss before getting an average below 50%?

5. A series of football punts is studied in an experiment. The initial speed of the football and the length of the punt are recorded.

Speed (m/s)	10	17	18	21	25
Distance (m)	10	28	32	43	61

Use a curve of best fit to estimate the length of a punt with an initial speed of 29 m/s and the initial speed of a punt with a length of 55 m.

A–15 Trigonometry of Right Triangles

By the Pythagorean relationship, $a^2 + b^2 = c^2$ for any right triangle, where c is the length of the hypotenuse and a and b are the lengths of the other two sides.

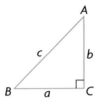

In any right triangle, there are three primary trigonometric ratios that associate the measure of an angle with the ratio of two sides. For example, for $\angle ABC$, in Figure 1,

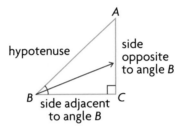

For $\angle B$

$$\sin B = \frac{opposite}{hypotenuse}$$

$$\cos B = \frac{adjacent}{hypotenuse}$$

$$\tan B = \frac{opposite}{adjacent}$$

Figure 1

Similarly, in Figure 2,

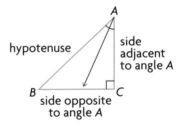

For $\angle A$

$$\sin A = \frac{opposite}{hypotenuse}$$

$$\cos A = \frac{adjacent}{hypotenuse}$$

$$\tan A = \frac{opposite}{adjacent}$$

Figure 2

Note how the opposite and adjacent sides change in Figures 1 and 2 with angles A and B.

EXAMPLE **1**

State the primary trigonometric ratios of $\angle A$.

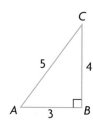

Solution

Sketch the triangle. Then label the opposite side, the adjacent side, and the hypotenuse.

$\sin A = \dfrac{\text{opposite}}{\text{hypotenuse}}$

$ = \dfrac{4}{5}$

$\cos A = \dfrac{\text{adjacent}}{\text{hypotenuse}}$

$ = \dfrac{3}{5}$

$\tan A = \dfrac{\text{opposite}}{\text{adjacent}}$

$ = \dfrac{4}{3}$

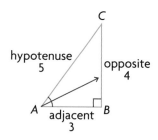

EXAMPLE **2**

A ramp must have a rise of one unit for every eight units of run. What is the angle of inclination of the ramp?

Solution

The slope of the ramp is $\dfrac{\text{rise}}{\text{run}} = \dfrac{1}{8}$. Draw a labelled sketch.

Calculate the measure of $\angle B$ to determine the angle of inclination.

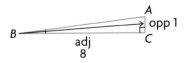

The trigonometric ratio that associates $\angle B$ with the opposite and adjacent sides is the tangent. Therefore,

$$\tan B = \frac{\text{opposite}}{\text{adjacent}}$$

$$B = \frac{1}{8}$$

$$B = \tan^{-1}\left(\frac{1}{8}\right)$$

$$B \doteq 7°$$

The angle of inclination is about $7°$.

EXAMPLE 3

Determine x to the nearest centimetre.

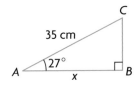

Solution

Label the sketch. The cosine ratio associates $\angle A$ with the adjacent side and the hypotenuse.

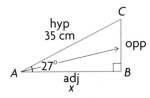

Then,

$$\cos A = \frac{\text{adjacent}}{\text{hypotenuse}}$$

$$\cos 27° = \frac{x}{35}$$

$$x = 35\cos 27°$$

$$x \doteq 31$$

So x is about 31 cm.

Practising

1. A rectangular lot is 15 m by 22 m. How long is the diagonal, to the nearest metre?

2. State the primary trigonometric ratios for ∠A.

a)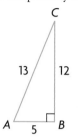

b)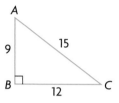

c)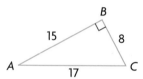

d)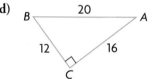

3. Solve for x to one decimal place.

a) $\sin 39° = \dfrac{x}{7}$

c) $\tan 15° = \dfrac{x}{22}$

b) $\cos 65° = \dfrac{x}{16}$

d) $\tan 49° = \dfrac{31}{x}$

4. Solve for ∠A to the nearest degree.

a) $\sin A = \dfrac{5}{8}$

c) $\tan B = \dfrac{19}{22}$

b) $\cos A = \dfrac{13}{22}$

d) $\cos B = \dfrac{3}{7}$

5. Determine x to one decimal place.

a)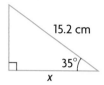

b)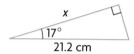

c)

d)

6. In △ABC, ∠B = 90° and AC = 13 cm. Determine

a) BC if ∠C = 17°

b) AB if ∠C = 26°

c) ∠A if BC = 6 cm

d) ∠C if BC = 9 cm

7. A tree casts a shadow 9.3 m long when the angle of the sun is 43°. How tall is the tree?

8. Janine stands 30.0 m from the base of a communications tower. The angle of elevation from her eyes to the top of the tower is 70°. How high is the tower if her eyes are 1.8 m above the ground?

9. A surveillance camera is mounted on the top of a building that is 80 m tall. The angle of elevation from the camera to the top of another building is 42°. The angle of depression from the camera to the same building is 32°. How tall is the other building?

PART 1 USING THE TI-83 PLUS AND TI-84 GRAPHING CALCULATORS

B–1 Preparing the Calculator

Before you graph any function, be sure to clear any information left on the calculator from the last time it was used. You should always do the following:

1. Clear all data in the lists.

Press [2nd] [+] [4] [ENTER].

2. Turn off all stat plots.

Press [2nd] [Y=] [4] [ENTER].

3. Clear all equations in the equation editor.

Press [Y=], then press [CLEAR] for each equation.

4. Set the window so that the axes range from −10 to 10.

Press [ZOOM] [6]. Press [WINDOW] to verify.

B–2 Entering and Graphing Functions

Enter the equation of the function into the equation editor. The calculator will display the graph.

1. Graph.

To graph $y = 2x + 8$, press [Y=] [2] [X, T, Θ, n] [+] [8]

[GRAPH]. The graph will be displayed as shown.

2. Enter all linear equations in the form $y = mx + b$.

If m or b are fractions, enter them between brackets. For example, enter
$2x + 3y = 7$ in the form $y = -\dfrac{2}{3}x + \dfrac{7}{3}$, as shown.

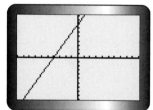

3. Press [GRAPH] to view the graph.

4. Press [TRACE] to find the coordinates of any point on a graph.

Use the left and right arrow keys to cursor along the graph.
Press [ZOOM] [8] [ENTER] [TRACE] to trace using integer intervals. If you are working with several graphs at the same time, use the up and down arrow to scroll between graphs.

B–3   Evaluating a Function

1. **Enter the function into the equation editor.**

   To enter $y = 2x^2 + x - 3$, press 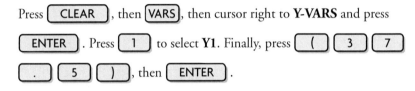.

2. **Use the value operation to evaluate the function.**

   To find the value of the function at $x = -1$, press [2nd] [TRACE] [ENTER], enter [(−)] [1] at the cursor, then press [ENTER].

3. **Use function notation and the Y-VARS operation to evaluate the function.**

   There is another way to evaluate the function, say at $x = 37.5$.

   Press [CLEAR], then [VARS], then cursor right to **Y-VARS** and press [ENTER]. Press [1] to select **Y1**. Finally, press [( ] [3] [7] [.] [5] [ )], then [ENTER].

B–4   Changing Window Settings

The window settings can be changed to show a graph for a given domain and range.

1. **Enter the function $y = x^2 - 3x + 4$ in the equation editor.**

2. **Use the WINDOW function to set the domain and range.**

   To display the function over the domain $\{x \mid -2 \le x \le 5\}$ and range

   $\{y \mid 0 \le y \le 14\}$, press [WINDOW] [(−)] [2] [ENTER], then [5] [ENTER], then [1] [ENTER], then [0] [ENTER], then [1] [4] [ENTER], then [1] [ENTER] and [1] [ENTER].

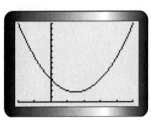

3. **Press [GRAPH] to show the function with this domain and range.**

B–5 Using the Split Screen

To see a graph and the equation editor at the same time, press ⸢ MODE ⸣ and cursor to **Horiz**. Press ⸢ ENTER ⸣ to select this, then press ⸢ 2nd ⸣ ⸢ MODE ⸣ to return to the home screen. Enter $y = x^2$ in **Y1** of the equation editor, then press ⸢ GRAPH ⸣.

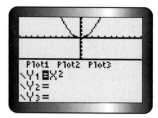

To see a graph and a table at the same time: press ⸢ MODE ⸣ and cursor to **G–T** (Graph-Table). Press ⸢ ENTER ⸣ to select this, then press ⸢ GRAPH ⸣.

It is possible to view the table with different increments. For example, to see the table start at $x = 0$ and increase in increments of 0.5, press ⸢ 2nd ⸣ ⸢ WINDOW ⸣ and adjust the settings as shown.

Press ⸢ GRAPH ⸣.

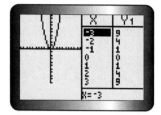

B–6 Using the TABLE Feature

A function such as $y = -0.1x^3 + 2x + 3$ can be displayed in a table of values.

1. Enter the function into the equation editor.

To enter $y = -0.1x^3 + 2x + 3$, press ⸢ Y= ⸣ ⸢ (−) ⸣ ⸢ . ⸣ ⸢ 1 ⸣ ⸢ X, T, Θ, n ⸣ ⸢ ^ ⸣ ⸢ 3 ⸣ ⸢ + ⸣ ⸢ 2 ⸣ ⸢ X, T, Θ, n ⸣ ⸢ + ⸣ ⸢ 3 ⸣.

2. Set the start point and step size for the table.

Press ⸢ 2nd ⸣ ⸢ WINDOW ⸣. The cursor is alongside "TblStart=." To start at $x = -5$, press ⸢ (−) ⸣ ⸢ 5 ⸣ ⸢ ENTER ⸣. The cursor is now alongside "ΔTbl=" (Δ, the Greek capital letter delta, stands for "change in.") To increase the x-value in steps of 1, press ⸢ 1 ⸣ ⸢ ENTER ⸣.

3. To view the table, press ⸢ 2nd ⸣ ⸢ GRAPH ⸣.

Use ⸢ ▲ ⸣ and ⸢ ▼ ⸣ to move up and down the table. Notice that you can look at higher or lower x-values than the original range.

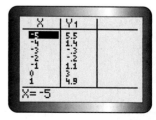

B–7   Making a Table of Differences

To create a table with the first and second differences for a function, use the STAT lists.

1. **Press STAT 1 and enter the x-values into L1.**

    For the function $f(x) = 3x^2 - 4x + 1$, use x-values from -2 to 4.

2. **Enter the function.**

    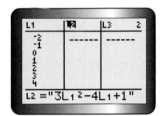

    Scroll right and up to select **L2**. Enter the function $f(x)$, using **L1** as the variable x. Press ALPHA + 3 2nd 1 x^2 —
    4 2nd 1 + 1 ALPHA + .

3. **Press ENTER to display the values of the function in L2.**

4. **Find the first differences.**

    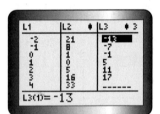

    Scroll right and up to select **L3**. Then press 2nd STAT . Scroll right to **OPS** and press 7 to choose Δ**List(**. Enter **L2** by pressing
    2nd 2) . Press ENTER to see the first differences displayed in **L3**.

5. **Find the second differences.**

    Scroll right and up to select **L4**. Repeat step 4, using **L3** in place of **L2**. Press
    ENTER to see the second differences displayed in **L4**.

B–8 Finding the Zeros of a Function

To find the zeros of a function, use the **zero** operation.

1. **Start by entering** $y = -(x + 3)(x - 5)$ **in the equation editor, then press**
   [GRAPH] [ZOOM] [6].

2. **Access the zero operation.**

   Press [2nd] [TRACE] [2].

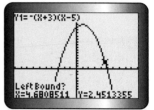

3. **Use the left and right arrow keys to cursor along the curve to any point to the left of the zero.**

   Press [ENTER] to set the left bound.

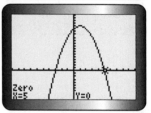

4. **Cursor along the curve to any point to the right of the zero.**

   Press [ENTER] to set the right bound.

5. **Press** [ENTER] **again to display the coordinates of the zero (the** *x*-intercept**).**

6. **Repeat to find the second zero.**

B–9 Finding the Maximum or Minimum Values of a Function

The least or greatest value can be found using the **minimum** operation or the **maximum** operation.

1. **Enter** $y = -2x^2 - 12x + 30.$
   Graph it and adjust the window as shown. This graph opens downward, so it has a maximum.

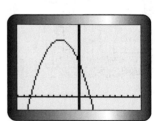

2. **Use the maximum operation.**

   Press [2nd] [TRACE] [4]. For parabolas that open upward, press

   [2nd] [TRACE] [3] to use the **minimum** operation.

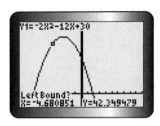

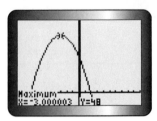

3. Use the left and right arrow keys to cursor along the curve to any point to the left of the maximum value.

   Press [ENTER] to set the left bound.

4. Cursor along the curve to any point right of the maximum value.

   Press [ENTER] to set the right bound.

5. Press [ENTER] again to display the coordinates of the optimal value.

B–10 Creating Scatter Plots and Determining Lines and Curves of Best Fit Using Regression

This table gives the height of a baseball above ground, from the time it was hit to the time it touched the ground.

Time (s)	0	1	2	3	4	5	6
Height (m)	2	27	42	48	43	29	5

Create a scatter plot of the data.

1. **Enter the data into lists.** To start press [STAT] [ENTER]. Move the cursor over to the first position in L_1 and enter the values for time. Press [ENTER] after each value. Repeat this for height in L_2.

2. **Create a scatter plot.** Press [2nd] [Y=] and [1] [ENTER]. Turn on Plot 1 by making sure the cursor is over **On**, the **Type** is set to the graph type you prefer, and L_1 and L_2 appear after **Xlist** and **Ylist**.

3. **Display the graph.** Press [ZOOM] [9] to activate **ZoomStat**.

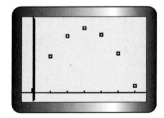

4. **Apply the appropriate regression analysis.** To determine the equation of the line or curve of best fit press (STAT) and scroll over to **CALC**. Press:

(4) to enable **LinReg(ax+b)**

(5) to enable **QuadReg**.

(0) to enable **ExpReg**.

(ALPHA) (C) to enable **SinReg**.

Press (2nd) (1) (,) (2nd) (2) (,) (VARS). Scroll over to **Y-VARS**. Press (1) twice. This action stores the equation of the line or curve of best fit into **Y1** of the equation editor.

5. **Display and analyze the results.**

Press (ENTER). In this case, the letters a, b, and c are the coefficients of the general quadratic equation $y = ax^2 + bx + c$ for the curve of best fit. R^2 is the percent of data variation represented by the model. In this case, the equation is about $y = -4.90x^2 + 29.93x + 1.98$.

Note: In the case of linear regression if r is not displayed, turn on the diagnostics function. Press (2nd) (0) and scroll down to **DiagnosticOn**. Press (ENTER) twice. Repeat steps 4 to 6.

6. **Plot the curve.**

Press (GRAPH)

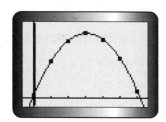

Appendix B

B–11 Finding the Points of Intersection of Two Functions

1. **Enter both functions into the equation editor.** In this case we will use $y = 5x + 4$ and $y = -2x + 18$.

2. **Graph both functions**. Press [GRAPH]. Adjust the window settings until the point(s) of intersection are displayed.

3. **Use the intersect operation.**

   Press [2nd] [TRACE] [5].

4. **Determine a point of intersection.** You will be asked to verify the two curves and enter a guess (optional) for the point of intersection. Press [ENTER] after each screen appears. The point of intersection is exactly (2, 14).

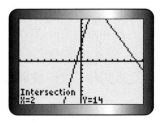

5. **Determine any additional points of intersection.** Press [TRACE] and move the cursor close to the other point you wish to identify. Repeat step 4.

B–12 Evaluating Trigonometric Ratios and Finding Angles

1. **Put the calculator in degree mode.**

   Press [MODE]. Scroll down and across to Degree. Press [ENTER].

2. **Use the [SIN], [COS], or [TAN] key to calculate trigonometric ratios.**

   To find the value of sin 54°, press [SIN] [5] [4] [ )] [ENTER].

3. **Use SIN⁻¹, COS⁻¹, or TAN⁻¹ to calculate angles.**

   To find the angle whose cosine is 0.6, press [2nd] [COS] [.] [6] [ )] [ENTER].

B–13 Graphing Trigonometric Functions

You can graph trigonometric functions in degree measure using the TI-83 Plus or TI-84 calculator. Graph the function $y = \sin x$ for $0° \leq x \leq 360°$.

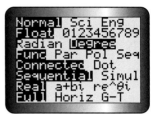

step 1

1. **Put the calculator in degree mode.**

   Press [MODE]. Scroll down and across to **Degree**. Press [ENTER].

2. **Enter $y = \sin x$ into the equation editor.**

   Press [Y=] [SIN] [X, T, Θ, n] [ )].

3. **Adjust the window to correspond to the given domain.**

   Press [WINDOW]. Set **Xmin = 0**, **Xmax = 360**, and **Xscl = 90**. These settings display the graph from 0° to 360°, using an interval of 90° on the x-axis. In this case, set **Ymin = −1** and **Ymax = 1**, since the sine function lies between these values. However, if this fact is not known, this step can be omitted.

step 3

4. **Graph the function using ZoomFit.**

   Press [ZOOM] [0]. The graph is displayed over the domain and the calculator determines the best values to use for **Ymax** and **Ymin** in the display window.

 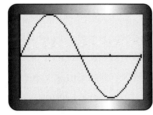

   step 4

   Note: You can use **ZoomTrig** (press [ZOOM] [7]) to graph the function in step 4. **ZoomTrig** will always display the graph in a window where **Xmin = −360°**, **Xmax = 360°**, **Ymin = −4**, and **Ymax = 4**.

B–14 Evaluating Powers and Roots

1. **Evaluate the power $(5.3)^2$.**

   Press [5] [.] [3] [x^2] [ENTER].

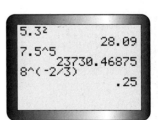

2. **Evaluate the power 7^5.**

   Press [7] [∧] [5] [ENTER].

3. **Evaluate the power $8^{-\frac{2}{3}}$.**

   Press [8] [∧] [( ] [−] [2] [÷] [3] [ )] [ENTER].

4. Evaluate the square root of 46.1

Press [2nd] [x^2] [4] [6] [.] [1] [)] [ENTER] .

5. Evaluate $\sqrt[4]{256}$

Press [4] [MATH] [5] [2] [5] [6] [ENTER] .

B–15 Analyzing Financial Situations Using the TVM Solver

Part 1: Introducing the TVM Solver

Press [MODE] and change the fixed decimal mode to 2, because most of the values that you are working with here represent dollars and cents. Scroll down to **Float**, across to **2**, and press [ENTER] .

Press [APPS] and then select **1:Finance**. From the Finance CALC menu, select **1:TVM Solver**. The screen that appears should be similar to the second one shown, but the values may be different.

You will notice eight variables on the screen.

N	total number of payment periods, or the number of interest conversion periods for simple annuities
I%	annual interest rate as a percent, not as a decimal
PV	present or discounted value
PMT	regular payment amount
FV	future or accumulated value
P/Y	number of payment periods per year
C/Y	number of interest conversion periods per year
PMT	Choose **BEGIN** if the payments are made at the beginning of the payment intervals. Choose **END** if the payments are made at the end of the payment intervals.

You may enter different values for the variables. Enter the value for money that is *paid* as a negative number, since the investment is a cash outflow; enter the value of money that is *received* as a positive number, since the money is a cash inflow. When you enter a whole number, you will see that the calculator adds the decimal and two zeros.

To solve for a variable, move the cursor to that variable and press [ALPHA] [ENTER] , and the calculator will calculate this value. A small shaded box to the left of the line containing the calculated value will appear.

Part 2: Determining Future Value and Present Value

EXAMPLE 1

Find the future value or amount of $7500 invested for nine years at 8%/a, compounded monthly.

Solution

The number of interest conversion periods, **N**, is $9 \times 12 = 108$, **I%** = 8, and **PV** = -7500. The value for present value, **PV**, is negative, because the investment represents a cash outflow. **PMT** = 0 and **FV** = 0. The payments per year, **P/Y**, and the compounding periods per year, **C/Y**, are both 12. Open the **TVM Solver** and enter these values. Scroll to the line containing **FV**, the future value, and press [ALPHA] [ENTER].

The investment will be worth $15 371.48 after nine years.

EXAMPLE 2

Maeve would like to have $3500 at the end of five years, so she can visit Europe. How much money should she deposit now in a savings account that pays 9%/a, compounded quarterly, to finance her trip?

Solution

Open the **TVM Solver** and enter the values shown in the screen, except the value for **PV**. The value for **FV** is positive, because the future value of the investment will be "paid" to Maeve, representing a cash inflow. Scroll to the line containing **PV** and press [ALPHA] [ENTER] to get $-$2242.86. The solution for **PV** is negative, because Maeve must pay this money and the payment is a cash outflow.

Part 3: Determining the Future or Accumulated Value of an Ordinary Simple Annuity

EXAMPLE 3

Celia deposits $1500 at the end of each year in a savings account that pays 4.5%/a, compounded annually. What will be the balance in the account after five years?

Solution

N = 5 and **I%** = 4.5. Because there is no money in the account at the beginning of the term, **PV** = 0. **PMT** = −1500. The payment, **PMT**, is negative, because Celia makes a payment, which is a cash outflow. **P/Y** = 1 and **C/Y** = 1. Open the **TVM Solver** and enter these values.

Scroll to the line containing **FV** and press [ALPHA] [ENTER].

The balance in Celia's account at the end of the year will be $8206.06.

EXAMPLE 4

Mr. Bartolluci would like to have $150 000 in his account when he retires in 15 years. How much should he deposit at the end of each month in an account that pays 7%/a, compounded monthly?

Solution

Open the **TVM Solver** and enter the values shown, except for **PMT**. Note that **N** = 12 × 15 = 180, and the future value, **FV**, is positive, since he will receive the money at some future time. Scroll to the line containing **PMT** and press [ALPHA] [ENTER].

Mr. Bartolluci must deposit $473.24 at the end of each month for 15 years. The payment appears as −473.24, because it is a cash outflow.

Part 4: Determining Present or Discounted Value of an Ordinary Simple Annuity

EXAMPLE 5

Northern Lights High School wishes to establish a scholarship fund. A $500 scholarship will be awarded at the end of each school year for the next eight years. If the fund earns 9%/a, compounded annually, what does the school need to invest now to pay for the fund?

Solution

Open the **TVM Solver** and enter 8 for **N**, 9 for **I%**, and 500 for **PMT**. The value for **PMT** is positive, because someone will receive $500 each year. Enter 0 for **FV**, since the fund will be depleted at the end of the term. Enter 1 for both **P/Y** and **C/Y**. Scroll to the line containing **PV** and press ⟨ ALPHA ⟩ ⟨ ENTER ⟩ .

The school must invest $2767.41 now for the scholarship fund. The present value appears as −2767.41, because the school must pay this money to establish the fund. The payment is a cash outflow.

EXAMPLE 6

Monica buys a snowboard for $150 down and pays $35 at the end of each month for 1.5 years. If the finance charge is 16%/a, compounded monthly, find the selling price of the snowboard.

Solution

Open the **TVM Solver** and enter the values as shown in the screen, except the value for **PV**. The payment, **PMT**, is positive, because the payments are a cash inflow for the snowboard's seller. Scroll to the line containing **PV** and press ⟨ ALPHA ⟩ ⟨ ENTER ⟩ .

The present value is $556.82. The present value appears as a negative value on the screen, because it represents what Monica would have to pay now if she were to pay cash.

The selling price is the sum of the positive present value and the down payment. Since the down payment is also a payment, add both numbers. The total cash price is $PV + \$150 = \$706.82$. Under this finance plan, Monica will pay $\$35 \times 18 + \$150 = \$780$.

B–16   Creating Repayment Schedules

In this section, you will create a repayment schedule, by applying some of the financial functions from the Finance CALC menu.

Part 1: Introducing Other Finance CALC Menu Functions

You have used the **TVM Solver** to find, for example, future value and present value. The calculator can use the information that you have entered into the **TVM Solver** to perform other functions. Here are three other functions:

Σ**Int**(*A, B, roundvalue*)	calculates the sum of the interest paid from period *A* to period *B*
Σ**Prn**(*A, B, roundvalue*)	calculates the sum of the principal paid from period *A* to period *B*
bal(*x, roundvalue*)	calculates the balance owing after period *x*

The calculator rounds as it calculates. You will need to tell the calculator the value for rounding, *roundvalue*. The greater *roundvalue* is, the greater the accuracy of the calculations. In this section, the *roundvalue* is 6, which is also the value that banks use.

Part 2: Using the TVM Solver and Other Finance CALC Menu Functions

Press ⬚ MODE ⬚ and change the fixed decimal mode to 2, because most of the values in this section represent dollars and cents.

EXAMPLE 1

Eleanor finances the purchase of a new pickup truck by borrowing $18 000. She will repay the loan with monthly payments. The term of the loan is five years. The interest rate is 14%/a, compounded monthly.

a)   How much is the monthly payment?
b)   How much will she pay in interest?
c)   How much will she still owe on the loan after the 30th payment, that is, at the halfway point in repaying the loan?
d)   What portion of the 30th payment reduces the principal?

Solution

a)   Press ⬚ APPS ⬚ and select **1:Finance**. Then press ⬚ ENTER ⬚ to select **1:TVM Solver** from the Finance CALC menu. Enter **N** = 60, because $12 \times 5 = 60$. Enter **I%** = 14, **PV** = 18 000, **FV** = 0, **P/Y** = 12, and **C/Y** = 12. Notice that the present value, **PV**, is a positive number because Eleanor receives (a cash inflow) $18 000 from the bank.

Scroll to the line containing **PMT** and press ⬚ ALPHA ⬚ ⬚ ENTER ⬚ . The monthly payment is $418.83.

The payment appears as a negative value, because Eleanor pays this amount each month. The actual value is $-418.828\ 515\ 3$, which the calculator rounded to -418.83.

b) Use Σ**Int**(A, B, *roundvalue*) to calculate the total interest that Eleanor will pay.

Press [2nd] [MODE] to return to the home screen.

Press [APPS] and the select **1:Finance** from the Finance CALC menu.

Select Σ**Int** by scrolling down or by pressing [ALPHA] [MATH].

Press [ENTER].

Press [1] [,] [6] [0] [,] [6] [ )] [ENTER].

The sum of the interest paid from the first period to the 60th period is calculated.

By the end of the loan, Eleanor will have paid $7129.71 in total interest. Eleanor will have paid $7129.71 + $18 000 = $25 129.71 in interest and principal for the truck. Note that the product of the payment, $418.83, and the total number of payments, 60, is $25 129.80. The difference of $0.09 is due to rounding, because 418.828 515 3 was rounded to 418.83.

c) Find the balance on the loan after the 30th payment. *roundvalue* must be consistent, that is, 6.

From the Finance CALC menu, select **bal** by scrolling or by pressing [9].

Press [3] [0] [,] [6] [ )] [ENTER].

Eleanor still owes $10 550.27 after the 30th payment. (Why is this amount not $9000?)

d) Find the portion of the 30th payment that reduces the principal by calculating the sum of the principal paid from the 30th payment to the 30th payment. In the words, you are calculating the sum of only one item, the 30th payment. *roundvalue* is again 6. From the Finance CALC menu,

select Σ**Prn** by scrolling down or by pressing [0]. Press [3] [0]

[,] [3] [0] [,] [6] [ )] [ENTER].

The portion of the 30th payment that reduces the principal is $292.33. The other portion of this payment, $126.50, is interest.

CALC VARS
5↑tvm_N
6:tvm_FV
7:npv(
8:irr(
9:bal(
0:ΣPrn(
A↓ΣInt(

ΣInt(1,60,6)
       -7129.71

bal(30,6)
      10550.27

ΣPrn(30,30,6)
      -292.33

Part 3: Using the Finance Functions to Create Repayment Schedules

Use the functions described in parts 1 and 2 to create repayment schedules or amortization tables.

EXAMPLE 2

Recall that Eleanor borrows $18 000 to purchase a pickup truck. She will repay the loan with monthly payments. The term of the loan is five years. The interest rate is 14%/a compounded monthly.

a) What will be the monthly outstanding balance on the loan after each of the first seven months?

b) Create a repayment schedule for the first seven months of the loan.

c) Use the repayment schedule to verify that the loan is completely paid after five years or 60 payments.

Solution

a) Find the outstanding balance after each payment for the first seven months. You will combine **sequence** (List OPS menu) and **bal**.

From the home screen, press [2nd] [STAT] [▶] [5] to select sequence.

Then press [APPS] [ENTER] [9] to select bal. Press [X, T, Θ, n] [,] [6] [)] [,] [X, T, Θ, n] [,] [1] [,] [7] [,] [1] [)]. Press [ENTER] to

calculate the sequence of balances, beginning with the first month, 1, and ending with the last month, 7. The increment for this sequence is 1, which is the last value entered. Recall that the increment is the change from payment number to payment number. Scroll right ([▶]) to see the other balances.

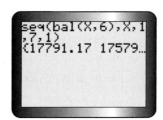

b) Create a repayment, or an amortization, schedule for the first seven months by comparing the interest, principal, and balance in a table.

Begin by opening the equation editor. Press [Y=]. Clear **Y1** to **Y3**, if necessary. Store the interest portion of each payment in **Y1**. Move the cursor to the right of **Y1=**. Press [APPS] [ENTER] [ALPHA] [MATH] to select **ΣInt**. Press [X, T, Θ, n] [,] [X, T, Θ, n] [,] [6] [ )] [ENTER].

To store the principal portion of each payment in Y2, press [APPS] [ENTER] [0] to select **ΣPrn**. Press [X, T, Θ, n] [,] [X, T, Θ, n] [,] [6] [ )] [ENTER].

To store the outsanding balance after each payment in **Y3**, press [APPS] [ENTER] [9] to select **bal**. Press [X, T, Θ, n] [,] [6] [ )] [ENTER].

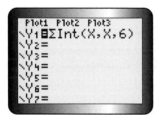 

Before viewing the table, press [2nd] [WINDOW]. Set **TblStart** to 1 and **ΔTbl** to 1. The table will start with payment 0 and the payment number will increase by 1 at each step.

Press [2nd] [GRAPH] to see the amortization table. Notice that the interest portion and the principal portion of each payment appear as negative values. Each payment, which is a combination of interest and principal, is a cash outflow for Eleanor.

Scroll right to see the values for **Y3**, the outstanding balance.

The outstanding balance after seven payments is $16 486.03.

c) Scroll or reset the tables's start value to see other entries in the amortization table. Scroll up to the beginning of the table. Notice that a substantial portion of the $418.83 payment is interest. Scroll down the table. At the end of the amortization, none of the payment is applied to the principal. The final outstanding balance is **2.5E⁻5**. As a decimal, this value is $0.000 25. Therefore, the amortization period of 60 payments is correct. The loan will be paid completely after five years or 60 payments.

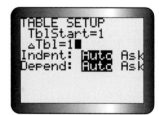

PART 2 USING A SPREADSHEET

B–17 Introduction to Spreadsheets

A spreadsheet is a computer program that can be used to create a table of values and then graph the values. It is made up of cells that are identified by column letter and row number, such as A2 or B5. A cell can hold a label, a number, or a formula.

Creating a Table

Use spreadsheets to solve problems like this:

How long will it take to double your money if you invest $1000 at 5%/a compounded quarterly?

To create a spreadsheet, label cell A1 as Number of Quarters, Cell B1 as Time (years), and cell C1 as Amount ($). Enter the initial values of 0 in A2, 0 in B2, and 1000 in C2. Enter the formulas =A2+1 in A3, =A3/4 in B3, and =1000*(1.0125)^A3 in C3 to generate the next values in the table.

	A	B	C
1	Number of Quarters	Time (years)	Amount ($)
2	0	0	1000
3	=A2+1	=A3/4	-1000*(1.0125^A3)
4			

Notice that an equal sign is in front of each formula, an asterisk (*) is used for multiplication, and a caret (^) is used for exponents. Next, use the cursor to select cell A3 to C3 and several rows of cells below them. Use the **Fill Down** command. This command inserts the appropriate formula into each selected cell.

	A	B	C
1	Number of Quarters	Time (years)	Amount ($)
2	0	0	1000
3	1	0.25	
4			

When the **Fill Down** command is used, the computer automatically calculates and enters the values in each cell, as shown below in the screen on the left.

Continue to select the cells in the last row of the table and use the **Fill Down** command to generate more values until the solution appears, as shown below in screen on the right.

	A	B	C
1	Number of Quarters	Time (years)	Amount ($)
2	0	0	1000
3	1	0.25	1012.50
4	2	0.5	1025.16
5	3	0.75	1037.97
6	4	1	1050.94

	A	B	C
1	Number of Quarters	Time (years)	Amount ($)
2	0	0	1000
3	1	0.25	1012.50
4	2	0.5	1025.1563
⋮	⋮	⋮	⋮
56	54	13.5	1955.8328
57	55	13.75	1980.2807
58	56	14	2005.0342

Creating a Graph

Use the spreadsheet's graphing command to graph the results. Use the cursor to highlight the portion of the table you would like to graph. In this case, Time versus Amount.

	A	B	C
1	**Number of Quarters**	**Time (years)**	**Amount ($)**
2	0	0	1000
3	1	0.25	1012.50
4	2	0.5	1025.1563
⋮	⋮	⋮	⋮
56	54	13.5	1955.8328
57	55	13.75	1980.2807
58	56	14	2005.0342

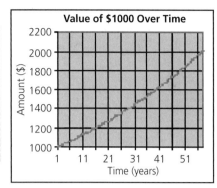

Different spreadsheets have different graphing commands. Check your spreadsheet's instructions to find the proper command. This graph appears.

Determining the Equation of the Curve of Best Fit

Different spreadsheets have different commands for finding the equation of the curve of best fit using regression. Check your spreadsheet's instructions to find the proper command for the type of regression that suits the data.

B–18   Creating an Amortization Table

You have decided to purchase $4000 in furniture for your apartment. Your monthly payment is $520 and the interest rate 12%/a, compounded monthly. A spreadsheet can create a table of the outstanding balance and the graph.

Create a Table of the Outstanding Balance

You may find that the spreadsheet software that you are using may have different commands or techniques for achieving the same results.

1. Create a spreadsheet with five columns, labeled Payment Number, Payment, Interest Paid, Principal Paid and Outstanding Balance in cells A1, B1, C1, D1 and E1, respectively.

2. In cells A2 and E2, enter 0 and 4000, respectively. Leave cell B2, C2 and D2 empty.

3. In cell A3, enter the expression =A2+1 and press (ENTER). In cell B3, enter 520. In cell C3, enter the expression for interest, =E2*0.01 and press (ENTER). In cell D3, enter the expression =B3-C3 and press (ENTER). In cell E3 enter =E2-D3 and press (ENTER).

	A	B	C	D	E
1	Payment Number	Payment	Interest Paid	Principal Paid	Outstanding Balance
2	0				4000
3	=A2 + 1	520	=E2*0.01	=B3 – C3	=E2 – D3

4. Select cells A3 across to E3 and down to E11. Use the Fill Down command to complete the table. You can display financial data in different ways. Select cells B3 to E11 and click the **$** button or icon. Each number will appear with a dollar sign, a comma, and will be rounded to two decimal places.

	A	B	C	D	E
1	Payment Number	Payment	Interest Paid	Principal Paid	Outstanding Balance
2	0				$4,000.00
3	1	$520.00	$40.00	$480.00	$3,520.00
4	2	$520.00	$35.20	$484.80	$3,035.20
5	3	$520.00	$30.35	$489.65	$2,545.55
6	4	$520.00	$25.46	$494.54	$2,051.01
7	5	$520.00	$20.51	$499.49	$1,551.52
8	6	$520.00	$15.52	$504.48	$1047.03
9	7	$520.00	$10.47	$509.53	$537.50
10	8	$520.00	$5.38	$514.62	$22.88
11	9	$520.00	$0.23	$519.77	$(496.89)

PART 3 USING THE GEOMETER'S SKETCHPAD

B–19 Graphing Functions

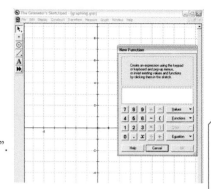

1. **Turn on the grid.**
   From the **Graph** menu, choose **Show Grid**.

2. **Enter the function.** From the **Graph** menu, choose **Plot New Function**. The function calculator should appear.

3. **Graph the function $y = x^2 - 3x + 2$.**
   Use either the calculator keypad or the keyboard to enter "x ^ 2 - 3 * x + 2". Then press the "OK" button on the calculator keypad. The graph of $y = x^2 - 3x + 2$ should appear on the grid.

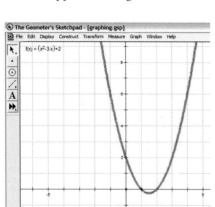

4. **Adjust the origin and/or scale.**
   To adjust the origin, left-click on the point at the origin to select it. Then left-click and drag the origin as desired. To adjust the scale, left-click in blank space to deselect, then left-click on the point at (1, 0) to select it. Left-click and drag this point to change the scale.

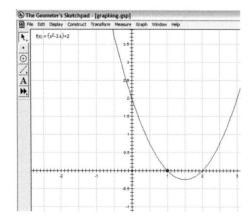

B–20 Graphing Trigonometric Functions

1. **Turn on the grid.**
   From the **Graph** menu, choose **Show Grid**.

2. **Graph the function $y = 2 \sin (30x) + 3$.**
   - From the **Graph** menu, choose **Plot New Function**. The function calculator should appear.
   - Use either the calculator keypad or the keyboard to enter "2 * sin (30 * x) + 3". To enter "sin," use the pull-down "Functions" menu on the calculator keypad.
   - Click on the "OK" button on the calculator keypad.
   - Click on "No" in the pop-up panel to keep degrees as the angle unit. The graph of $y = 2 \sin (30x) + 3$ should appear on the grid.

3. **Adjust the origin and/or scale.**
   Left-click on and drag either the origin or the point (1, 0).

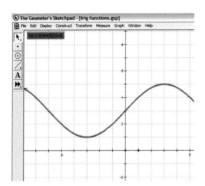
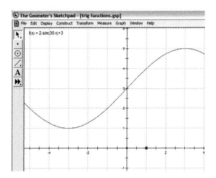

B–21 Creating Geometric Figures

To draw a triangle, follow these steps.

1. **Plot the vertices of the figure.**

   Choose the **Point** tool .

2. **Select the vertices.**

   Choose the **Arrow Selection** tool . Left-click and drag around the three vertices to select them.

3. **Draw the sides of the figure.**
   From the **Construct** menu, choose **Segments**. The sides of the triangle should appear.

B–22 Measuring Sides and Angles, and Using the Calculator

As an example, draw a triangle and verify that the sine law holds for it.

1. **Draw a triangle.**

   See "Creating Geometric Figures" (B-21).

2. **Find the measure of each side.**
   - Choose the **Arrow Selection** tool.
   - Left-click on any blank area to deselect.
   - Left-click on a side to select it.
   - From the **Measure** menu, choose **Length**.
   - Repeat for the other two sides.

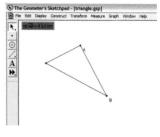

3. **Find the measure of each angle.**
   - Left-click on any blank area to deselect.
   - To measure *BAC*, select the vertices *in that order*: first *B*, then *A*, then *C*.
   - From the **Measure** menu, choose **Angle**.
   - Repeat for the other two angles.

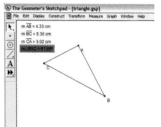

4. **For side *AB* and the opposite *ACB*, calculate the ratio $\dfrac{AB}{\sin ACB}$.**
   - From the Measure menu, choose Calculate… The calculator keypad should appear.
   - Left-click on the measure "m$\overline{AB}$" to select it.
   - Use the calculator keypad to enter "÷".
   - Use the "Functions" menu on the keypad to select "sin".
   - Select the measure "m∠ACB".
   - Press the "OK" button on the calculator keypad. The ratio should appear as a new measure.

5. **Repeat step 4 for the other two pairs of corresponding sides and angles.**

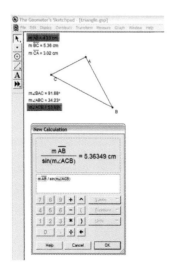

Appendix B

PART 4 USING FATHOM

B–23 Creating a Scatter Plot and Determining the Equation of a Line or Curve of Good Fit

1. **Create a case table.** Drag a case table from the object shelf, and drop it in the document.

2. **Enter the Variables and Data.** Click <new>, type a name for the new variable or attribute and press Enter. (If necessary, repeat this step to add more attributes; pressing Tab instead of Enter moves you to the next column.) When you name your first attribute, Fathom creates an empty collection to hold your data (a little, empty box). The collection is where your data are actually stored. Deleting the collection deletes your data. When you add cases by typing values, the collection icon fills with gold balls. To enter the data click in the blank cell under the attribute name and begin typing values. (Press Tab to move from cell to cell.)

3. **Graph the data.** Drag a new graph from the object shelf at the top of the Fathom window, and drop it in a blank space in your document. Drag an attribute from the case table, and drop it on the prompt below and/or to the left of the appropriate axis in the graph.

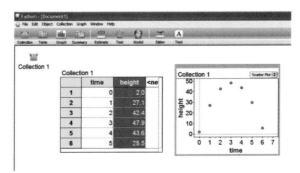

4. **Create a function.** Right click the graph and select **Plot Function.** Enter your function using a parameter that can be adjusted to fit the curve to the scatter plot. (a was used in this case).

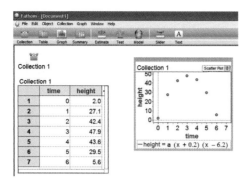

5. **Create a slider for the parameter(s) in your equation.** Drag a new slider from the object shelf at the top of the Fathom window, and drop it in a blank space below your graph. Type in the letter of the parameter used in your function in step 4 over V1. Click on the number then adjust the value of the slider until you are satisfied with the fit.

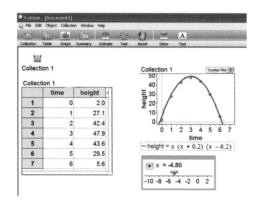

The equation of a curve of good fit is $y = -4.8(x + 0.2)(x - 6.2)$.

Glossary

Instructional Words

C

calculate: Figure out the number that answers a question; compute

clarify: Make a statement easier to understand; provide an example

classify: Put things into groups according to a rule and label the groups; organize into categories

compare: Look at two or more objects or numbers and identify how they are the same and how they are different (e.g., Compare the numbers 6.5 and 5.6. Compare the size of the students' feet. Compare two shapes.)

conclude: Judge or decide after reflection or after considering data

construct: Make or build a model; draw an accurate geometric shape (e.g., Use a ruler and a protractor to construct an angle.)

create: Make your own example or problem

D

describe: Tell, draw, or write about what something is or what something looks like; tell about a process in a step-by-step way

determine: Decide with certainty as a result of calculation, experiment, or exploration

draw: 1. Show something in picture form (e.g., Draw a diagram.)
2. Pull or select an object (e.g., Draw a card from the deck. Draw a tile from the bag.)

E

estimate: Use your knowledge to make a sensible decision about an amount; make a reasonable guess (e.g., Estimate how long it takes to cycle from your home to school. Estimate how many leaves are on a tree. What is your estimate of 3210 + 789?)

evaluate: 1. Determine if something makes sense; judge
2. Calculate the value as a number

explain: Tell what you did; show your mathematical thinking at every stage; show how you know

explore: Investigate a problem by questioning, brainstorming, and trying new ideas

extend: 1. In patterning, continue the pattern
2. In problem solving, create a new problem that takes the idea of the original problem further

J

justify: Give convincing reasons for a prediction, an estimate, or a solution; tell why you think your answer is correct

M

measure: Use a tool to describe an object or determine an amount (e.g., Use a ruler to measure the height or distance around something. Use a protractor to measure an angle. Use balance scales to measure mass. Use a measuring cup to measure capacity. Use a stopwatch to measure the time in seconds or minutes.)

model: Show or demonstrate an idea using objects and/or pictures (e.g., Model addition of integers using red and blue counters.)

P

predict: Use what you know to work out what is going to happen (e.g., Predict the next number in the pattern 1, 2, 4, 7,....)

R

reason: Develop ideas and relate them to the purpose of the task and to each other; analyze relevant information to show understanding

relate: Describe how two or more objects, drawings, ideas, or numbers are similar

represent: Show information or an idea in a different way that makes it easier to understand (e.g., Draw a graph. Make a model.)

S

show (your work): Record all calculations, drawings, numbers, words, or symbols that make up the solution

sketch: Make a rough drawing (e.g., Sketch a picture of the field with dimensions.)

solve: Develop and carry out a process for finding a solution to a problem

sort: Separate a set of objects, drawings, ideas, or numbers according to an attribute (e.g., Sort 2-D shapes by the number of sides.)

V

validate: Check an idea by showing that it works

verify: Work out an answer or solution again, usually in another way; show evidence of

visualize: Form a picture in your head of what something is like; imagine

Mathematical Words

A

acute angle: An angle greater than $0°$ and less than $90°$

adjacent sides: Two sides in a triangle or polygon that share a vertex with each other

amortization: The process of gradually reducing a debt through instalment payments of principal and interest

amount: The sum of the original principal and the interest; given by $A = P + I$, where A is the amount, P is the principal, and I is the interest

amplitude: The distance from the function's equation of the axis to either the maximum or the minimum value

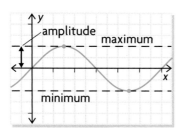

angle of depression: The angle between the horizontal and the line of sight when one is looking down at an object

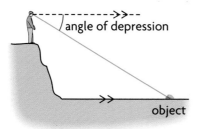

angle of elevation: The angle between the horizontal and the line of sight when one is looking up at an object

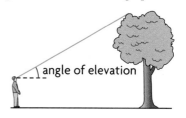

annuity: A series of equal deposits or payments made at regular intervals

asymptote: A line that a curve approaches, but never reaches on some part of its domain

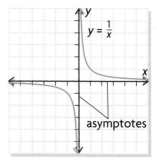

axis of symmetry: A line in a 2-D figure such that, if a perpendicular is constructed, any two points lying on the perpendicular and the figure are at equal distances from this line

B

base: The number that is used as a factor in a power (e.g., In the power 5^3, 5 is the base.)

bearing: The direction in which you have to move in order to reach an object. A bearing is a clockwise angle from magnetic north. For example, the bearing of the lighthouse shown is 335°

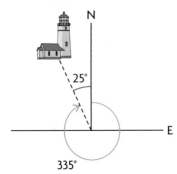

binomial: An algebraic expression with two term. For example, $3x + 2$

C

circumference: The boundary of a circle; the length of this boundary. The formula to calculate the length is $C = 2\pi r$, where r is the radius, or $C = \pi d$, where d is the diameter

coefficient: The factor by which a variable is multiplied. For example, in the term $3x$, the coefficient of x is 3. In the term $0.25y$, the coefficient of y is 0.25

completing the square: The process of adding a constant to a given quadratic expression to form a perfect trinomial square. For example, $x^2 + 6x + 2$ is not a perfect square, but if 7 is added to it, it becomes $x^2 + 6x + 9$, which is $(x + 3)^2$

compounding period: Each period over which compound interest is earned or charged in an investment or loan

compound interest: Interest calculated at regular periods and added to the principal for the next period

contained angle: The angle between two known sides

cosine law: In any acute $\triangle ABC$,
$c^2 = a^2 + b^2 - 2ab \cos C$

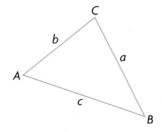

curve of best fit: The curve that best describes the distribution of points in a scatter plot. Typically found using regression analysis

curve of good fit: A curve that approximates or is close to the distribution of points in a scatter plot. Typically found using an informal process

cycle: A series of events that are regularly repeated; a complete set of changes, starting from one point and returning to the same point in the same way

D

decomposing: Breaking a number or expression into parts that make it up

decreasing function: A function whose y-values decrease as the x-values increase. The graph falls from left to right

degree: The degree of a polynomial with a single variable, say, x, is the value of the highest exponent of the variable. For example, for the polynomial $5x^3 - 4x^2 + 7x - 8$, the highest power or exponent is 3; the degree of the polynomial is 3

dependent variable: In an algebraic relation, the variable whose value depends on the value of another variable. Often represented by y

difference of squares: An expression of the form $a^2 - b^2$, which involves the subtraction of two squares

discrete: Consisting of separate and distinct parts. Discrete variables measure things that can be counted using whole numbers, such as books on a shelf, people in a room, or letters in a word. Continuous variables are not discrete; for example, the temperature of a room

discriminant: The expression $b^2 - 4ac$ in the quadratic formula

domain: The set of all values for which the independent variable is defined

down payment: The partial amount of a purchase paid at the time of purchase

E

equation of the axis: The equation of the horizontal line halfway between the maximum and the minimum is determined by

$$y = \frac{(\text{maximum value} + \text{minimum value})}{2}$$

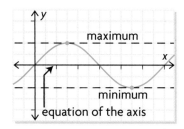

expand: To write an expression in extended but equivalent form. For example, $3(5x + 2) = 15x + 6$

exponent: The number that tells how many equal factors are in a power

extrapolate: To predict a value by following a pattern beyond known values

F

factor: To express a number as the product of two or more numbers, or an algebraic expression as the product of two or more other algebraic expressions. Also, the individual numbers or algebraic expressions in such a product

factored form: A quadratic function in the form $f(x) = a(x - r)(x - s)$

function: A relation in which there is only one value of the dependent variable for each value of the independent variable (i.e., for every x-value, there is only one y-value)

function notation: $f(x)$ is called function notation and is used to represent the value of the dependent variable for a given value of the independent variable, x

future value: The final amount (principal plus interest) of an investment or loan when it matures at the end of the investment or loan period

G

Guaranteed Investment Certificate (GIC): A Guaranteed Investment Certificate is a secure investment that guarantees to preserve your money. Your investment earns interest, at either a set or a variable rate

H

hypotenuse: The longest side of a right triangle; the side that is opposite the right angle

I

increasing function: A function whose y-values increase as the x-values increase. The graph rises from left to right

independent variable: In an algebraic relation, a variable whose values may be freely chosen and upon which the values of the other variables depend. Often represented by x

integers: The set of integers, **I,** is the set consisting of the numbers ..., $-3, -2, -1, 0, 1, 2, 3, ...$ This statement is expressed mathematically within the set notation $\{x \in \mathbf{I}\}$

intercept: See *x-intercept*, *y-intercept*

interest: The cost of borrowing money or the money earned from an investment

interest rate: A percent of money borrowed or invested that is paid for a specified period of time

interpolate: To estimate a value between two known values

investment: A sum of money that is deposited into a financial institution that earns interest over a specified period of time

K

key points: Points of any function that define its general shape

Key Points of $f(x) = x^2$

x	$f(x) = x^2$
-3	9
-2	4
-1	1
0	0
1	1
2	4
3	9

L

length: The measurment of the extent of an object or shape along its greatest dimension

like terms: Algebraic terms that have the same variables and exponents apart from their numerical coefficients. Like terms can be combined by adding or subtracting their numerical coefficents. For example, $2y + 8y = 10y$, $3x^2 - 5x^2 = -2x^2$, $3mr^2 + 6mr^2 = 9mr^2$

linear equation: An equation of the form $ax + b = 0$, or an equation that can be rewritten in this form. The algebraic expression involved is a polynomial of degree 1 (e.g., $2x + 3 = 6$ or $y = 3x - 5$)

line of best fit: The straight line that best describes the distribution of points in a scatter plot. Typically found using linear regression analysis

line of good fit: The straight line that reasonably describes the distribution of points in a scatter plot. Typically found using an informal process

loan: A sum of money that is borrowed from a financial institution that must be repaid with interest in a specified period of time

M

mapping diagram: A drawing with arrows to show the relationship between each value of x and the corresponding values of y

maturity: The final payment date of a loan or investment, after which point no further interest or principal is paid

maximum value: The greatest value taken by the dependent variable in a relation or function

minimum value: The least value taken by the dependent variable in a relation or function

N

negative correlation: This indicates that as one variable in a linear relationship increases, the other decreases and vice versa

O

oblique triangle: A triangle (acute or obtuse) that does not contain a right angle

opposite side: In a triangle or polygon, the side that is located opposite a specific angle

optimal value: The maximum or minimum value of the dependent variable

P

parabola: The graph of a quadratic relation of the form $y = ax^2 + bx + c$ ($a \neq 0$). The graph, which resembles the letter U, is symmetrical

peak: The highest point(s) on a graph

perfect square trinomial: A trinomial that has two identical binomial factors; for example, $x^2 + 6x + 9$ has the factors $(x + 3)(x + 3)$

period: The interval of the independent variable (often time) needed for a repeating action to complete one cycle

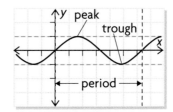

periodic function: A function whose values are repeated at equal intervals of the independent variable

positive correlation: This indicates that both variables in a linear relationship increase or decrease together

power: A numerical expression that shows repeated multiplication (e.g., The power 5^3 is a shorter way of writing $5 \times 5 \times 5$). A power has a base and an exponent; the exponent tells the number of equal factors there are in a power

present value: The principal that must be invested today to obtain a given amount in the future

principal: A sum of money that is borrowed or invested

Pythagorean theorem: The conclusion that, in a right triangle, the square of the length of the longest side is equal to the sum of the squares of the lengths of the other two sides

Q

quadratic equation: An equation that contains a polynomial whose highest degree is 2. For example, $x^2 + 7x + 10 = 0$

quadratic formula: A formula for determining the roots of a quadratic equation of the form $ax^2 + bx + c = 0$. The formula uses the coefficients of the terms in the quadratic equation:

$$x = \frac{-b \pm \sqrt{b^2 - 4ac}}{2a}$$

quadratic function: A function that contains a polynomial whose highest degree is 2. Its graph is a parabola. For example, $f(x) = 2x^2 + 3x - 5$

quadratic regression: A process that fits the second-degree polynomial $ax^2 + bx + c$ to the data

R

radical: The indicated root of a quantity. For example, $\sqrt[3]{8} = 2$ since $2 \times 2 \times 2 = 2^3 = 8$

range: The set of all values of the dependent variable. All such values are determined from the values in the domain

rational number: A number that can be expressed as the quotient of two integers where the divisor is not 0

real numbers: The set of real numbers, **R**, is the set of all decimals—positive, negative, and 0, terminating and nonterminating. This statement is expressed mathematically with the set notation $\{x \in \mathbf{R}\}$

reflection: A transformation in which a 2-D shape is flipped. Each point in the shape flips to the opposite side of the line of reflection, but stays the same distance from the line

relation: A relationship between two variables; values of the independent variable are paired with values of the dependent variable

right angle: An angle that measures $90°$

root of an equation: A number that, when substituted for the unknown, makes the equation a true statement. For example, $x = 2$ is a root of the equation $x^2 - x - 2 = 0$ because $2^2 - 2 - 2 = 0$. The root of an equation is also known as a *solution* to that equation

S

scatter plot: A graph that attempts to show a relationship between two variables by means of points plotted on a coordinate grid

set notation: A way of writing a set of items or numbers within curly brackets, $\{\ \}$

simple interest: Interest earned or paid only on the original sum of money invested or borrowed

sine function: A sine function is the graph of $f(x) = \sin x$, where x is an angle measured in degrees. It is a periodic function

sine law: In any triangle, the ratios of each side to the sine of its opposite angle are equal

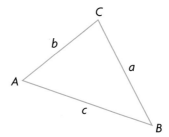

$$\frac{\sin A}{a} = \frac{\sin B}{b} = \frac{\sin C}{c}$$

sinusoidal function: A type of periodic function created by transformations of $f(x) = \sin x$

solution to a quadratic equation: The value of a variable that makes the equation true

standard form: A quadratic function in the form $f(x) = ax^2 + bx + c$

surface area: The total area of all the faces of any 3-D shape

T

term deposit: An investment purchased from a bank, trust company, or credit union for a fixed period or term

transformations: Transformations are operations performed on functions to change the position or shape of the associated curves or lines

transformed functions: The resulting function when the shape and/or position of the graph of $f(x)$ is changed

translation: Two types of translations can be applied to the graph of a function $y = f(x - h) + k$:

- Horizontal translations—all points on the graph move to the right when $h > 0$ and to the left when $h < 0$

- Vertical translations—all points on the graph move up when $k > 0$ and down when $k < 0$

trend: A relationship between two variables for which the independent variable is time

trinomial: An algebraic expression with three terms. For example, $2x^2 - 6x + 7$

trough: The lowest point(s) on a graph

U

unlike terms: Algebraic terms that have diffent variables and/or exponents apart from their numerical coefficients. Unlike terms cannot be combined by adding or subtracting their numerical coefficients. For example, $3x^3$, $5y$, and $2x$

V

variable: A letter or symbol, such as a, b, x, or n, that represents a number (e.g., In the formula for the area of a rectangle, the variables A, l, and w represent the area, length, and width of the rectangle.)

vertex (plural vertices): The point at the corner of an angle or shape (e.g., A cube has eight vertices. A triangle has three vertices. An angle has one vertex.)

vertex form: A quadratic function in the form $f(x) = a(x - h)^2 + k$, where the vertex is (h, k)

vertical compression: When $0 < a < 1$, the graph of the function $af(x)$ is compressed vertically

vertical-line test: A test to determine whether the graph of a relation is a function. The relation is not a function if any vertical line drawn through the graph of the relation passes through two or more points

vertical reflection: When $a < 0$, the graph of the function $af(x)$ is reflected in the x-axis

vertical stretch: When $a > 1$, the graph of the function $af(x)$ is stretched vertically

vertical translation: When $d > 0$, the graph of the function $f(x) + d$ is shifted d units up. When $d < 0$, the graph of the function $f(x) + d$ is shifted d units down

X

x-intercept: The value at which a graph meets the x-axis. The value of y is 0 for all x-intercepts

Y

y-intercept: The value of the dependent variable when the independent variable is zero; sometimes called the initial value

Z

zeros of a relation: The values of x for which a relation has the value zero. The zeros of a relation correspond to the x-intercepts of its graph

Answers

Chapter 1

Getting Started, pp. 2–4

1. a) (ii)   **c)** (iv)   **e)** (vi)
    b) (v)   **d)** (i)   **f)** (iii)

2. a) -2 **b)** -1 **c)** -3 **d)** -12

3. a)

x	y
-2	8
-1	2
0	0
1	2
2	8

b)

x	y
-2	-19
-1	-7
0	-3
1	-7
2	-19

c)

x	y
-2	7
-1	5.5
0	5
1	5.5
2	7

d)

x	y
-2	-15
-1	0
0	5
1	0
2	-15

4. a) $y = -3x + 5$; -1 **b)** $y = 3 - 3x$; -3

5. a) yes   **b)** yes

6. a) Yes. When I substitute 2 for x and -1 for y into $2x - y$, the equation $2x - y = 5$ is true.
    b) No. When I substitute -1 for x and 29 for y, the equation $y = -2x^2 - 5x + 22$ is false.

7. a) 6; 4; $-\dfrac{2}{3}$ **b)** -8; 2; $\dfrac{1}{4}$

8. a) 2; $x = 0$ **b)** -4; $x = 0$

9. a) B   **b)** C   **c)** D   **d)** A

10. a)

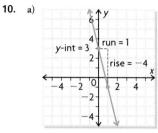

b)

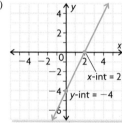

c)

x	y
-3	-1
-1	0
1	1
3	2

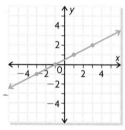

11. slope $= -2$, y-int $= 5$, equation: $y = -2x + 5$

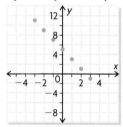

12.

Definition:	Characteristics:
A quadratic relations is a relation that can be described by an equation that contains a polynomial whose highest degree is 2.	• axis of symmetry • single maximum or minimum value

Quadratic Relation

Examples:	Non-examples:
$y = x^2 + 7x + 10$ $y = x^2 - 8x + 7$	$y = 4x + 3$ $y = x^3 - 1$

Lesson 1.1, pp. 13–16

1. a) i) D = {1, 3, 4, 7}, R = {2, 1}

   ii) Function since there is only one *y*-value for each *x*-value.

   b) i) D = {1, 4, 6}, R = {2, 3, 5, 1}

   ii) Not a function since there are two *y*-values for *x* = 1.

   c) i) D = {1, 0, 2, 3}, R = {0, 1, 3, 2}

   ii) Function since there in only one *y*-value for each *x*-value.

   d) i) D = {1}, R = {2, 5, 9, 10}

   ii) Not a function since there are four *y*-values for *x* = 1.

2. a) i) D = {1, 3, 4, 6}, R = {1, 2, 5}

   ii) Function since there in only one value in the range for each value in the domain.

   b) i) D = {1, 2, 3}, R = {4, 5, 6}

   ii) Not a function since there are two values in the range for the value of 1 in the domain.

   c) i) D = {1, 2, 3}, R = {4}

   ii) Function since there is only one value in the range for each value in the domain.

   d) i) D = {2}, R = {4, 5, 6}

   ii) Not a function since there are three values in the range for the value of 2 in the domain.

3. a) i) D = {*x* | 2 ≤ *x* ≤ 13}, R = {*y* | 2 ≤ *y* ≤ 7}

   ii) Function since a vertical line passes through only one *y*-value at any point.

   b) i) D = {*x* | 0 ≤ *x* ≤ 6}, R = {*y* | 0 ≤ *y* ≤ 9}

   ii) Not a function since a vertical line passes through two points when *x* = 1 and when *x* = 5.

   c) i) D = {*x* | −2 ≤ *x* ≤ 4}, R = {*y* | −4 ≤ *y* ≤ 5}

   ii) Not a function since a vertical line passes through two points at several values of *x*.

   d) i) D = {*x* | *x* ∈ **R**}, R = {*y* | *y* ≥ 2}

   ii) Function since a vertical line passes through only one *y*-value at any point.

4. a) Function since there is only one *y*-value for each *x*-value.

   b) Not a function since there are two *y*-values for *x* = 1.

   c) Function since there is only one *y*-value for each *x*-value.

   d) Not a function since there are two *y*-values for *x* = 1.

5. a) function

   b) not a function; (−5, −7) or (−5, −2)

6. a) Function since a vertical line will pass through only one *y*-value for any *x*-value.

   D = {0, 1, 2, 3, 4}, R = {2, 4, 6, 8, 10}

   b) Not a function since a vertical line passes through two points at *x* = 0 and *x* = 1.

   D = {0, 1, 2}, R = {2, 4, 6, 8, 10}

   c) Not a function since a vertical line passes through two points at *x* = 1 and *x* = 3.

   D = {1, 3, 5}, R = {1, 2, 3, 4, 5}

   d) Function since a vertical line will pass through only one *y*-value for any *x*-value.

   D = {2, 4, 6, 8, 10}, R = {1}

7. a) Function since a vertical line will pass through only one *y*-value for any *x*-value.

   D = {*x* ∈ **R**}, R = {*y* ∈ **R** | *y* ≥ 1}

   b) Not a function since a vertical line passes through two points at several values of *x*.

   D = {*x* ∈ **R** | *x* ≥ 1}, R = {*y* ∈ **R**}

c) Function since a vertical line will pass through only one *y*-value for any *x*-value.

   D = {*x* ∈ **R**}, R = {*y* ∈ **R**}

d) Not a function since a vertical line passes through two points at several values of *x*.

   D = {*x* ∈ **R** | −3 ≤ *x* ≤ 3}, R = {*y* ∈ **R** | −3 ≤ *y* ≤ 3}

8. a)

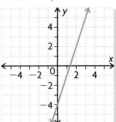

b)

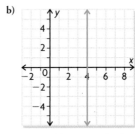

c) A vertical line cannot be the graph of a linear function because there are infinitely many *y*-values for the one *x*-value. Also, it doesn't pass the vertical-line test.

9. a) (Sonia, 8), (Jennifer, 10), (Afzal, 9), (Tyler, 8)

   b) D = {Sonia, Jennifer, Afzal, Tyler}, R = {7, 8, 9, 10}

   c) Yes, because there is only one arrow from each student to one mark.

10. a) Yes, because each mammal has only one resting pulse rate.

   b) Yes, because each mammal has only one resting pulse rate.

11. a) outdoor temperature: domain; heating bill: range

   b) time spent doing homework: domain; report card mark: range

   c) person: domain; date of birth: range

   d) number of cuts: domain; number of slices of pizza: range

12. a) This represents a function because for each size and type of tire there is only one price.

   b) This might not represent a function because more than one tire size could have the same price.

13. a) The date is the independent variable. The temperature is the dependent variable.

   b) The domain is the set of dates for which a temperature was recorded.

   c) The range is the set of outdoor temperatures in degrees Celsius.

   d) One variable is not a function of the other because during a certain date the outdoor temperature could vary over several degrees, and given any temperature, there could have been multiple days for which that temperature was recorded.

14. For example,

Definition:	Rules:
A function is a relation in which there is only one *y*-value (the dependent variable) for every *x*-value (the independent variable).	• A relation in which each element of the domain corresponds to only one element of the range. • Use a vertical line to check whether the graph of the relation is a function. If the line crosses only one point along the graph, then the relation is a function.

Function

Examples:

x	*y*
−3	2
−2	1
−1	0
0	−1
1	−2
2	−3

Non-examples:

x	*y*
3	5
2	3
1	1
3	−5
2	−3
1	−1

Student   Height (cm)

Sheila, Geoff, Meaghan → 150, 165, 158

Herb → Parsley, Thyme, Oregano

15. 0 to 66 best represents the range in the relationship. Since height is the range, the height would start at 66 m, when the rock rolls off the cliff, and would end at 0 m, when the rock hits the ground.

16. The most reasonable set of values for the domain is positive integers. The domain is the set of items sold, and thus no negative integer could appear in the domain.

Lesson 1.2, pp. 24–25

1.

Time (s)	0	0.1	0.2	0.3	0.4	0.5	0.6	0.7	0.8
Height (m)	10	9.84	9.36	8.56	7.44	6.00	4.24	2.16	0.00
First Differences		0.16	0.48	0.80	1.21	1.44	1.76	2.08	2.16
Second Differences			0.32	0.32	0.41	0.23	0.32	0.32	0.08

a) Distance as a function of time, $d(t)$, is a quadratic function of time since most of the second differences are close to or equal to 0.32, but the first distances are not close to a fixed number.

b) $D = \{t \mid 0 \le t \le 0.8, t \in \mathbf{R}\}$, $R = \{d \mid 0 \le d \le 0, d \in \mathbf{R}\}$

2. a) $f(x) = 60x$

b) The degree is 1. The function is linear.

c) $1800

d) $D = \{x \mid 0 \le x \le 60, x \in \mathbf{W}\}$, $R = \{f(x) \mid 0 \le f(x) \le 3600, f(x) \le \mathbf{W}\}$

3. a) 1; linear   c) 2; quadratic

b) 2; quadratic   d) 1; linear

4.

Time (s)	0	1	2	3	4	5
Height (m)	0	15	20	20	15	0
First Differences		15	5	0	−5	−15
Second Differences			−10	−5	−5	−10

Since the second differences are not constant, the relationship is neither linear nor quadratic.

5.

Time (h)	0	1	2	3
Bacteria Count	12	23	50	100
First Differences		11	27	50
Second Differences			16	23

Based on the data given and the differences, the data are neither linear nor quadratic.

6. a) 2; quadratic   c) 2; quadratic

b) 2; quadratic   d) 1; linear

7. a)

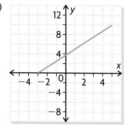

b)

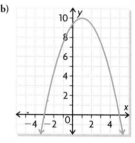

8. a) $f(t) = 29.4t - 4.9t^2$

b) The degree is 2, so the function is quadratic.

c)

Time (s)	0	1	2	3	4	5	6
Height (m)	0	24.5	39.2	44.1	39.2	24.5	0
First Differences		24.4	14.7	4.9	−4.9	−14.7	−24.5
Second Differences			−9.7	−9.8	−9.8	−9.8	−9.8

Since the second differences are constant, then the function is quadratic.

d)

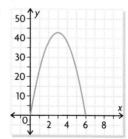

e) The ball is at its greatest height at 3 s. $f(3) = 44.1$

f) The ball is on the ground at 0 s and 6 s. $f(0) = 0$ and $f(6) = 0$

9. Table of values: A table of values to check the differences. If the first differences are constant, then it is a linear function. If the second differences are constant, then it is a quadratic function.

Graph: The graph of a linear function looks like a line, and the graph of a quadratic function looks like a parabola.

Equation: The degree of an equation will determine the type of function. If the degree is 1, then the function is linear. If the degree is 2, then the function is quadratic.

10. a) $I(x) = \$50 + \$0.05x$, x is the number of bags of peanuts sold.
b) $R(x) = \$2.50x$, x is the number of bags of peanuts sold.
c) 21

Lesson 1.3, pp. 32–35

1. The y-value is $\frac{1}{2}$ when the x-value is 3.
2. a) 1 **b)** 4 **c)** 2
3. a) $f(-2) = -7; f(0) = 5; f(2) = -7;$
    $f(2x) = -3(2x)^2 + 5 = -12x^2 + 5$
b) $f(-2) = 21; f(0) = 1; f(2) = 13;$
    $f(2x) = 4(2x)^2 - 2(2x) + 1 = 16x^2 - 4x + 1$
4. a) 72 cm; This represents the height of the stone above the river when the stone was released.
b) 41.375 cm; This represents the height of the stone above the river 2.5 s after the stone was released.
c) The stone is 27.9 cm above the river 3 s after it was released.
5. a) 1 **b)** 8 **c)** 41 **d)** -7
6. a) D = $\{-2, 0, 2, 3, 5, 6\}$, R = $\{1, 2, 3, 4, 5\}$
b) i) 4 **ii)** 2 **iii)** 5 **iv)** -2
c) They are not the same function because of order of operations.
d) $f(2) = 5$ corresponds to (2, 5). 2 is the x-coordinate of the point. $f(2)$ is the y-coordinate of the point.
7. 6; the y-value when $x = 2$ is 6
8. The point on the graph is $(-2, 6)$ because the y-value is 6 when $x = -2$.
9. a) 1; 19 **b)** $-1; -9$ **c)** 13; 23 **d)** $-4; 44$
10. a) i) 5 **iii)** 11 **v)** 3
     ii) 8 **iv)** 14 **vi)** 3
b) first differences
11. a) i) 9 **iii)** 1 **v)** 2
     ii) 4 **iv)** 0 **vi)** 2
b) second differences
12. a) 17
b) y-coordinate of the point on the graph with x-coordinate 2
c) $x = 1$ or $x = 5$
d) No, $f(3) = -1$
13. a) x represents one of the numbers. $(10 - x)$ represents the other number. $P(x)$ represents the product of the two numbers.
b) The domain is the set of all whole numbers between 0 and 10.
c)

x	0	1	2	3	4	5	6	7	8	9	10
P(x)	0	9	16	21	24	25	24	21	16	9	0

d) 5 and 5; 25; The largest product would occur when both numbers are the same.
14. a)

Fertilizer, x (kg/ha)	0	0.25	0.50	0.75	1.00
Yield, y(x) (tonnes)	0.14	0.45	0.70	0.88	0.99

Fertilizer, x (kg/ha)	1.25	1.50	1.75	2.00
Yield, y(x) (tonnes)	1.04	1.02	0.93	0.78

b) 1.25 kg/ha
c) The answer changes because the table gives only partial information about the function.

15. For example, for the function $f(x) = -x^2 + 8x - 7$, if I substitute $x = 3$ into the function, then $f(3) = -(3)^2 + 8(3) - 7 = 8$. This means that the y-value is 8 when the x-value is 3. This also corresponds to the point (3, 8) on the graph of the function $f(x)$, where 3 is the x-coordinate and 8 is the y-coordinate.

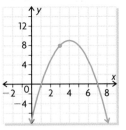

16. a) $h(0)$ represents the height of the glider when it is launched, at 0 s.
b) $h(3)$ represent the height of the glider at 3 s.
c) The glider is at its lowest point between 7 and 8 s, about 7.5 s. The vertical distance between the top of the tower and the glider at this time is 14.0625 m.
17. a)

Time (s)	0	0.5	1	1.5	2	2.5	3
Height (m)	1	14.75	26	34.75	41	44.75	46

Time (s)	3.5	4	4.5	5	5.5	6
Height (m)	44.75	41	34.75	26	14.75	1

Time (s)	6.5	7	7.5	8
Height (m)	-15.25	-34	-55.25	-79

b)

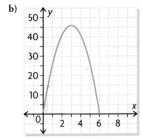

c) 3 s
d) 46 m
e) about 6.1 s

Mid-Chapter Review, p. 37

1. a) Not a function because there are two y-values for the x-value of 1.
b) Function because for every x-value there is only one y-value.
c) Not a function because there are two y-values for the x-value of 7.
d) Function because for every x-value there is only one y-value.
e) Not a function because the vertical line test isn't passed.
2. a) For f: D = $\{1, 2, 3\}$, R = $\{2, 3, 4\}$. For g: D = $\{1, 2, 3\}$, R = $\{0, 1, 2, 3, 4\}$.
b) f is a function because there is only one y-values for each x-value. g is not a function because there are two y-values for an x-value of 2, and there are two y-values for an x-value of 3.

3. a) $D = \{x \mid -3 \le x \le 3, x \in \mathbf{R}\}$, $R = \{y \mid -3 \le y \le 3, y \in \mathbf{R}\}$, not a function because it fails the vertical-line test

b) $D = \{x \mid x \le 0, x \in \mathbf{R}\}$, $R = \{y \mid y \in \mathbf{R}\}$, not a function because it fails the vertical-line test

4. a) quadratic (second differences are all 12.4)

b) between 1 and 1.5

c) between 99.2 and 155

5. a)

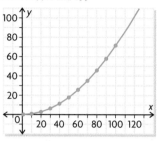

b) 120 km/h

c) A quadratic relation can model the data since the second differences are almost constant.

6. a) -9　　**b)** $4m - 5$　　**c)** $2n - 4$

7. a) 6　　**b)** $18m^2 - 9m + 1$　　**c)** 1

Lesson 1.4, p. 40

1. a) $a = 1, h = 4, k = 0$　　**e)** $a = -1, h = 0, k = 0$

b) $a = 1, h = 0, k = 5$　　**f)** $a = 2, h = 0, k = 0$

c) $a = 1, h = -2, k = 0$　　**g)** $a = -\dfrac{1}{2}, h = 0, k = 0$

d) $a = 1, h = 0, k = -3$

Lesson 1.5, pp. 47–50

1. a) (vi) because it opens down and the vertex is at $(2, -3)$

b) (iv) because it opens down and the vertex is at $(0, -4)$

c) (i) because it opens up and the vertex is at $(0, 5)$

d) (v) because it opens up and the vertex is at $(-2, 0)$

e) (iii) because it opens up and the vertex is at $(2, 0)$

f) (ii) because it opens down and the vertex is at $(-4, 2)$

2. a) $a = -3, h = 0, k = 0$; opens down, vertically stretched by a factor of 3

x	y
−3	−27
−2	−12
−1	−3
0	0
1	−3
2	−12
3	−27

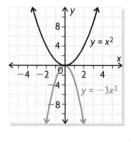

b) $a = 1, h = -3, k = -2$; opens up, move 3 units to the left and 2 units down

x	y
−3	−2
−2	−1
−1	2
0	7
1	14
2	23
3	34

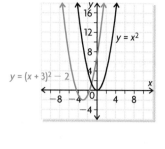

c) $a = 1, h = 1, k = 1$; opens up, move 1 unit to the right and 1 unit up

x	y
−3	17
−2	10
−1	5
0	2
1	1
2	2
3	5

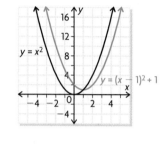

d) $a = -1, h = 0, k = -2$; opens down, move 2 units down

x	y
−3	−11
−2	−6
−1	−3
0	−2
1	−3
2	−6
3	−11

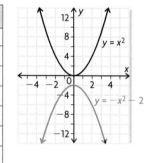

e) $a = -1, h = 2, k = 0$; opens down, move 2 units to the right

x	y
−3	−25
−2	−16
−1	−9
0	−4
1	−1
2	0
3	−1

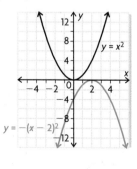

f) $a = \dfrac{1}{2}$; $h = -3$, $k = 0$; opens up, move 3 units to the left, and vertically compress by a factor of 2

x	y
−3	0
−2	0.5
−1	2
0	4.5
1	8
2	12.5
3	18

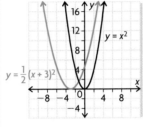

3. a) (iv) **b)** (i) **c)** (iii) **d)** (ii)
4. a) Move 5 units to the right and 3 units up.

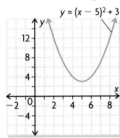

b) Move 1 unit to the left and 2 units down.

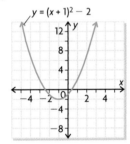

c) Vertically stretch by a factor of 3 and move 4 units down.

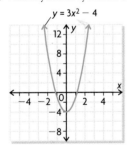

d) Reflect graph in x-axis, vertically compress by a factor of 3, and move 4 units to the left.

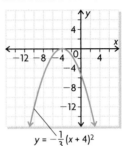

5. a) $y = 5x^2$
  b) $y = \dfrac{1}{2}x^2$
  c) $y = -(x-2)^2$
  d) $-\dfrac{1}{3}x^2 + 2$

6. a) $y = (x-2)^2 - 6$
  b) $y = (x+2)^2 - 4$
  c) $y = x^2 - 1$
  d) $y = (x-5)^2 - 5$

7. a) $y = x^2 + 4$
  b) $y = (x-5)^2$
  c) $y = -(x-5)^2$
  d) $y = 2(x-2)^2$
  e) $y = -0.5(x+2)^2$
  f) $y = -0.5(x-1)^2$

8. a) i) The shape of the graph $f(x) = -(x-2)^2$ is the same shape as the graph of $f(x) = x^2$.
  ii) The vertex is at $(2, 0)$ and the axis of symmetry is $x = 2$.
  iii), iv)

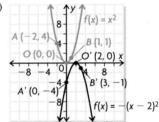

b) i) The shape of the graph of $f(x) = \dfrac{1}{2}x^2 + 2$ is the same as the graph of $f(x) = x^2$ compressed vertically by a factor of 2.
  ii) The vertex is at $(0, 2)$ and the axis of symmetry is $x = 0$.
  iii), iv)

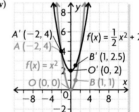

c) i) The graph of $f(x) = (x + 2)^2 - 2$ is the same shape as the graph of $f(x) = x^2$.
   ii) The vertex is at $(-2, -2)$ and the axis of symmetry is $x = -2$.
   iii), iv) $f(x) = (x + 2)^2 - 2$

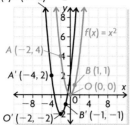

9. a) $y = (x - 2)^2$
   b) $y = (x + 4)^2$
   c) $y = (x + 4)^2 - 5$
   d) $y = \frac{1}{4}x^2$
   e) $y = 2(x + 4)^2$
   f) $y = 3(x - 2)^2 - 1$

10. a) Either $a > 0$ and $k < 0$ or $a < 0$ and $k > 0$
    b) $a \in \mathbf{R}$ and $k = 0$
    c) Either $a > 0$ and $k > 0$ or $a < 0$ and $k > 0$

11. a) If the graph of the function for Earth, $h(t) = -4.9t^2 + 100$, is the base graph, then the graph for Mars is wider, the graph for Saturn is slightly narrower, and the graph for Neptune is slightly wider.
    b) Neptune
    c) Mars

12. a) The x-coordinates are decreased by 7 and the y-coordinates are unchanged.
    b) The x-coordinates are unchanged and the y-coordinates are increased by 7.
    c) The x-coordinates are increased by 4 and the y-coordinates are multiplied by -2.
    d) The x-coordinates are unchanged and the y-coordinates are multiplied by $-\frac{1}{2}$ and decreased by 4.

13. a) The graphs get narrower and narrower.
    b) The graphs get wider and wider.

14. a) $y = -2x^2, y = -2(x - 2)^2, y = -2(x - 4)^2, y = -2(x + 2)^2, y = -2(x + 4)^2$
    b) Answers may vary. For example, $y = 0.5x^2, y = 0.5(x - 6)^2, y = 0.5(x - 3)^2, y = 0.5(x + 3)^2, y = 0.5(x + 6)^2$.

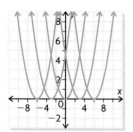

Lesson 1.6, pp. 56–58

1. a) horizontal translation 2 units to the left and vertical stretch by a factor of 3
   b) horizontal translation 3 units to the right, reflection in x-axis, vertical stretch by a factor of 2, and vertical translation 1 unit up
   c) vertical compression by a factor of 3 and vertical translation 3 units down

d) horizontal translation 2 units to the left, reflection in x-axis, vertical compression by a factor of 2, and vertical translation 4 units up

2. a)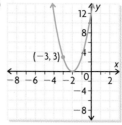

   $f(-3) = 3(-3 + 2)^2 = 3(-1)^2 = 3$

   b)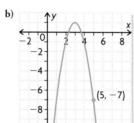

   $f(5) = -2(5 - 3)^2 + 1 = -2(2)^2 + 1 = -7$

   c)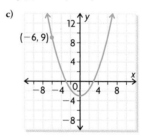

   $f(-6) = \frac{1}{3}(-6)^2 - 3 = 12 - 3 = 9$

   d)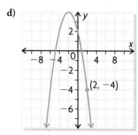

   $f(2) = -\frac{1}{2}(2 + 2)^2 + 4 = -8 + 4 = -4$

3. a) (iv)   b) (i)   c) (iii)   d) (ii)

4. a) vertical compression by a factor of 2 and a vertical translation 2 units down; $y = \frac{1}{2}x^2 - 2$
   b) horizontal translation 4 units to the right and reflection in x-axis; $y = -(x - 4)^2$
   c) reflection in x-axis, vertical stretch by a factor of 2, and a vertical translation 3 units down; $y = -2x^2 - 3$
   d) horizontal translation 4 units to the left, reflection in x-axis, and a vertical translation 2 units down; $y = -(x + 4)^2 - 2$

5. a) $y = 2(x - 4)^2 - 1$ **d)** $y = \frac{1}{2}(x - 2)^2 - 2$

b) $y = -\frac{1}{3}(x + 2)^2 + 3$ **e)** $y = -2(x + 1)^2 + 4$

c) $y = -\frac{1}{2}(x + 3)^2 + 2$ **f)** $y = (x - 4)^2 + 5$

6. a) $y = 5(x - 2)^2 - 4$ **d)** $y = (x - 2)^2 - 1$

b) $y = \frac{1}{2}(x - 2)^2 - 4$ **e)** $y = -(x - 4)^2 - 8$

c) $y = x^2 - 4$

7. a) horizontal translation 5 units to the right, vertical stretch by a factor of 4, vertical reflection in the x-axis, and vertical translation 3 units up

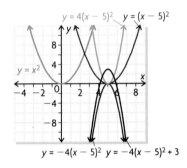

$y = -4(x - 5)^2$ $y = -4(x - 5)^2 + 3$

b) horizontal translation 1 unit to the left, vertical stretch by a factor of 2, and vertical translation 8 units down

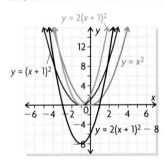

c) horizontal translation 2 units to the left, vertical compression by a factor of $\frac{3}{2}$, and vertical translation 1 unit up

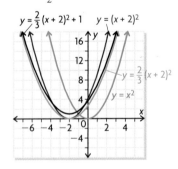

d) horizontal translation 1 unit to the right, vertical compression by a factor of 2, vertical reflection in x-axis, and vertical translation 5 units down

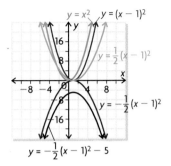

$y = -\frac{1}{2}(x - 1)^2 - 5$

e) horizontal translation 3 units to the right, vertical reflection in the x-axis, and vertical translation 2 units up

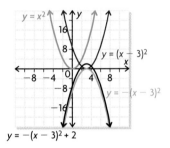

$y = -(x - 3)^2 + 2$

f) horizontal translation 1 unit to the left, vertical stretch by a factor of 2, and vertical translation 4 units up

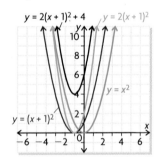

8.

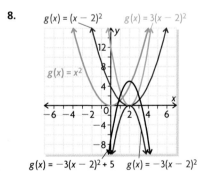

$g(x) = -3(x - 2)^2 + 5$ $g(x) = -3(x - 2)^2$

9. a) $y = x^2 + 2$, $y = 2x^2 + 2$, $y = \frac{1}{2}x^2 + 2$, $y = -x^2 + 2$,

$y = -2x^2 + 2$, $y = -\frac{1}{2}x^2 + 2$

b) $y = x^2 + 6$, $y = x^2 + 4$, $y = x^2 + 2$, $y = x^2$, $y = x^2 - 2$,
$y = x^2 - 4$, $y = x^2 - 6$

10. a) horizontal translation 6 units to the left, vertical compression by a factor of 4, vertical reflection in the x-axis, and vertical translation 2 units up

b) $y = -\frac{1}{4}(x + 6)^2 + 2$

11. a) horizontal translation 7 units to the left, vertical stretch by a factor of 2, and vertical translation 3 units down

b) vertical stretch by a factor of 2 and vertical translation 7 units up

c) horizontal translation 4 units to the right, vertical stretch by a factor of 3, vertical reflection in the x-axis, and vertical translation 2 units up

d) vertical stretch by a factor of 3, vertical reflection in the x-axis, and vertcial translation 4 units down

12. a) constant: a, k; changed: h

b) constant: a, h; changed: k

c) constant: h, k; changed: a

Lesson 1.7, pp. 63–65

1. a) Domain is all real numbers and range is all real numbers.

b) horizontal lines: domain is all real numbers, range is the y-value of the horizontal line

vertical lines: domain is the x-value of the vertical line, range is all real numbers

2. a) $D = \{x \in \mathbf{R}\}$, $R = \{y \in \mathbf{R} \mid y \le 5\}$; $f(x)$ is a quadratic function that opens down and the vertex is at $(-3, 5)$, which is a maximum.

b) $D = \{x \in \mathbf{R}\}$, $R = \{y \in \mathbf{R} \mid y \ge 5\}$; $f(x)$ is a quadratic function that opens up and the vertex is at $(-1, 5)$, which is a minimum.
$f(x) = 2x^2 + 4x + 7$

c) $D = \{x \in \mathbf{R}\}$, $R = \{y \in \mathbf{R}\}$; $f(x)$ is a linear function with nonzero slope.

d) $D = \{x = 5\}$, $R = \{y \in \mathbf{R}\}$; $x = 5$ is a vertical line.

3. $D = \{t \in \mathbf{R} \mid 0 \le t \le 20\}$, $R = \{f(t) \in \mathbf{R} \mid 0 \le f(t) \le 500\}$; Since t represents time, t cannot be negative or greater than 20 (after $t = 20$ the height is negative). Since $f(t)$ represents the height, the height is positive between 0 and 500.

4. a) $D = \{x \in \mathbf{R}\}$, $R = \{y \in \mathbf{R} \mid y \le 2\}$

b) $D = \{x \in \mathbf{R}\}$, $R = \{y \in \mathbf{R}\}$

c) $D = \{x \in \mathbf{R}\}$, $R = \{y = 8\}$

d) $D = \{x = 4\}$, $R = \{y \in \mathbf{R}\}$

e) $D = \{x \in \mathbf{R}\}$, $R = \{y \in \mathbf{R} \mid y \ge -4\}$

f) $D = \{x \in \mathbf{R}\}$, $R = \{y \in \mathbf{R}\}$

5. $D = \{x \in \mathbf{R} \mid 1 \le x \le 11\}$, $R = \{y \in \mathbf{R} \mid 12 \le y \le 36\}$

6. a)

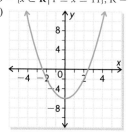

$D = \{x \in \mathbf{R}\}$, $R = \{y \in \mathbf{R} \mid y \ge -6\}$

b)

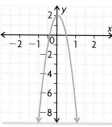

$D = \{x \in \mathbf{R}\}$, $R = \left\{ y \in \mathbf{R} \mid y \le \frac{81}{40} \right\}$

c)

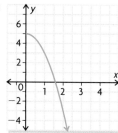

$D = \{x \in \mathbf{R} \mid x \ge 0\}$, $R = \{y \in \mathbf{R} \mid y \le 5\}$

d)

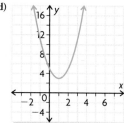

$D = \{x \in \mathbf{R}\}$, $R = \{y \in \mathbf{R} \mid y \ge 3\}$

e)

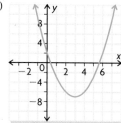

$D = \{x \in \mathbf{R}\}$, $R = \{y \in \mathbf{R} \mid y \ge -7\}$

f)

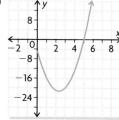

$D = \{x \in \mathbf{R} \mid x \ge 0\}$, $R = \{y \in \mathbf{R} \mid y \ge -21.33\}$

Answers

7. a)

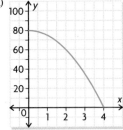

b) Since t represents time, $t \geq 0$. Since the height of the pebble is negative after $t = 4$, $t \leq 4$.

c) The bridge is 80 m high since the maximum height of 80 m occurs when $t = 0$ s, when the pebble is dropped.

d) It takes 4 s for the pebble to hit the water since the height at $t = 4$ s is 0 m.

e) $D = \{t \in \mathbf{R} \mid 0 \leq t \leq 4\}$, $R = \{h(t) \in \mathbf{R} \mid 0 \leq h(t) \leq 80\}$

8. a)

Number of People, n	Cost, $C(n)$
0	$550
1	$568
2	$586
3	$604
4	$622
5	$640

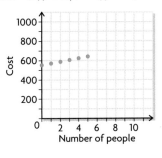

b) $C(n) = 550 + 18n$

c) $D = \{n \in \mathbf{R} \mid n \geq 0\}$, $R = \{C(n) \in \mathbf{R} \mid C(n) \geq 550\}$

9. a) $d(t) = -10t$

b) 5 h

c) $D = \{t \in \mathbf{R} \mid t \geq 0\}$, $R = \{d(t) \in \mathbf{R} \mid 3000 \geq d(t) > 0\}$

10. a)

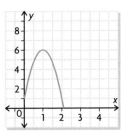

b) t represents time, which is never negative and is between 0 and 2.095 s, when the ball hits the ground.

c) 6 m

d) 1 s

e) 2.095 s

f) never

g) $D = \{t \in \mathbf{R} \mid 0 \leq t \leq 2.095\}$, $R = \{h(t) \in \mathbf{R} \mid 0 \leq h(t) \leq 6\}$

11. a)

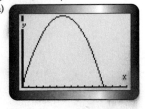

b) $0.58

c) $D = \{x \in \mathbf{R} \mid 0 \leq x \leq 1.17\}$, $R = \{R(x) \in \mathbf{R} \mid 0 \leq R(x) \leq 102.08\}$

12. a) The values of the domain and range must make sense for the situation.

b) Restrictions are necessary in order for the situation to make sense. For example, it doesn't make sense to have a negative value for time or a negative value for height in most situations.

13. a) $D = \{r \in \mathbf{R} \mid r \geq 0\}$

b) $R = \{A(r) \in \mathbf{R} \mid A(r) \geq 0\}$

Chapter Review, pp. 68–69

1. a) $D - \{a \in \mathbf{R} \mid 1 \leq a \leq 12\}$

b) $R = \{m \in \mathbf{R} \mid 11.5 \leq m \leq 49.5\}$

c) function

2. quadratic function

3. a) 1; linear

b) 2; quadratic

c) 3; neither linear nor quadratic

4. a) $f(-1) = 7$

b) $f(3) = 19$

c) $f(0.5) = 0.25$

5. a) 5   **c)** 31

b) 4   **d)** 4

6. a) vertex at $(0, -7)$, $x = 0$, opens up

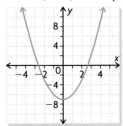

b) vertex at $(-1, 10)$, $x = -1$, opens down

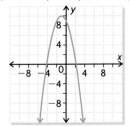

c) vertex at $(-2, -3)$, $x = -2$, opens down

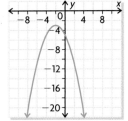

d) vertex at $(5, 0)$, $x = 5$, opens up

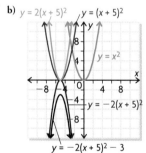

7. a) vertical translation 7 units down
b) horizontal translation 1 unit to the left, vertical reflection in x-axis, and vertical translation 10 units up
c) horizontal translation 2 units to the left, vertical compression by a factor of 2, vertical reflection in x-axis, and vertical translation 3 units down
d) horizontal translation 5 units to the right and vertical stretch by a factor of 2

8. a) i) vertical stretch by a factor of 5 and vertical translation 4 units down
ii) horizontal translation 5 units to the right and vertical compression by a factor of 4
iii) horizontal translation 5 units to the left, vertical stretch by a factor of 3, vertical reflection in the x-axis, and vertical translation 7 units down
b) i) $D = \{x \in \mathbf{R}\}$, $R = \{y \in \mathbf{R} \mid y \geq -4\}$
ii) $D = \{x \in \mathbf{R}\}$, $R = \{y \in \mathbf{R} \mid y \geq 0\}$
iii) $D = \{x \in \mathbf{R}\}$, $R = \{y \in \mathbf{R} \mid y \geq -7\}$

9. a) horizontal translation 5 units to the left, vertical stretch by a factor of 2, vertical reflection in the x-axis, and vertical translation 3 units down.
b)

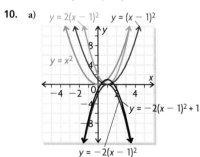

10. a)

b) $y = -2(x - 1)^2 + 1$
11. a) $y = 2x^2 - 8$
b) $(0, -4)$
c) The graphs are different because the operations of multiplying by 2 and subtracting 4 are done in different orders for (a) and (b).
d) vertical stretch by a factor of 2, vertical translation 4 units down

12. a) 1.34 m **d)** no
b) 9 m **e)** 2.6052 s
c) 1.343 75 m
13. a) 57 m
b) 6.4 s
c) $D = \{t \in \mathbf{R} \mid 0 \leq t \leq 6.4\}$, $R = \{h(t) \in \mathbf{R} \mid 0 \leq h(t) \leq 57\}$
14. 11.75 s

Chapter Self-Test, p. 70

1. a) $D = \{1, 3, 4, 7\}$, $R = \{1, 2\}$; function since there is only one y-value for each x-value
b) $D = \{1, 3, 4, 6\}$, $R = \{1, 2, 5\}$; function since there is only one value in $g(x)$ for each value in x
c) $D = \{1, 2, 3\}$, $R = \{2, 3, 4, 5\}$; not a function since there are two y-values for $x = 1$ and it fails the vertical-line test
2. A function is a relation in which there is only one y-value for each x-value. For example, $y = x^2$ is a function, but $y = \sqrt{x}$ is not a function.

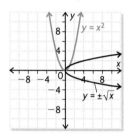

3.

Time (s)	0	1	2	3	4	5
Height (m)	0	30	40	40	30	0
First Differences		30	10	0	-10	-30
Second Differences			-20	-10	-10	-20

Since the first differences and the second differences are not constant, the data does not represent a linear or a quadratic function.
4. a) $f(2) = 3(2)^2 - 2(2) + 6 = 14$
b) $f(x - 1) = 3(x - 1)^2 - 2(x - 1) + 6 = 3x^2 - 8x + 11$
5. a) 28
b) the y-coordinate when the x-coordinate is 1
c) $D = \{x \in \mathbf{R}\}$, $R = \{y \in \mathbf{R} \mid y \geq 1\}$
d) It passes the vertical line test.
e) It's a quadratic that opens up.
6. a) horizontal translation 3 units to the right, vertical stretch by a factor of 5, and vertical translation 1 unit up
b) minimum value is 1 when $x = 3$; there is no maximum value.
c) $D = \{x \in \mathbf{R}\}$, $R = \{y \in \mathbf{R} \mid y \geq 1\}$
d)

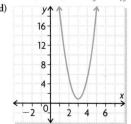

7. a)

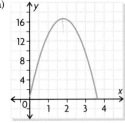

b) t represents time, which can never be negative.
c) 16.7 m
d) 1.8 s
e) 3.6 s
f) $D = \{t \in \mathbf{R} \mid 0 \le t \le 3.6\}$, $R = \{h(t) \in \mathbf{R} \mid 0 \le h(t) \le 16.7\}$

Chapter 2

Getting Started, pp. 74–76

1. a) (vii)   **c)** (viii)   **e)** (i)   **g)** (ii)
    b) (iv)   **d)** (vi)   **f)** (iii)   **h)** (v)
2. a) $-6x + y$ **c)** $3x - 3y + 3$
    b) xy **d)** $-9a - 7b - 7ab$
3. a) x^5 **b)** $10x^3$ **c)** x^2 **d)** $2x^2$
4. a) $(x)(x) = x^2$ **b)** $(2x)(4x) = 8x^2$
5. a) $9x - 24$ **d)** $-5d^3 - 11d^2 - 6d + 36$
    b) $-32x^2 + 8x - 4$ **e)** $6x^3 + 10x^2$
    c) $14x^2 - 10x + 8$ **f)** $-5x^4 + 15x^3 - 20x^2$
6. a) $2(x - 5)$ **c)** $5(5x^2 + 4x - 20)$
    b) $6(x^2 + 4x + 5)$ **d)** $x^3(7x + 12 - 9x^2)$
7. a) x^5 **b)** $-30x^7$ **c)** $15x^7$ **d)** $2x^3$
8. a) (i) and (v)   **c)** (iii)
    b) (ii), (iv), and (vi) **d)** (iii) and (iv)
9. a) 8   **c)** 18   **e)** $3x + 2$
    b) 8   **d)** x **f)** $5x$
10. a) $2 \times 3 \times 13$ **b)** 7×3^2 **c)** $5^2 \times 11^2$ **d)** 41
11. a)

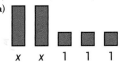

b)

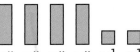

c)

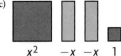

d)

e)

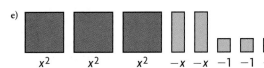

f)

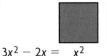

12. a) (iii)   **b)** (ii)   **c)** (i)
13. a)

$3x^2 - 2x =$

b)

$2x + 4 =$

c)

$-2x^2 - x =$

d)

$x^2 + 3x =$

14. a) Agree: Area = length $\times$ width, in which length and width are factors of the product Area.
    b) Disagree: factors are integers; Agree: factors are terms that multiply together to make the product.
    c) Disagree: -2 and $-x - 3$ are other factors of $2x + 6$.

Lesson 2.1, pp. 85–87

1. a) $A = (2x + 1)(x + 3) = 2x^2 + 7x + 3$
    b) $A = (2x + 3)(3x - 2) = 6x^2 + 5x - 6$
2. a) $x^2 + 4x - 21$ **c)** $4x^2 - 20x + 25$
    b) $a^2 + 12a + 36$ **d)** $m^2 - 81$
3. a) $3x^2 - 3x - 90$ **c)** $-3n^2 + 2n + 4$
    b) $a^2 - 2a + 7$ **d)** $-6x^2 + 24x + 1$
4. a) $(x + 1)(2x + 2) = 2x^2 + 4x + 2$
    b) $(2x - 1)(x + 3) = 2x^2 + 5x - 3$
5. a) $12x^2 + 7x - 10$ **d)** $4a^2 + 29a + 32$
    b) $45x^2 + 60x + 20$ **e)** $-20n^2 - 2n$
    c) $-14x^2 - 12x + 19$ **f)** $4x^2 + 75$
6. a) 11   **c)** $-2x^2 - x + 6$
    b) -2 **d)** -22
    e) $-2x^2 - x + 6$ evaluated for $x = -4$ is -22.
    It was shown in part (c) that the factors of $-2x^2 - x + 6$ are $(3 - 2x)$ and $(x + 2)$. Parts (a) and (b) showed that $(3 - 2x)$ and $(x + 2)$ evaluated for $x = 4$ are 11 and -2, respectively, the

product of which is -22. So, parts (a)–(b) show that if you evaluate two expressions for a specific number, x, and multiply the results of the evaluation together, it is equal to the product of those two expressions evaluated for that same number, x.

7. a) $2x^2 - 10x$ **c)** $18x^2 + 27x - 35$
b) $a^2 - 2a - 63$ **d)** $11m^2 - 15m - 12$

8. The highest exponent comes from $2x$ times $3x$, or $6x^2$.

9. Answers may vary. E.g., 4 by $x^2 + 2x$ or $4x$ by $x + 2$

$x^2 + 2x$ $x + 2$

4           $4x$

10. a) $(2x + 3)(2x - 4) = 4x^2 - 2x - 12$
b) $\frac{1}{2}(2x - 1)(4x + 2) = 4x^2 - 1$

11. a) πx^2 **b)** $\pi(x + 5)^2$ **c)** $10\pi x + 25\pi$

12. a) Answers may vary. E.g.,
$(6x^2 - 8x) + (-15x^2 - 18x) = -9x^2 - 26x$
b) Answers may vary. E.g., $(6x^2 - 8x) + (-15x^2 + 8x) = -9x^2$

13. a) Answers may vary. E.g., $(2x + 3)(2x - 4) = 4x^2 - 2x - 12$
b) Answers may vary. E.g., $(2x + 3)(2x - 3) = 4x^2 - 9$

14. a) $6x^2 - xy - y^2$ **c)** $25m^2 - 49n^2$
b) $9a^2 - 30ab + 25b^2$ **d)** $-4x^2 - 10xy + 6y^2$

15. a) $(n + 1)^2 - n^2 = n^2 + 2n + 1 - n^2$
b) 12
c) 5, 12, 13
d) 7, 24, 25; 9, 40, 41; 11, 60, 61

Lesson 2.2, pp. 92–94

1. a) $6x^2 + 4x; 2x$ **b)** $16x - 12x^2; 4x$

2. a) 3   **b)** x

3. a) $2(2x^2 - 3x + 1)$ **c)** $(a + 7)(5a + 2)$
b) $5x(x - 4)$ **d)** $(3m - 2)(4m - 1)$

4. a) $-6x^2; 6x$ **b)** $4x^2 + 16x; 4x$

5. a) 6   **b)** $2x$ **c)** 2   **d)** -5

6. a) $9x(3x - 1)$ **d)** $-2(a^2 + 2a - 3)$
b) $-4m(2m - 5)$ **e)** $(x + 7)(3x - 2)$
c) $5(2x^2 - x + 5)$ **f)** $(3x - 2)(2x + 1)$

7. a) $2x - 1$ **b)** $m^2 - 2m + 2$

8. $2\pi r(r + 10)$

9. $2a(3a - 2) + 7(2 - 3a) = 2a(3a - 2) + 7(-1)(3a - 2)$
$= 2a(3a - 2) - 7(3a - 2)$
$= (3a - 2)(2a - 7)$

10. a) Answers may vary. E.g., $6x^2 + 6x = 6x(x + 1); 12x^2 + 18x$
$= 6x(2x + 3); 6x^2 - 30x = 6x(x - 5)$
b) Answers may vary. E.g., $7x^2 + 7x + 7 = 7(x^2 + x + 1);$
$7x^2 + 14x + 70 = 7(x^2 + 2x + 10);$
$14x^2 + 7x + 7 = 7(2x^2 + x + 1)$

11. Both common factors divide evenly into the quadratic, but the first term is usually positive after factoring.

12. k must be a number that has a common factor with 6 and 12 but not with 6 and 4. So k can be any odd multiple of 3, such as $\pm 3, \pm 9, \pm 15, \ldots$

13. a) $n + 2$ **c)** $2n^2 + 4n + 4 = 2(n^2 + 2n + 2)$
b) $n^2 + (n + 2)^2$

14. If a polynomial has two terms and each term has a common factor, then this can be divded out of both terms. For example, $6x^2 + 8x = 2x(3x + 4)$.

15. a) $5xy(x - 2y)$ **c)** $(x + y)(3x - y)$
b) $5a^2b(2b^2 - 3b + 4)$ **d)** $(x - 2)(5y + 7)$

16. a) $(3a + b)(3x + 2)$ **c)** $(x + y)(x + y + 1)$
b) $(2x - 1)(5x - 3y)$ **d)** $(1 + x)(1 + y)$

Lesson 2.3, pp. 99–100

1. a) $x^2 + x - 6 = (x + 3)(x - 2)$
b) $x^2 - x - 12 = (x - 4)(x + 3)$

2. a) $(x + 5)$ **c)** $(x + 8)$
b) $(x + 4)$ **d)** $(x + 2)$

3. a) $(x + 5)(x + 4)$ **c)** $(m + 3)(m - 2)$
b) $(a - 5)(a - 6)$ **d)** $2(n + 5)(n - 7)$

4. a) $x^2 + 4x + 3 = (x + 3)(x + 1)$
b) $x^2 - 3x - 10 = (x - 5)(x + 2)$

5. a) $(x + 6)$ **b)** $(x - 7)$ **c)** $(x + 5)$ **d)** $(x - 3)$

6. a) $(x - 5)(x - 2)$ **d)** $(w + 3)(w - 6)$
b) $(y - 5)(y + 11)$ **e)** $(x - 11)(x - 3)$
c) $(x - 5)(x + 2)$ **f)** $(n - 10)(n + 9)$

7. Answers may vary. E.g., $x^2 + 6x + 5; x^2 + 7x + 10;$
$2x^2 + 16x + 30$

8. $(2 - x) = -(x - 2)$ and $(5 - x) = -(x - 5)$. When multiplied, the two minus signs become a positive sign.

9. a) $(a + 5)(a + 2)$ **d)** $(x - 4)(x + 15)$
b) $-3(x + 3)(x + 6)$ **e)** $(x + 5)(x - 2)$
c) $(z - 5)^2$ **f)** $(y + 6)(y + 7)$

10. $f(n) = (n - 3)(n + 1)$ when factored. If $n = 4$, the first factor becomes 1. Therefore, prime. Other numbers larger than 4 always produce two factors and are not prime.

11. We are looking for numbers that add to 0 and multiply to -16 to be able to factor.

12. a) Answers may vary. E.g., $b = 3$ and $c = 2$.
b) Answers may vary. E.g., $b = 5$ and $c = 6$.
c) Answers may vary. E.g., $b = 5$ and $c = 9$.

13. a) $k = 4$ or 5. These are the possible sums of the factors of 4.
$2 + 2 = 4$ and $1 + 4 = 5$
b) $k = 4, 3,$ or 0. These are the possible products of the addends of 4.
$2 \times 2 = 4, 1 \times 3 = 3,$ and $0 \times 4 = 0$
c) $k = 4$ or 0. These are the possible numbers whose factors when added together give the same value. $2 \times 2 = 4$ and $2 + 2 = 4.$
$0 \times 0 = 0$ and $0 + 0 = 0$

14. a) $(x + 5y)(x - 2y)$
b) $(a + 3b)(a + b)$
c) $-5(m - n)(m - 2n)$
d) $(x + y - 2)(x + y - 3)$

15. $\left(1 + \frac{3}{x}\right)(x + 4)$ or $\left(1 + \frac{4}{x}\right)(x + 3)$

Mid-Chapter Review, p. 103

1. a) $2x^2 - 18x + 15$ **c)** $-5x^2 - 15x$
b) $18n^2 + 8$ **d)** $-18a^2 + 2b^2$

2. $(2x + 1)$ and $(3x - 2); 6x^2 - x - 2$

3. $6x^2 + 5x - 4$

4. $5x$

5. a) $-4x(2x - 1)$ **c)** $5(m^2 - 2m - 1)$
b) $3(x^2 - 2x + 3)$ **d)** $(2x - 1)(3x + 5)$

6. $-8x^2 + 4x; 4x$

7. a) $6x^2 + 24x + 24$ **c)** 6 is the common factor and $6 = 2 \times 3$.
b) yes

8. $x^2 + 3x - 10; (x - 2)(x + 5)$

9. a) $(x - 3)(x + 5)$ **c)** $(x - 7)(x - 5)$
b) $(n - 2)(n - 6)$ **d)** $2(a - 4)(a + 3)$

10. 4 does not divide evenly into 6.

11. Yes, because the positive factors of c that add to b can both be made negative so that they add to $-b$.

Lesson 2.4, pp. 109–110

1. a) $8x^2 + 14x + 3$; $(2x + 3)(4x + 1)$
   b) $3x^2 + x - 2$; $(3x - 2)(x + 1)$

2. a) $(2a - 1)$ **c)** $(x + 2)$
   b) $(5x - 2)$ **d)** $4(n - 3)$

3. $3x^2 + 16x + 5$; $(x + 5)(3x + 1)$

4. a) $(x - 4)(2x + 1)$ **d)** $2(x + 4)(x + 1)$
   b) $3(x + 1)(x + 5)$ **e)** $3(x - 3)(x + 7)$
   c) $(x + 3)(5x + 2)$ **f)** $(x - 7)(2x - 1)$

5. a) $(2x + 1)(4x + 3)$ **d)** $(3x - 2)(5x + 2)$
   b) $3(m - 1)(2m + 1)$ **e)** $2(n + 5)(3n - 2)$
   c) $(a - 4)(2a - 3)$ **f)** $2(4x + 3)(2x - 1)$

6. Answers may vary. E.g., $4x^2 - 8x - 5$; $4x^2 - 4x - 15$; $4x^2 - 25$

7. a) $k = 4, 3$ **c)** $k = 3, 6$
   b) $k = 18, 6, -6, -18, -39$

8. Yes. We want the product to be c and the sum to be b.

9. a) $(2x - 3)(3x + 4)$ **d)** $3n(n - 4)(4n - 9)$
   b) $(k + 5)(8k + 3)$ **e)** $3(k - 4)(k + 2)$
   c) $5(2r - 7)(3r + 2)$ **f)** $(4y - 5)(6y + 5)$

10. a) $(x + 3)(x + 2)$ **d)** $(a - 4)(a + 3)$
   b) $(x - 6)(x + 6)$ **e)** $4(x + 6)(x - 2)$
   c) $(5a + 2)(a - 3)$ **f)** $(2x - 3)(3x + 1)$

11. Yes. $n = 16, 24, 41, 49, \ldots$. The factors are $2(n + 1)(3n + 2)$. If either factor is a multiple of 25, the expression is a multiple of 50.

12. Once you have found the greatest common factor you can ask yourself what must you multiply the common factor by to get each term of the original polynomial. This helps you find the other factor. For $-4x^2 + 38x - 48$, the greatest common factor is -2. So the factors are $-2(2x^2 - 19x + 24)$. Now try to find two binomials that multiply to give the trinomial in the brackets, $-4x^2 + 38x - 48 = -2(2x - 3)(x - 8)$

13. a) $(2x + 3y)(3x + y)$ **c)** $(2x - 3y)(4x - y)$
   b) $(a - 2b)(5a + 3b)$ **d)** $4(a + 5)(3a - 2)$

14. No. If a and c are odd, their product is odd. So we look for a pair of odd numbers that multiply to ac and add to b. But two odd numbers add to an even number, so b is not odd.

Lesson 2.5, pp. 115–116

1. a) $4x^2 - 1$; $(2x + 1)(2x - 1)$
   b) $9x^2 + 6x + 1$; $(3x + 1)(3x + 1)$

2. a) $(x - 5)$ **c)** $(5a + 6)$ **e)** $(2m - 3)$
   b) $(n + 4)$ **d)** 7   **f)** $(3x + 1)$

3. a) $(x + 6)(x - 6)$ **c)** $(x + 8)(x - 8)$ **e)** $(x + 10)(x - 10)$
   b) $(x + 5)^2$ **d)** $(x - 12)^2$ **f)** $(x + 2)^2$

4. a) $(7a + 3)^2$ **c)** $-2(2x - 3)^2$ **e)** $4(2 - 3x)(2 + 3x)$
   b) $(x + 11)(x - 11)$ **d)** $20(a + 3)(a - 3)$ **f)** $(x + 3)^2$

5. a) 400   **b)** 580

6. a) 5 is not a perfect square.
   b) $b(-b)$ is not $+9$ and 5 is not a perfect square.

7. $(x - 3)(x - 2)(x + 2)(x + 3)$

8. 7, 5; $-7, -5$; $-7, 5$; 7, -5; 5, 1; $-5, -1$; 5, -1; $-5, 1$

9. $x^2 - (x - 2)^2 = 4(x - 1)$; average of x and $x - 2$ is $x - 1$

---

10. A perfect-square trinomial has two of the three terms perfect squares and the non-square term is 2 times the product of the square roots of the two square terms. A difference of squares polynomial has two perfect square terms separated by a minus sign.

11. a) $(10x + 3y)(10x - 3y)$ **c)** $(2x - y + 3)(2x - y - 3)$
   b) $(2x + y)^2$ **d)** $10(3x - 2y)^2$

12. a) $(2x - 5y - 2z)(2x - 5y + 2z)$
   b) $-(x - 16)(x + 2)$

Chapter Review, pp. 120–121

1. a) $7x^2 + 11x - 9$ **c)** $-54x^2 + 63x - 10$
   b) $-144a^2 + 225$ **d)** $-20n^2 + 100n - 125$

2. $(4x - 1)(x + 4)$; $4x^2 + 15x - 4$

3. a) $(x - 2)$ **c)** $(b^2 - 2)$
   b) 3   **d)** $4x + 5$

4. a) $5x(2x - 1)$ **c)** $-2(x - 1)(x + 4)$
   b) $12(n^2 - 2n + 4)$ **d)** $(3a + 2)(5 - 7a)$

5. a) 2 by $3x^2 - 4$ **b)** No. $3x^2 - 4$ does not factor.

6. a) Answers may vary. E.g., $7x^2 - 7x$; $14x^2 - 21x$; $7x^3 + 21x^2 - 35x$
   b) $7x(x - 1)$; $7x(2x - 3)$; $7x(x^2 + 3x - 5)$

7. a) $6x - 2x^2 = 2x(3 - x)$
   b) $x^2 + 2x - 15 = (x - 3)(x + 5)$

8. a) $(x + 7)$ **c)** $(b + 4)$
   b) $(a - 4)$ **d)** $(x - 5)$

9. a) $(x + 5)(x + 2)$ **c)** $(x + 7)(x - 6)$
   b) $(x - 3)(x - 9)$ **d)** $(x - 10)(x + 9)$

10. Answers may vary. E.g.,
   $x^2 + 5x + 6 = (x + 2)(x + 3)$; $x^2 + 3x + 2 = (x + 1)(x + 2)$

11. a) $3x - 2x$ **c)** $6x - 56x$
   b) $24x - 15x$ **d)** $6x - 15x$

12. a) $(2x + 5)$ **c)** $(2b + 3)$
   b) $(a + 4)$ **d)** $(3x + 4)$

13. a) $(2x - 5)(3x - 2)$ **c)** $(4x - 3)(5x + 6)$
   b) $(2a - 3)(5a + 2)$ **d)** $(n + 1)(6n + 7)$

14. a) $9x^2 + 12x + 4$; $(3x + 2)(3x + 2)$
   b) $9x^2 - 4$; $(3x - 2)(3x + 2)$

15. a) $(x - 5)$ **c)** $(2b - 5)$
   b) $(3a + 1)$ **d)** $(3x - 8)$

16. a) $(2x + 3)(2x - 3)$ **c)** $(x - 2)(x + 2)(x^2 + 4)(x^4 + 16)$
   b) $(4a - 3)^2$ **d)** $(x + 1)^2$

17. $x^2 + 0x + 1$ cannot be factored as no numbers that multiply to $+1$ add to 0.

18. a) $(x + 5)(x - 3)$ **d)** $3(6x - 1)(x + 1)$
   b) $5(m + 4)(m - 1)$ **e)** $4(3x + 2)(3x + 2)$
   c) $2(x + 3)(x - 3)$ **f)** $5c^2(3c + 5)$

19. Factoring can be the opposite of expanding, for example:
   $(x + 1)(x + 2) = x^2 + 3x + 2$
   [factored] ⟷ [expanded]

Chapter Self-Test, p. 122

1. a) $-7x^2 + 2x$ **c)** $22x^2 - 105x + 119$
   b) $-200n^2 - 80n - 128$ **d)** $-78a^2 + 21a + 48$

2. $(3x - 3)(2x + 5)$; $6x^2 + 9x - 15$

3. a) $6x^2 - 7x - 3$ **c)** $12x^2 - 6x - 18$
   b) $4x - 6$ by $3x + 3$

4.

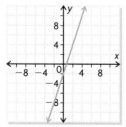

-2x is the common factor. It divides into $-2x^2$, x times and into $8x$, -4 times. The factors are $-2x(x-4)$.

5. a) $(x+4)(x-3)$ **c)** $-5(x-8)(x-7)$
   b) $(a+7)(a+9)$ **d)** $(y-6)(y+9)$

6. a) $(x-5)(2x+1)$ **c)** $3(x-2)(2x-1)$
   b) $(3n-16)(4n-1)$ **d)** $(4a+3)(2a-5)$

7. 3 by $4x^2-x-5$; $(3x+3)$ by $(4x-5)$; $(x+1)$ by $(12x-15)$

8. $m=12, -12$

9. a) $(11x-5)(11x+5)$ **c)** $(x-3)(x+3)(x^2+9)$
   b) $(6a-5)^2$ **d)** $(n+3)^2$

10. 7, 2; $-7, -2$; $-7, 2$; 7, -2; 9, 6; $-9, -6$; $-9, 6$; 9, -6; 23, 22; $-23, -22$; $-23, 22$; 23, -22

Chapter 3

Getting Started, pp. 126–128

1. a) (v)   **c)** (iii)   **e)** (ii)
   b) (vi)   **d)** (iv)   **f)** (i)

2. a) $y=6$ **c)** $a=\dfrac{9}{2}$

   b) $x=\dfrac{16}{3}$ **d)** $c=\dfrac{7}{4}$

3. a) vertex: $(0, 2)$; axis: $x=0$; domain: $\{x \in \mathbf{R}\}$;
   range: $\{y \in \mathbf{R} \mid y \geq 2\}$
   b) vertex: $(3, -5)$; axis: $x=3$; domain: $\{x \in \mathbf{R}\}$;
   range: $\{y \in \mathbf{R} \mid y \leq -5\}$

4. a) $5y-5x$ **c)** $6x^4+10x^3$
   b) $8m-4$ **d)** $-x^2-2x+12$

5. a) x-intercept: $\dfrac{7}{3}$; y-intercept: -7

   b) x-intercept: 6; y-intercept: -2
   c) x-intercept: 2; y-intercept: 5
   d) x-intercept: 4; y-intercept: 3

6. a) $5(x^2-x+3)$ **d)** $(2x+3)(3x-1)$
   b) $(x-10)(x-1)$ **e)** $2(x-1)(x+3)$
   c) $(x+5)(2x-3)$ **f)** $(x-11)(x+11)$

7. a)

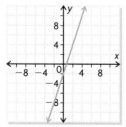

b)

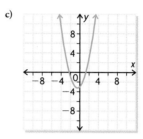

c)

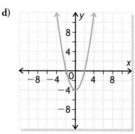

d)

8. a) x-intercepts: 3, -2; y-intercept: -6; min -6, the parabola opens up and $a > 0$
   b) x-intercepts: 3, -3; y-intercept: 9; max 9, the parabola opens down and $a < 0$

9. a) linear, the graph shows a straight line or first difference is -1
   b) nonlinear, the graph does not show a straight line or first difference is not constant

10.

Factoring atrategies:
Common factor to be done first
Difference of squares for 2 terms separated by a $-$ sign
Simple trinomials for 3 terms starting with 1 or a prime number times x^2
Complex trinomials for 3 terms not starting with 1 or a prime number times x^2

Factoring Quadratics

Examples:	Non-examples:
$3x^2+6x+9 = 3(x^2+2x+3)$	$3x+6$ is not quadratic
$4x^2-25 = (2x+5)(2x-5)$	x^3+1 is not quadratic
$x^2+x-20 = (x+5)(x-4)$	x^2+x+1 does not factor
$4x^2+16x+15 = (2x+3)(2x+5)$	

Lesson 3.1, p. 131

1. Yes; it matches the minimum number of moves I found when I played the game, and when I substituted 6 into the model, I got 48.

2. Answers may vary. E.g.: R = red, B = blue, S = slide, J = jump

N																
1	RS	BJ	RS													
2	RS	BJ	BS	RJ	RJ	BS	BJ	RS								
3	RS	BJ	BS	RJ	RJ	RS	BJ	BJ	BJ	RS	RJ	RJ	BS	BJ	RS	

It is very symmetric. Each row of the table starts and ends with a slide.

3. Yes; number of moves: 5, 11, 19, ...; $f(x) = x^2 + 3x + 1$, when graphed it appears quadratic

Lesson 3.2, pp. 139–142

1. a) $-5, 3$ **b)** $f(x) = (x + 5)(x - 3)$

2. a) $f(x) = 2x(x + 6)$, zeros: $0, -6$; axis: $x = -3$; vertex: $(-3, -18)$

b) $f(x) = (x - 4)(x - 3)$, zeros: $4, 3$; axis: $x = \frac{7}{2}$; vertex: $\left(\frac{7}{2}, -\frac{1}{4}\right)$

c) $f(x) = -(x + 10)(x - 10)$, zeros: $-10, 10$; axis: $x = 0$; vertex: $(0, 100)$

d) $f(x) = (x + 3)(2x - 1)$, zeros: $-3, \frac{1}{2}$; axis: $x = -\frac{5}{4}$; vertex: $\left(-\frac{5}{4}, -\frac{49}{8}\right)$

3. a) $f(x) = 3x^2 - 12x$; 0
b) $f(x) = x^2 + 2x - 35$; -35
c) $f(x) = 6x^2 - 20x - 16$; -16
d) $f(x) = 6x^2 + 7x - 20$; -20

4. a) zeros: $0, -6$; axis of symmetry: $x = -3$; vertex: $(-3, -18)$
b) zeros: $8, -4$; axis: $x = 2$; vertex: $(2, -36)$
c) zeros: $10, 2$; axis: $x = 6$; vertex: $(6, 16)$

d) zeros: $-\frac{5}{2}, \frac{9}{2}$; axis: $x = 1$; vertex: $(1, 49)$

e) zeros: $-\frac{3}{2}, 2$; axis: $x = \frac{1}{4}$; vertex: $\left(\frac{1}{4}, -\frac{49}{8}\right)$

f) zeros: $5, -5$; axis $x = 0$; vertex: $(0, 25)$

5. a) $g(x) = 3x(x - 2)$; zeros: $0, 2$; axis: $x = 1$; vertex: $(1, -3)$
b) $g(x) = (x + 3)(x + 7)$; zeros: $-7, -3$; axis: $x = -5$; vertex: $(-5, -4)$

c) $g(x) = (x + 2)(x - 3)$; zeros: $-2, 3$; axis: $x = \frac{1}{2}$; vertex: $\left(\frac{1}{2}, -6.25\right)$

d) $g(x) = 3(x - 1)(x + 5)$; zeros: $-5, 1$; axis: $x = -2$; vertex: $(-2, -27)$

e) $g(x) = (x - 7)(2x + 1)$; zeros: $-\frac{1}{2}, 7$; axis: $x = 3.25$; vertex: $(3.25, -28.125)$
f) $g(x) = -6(x - 2)(x + 2)$; zeros: $-2, 2$; axis: $x = 0$; vertex: $(0, 24)$

6. a) (iii)   **c)** (iv)   **e)** (v)
b) (ii)   **d)** (i)
I expanded or graphed.

7. a) 20.25, max   **c)** 45.125, max   **e)** 25, max
b) -49, min   **d)** -2.25, min   **f)** -4, min

8. a)

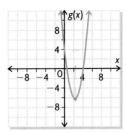

b)

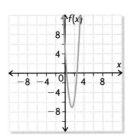

c)

d)

e)

f)

9. a) the y-intercept, the direction the parabola opens; in $f(x) = x^2 - x - 20$, the y-intercept is -20, and the parabola opens upward because a is $+1$, which is greater than 0
b) the x-intercept(s), the axis of symmetry, the direction the parabola opens; in $f(x) = (x - 5)(x + 7)$, the zeros are 5 and -7, the axis of symmetry is $x = -1$, and the parabola opens upward because a is $+1$, which is greater than 0

10. a) zeros: 40, −10; axis: $x = 15$; y-intercept: 400; vertex: (15, 625); max: 625

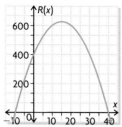

b) $f(x) = x^2 - 2x - 8$; zeros: 4, −2; axis: $x = 1$; y-intercept: −8; vertex: (1, −9); min: −9

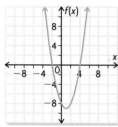

c) $f(x) = (4 - x)(x + 2)$; zeros: 4, −2; axis: $x = 1$; y-intercept: 8; vertex: (1, 9); max: 9

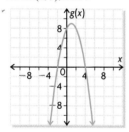

d) $p(x) = -x^2 + 2x + 3$; zeros: 3, −1; axis: $x = 1$; y-intercept: 3; vertex: (1, 4); max: 4

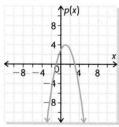

e) $j(x) = (2x - 11)(2x + 11)$; zeros: $\dfrac{11}{2}$, $-\dfrac{11}{2}$; axis: $x = 0$; y-intercept: −121; vertex: (0, −121); min: −121

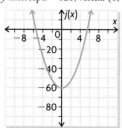

11. $h(t) = -5t(t - 3)$; 11.25 m

12. a) $f(x) = (x + 2)(x - 7)$; $f(x) = x^2 - 5x - 14$
b) $f(x) = (7 - x)(x + 5)$; $f(x) = -x^2 + 2x + 35$
c) $f(x) = \dfrac{1}{2}(x + 3)(x + 6)$; $f(x) = \dfrac{1}{2}x^2 + \dfrac{9}{2}x + 9$
d) $f(x) = -\dfrac{2}{9}x(x - 6) = -\dfrac{2x^2}{9} + \dfrac{4x}{3}$

13. a) $x = 5$; (5, 6); $f(x) = -\dfrac{2}{3}(x - 2)(x - 8)$;
$f(x) = -\dfrac{2}{3}x^2 + 20x - \dfrac{32}{3}$
b) $x = -\dfrac{9}{2}$; $\left(-\dfrac{9}{2}, -2\right)$; $f(x) = \dfrac{8}{25}(x + 7)(x + 2)$;
$f(x) = \dfrac{8x^2}{25} + \dfrac{72x}{25} + \dfrac{112}{25}$
c) $x = 4$; (4, 5); $f(x) = -\dfrac{1}{5}(x + 1)(x - 9)$;
$f(x) = -\dfrac{x^2}{5} + \dfrac{8x}{5} + \dfrac{9}{5}$
d) $x = -4$; (−4, −5); $f(x) = \dfrac{5}{16x}(x + 8)$;
$f(x) = \dfrac{5}{16}x^2 + \dfrac{5}{2}x$

14. a) $t = 7$ s
b) $t = 5$ s
15. 75 km/h
16. Find the zeros and then the axis of symmetry. Find the vertex and use all those points to graph the function.
17. $y = \left(-\dfrac{1}{48}\right)x^2 + 192$
18. $h(t) = -t(4.9t - 30)$
19. a) $h(d) = -0.0502(d - 21.9)(d + 1.2)$
b) r and s are the points where the shot is on the ground.

Lesson 3.3, pp. 149–152

1. a) zeros: −2, 7

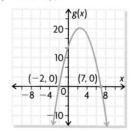

b) zeros: −3, −5

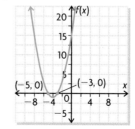

2. a) zeros: $-4, 5$ **b)** zeros: $-1.4, 6.4$

3. a) $x^2 - 2x - 35 = 0$; zeros: $-5, 7$

b) $-x^2 + 3x + 4 = 0$; zeros: $-1, 4$

c) $x^2 + 3x - 5 = 0$; zeros: $-4.2, 1.2$

d) $-6x^2 - x + 2 = 0$; zeros: $-\dfrac{2}{3}, \dfrac{1}{2}$

4. a) zero: 4

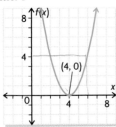

b) zeros: $-4, \dfrac{5}{2}$

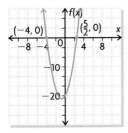

c) zeros: 1, 2

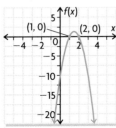

d) zeros: $-5.5, 1.5$

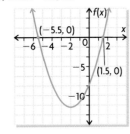

e) no solution

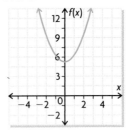

f) zeros: $-4, 4$

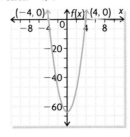

5. a) zeros: $-2.0, \dfrac{1}{3}$

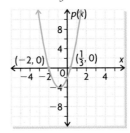

b) zeros: $-\dfrac{3}{2}, 7$

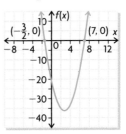

c) zeros: $-\dfrac{3}{4}, \dfrac{1}{2}$

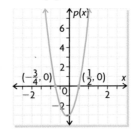

d) no solution

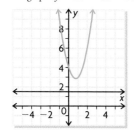

6. a) 71 900 people **b)** partway through 2019
  c) 1983
7. 26.9 s
8. 3.53 s
9. 2
10. a) $b = 5$ or $b = 2$, so to break even, the company must produce 5000 or 2000 skateboards
  b) 3500
11. a) 45 m **b)** 4 s
12. a) 15 min **b)** \$160
13. a) i) Answers may vary. E.g., $f(x) = 8x^2 + 2x - 3$
    ii) Answers may vary. E.g., $f(x) = x^2 - 6x + 9$
    iii) Answers may vary. E.g., $f(x) = x^2 + 2$
  b) There can only be no zeros, one zero, or two zeros because a quadratic function either decreases and increases (or increases and decreases), so it will not be able to cross the x-axis a third time.
14. a) $f(x) = 3x^2 - 2x + 1$
  b) Graph the function and find where the graph crosses the x-axis.
15. $y = -0.3x^2 + 150$
16. a) $(1.5, 5)$ and $(-2, -9)$
  b) $(-1.5, -3.75)$ and $(5, 6)$
17. The graphs $y = 2x^2 - 3x + 4$ and $y = 1.5$ do not intersect.

Mid-Chapter Review, p. 155

1. a) $f(x) = 2x^2 + 17x + 21$
  b) $g(x) = -3x^2 + 16x + 12$
  c) $f(x) = -8x^2 - 2x + 15$
  d) $g(x) = -6x^2 + 13x - 5$
2. a) (ii) **c)** (i) **e)** (iii)
  b) (i) **d)** (ii)
3. a) minimum -36 **c)** minimum -28.125
  b) maximum 40.5 **d)** maximum 15.125
4. Answers may vary. E.g., Factored form is most useful because it gives the zeros, and the midpoint of the zeros gives the vertex. In $f(x) = (x - 2)(x - 6)$, the zeros are 2 and 6, and the vertex is $(4, -4)$.

5. a) zeros: $-4, 2$

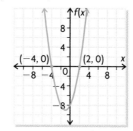

  b) zeros: $\dfrac{1}{3}, -\dfrac{1}{5}$

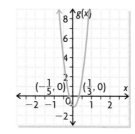

  c) zeros: $-\dfrac{1}{4}, -\dfrac{1}{2}$

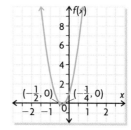

  d) zeros: $2.5, -1$

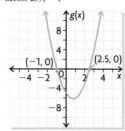

6. a) zeros: $-5, 3$

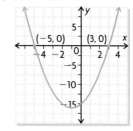

b) zeros: $-3, -\dfrac{5}{2}$

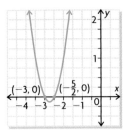

c) zeros: $2, -\dfrac{3}{2}$

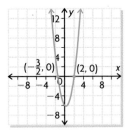

d) zeros: $-3, -4$

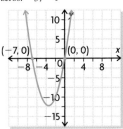

7. It is a dangerous ball. It is above 1 m from $0.11 < t < 1.9$.

8. No. $(x - 3)(x + 4) = x^2 - x + 12$

Lesson 3.4, pp. 161–163

1. a) $x = -3, 5$ **c)** $x = -\dfrac{1}{2}, \dfrac{5}{3}$

   b) $x = 6, 9$ **d)** $x = 0, 3$

2. a) $(x + 5)(x - 4); x = -5, 4$

   b) $(x - 6)(x + 6), x = -6, 6$

   c) $(x + 6)(x + 6); x = -6$

   d) $x(x - 10); x = 0, 10$

3. a) yes   **c)** yes

   b) no   **d)** no

4. a) $x = -6, 9$ **c)** $x = -7$ **e)** $x = -\dfrac{1}{2}, 5$

   b) $x = 13, -13$ **d)** $x = 14, 3$ **f)** $x = \dfrac{1}{3}, -4$

5. a) $x = 17, -17$ **c)** $x = -3, 5$ **e)** $x = -3, 5$

   b) $x = \dfrac{5}{3}$ **d)** $x = -\dfrac{1}{2}, 7$ **f)** $x = -\dfrac{3}{2}, 5$

6. a) $x = -10, 0$ **c)** $x = -7, -2$ **e)** $x = 5, \dfrac{16}{3}$

   b) $x = -\dfrac{1}{3}, 0$ **d)** $x = -3, 9$ **f)** $x = -\dfrac{5}{2}, -\dfrac{1}{3}$

7. a) $x = -6, 7$ **c)** $x = 4, -9$ **e)** $x = -\dfrac{1}{4}, -\dfrac{3}{2}$

   b) $x = -4, 1$ **d)** $x = 2, \dfrac{8}{3}$ **f)** $x = \dfrac{3}{2}, -\dfrac{1}{5}$

8. 5 s

9. a) 0, 24   **b)** 288 m^2

10. a) at 0 and 6000 snowboards

   b) when they sell between 0 and 6000 snowboards

11. $t = 3\,s$

12. 9 s

13. $30\,025

14. a) $P(x) = -5x^2 + 19x - 10$

   b) between 631 and 3169 pairs of shorts

15. No. Some equations do not factor or are too difficult to factor.

16. It is faster and helps find the maximum or minimum value, but it can be difficult or impossible to factor the equation.

17. 14.1 s

18. $2 < t < 10$

Lesson 3.5, pp. 168–169

1. a) between 400 and 900 games

   i) table of values

x	$P(x)$
3	-6
4	0
5	4
6	6
7	6
8	4
9	0

   ii) factoring: $P(x) = -(x - 9)(x - 4)$

2. $-5(t - 10)(t + 1) = 0; t = 10$ or $t = -1$; the ball hits the ground after 10 s.

3. $0.0056s^2 + 0.14s = 7; 0.5$ km/h

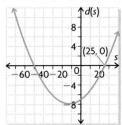

4. Beverly used an appropriate method, but she should have substituted 20 instead of 2020 because $t = 0$ corresponds to the year 2000 to get a population of 74 000.

5. Solution 1: Using a table of values
$x = 4$

x	y
−2	12
−1.5	0
−1	−10
0	−24
1	−30
2	−28
3	−18
4	0

Solution 2: Using a graph
$x = 4$ and $x = −1.5$

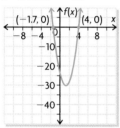

Solution 3: By factoring
$x = −\dfrac{3}{2}$ and $x = 4$
$2(2x + 3)(x − 4) = 0$

6. The population will be 312 000 in 2008 and 1972.
7. You can use graphing and factoring. Your answers will be $x = −3, 5$.
8. approximately 15.77 m
9. $450 000
10. 1 s
11. **a)** Answers may vary. E.g., $16 = −2x^2 + 32x + 110$
$0 = −2x^2 + 32x − 126$
$0 = −2(x^2 + 16x − 63)$
$0 = −2(x + 7)(x − 9)$
$x = 7$ and $x = 9$
They must sell either 7000 or 9000 games.
b) Answers may vary. E.g., I let $P(x) = 16$ in the function, since P is profit in thousands of dollars. I rearranged the equation to get 0 on the left side, then factored the right side. I determined values for x where each of the factors were zero. These were the solutions to the equation. I multipled these numbers by 1000, since they represent the number of games in thousands.
12. She noticed that when $t = −30$ and 10, $P(t) = 35 000$. Since $t = 0$ corresponds to 2000, then $t = −30$ corresponds to 1970.
13. 5 and 5; product is 25
14. 6 m × 6 m

Lesson 3.6, pp. 176–179

1. **a)** $x = 3, x = −5$; $a = 2$; $y = 2(x − 3)(x + 5)$; $y = 2x^2 + 4x − 30$
b) $x = 1.5, x = −3$; $a = −2$; $y = (−2x + 3)(x + 3)$;
$y = −2x^2 − 3x + 9$

2. **a)** not quadratic because the graph does not have a maximum or minimum because the data does not increase and decrease or vice versa

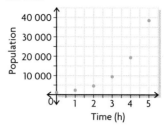

b) quadratic, the graph looks parabolic

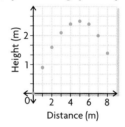

3. **a)** $y = 0.5(x − 2)(x + 1)$; $y = 0.5x^2 − 0.5x − 1$
b) $y = −0.5(x + 2)(x − 3)$; $y = −0.5x^2 + 0.5x + 3$
4. **a)** $y = −2x^2 + 2x + 24$
b) $y = x^2 + 7x + 10$
c) $y = 3x^2 + 6x − 105$
d) $y = −3x^2 + 18x − 24$
5. $h(x) = −0.37x^2 + 1.48x + 0.52$; 3.6 m
6. $h(t) = −5t^2 + 35$, $t = 2.45$ s
7. $h(t) = −4.9t^2 + 9.7t + 1$, $t = 1.7$ s, 0.2 s
8. $h(t) = −4.9t^2 + 17.9t + 0.5$, $t = 3.7$ s
9. **a)** $y = 0.21(x − 3)(x − 12)$
b) −4.25 m
10. **a)** $y = −x^2 + 7x + 50$
b) They will sell no more shoes in month 12.
11. **a)** 0, 0, 1, 3, 6, 10, 15
b) number of lines $= \dfrac{n(n − 1)}{2}$, where n is the number of dots and $n ≥ 2$
12. The zeros are a and b. So, $f(x) = k(x − a)(x − b)$. To determine k, use another point on the parabola.
13. $y = 0.000\ 36x^2 + 4$
14. **a)**

Time (s)	2.5	3.0	3.5	4.0	4.5	5.0	5.06
Height (m)	32.125	30.9	27.225	21.1	12.525	1.5	0

b) $y = −4.9t^2 + 24.5t + 1.5$
c) 32.125 m

Chapter Review, pp. 182–183

1. **a)** (ii) **c)** (iv)
b) (iii) **d)** (i)
2. **a)** min −36 **b)** min −3.125 **c)** max 15.125
3. **a)** $x = −5$ or $x = 3$, min −16
b) $x = 7$ or $x = 1$, max 9
c) $x = −8$ or $x = −1$, min −24.5

d) $x = -\dfrac{1}{2}$ or $x = -3$, min -3.125

e) $x = -\dfrac{3}{2}$ or $x = \dfrac{1}{3}$, min $-\dfrac{121}{24}$

f) $x = -7$ or $x = 7$, max 49

4. a) 3.15 m   **b)** 2.38 s

5. a) zeros: $-3, 5$; vertex: $(1, 32)$; y-int: 30
  b) zeros: $4, -2$; vertex: $(1, -18)$; y-int: -16

6. a) $x = -7$ or $x = 5$
  b) $x = 3$ or $x = -8$

c) $x = \dfrac{1}{3}$

d) $x = -\dfrac{1}{2}$ or $x = \dfrac{5}{3}$

7. $t = 1$ and 9 s

8. a) 2014   **b)** 87 850 people

9. $x = 2$; No, you cannot just change the 17 to be positive. The answer should be $x = 2$.

10. a) $y = -4.9x^2 + 37.6x + 14$
  b) It is close to most of the points on the graph.

11. $y = -3(x + 2)(x - 4)$ or $y = -3x^2 + 6x + 24$

Chapter Self-Test, p. 184

1. a) $f(x) = 6x^2 - 37x + 45$
  b) $f(x) = -5x^2 - 4x + 12$

2. a) $f(x) = (x - 9)(x + 9)$
  b) $f(x) = (2x - 1)(3x + 4)$

3. a) $x = 5, -7; x = -1$; minimum -36
  b) $x = \dfrac{1}{2}, -\dfrac{7}{2}; x = -1.5$; maximum 16

4. No. Some may not factor at all, while for others it may not be as obvious what the factors are.

5. a) $x = 0$ and $x = -6$

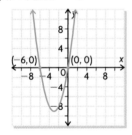

b) $x = -1$ and $x = 3.5$

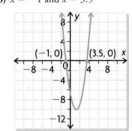

6. a) $x = -6$ and $x = \dfrac{1}{2}$
  b) $x = -3$ and $x = 7$

7. a) 2015
  b) 11 700 people

8. a) 6.4 s
  b) 57 m

9. a) $y = -343x^2 + 965x - 243$
  b) It is close to most of the data points.
  c) 36491 kg/ha

10. It makes it easier to answer questions about the data.

Cumulative Review Chapters 1–3, pp. 186–189

1. (a)   **5.** (c)   **9.** (d)   **13.** (c)   **17.** (a)
2. (c)   **6.** (b)   **10.** (b)   **14.** (a)   **18.** (b)
3. (a)   **7.** (d)   **11.** (d)   **15.** (c)
4. (d)   **8.** (c)   **12.** (c)   **16.** (c)

19. a) Domain $\{x \in \mathbf{R}\}$, Range $\{y \in \mathbf{R} \mid y \geq 4\}$
  b) Transformations: vertical stretch by a factor of 3, horizontal translation 4 to the right, vertical translation 5 up
  c)

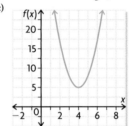

20. a)

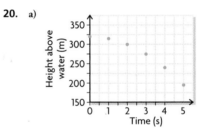

b) quadratic; the graph appears to have a shape of part of a parabola. The second differences are also constant.

c)

$t(s)$	6	7	8
$h(m)$	140	75	0

d)

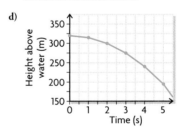

e) vertex: $(0, 320)$; axis: $x = 0$; zeros: $x = 8$ and $x = -8$
  $h(t) = -5(t - 8)(t + 8)$ or $h(t) = -5t^2 + 320$
  f) domain $\{t \in \mathbf{R} \mid 0 \leq t \leq 8\}$;
  range $\{h(t) \in \mathbf{R} \mid 0 \leq h(t) \leq 320\}$
  g) i) 38.75 m **ii)** 7.35 s

Chapter 4

Getting Started, pp. 192–194

1. **a)** (iv)   **c)** (v)   **e)** (i)
   b) (iii)   **d)** (ii)   **f)** (vi)

2. **a)** $(x + 10)(x - 3); x = -10$ or $x = 3$
   b) $(x + 5)(x + 3); x = -5$ or $x = -3$
   c) $(x + 2)(x - 3); x = -2$ or $x = 3$
   d) $(x - 2)(x - 3); x = 2$ or $x = 3$

3. **a)** $(x + 3)^2$ **c)** $(3x + 1)^2$
   b) $(x - 4)^2$ **d)** $(2x - 3)^2$

4. **a)** $-15x^2 + 7x + 8$ **c)** $12x^2 - x - 35$
   b) $2x^2 + 6$ **d)** $4x^2 - 1$

5. **a)** $(x - 5)(x + 8)$ **c)** $(9x - 7)(9x + 7)$
   b) $(2x + 3)(3x - 2)$ **d)** $(3x + 1)^2$

6. **a)** 9-   **c)** $28x$
   b) $10x$ **d)** 16

7. **a)** $f(x) = (x - 9)(x + 2)$ **c)** $h(x) = (2x - 5)(2x + 5)$
   b) $g(x) = -(x - 8)(2x - 1)$ **d)** $y = (3x - 1)(2x + 5)$

8. **a)** vertex: $(-3, -4)$; domain: $\{x \in \mathbf{R}\}$; range: $\{y \in \mathbf{R} \mid y \geq -4\}$
   b) vertex: $\left(\dfrac{5}{4}, -\dfrac{121}{8}\right)$; domain: $\{x \in \mathbf{R}\}$; range: $\left\{y \in \mathbf{R} \mid y \geq -\dfrac{121}{8}\right\}$
   c) vertex: $\left(-\dfrac{7}{12}, \dfrac{121}{24}\right)$; $\{x \in \mathbf{R}\}$; range: $\left\{y \in \mathbf{R} \mid y \leq \dfrac{121}{24}\right\}$
   d) vertex: $\left(\dfrac{3}{2}, \dfrac{147}{4}\right)$; $\{x \in \mathbf{R}\}$; range: $\left\{y \in \mathbf{R} \mid y \leq \dfrac{147}{4}\right\}$

9. **a)** $f(x) = x^2$ vertically shifted down 5 units

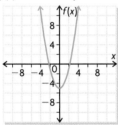

   b) $f(x) = x^2$ translated 2 units right and 1 unit up

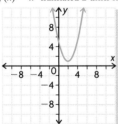

   c) $f(x) = x^2$ translated 3 units left, stretched vertically by a factor of 2 and translated 4 units down

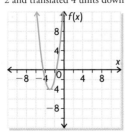

   d) $f(x) = x^2$ translated 4 units right, compressed vertically by a factor of 2, reflected in the x-axis and translated 2 units up

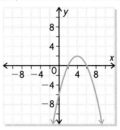

10. 0.5 s

11.

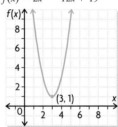

Essential characteristics:		Non-essential characteristics:
	Does not have to = 0.	
A polynomial equation containing one variable with highest degree 2.	Can have fractional or decimal coefficients.	
	Can have variables on both sides of the = sign.	
	Does not have to have a solution.	
Examples:	**Quadratic Equation**	Non-examples:
$x^2 - 9 = 0$		$x^2 + x^2 = 0$
$x^2 + 4x - 21 = 0$		$x^2 + \frac{1}{x} = 0$
$(3x - 1)(2x + 5) = 0$		

Lesson 4.1, pp. 203–205

1. **a)** $(3, -5)$; min   **c)** $(-1, 6)$; max
   b) $(5, -1)$; max   **d)** $(-5, -3)$; min

2. **a)** $x = 3$, domain: $\{x \geq \mathbf{R}\}$; range: $\{y \geq \mathbf{R} \mid y \geq -5\}$
   b) $x = 5$, $\{x \geq \mathbf{R}\}$; range: $\{y \geq \mathbf{R} \mid y < -1\}$
   c) $x = -1$, $\{x \geq \mathbf{R}\}$; range: $\{y \geq \mathbf{R} \mid y < 6\}$
   d) $x = -5$, $\{x \geq \mathbf{R}\}$; range: $\{y \geq \mathbf{R} \mid y \geq -3\}$

3. **a)** $f(x) = 2x^2 - 12x + 19$

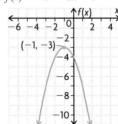

   b) $f(x) = -x^2 - 2x - 4$

4.

	Function	Vertex	Axis of Symmetry	Opens Up/Down	Range
a)	$f(x) = (x - 3)^2 + 1$	$(3, 1)$	$x = 3$	up	$y \geq 1$
b)	$f(x) = -(x + 1)^2 - 5$	$(-1, -5)$	$x = -1$	down	$y \leq -5$
c)	$y = 4(x + 2)^2 - 3$	$(-2, -3)$	$x = -2$	up	$y \geq -3$
d)	$y = -3(x + 5)^2 + 2$	$(-5, 2)$	$x = -5$	down	$y \leq 2$
e)	$f(x) = -2(x - 4)^2 + 1$	$(4, 1)$	$x = 4$	down	$y \leq 1$
f)	$y = \frac{1}{2}(x - 4)^2 + 3$	$(4, 3)$	$x - 4$	up	$y \geq 3$

Sketches (from table above)

a)
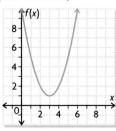

b)
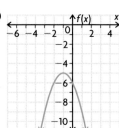

c)
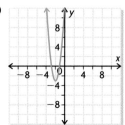

d)
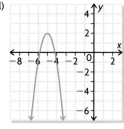

e)
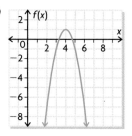

f)
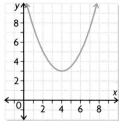

5. **a)** $x = -8, 14$
   b) $x = -12, 2$
   c) $x = 1$
   d) $x = -6.873$ or -0.873
   e) $x = 0.192$ or 7.808
   f) no zeros

6. $f(x) = (x - 7)^2 - 25$; vertex form: vertex $(7, -25)$; axis: $x = 7$; direction of opening: up
   $f(x) = x^2 - 14x + 24$; standard form: y intercept is 24; direction of opening: up
   $f(x) = (x - 12)(x - 2)$; factored form: zeros: $x = 12$ and $x = 2$; direction of opening: up

7. **a)** after 0.3 s; maximum since $a < 0$ so the parabola opens down
   b) 110 m
   c) 109.55 m

8. **a)** $y = -3(x + 4)^2 + 8$, $y = -3x^2 - 24x - 40$
   b) $y = -(x - 3)^2 + 5$, $y = -x^2 + 6x - 4$
   c) $f(x) = 4(x - 1)^2 - 7$, $y = 4x^2 - 8x - 3$
   d) $y = (x + 6)^2 - 5$, $y = x^2 + 12x + 31$

9. **a)** $f(x) = (x + 3)^2 - 4$
   b) $f(x) = (x - 2)^2 + 1$
   c) $f(x) = -(x - 4)^2 - 2$
   d) $f(x) = -(x + 1)^2 + 4$

10. **a)** 23 m   **b)** 2 s   **c)** 1 s or 3 s

11. **a)** vertex: $(4, 0)$; two possible points $(3, 1)$, $(5, 1)$

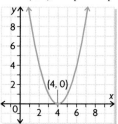

b) vertex: $(0, 4)$; two possible points: $(2, 8)$, $(-2, 8)$

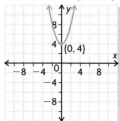

c) vertex: $(1, -3)$; two possible points: $(-1, 5)$, $(3, 5)$

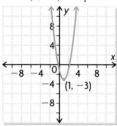

d) vertex: $(-3, 5)$; two possible points: $(-2, 3)$, $(-4, 3)$

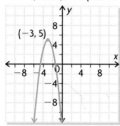

12. $y = 2(x + 1)^2 - 8$

13. Expand and simplify. For example,
$$f(x) = (x + 2)^2 - 4$$
$$= x^2 + 4x + 4 - 4$$
$$= x^2 + 4x$$

14. a) y-intercept, opens upward or downward
   b) vertex, opens upward or downward, max/min

15. $y = -0.88(x - 1996)^2 + 8.6$

16. $f(x) = (x - 1)^2 - 36$

Lesson 4.2, pp. 213–215

1. a) 16   **b)** 36   **c)** 25   **d)** $\dfrac{25}{4}$

2. a) $m = 10, n = 25$ **c)** $m = 12, n = 36$

   b) $m = 6, n = 9$ **d)** $m = \dfrac{7}{2}, n = \dfrac{49}{4}$

3. a) $(x + 7)^2$ **c)** $(x - 10)^2$
   b) $(x - 9)^2$ **d)** $(x + 3)^2$

4. a) $(x + 6)^2 + 4$ **c)** $(x - 5)^2 + 4$

   b) $(x - 3)^2 - 7$ **d)** $\left(x - \dfrac{1}{2}\right)^2 - \dfrac{13}{4}$

5. a) $a = 3, h = 2, k = 5$
   b) $a = -2, h = -5, k = -3$
   c) $a = 2, h = 3, k = 5$
   d) $a = \dfrac{1}{2}, h = -3, k = -5$

6. a) $f(x) = (x + 4)^2 - 13$ **d)** $f(x) = -(x - 3)^2 + 16$

   b) $f(x) = (x - 6)^2 - 1$ **e)** $f(x) = -\left(x - \dfrac{3}{2}\right)^2 + \dfrac{1}{4}$

   c) $f(x) = 2(x + 3)^2 - 11$ **f)** $f(x) = 2\left(x + \dfrac{3}{4}\right)^2 - \dfrac{1}{8}$

7. a) $f(x) = (x - 2)^2 + 1$; domain: $\{x \in \mathbf{R}\}$; range: $\{y \in \mathbf{R} \mid y \geq 1\}$

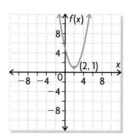

   b) $f(x) = (x + 4)^2 - 3$; domain: $\{x \in \mathbf{R}\}$; range: $\{y \in \mathbf{R} \mid y \geq -3\}$

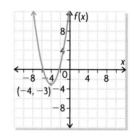

   c) $f(x) = 2(x + 3)^2 + 1$; domain: $\{x \in \mathbf{R}\}$; range: $\{y \in \mathbf{R} \mid y \geq 1\}$

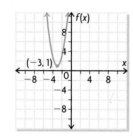

   d) $f(x) = -(x - 1)^2 - 6$; domain: $\{x \in \mathbf{R}\}$; range: $\{y \in \mathbf{R} \mid y \leq -6\}$

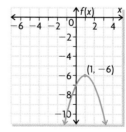

e) $f(x) = -3(x + 2)^2 + 1$; domain: $\{x \in \mathbf{R}\}$; range: $\{y \in \mathbf{R} \mid y \leq 1\}$

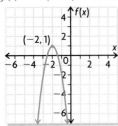

f) $f(x) = \frac{1}{2}(x + 3)^2 - \frac{1}{2}$, domain: $\{x \in \mathbf{R}\}$; range: $\{y \subset \mathbf{R} \mid y \geq -0.5\}$

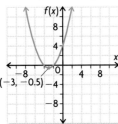

8. **a)** $g(x) = 4(x - 3)^2 - 5$
  b) $x = 3$
  c) $(3, -5)$
  d) minimum value of -5 because $a > 0$; parabola opens up
  e) domain: $\{x \in \mathbf{R}\}$
  f) $\{g(x) \in \mathbf{R} \mid g(x) \geq -5\}$
  g)

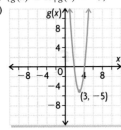

9. Colin should have taken the square root of both 9 and 4 to get $\frac{3}{2}$, not $\frac{3}{4}$.
10. 61 250 m²
11. $15
12. **a)** $y = 3(x - 5)^2 - 2$
  b) vertex, axis of symmetry, max/min
13. Reflection about the x-axis, a vertical stretch of 2, a horizontal shift of 4, a vertical shift of 3. I completed the square to determine the vertex.
14. Each form provides different information directly.
15. $f(x) = ax^2 + bx + c$
  $= x^2 + bx + c$
  $= x^2 + bx + \left(\frac{b}{2}\right)^2 - \left(\frac{b}{2}\right)^2 + c$
  $= \left(x + \frac{b}{2}\right)^2 + \left(c - \frac{b}{2}\right)^2$
  Example:
  $f(x) = x^2 + 6x + 4$
  $= x^2 + 6x + \left(\frac{6}{2}\right)^2 - \left(\frac{6}{2}\right)^2 + 4$
  $= x^2 + 6x + 9 - (9 + 4)$
  $= (x + 3)^2 - 5$

16. $y = -5(x - 1)^2 + 6$
17. $(1, 6)$

Lesson 4.3, pp. 222–223

1. **a)** $a = 3, b = -5, c = 2$ **c)** $a = 16, b = 24, c = 9$
  b) $a = 5, b = -3, c = 7$ **d)** $a = 2, b = -10, c = 7$

2. **a)** **a)** $x = \dfrac{-(-5) \pm \sqrt{(-5)^2 - 4(3)(2)}}{2(3)}$

  b) $x = \dfrac{-(-3) \pm \sqrt{(-3)^2 - 4(5)(7)}}{2(5)}$

  c) $x = \dfrac{-(24) \pm \sqrt{(24)^2 - 4(16)(9)}}{2(16)}$

  d) $x = \dfrac{-(-10) \pm \sqrt{(-10)^2 - 4(2)(7)}}{2(2)}$

  b) **a)** $x = 1$ or $x = \dfrac{2}{3}$ **c)** $x = -\dfrac{3}{4}$
  b) no solution   **d)** $x = 4.16$ or $x = 0.84$

3. **a)** no solution
  b) $x = -5$ or $x = 1.5$
  c) $x = 5.5$
  d) $x = -0.84$ or $x = 2.09$
  e) no solution
  f) $x = -0.28$ or $x = 0.90$
4. **a)** no solution
  b) $x = -5$ or $x = 1.5$
  c) $x = 5.5$
  d) $x = -0.84$ or $x = 2.09$
  e) no solution
  f) $x = -0.28$ or $x = 0.90$
5. For example:
  a) factoring, $x = 0$ or $x = 15$
  b) take the square root of both sides, $x = -10.72$ or $x = 10.72$
  c) factoring, $x = 8$ or $x = \dfrac{3}{2}$
  d) expand, then use quadratic formula, $x = 2.31$ or $x = 11.69$
  e) isolate the squared term, $x = -2$ or $x = 8$
  f) quadratic formula, $x = 0.09$ or $x = 17.71$
6. **a)** 10.9 s   **b)** approx. 9 s
7. 20 m × 20 m and 10 m × 40 m
8. 9.02 s
9. 91 km/h
10. Answers may vary. E.g., 2 solutions $2x^2 - 4x = 0$, $3x^2 - 4x - 2 = 0$; 1 solution $3x^2 - 6x + 3 = 0$, $x^2 - 2x + 1 = 0$; 0 solutions $x^2 - 6x + 10 = 0$
11. **a)** $x = 1.62$ or $x = 6.38$
  b) $x = 1.62$ or $x = 6.38$
  c) The method in part (a) was best because it required fewer steps to find the roots.
12. $(-1, -2)$ and $(3, 6)$
13. $(-1.21, -5.96)$ and $(2.21, 4.29)$
14. **a)** $x = 0$ and $x = 0.37$
  b) $x = -3$ and $x = -2.4$, and $x = 2.4$ and $x = 3$

Mid-Chapter Review, p. 226

1. **a)** $f(x) = x^2 - 16x + 68$
   b) $g(x) = -x^2 + 6x - 17$
   c) $f(x) = 4x^2 - 40x + 109$
   d) $g(x) = -0.5x^2 + 4x - 6$

2. **a)** vertex: $(3, 6)$; axis: $x = 3$; min. 6; domain: $\{x \in \mathbf{R}\}$;
   range: $\{y \in \mathbf{R} \mid y \geq 6\}$

   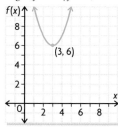

   b) vertex: $(-5, -7)$; axis: $x = -5$; max. -7; domain: $\{x \in \mathbf{R}\}$;
   range: $\{y \in \mathbf{R} \mid y \leq -7\}$

   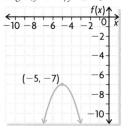

3. Answers may vary. E.g., the vertex form. From this form, I know the location of the vertex, which helps me sketch the graph. If $f(x) = (x + 2)^2 - 5$, then the vertex is $(-2, -5)$ and the parabola opens up.

4. **a)** $f(x) = 2(x - 2)^2 + 5$
   b) $f(x) = -3(x + 1)^2 - 4$

5. **a)** $f(x) = (x + 5)^2 - 13$

   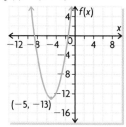

   b) $f(x) = 2(x + 3)^2 - 21$

   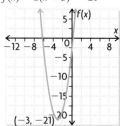

c) $f(x) = -(x + 4)^2 + 6$

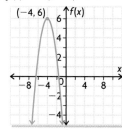

d) $g(x) = 2(x - 0.5)^2 + 7$

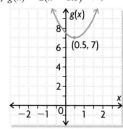

6. $205
7. 10 000 m²
8. **a)** $x = -5, x = 3$ **c)** $x = 5.27, x = 8.73$
   b) $x = \dfrac{1}{3}$ **d)** no solution
9. about 4 s
10. $1.67 and $10
11. **a)** 62 m   **b)** about 8 s   **c)** at 3 s and 5 s

Lesson 4.4, pp. 232–233

1. **a)** $(-5)^2 - 4(1)(7)$ **c)** $(12)^2 - 4(3)(-7)$
   b) $(11)^2 - 4(-5)(17)$ **d)** $(1)^2 - 4(2)(-11)$
2. **a)** two distinct   **c)** one   **e)** two distinct
   b) none   **d)** none   **f)** one
3. No, there are two solutions; $b^2 - 4ac = (-5)^2 - 4(1)(2) > 0$
4. **a)** two   **c)** one   **e)** none
   b) none   **d)** two   **f)** one
5. $k > 4$ or $k < -4$
6. $m = 4$
7. $k < \dfrac{1}{2}, k = \dfrac{1}{2}, k > \dfrac{1}{2}$
8. **a)** Put a, b, and c into the discriminant and set the discrimant equal to zero. Solve for k.
   b) $(-5)^2 - 4(3)(k) = 0, k = \dfrac{25}{12}$
9. $k = -4, k = 8$
10. **a)** $-50 < k < 50$
    b) $k = \pm 50$
    c) $k < -50, k > 50$
11. Answers may vary. E.g., if $a = 2$, $b = 3$, and $c = -1$, then $0 = 2x^2 + 3x - 1$. In this case, the discriminant $(b^2 - 4ac)$ has a value of 17, so there are 2 solutions.
12. Yes. The discriminant will be greater than zero. $x = 99.9$
13. Answers may vary. E.g., discriminant, quadratic formula
14. Yes. Set the expression equal to each other and solve. You would get $p = 1$ or $p = -13$.
15. two distinct zeros for all values of k

Lesson 4.5, pp. 239–241

1. Complete the square to put the function in vertex form. The y-coordinate of the vertex will be the maximum revenue.

2. $3025

3. Solve the equation $-4.9t^2 + 1.5t + 17 = 5$ for $t > 0$.

4. 1.73 s

5. $1865; $6

6. **a)** 8600
   b) approx. 2011
   c) no; the graph does not cross the horizontal axis

7. when selling between 1000 and 7000 pairs of shoes

8. 128 m²

9. **a)** 8.4 m
   b) 20 km/h

10. 2 m

11. 2017

12. **a)** $6.50   **b)** $8.00

13. **a)** about 1.39 m   **c)** no, the ball hits the ground at 4.3 s
   b) 23 m   **d)** $t = 0.47$ s and $t = 3.73$ s

14. **a)** $f(x) = -60(x - 5.5)^2 + 1500$
   b) $2 or $9

15. An advantage of the vertex form is that is provides the minimum or maximum values of the function. A disadvantage is that you must expand and simplify to find the zeros of the function using the quadratic formula.

16. 5 cm × 12 cm

17. 10 cm

Lesson 4.6, pp. 250–252

1. **a)** no, the graph increases too quickly
   b) yes, the graph decreases and increases

2. **a)**

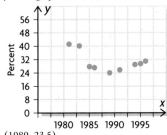

   b) (1989, 23.5)
   c) up, e.g., $a = 1$
   d) e.g., needs to be wider so use a smaller, positive value for a
   e) $y = 0.23(x - 1989)^2 + 23.5$, or
      $y = 0.23x^2 - 912.53x + 908\ 060.52$
   f) domain: $\{x \in \mathbf{R} \mid 1981 \leq x \leq 1996\}$;
      range: $\{y \in \mathbf{R} \mid 23.5 \leq y \leq 41.7\}$

3. **a)** $y = (x - 5)^2 - 3$
   b) $y = -0.5(x + 4)^2 + 6$

4. **a)** $y = 2x^2 - 8x + 11$ **c)** $y = -4x^2 + 24x - 43$
   b) $y = -2x^2 - 4x + 3$ **d)** $y = 6x^2 + 24x + 19$

5. **a)**

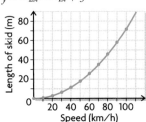

b) $y = 0.007x^2 - 0.0005x - 0.016$
c) Skid will be 100.7 m.
d) domain: $\{x \in \mathbf{R} \mid x \geq 0\}$; range: $\{y \in \mathbf{R} \mid y \geq 0\}$

6. **a)** $y = 0.000\ 83x^2 - 0.116x + 21.1$
   b) 21.1¢
   c) domain: $\{x \in \mathbf{R} \mid x \geq 0\}$; range: $\{y \in \mathbf{R} \mid y \geq 17\}$

7. **a)** $y = -15(x - 1987)^2 + 1075$
   b) not very well, because they would be selling negative cars
   c) domain: $\{x \in \mathbf{R} \mid x \geq 1982\}$; range: $\{y \in \mathbf{R} \mid y \leq 1075\}$

8. $y = -5(x - 2)^2 + 20.5$; $x \geq 0, y \geq 0$; 4 s

9. **a)** $y = -0.53x^2 + 1.39x + 0.13$
   b) domain: $\{x \in \mathbf{R} \mid x \geq 0\}$; range: $\{y \in \mathbf{R} \mid y \leq 1.1\}$
   c) 0.57 kg/ha or 2.05 kg/ha

10. **a)** $y = 0.03(x - 90)^2 + 16$
   b) 1096 mph, no a regular car can't drive that fast!

11. $f(x) = 2x^2 + 4x + 6$

12. If the zeros of the function can be determined, use the factored form $f(x) = a(x - r)(x - s)$. If not, then use graphing technology and quadratic regression.

13. 12.73 m

14. 15, 24

15. $6250, $12.50

Chapter Review, pp. 254–255

1. **a)** $f(x) = x^2 + 6x + 2$
   b) $f(x) = -x^2 - 14x - 46$
   c) $f(x) = 2x^2 - 4x + 7$
   d) $f(x) = -3x^2 + 12x - 16$

2. **a)** $y = (x - 3)^2 - 5$ **c)** $f(x) = \dfrac{1}{2}(x + 2)^2 - 3$
   b) $f(x) = -(x - 6)^2 + 4$ **d)** $f(x) = 2(x - 5)^2 + 3$

3. **a)** $f(x) = (x + 1)^2 - 16$ **d)** $f(x) = 3(x + 2)^2 + 7$
   b) $f(x) = -(x - 4)^2 + 9$ **e)** $f(x) = \dfrac{1}{2}(x - 6)^2 + 8$
   c) $f(x) = 2(x + 5)^2 - 34$ **f)** $f(x) = 2\left(x + \dfrac{1}{2}\right)^2 + \dfrac{7}{2}$

4. **a)** vertex: $(5, -2)$; axis: $x = 5$; opens up because $a > 0$; two zeros because the vertex is below the x-axis; parabola opens up

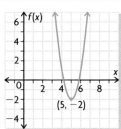

   b) vertex: $(-3, -1)$; axis: $x = -3$; opens down because $a < 0$; no zeros because the vertex is below the x-axis; parabola opens down

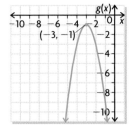

c) vertex: $(-1, 5)$; axis: $x = -1$; will have no zeros because the vertex is above the x-axis; parabola opens up

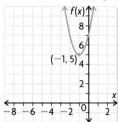

d) vertex: $(8, 0)$; axis: $x = 8$; $a < 0$, will have one zero because the vertex is on the x-axis; parabola opens down

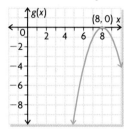

5. a) $x = -\dfrac{1}{2}, x = 8$

b) no solution

c) $x = \dfrac{1}{3}$

d) $x = 1.16, x = -2.40$

6. a) 2.7 m   **b)** 1.5 s

7. a) no solution
b) two distinct solutions
c) one solution

8. a) $\dfrac{16}{5} > k$ **b)** $\dfrac{16}{5} = k$ **c)** $\dfrac{16}{5} < k$

9. a) 25 cars   **b)** \$525

10. a) 41 472 m^2 **b)** 10 720 m^2 **c)** 36 m

11. a) $y = -4.9(x - 1.5)^2 + 10.5$
b) domain: $\{x \in \mathbf{R} \mid 0 \le x \le 2.96\}$;
range: $\{y \in \mathbf{R} \mid 0 \le y \le 10.5\}$
c) 2.8 s

12. a)

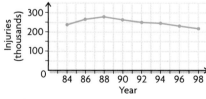

b) $y = -673.86x^2 + 2\,680\,816x - 2\,665\,999\,342$
c) Use the quadratic regression function on a graphing calculator to determine the curve's equation. $y = -673.86x^2 + 2\,680\,816x - 2\,665\,999\,342$
d) 160 715, using the above equation
e) 1989

Chapter Self-Test, p. 256

1. a) $f(x) = x^2 + 6x + 2$
b) $f(x) = -3x^2 - 30x - 73$
2. a) $f(x) = (x - 5)^2 + 8$
b) $f(x) = -5(x - 2)^2 + 8$
3. a) vertex: $(8, 3)$; axis: $x = 8$; min. 3

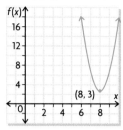

b) vertex: $(7, 6)$; axis: $x = 7$; max. 6

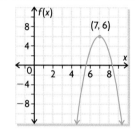

4. If you get a negative under the square root, you cannot solve. There are no solutions to these equations.
5. a) $x = -1.12, x = 7.12$
b) no solution
6. a) no solution   **b)** one solution
7. a) 20.1m   **b)** 15.2 m
8. a) \$13 812   **b)** \$24
9. a) $y = -1182(x - 2)^2 + 4180$
b) It fits the data pretty well.
c) domain: $\{x \in \mathbf{R} \mid x \ge 0\}$; range: $\{y \in \mathbf{R} \mid y \le 4180\}$
d) \$4106
10. It is easier to use this method to find the max/min and vertex.

Chapter 5

Getting Started, pp. 260–262

1. a) (v)   **c)** (ii)   **e)** (iv)
b) (vi)   **d)** (i)   **f)** (iii)

2. a) $\sin A = \dfrac{12}{13}, \cos A = \dfrac{5}{13}, \tan A = \dfrac{12}{5}$

b) $\sin A = \dfrac{8}{17}, \cos A = \dfrac{15}{17}, \tan A = \dfrac{8}{15}$

3. a) $\angle A \doteq 27°$ **b)** $b \doteq 3$ **c)** $\angle A \doteq 50°$ **d)** $d \doteq 16$
4. a) 12 cm   **b)** 10 m

5. a) $\sin A = \dfrac{4}{13}, \cos A = \dfrac{12}{13}, \tan A = \dfrac{1}{3}; \angle A \doteq 18°$

b) $\sin D = \dfrac{5}{7}, \cos D = \dfrac{5}{7}, \tan D = 1; \angle D \doteq 44°$

c) $\sin C = \dfrac{12}{13}, \cos C = \dfrac{4}{13}, \tan C = 3; \angle C \doteq 72°$

6. a) 0.7660 **b)** 0.9816 **c)** 3.0777
7. a) 47° **b)** 11° **c)** 80°
8. a) $\theta = 71°, \phi = 109°$ **b)** $\theta \doteq 72°, \phi \doteq 18°$
9. a) 27° **b)** 2.3 m
10. a) Answers may vary. E.g., Given the triangle below, calculate the length of h to the nearest metre.

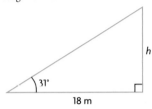

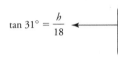

$\tan 31° = \dfrac{h}{18}$ ← I knew one angle and the side adjacent to that angle. I needed to determine the length of the opposite side, so I used tangent.

$18 \times \tan 31° = \overset{1}{\cancel{18}} \times \dfrac{h}{\underset{1}{\cancel{18}}}$ ← To solve for h, I multiplied both sides of the equation by 18.

$18 \times \tan 31° = h$

$11 \text{ m} \doteq h$ ← I rounded to the nearest metre.

The height of the triangle is about 11 m.

b) Answers may vary. E.g., Given the triangle below, calculate the measure of angle θ to the nearest degree.

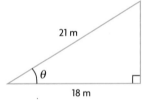

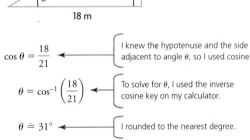

$\cos \theta = \dfrac{18}{21}$ ← I knew the hypotenuse and the side adjacent to angle θ, so I used cosine.

$\theta = \cos^{-1}\left(\dfrac{18}{21}\right)$ ← To solve for θ, I used the inverse cosine key on my calculator.

$\theta \doteq 31°$ ← I rounded to the nearest degree.

Angle θ is about 31°.

Lesson 5.1, pp. 271–273

1. a) 0.2588 **b)** 0.5736
2. a) 22° **b)** 45°
3. a) $\sin A = \dfrac{3}{5}, \cos A = \dfrac{4}{5}, \tan A = \dfrac{3}{4}; \angle A \doteq 37°$
b) $\sin D = \dfrac{6.9}{13}, \cos D = \dfrac{11}{13}, \tan D = \dfrac{6.9}{11}; \angle D \doteq 32°$
4. a) 8 cm **b)** 11 cm

5. a) $x \doteq 62°, y \doteq 28°, z \doteq 17$ **c)** $x \doteq 64°, y \doteq 26°, q \doteq 4$
b) $i \doteq 8, j \doteq 7$ **d)** $x \doteq 18°, l \doteq 3, j \doteq 10$
6. a) (Eiffel Tower) 254 m **c)** (Leaning Tower of Pisa) 56 m
b) (Empire State Building) 381 m **d)** (Big Ben's clock tower) 97 m
Order of heights (tallest to shortest): (b), (a), (d), (c)
7. 7.4 m
8. 26°
9. a) 0.3 **b)** 18°
10. 120 m
11. a) 30.91 m **b)** 29.86 m
12. a) nothing **b)** scaffolding **c)** planks
13. Answers may vary. E.g., They could place the pole into the ground on the opposite bank of the river (side B). Then they measure the angle of elevation from the other bank (side A). The width of the river can be calculated using tangent.

$\tan \theta = \dfrac{h}{x}$

$x = \dfrac{h}{\tan \theta}$, where h is the height of the pole and θ is the angle of elevation

14. 105 m
15. a) 63° **b)** 4.22 m **c)** 2.6 m
16. No, the height of the array will be 3.3 m.
17. 29 m
18. a) 18.4 m
b) (Jodi) 85.7 m, (Nalini) 135.0 m
c) (pole) 5.5 m, (Jodi) 19.4 m, (Nalini) 30.6 m

Lesson 5.2, pp. 280–282

1. a) 8.3 cm^2 **b)** 43.1 cm^2
2. In $\triangle ATB$, calculate AB using tangent. In $\triangle CTB$, calculate BC using tangent. Then add AB and BC. The length is about 897.8 m.
3. Karen; Karen's eyes are lower than Anna's. Thus, the angle of elevation is greater.
4. a) tangent; Tangent relates a known side and the length we must find.
b) 26 m
5. No. The height of a pyramid is measured from the very top of the pyramid to the centre of the base. I don't know how far that point on its base is from the person measuring the angle of elevation.
6. Darren can only store the 1.5 m plank in the garage.
7. a) 58.5 m **b)** 146 m **c)** 87.7 m
8. 12 m
9. 31 m
10. 22.2 m
11. a) 35° **b)** 0.8 m
12. a) 180 cm^2 **b)** (volume) 18 042 cm^3, (surface area) 5361 cm^2
13. a) 54° **b)** 24 703 m
14. Draw a perpendicular from the base to divide x into two parts, x_1 and x_2. Use cosine to solve for x_1 and x_2. Then, add x_1 and x_2 to determine x.

$\cos 24° = \dfrac{x_1}{148}$ and $\cos 19° = \dfrac{x_2}{181}$

$x = x_1 + x_2$

$x \doteq 306$

15. 12 cm
16. 0.7 m^2

Lesson 5.3, pp. 288–290

1. a)

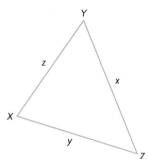

b) $\dfrac{\sin X}{x} = \dfrac{\sin Y}{y} = \dfrac{\sin Z}{z}$ or $\dfrac{x}{\sin X} = \dfrac{y}{\sin Y} = \dfrac{z}{\sin Z}$

2. a) $x \doteq 4.6$ **b)** $\theta \doteq 24°$

3. a) $b \doteq 12$ cm   **b)** $\angle D \doteq 49°$

4. $q \doteq 8$ cm

5. a) $75°$ **b)** 82 m

6. a) $\angle C \doteq 39°$, $b \doteq 4.7$ cm   **c)** $\angle C \doteq 53°$, $b \doteq 11.6$ cm

   b) $\angle C \doteq 74°$, $b \doteq 8.7$ cm   **d)** $\angle C \doteq 71°$, $b \doteq 10.8$ cm

7. a) 20.3 cm   **b)** 13.6 cm^2

8. a) $\angle A = 75°$, $a \doteq 13$ cm, $b \doteq 13$ cm

   b) $\angle M \doteq 58°$, $\angle N \doteq 94°$, $n \doteq 9$ cm

   c) $\angle Q \doteq 55°$, $q \doteq 8$ cm, $s \doteq 10$ cm

   d) $\angle D \doteq 57°$, $\angle F \doteq 75°$, $f \doteq 10$ cm

9. a) $\angle A \doteq 42°$, $\angle B \doteq 68°$, $b \doteq 14.6$

   b) $\angle D \doteq 94°$, $d \doteq 21.2$, $e \doteq 13.1$

   c) $\angle J \doteq 31°$, $\angle H \doteq 88°$, $h \doteq 6.1$

   d) $\angle K \doteq 61°$, $\angle M \doteq 77°$, $m \doteq 14.0$

   e) $\angle P \doteq 36°$, $\angle Q \doteq 92°$, $q \doteq 2.0$

   f) $\angle Z \doteq 61°$, $x \doteq 5.2$, $y \doteq 7.6$

10. (1.9 m chain) $52°$, (2.2 m chain) $42°$

11. Calculate $\angle B$, then use the sine law to determine b and c.

$$\dfrac{b}{\sin 64°} = \dfrac{289}{\sin 51°}$$

$$\dfrac{c}{\sin 65°} = \dfrac{289}{\sin 51°}$$

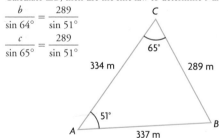

Phones A and B are farthest apart.

12. Answers may vary. E.g., The primary trigonometric ratios only apply when you have a right triangle. If you have a triangle that doesn't have a right angle, you would have to use the height of the triangle in order to use primary trigonometric ratios. This would mean you use two smaller triangles, instead of the original one. If you use the sine law, you can calculate either an angle or side directly without unnecessary calculations.

13.

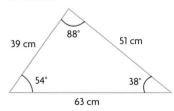

14. Answers may vary. E.g., Given any acute angle θ, the value of $\sin \theta$ ranges from 0 to 1. If θ gets very small, $\sin \theta$ approaches zero. But if θ gets close to $90°$, $\sin \theta$ approaches 1. Suppose in $\triangle XYZ$, $\angle X$ is $90°$.

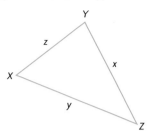

Suppose that $\angle X$ is the largest angle. Let's look at $\angle Y$. Since Y is a smaller angle than X in $\triangle XYZ$, we have $\sin Y < \sin X$. Thus, $\dfrac{1}{\sin X} < \dfrac{1}{\sin Y}$. Multiplying through by x yields $\dfrac{x}{\sin X} < \dfrac{x}{\sin Y}$. But by the sine law, $\dfrac{x}{\sin X} = \dfrac{y}{\sin Y}$. Thus, the previous inequality becomes $\dfrac{y}{\sin Y} < \dfrac{x}{\sin Y}$. Cancelling $\sin Y$, we obtain $y < x$. The same argument applies to side z.

15. 15.1 m

16. a) 4.0 m, 3.4 m   **b)** 85%

Mid-Chapter Review, pp. 292–293

1. 100 m

2. a) $23°$ **b)** 1.9 m

3. a) $x \doteq 4.4$ cm, $y \doteq 8.6$ cm, $\angle X = 27°$

   b) $\angle K \doteq 44°$, $\angle L \doteq 46°$, $j \doteq 99.8$ cm

   c) $m \doteq 13.2$ m, $n \doteq 24.2$ m, $\angle M = 33°$

   d) $i \doteq 20.9$ m, $\angle I \doteq 59°$, $\angle G \doteq 31°$

4. $84°$

5. 5.1 m^2

6. 3.9 km

7. $44°$

8. a) $56°$ **b)** 126 m

9. a) $\angle E = 71°$, $d \doteq 24$, $f \doteq 18$

   b) $\angle P = 51°$, $p \doteq 6$, $q \doteq 6$

   c) $\angle C = 58°$, $b \doteq 24$, $c \doteq 22$

   d) $\angle X = 103°$, $y \doteq 13$, $z \doteq 6$

10. a) 123 m   **b)** 6 m

11. a) $68°$ **b)** 69 cm

Lesson 5.4, pp. 299–301

1. i) (a); In order to apply the sine law, we would need to know $\angle A$ or $\angle C$.

   ii) $b^2 = a^2 + c^2 - 2ac \cos B$

2. a) $b \doteq 6.4$ cm   **b)** $b \doteq 14.5$ cm

3. a) $\angle B \doteq 83°$ **b)** $\angle D \doteq 55°$

4. a) 12.9 cm   **b)** 10.1 cm   **c)** 8.1 cm

5. a) $a \doteq 11.9$ cm, $\angle B \doteq 52°$, $\angle C \doteq 60°$

   b) $d \doteq 7.7$ cm, $\angle E \doteq 48°$, $\angle F \doteq 80°$

   c) $h \doteq 7.2$ cm, $\angle I \doteq 48°$, $\angle F \doteq 97°$

   d) $\angle P \doteq 37°$, $\angle Q \doteq 40°$, $\angle R \doteq 103°$

6. No, the interior angles are about $22°$, $27°$, and $130°$.

7. 18 km

8. No; E.g., Calculate $\angle G$ using the fact that the sum of the interior angles equals 180°. Use the sine law to calculate f and h.
$\angle G \doteq 85°, f \doteq 45.3$ m, $h \doteq 59.7$ m

9. 7.1 km

10. Answers may vary. E.g., Use the cosine law to calculate r. Then use the sine law to calculate angle θ. $\theta \doteq 29°$

11. **a)** 13 cm   **b)** 6 cm

12. **a)** Answers may vary. E.g., Two sailboats headed on two different courses left a harbour at the same time. One boat travelled 25 m and the other 35 m. The sailboats travelled on paths that formed an angle of 78° with the harbour. Calculate the separation distance of the sailboats.
$d^2 = 25^2 + 35^2 - 2(25)(35)\cos 78°$
$d \doteq 39$ m

13. **a)** (AX) 0.9 km, (AY) 2.0 km
b) (BX) 1.6 km, (BY) 0.6 km
c) 1.7 km

Lesson 5.5, pp. 309–311

1. **a)** primary trigonometric ratios, $\sin 39° = \dfrac{x}{43.0}$

b) the sine law, $\dfrac{\sin \theta}{3.1} = \dfrac{\sin 42°}{2.2}$

c) the cosine law, $\cos \theta = \dfrac{3.6^2 + 5.2^2 - 4.1^2}{2(3.6)(5.2)}$

2. **a)** $x \doteq 27.1$ **b)** $\theta \doteq 71°$ **c)** $\theta \doteq 52°$

3. 486 cm²

4. 1.4 m

5. 260.7 m

6. 50°

7. 27°

8. 0.004 s

9.

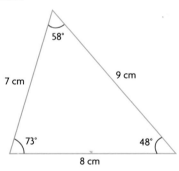

10. (perimeter) 32.7 cm, (area) 73.6 cm²

11. 6.3 m

12. 107.7 m

13. (if $\angle ABC = 61°$) 78°, (if $\angle ABC = 65°$) 72°

14.

the cosine law	3rd problem	$x^2 \doteq 1.7^2 + 3.8^2 - 2(1.7)(3.8)\cos 70°$ $x \doteq 3.6$ km
the sine law	1st problem	$\dfrac{b}{\sin 58°} = \dfrac{54}{\sin 51°}$ $b \doteq 59$ m $\dfrac{j}{\sin 71°} = \dfrac{54}{\sin 51°}$ $j \doteq 66$ m
primary trigonometric ratios	2nd problem	$\tan 41° = \dfrac{h}{11.4}$ $h \doteq 9.9$ m

15. 30.3 cm²

16. 61 m

17.

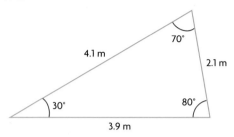

Chapter Review, pp. 314–315

1. **a)** 4 cm   **b)** 40°

2. 27°

3. **a)** 12 m   **b)** 6.4 m²

4. 84 m

5. **a)** $\angle A = 80°$, $a \doteq 19$ cm, $b \doteq 18$ cm
b) $\angle M \doteq 64°$, $\angle N \doteq 86°$, $n \doteq 20$ cm
c) $\angle Q = 45°$, $q \doteq 6$ cm, $s \doteq 9$ cm
d) $\angle D \doteq 40°$, $\angle F \doteq 88°$, $f \doteq 14$ cm

6. 59 m

7. **a)** $b \doteq 5$ cm, $\angle A \doteq 78°$, $\angle C \doteq 54°$
b) $\angle T = 90°$, $\angle U \doteq 53°$, $\angle V \doteq 37°$
c) $m \doteq 9$ cm, $\angle N \doteq 37°$, $\angle L \doteq 83°$
d) $y \doteq 6$ m, $\angle X = \angle C = 70°$

8. 47°

9. **a)**

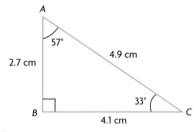

b)

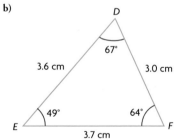

c)

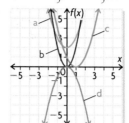

d)

J

51°

21.0 cm   18.0 cm

55°   74°
K   L
17.0 cm

10. **a)** 96°, 84°
   b) 10.8 cm

Chapter Self-Test, p. 316

1. 3 m
2. Yes, the angle of elevation is 6.4°, which is greater than 4.5°.
3. $x \doteq 111$ m, $y \doteq 108$ m
4. **a)** $\angle A \doteq 120°$, $\angle C \doteq 33°$, $a \doteq 29$ cm
   b) $\angle F \doteq 47°$, $d \doteq 31$ cm, $e \doteq 37$ cm
5. 2 m
6. 8 m
7. 795 m
8. 197 m

CHAPTER 6

Getting Started, pp. 320–322

1. **a)** (i)   **c)** (vi)   **e)** (iv)
   b) (iii)   **d)** (v)   **f)** (ii)
2. **a)** about 175 m
   b) about 30 s; Started fast for 8s; then slower; then top speed
   c) for first 8 s started off slower; then sped up; at 20 s full speed
3. **a)** Moved to the right 5 units, stretched by a factor of 3, and moved up 4 units
   b) Moved to the right 2 units, stretched by a factor of 2, and moved up 1 unit
   c) Moved to the left 1 unit, compressed by a factor of 0.5, and moved down 3 units
   d) Moved to the left 2 units, reflection in the x-axis and compressed by a factor of $\frac{1}{4}$, and moved down 4 units

4. **a)** **i)** 0.5; Truck is moving at a steady pace away from the detector.
      ii) (0, 1.5); Spot where the truck starts away from the detector.
      iii) 0; Truck is not moving for 2 s.
      iv) (8, 0); Truck stops and this is how long it took for the truck to go the entire distance (to and from the detector).
   b) domain: $\{t \in \mathbf{R} \mid 0 \le t \le 8\}$; range $\{t \in \mathbf{R} \mid 0 \le d \le 3\}$
5. **a)** $\sin A = \frac{3}{5}$; $\cos A = \frac{4}{5}$; $\tan A = \frac{3}{4}$
6.

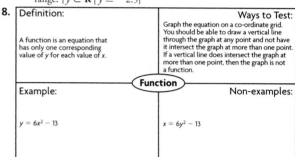

7. **a)** max: 4; zeros: (0, 0), (6, 0); domain: $\{x \in \mathbf{R}\}$;
      range: $\{y \in \mathbf{R} \mid y \le 4\}$
   b) min: −2.5; zeros: (−3, 0), (1, 0); domain: $\{x \in \mathbf{R}\}$;
      range: $\{y \in \mathbf{R} \mid y \ge -2.5\}$

8.

Definition:		Ways to Test:
A function is an equation that has only one corresponding value of y for each value of x.		Graph the equation on a co-ordinate grid. You should be able to draw a vertical line through the graph at any point and not have it intersect the graph at more than one point. If a vertical line does intersect the graph at more than one point, then the graph is not a function.
	Function	
Example:		Non-examples:
$y = 6x^2 - 13$		$x = 6y^2 - 13$

Lesson 6.1, p. 325

1. **a)**

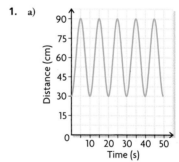

   b)

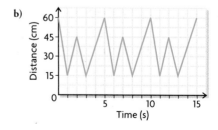

2. Explanation: Start the paddle at 20 cm from the sensor. For 2.5 s, move to 80 cm. For 2.5 s, do not move the paddle. For 2.5 s more, move to 20 cm from the sensor. Repeat this 3 times.

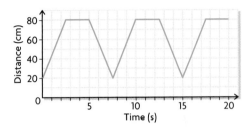

Lesson 6.2, pp. 330–334

1. a) yes; pattern is repeated
   b) yes; pattern is repeated
   c) no; not a function
2. days 89 and 119
3. 5
4. a)

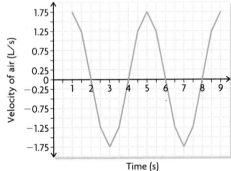

   b) 4 s
   c) breathing in, breathing out
   e) 10 s, 12 s, 14 s, 16 s, 18 s
5. a)

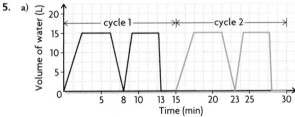

   b) 15 min; time for one dishwasher cycle
   c)

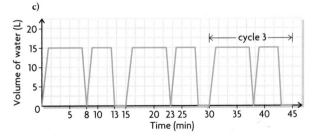

   d) 240 L
   e) domain: $\{t \in \mathbf{R} \mid 0 \le t \le 120\}$; range: $\{v \in \mathbf{R} \mid 0 \le v \le 15\}$

6. a) 8 s
   b) time to complete one full rotation
   c) 6 m; maximum height
   d) 4 m
   e) $t = 4$, $t = 12$, $t = 20$
   f) $t = 2$, $t = 6$, $t = 10$, $t = 14$, $t = 18$, $t = 22$
7. a) yes; repeating pattern   **d)** yes; repeating pattern
   b) no   **e)** no
   c) no   **f)** yes; repeating pattern
8. a) yes; heart beats regularly   **d)** yes; steady pace

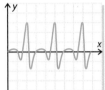   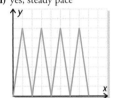

   b) no   **e)** yes; repeated rotation

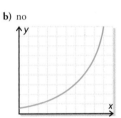   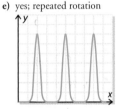

   c) no   **f)** no

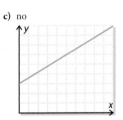   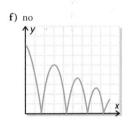

9. a) The wave pattern is repeating.
   b) The change is 10 m.
10. a) The washer fills up with water, washes the dishes but does not add water, dishwasher empties of water, fills up again with a small amount of water, rinses again, and fills up one more time to rinse dishes, and empties again, and then waits 2 min before starting this cycle all over again.
   b) 26 min
   c) 50 L, 10 L, 50 L
   d) 110 L
11. The paddle is moved in a steady back and forth motion in front of the detector.
12.

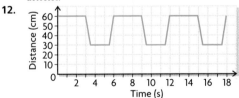

13. A periodic function is a function that produces a graph that has a regular repeating pattern over a constant interval. E.g. A search light on top of a lighthouse—this will go around in a circular period during the same time period and then keep repeating this pattern.

14. approximately periodic; speeds will vary slightly from lap to lap

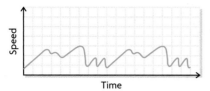

15. a) The light is green for 60 s, yellow for 20 s, red for 40 s, and keeps repeating this pattern.
   b) 120 s
   c) 140 s

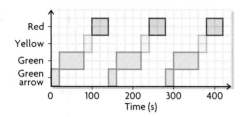

Lesson 6.3, pp. 339–343

1. a) yes; repeating pattern of waves
   b) no; not the same shape as a sine function
   c) no; not the same shape as a sine function
2. a) 2 s; takes time to get into rhythm
   b) 2 s; time to complete one jump
   c) $H = 2.25$
   d) amplitude $= 1.75$ m; amplitude is maximum or minimum distance above axis
3. a) $D = 9$; where the swing is location between the up swinging and down swinging
   b) $a = 5$
   c) 4 s; time to complete one full swing
   d) 4 m from the detector
   e) No. When she is initially starting to swing, the amplitude between the successive waves would be getting larger.
   f) yes, she's swinging away from the detector
4. a) 0.03
   b) $y = 0$
   c) 4.5
   d) period in seconds; axis: current in amperes; amplitude in amperes
5. a) period $= 360°$; axis $y = 3$; amplitude $= 2$
   b) period $= 360°$; axis $y = 1$; amplitude $= 3$
   c) period $= 720°$; axis $y = 2$; amplitude $= 1$
   d) period $= 180°$; axis $y = -1$; amplitude $= 1$
   e) period $= 1440°$; axis $y = 0$; amplitude $= 2$
   f) period $= 720°$; axis $y = 2$; amplitude $= 3$
6. a)

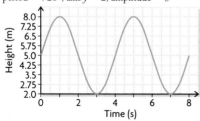

b)

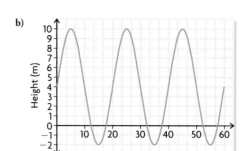

c)

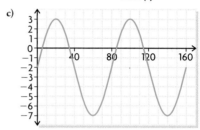

7. a) periodic because it's a regular, predictable cycle
   b) not sinusoidal; maximum values change from day to day, so do the minimum values
8. a)

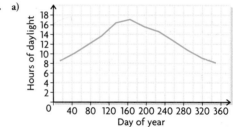

   b) 365 days
   c) $y = 12.6$ h; average number of hours of sunlight per day over the entire year
   d) $a = 4.5$ h; how many hours more or less one might expect to have from the average number of hours of sunlight per day
9. a) 20 m   **c)** 2 m
   b) 12 m   **d)** 12 m
10. a)

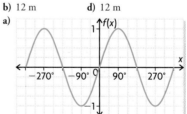

   b) period: $360°$; axis: $y = 0$; amplitude: 1
11. a)

Rotation $(\theta)°$	0	30	60	90	120	150	180
$\cos \theta$	1	0.866	0.5	0	−0.5	−0.866	−1

Rotation $(\theta)°$	210	240	270	300	330	360
$\cos \theta$	−0.866	−0.5	0	0.5	0.866	1

b)

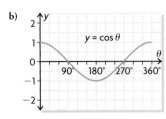

c) yes; it's the sine curve shifted left 90°
d) period: 360°; axis: $y = 0$; amplitude $= 1$
e) $\sin \theta° = \cos (\theta - 90°)$

12. a)

Rotation $(\theta)°$	0	30	60	90	120	150	180
$\tan \theta$	0	0.58	1.73	E	−1.7	−0.58	0

Rotation $(\theta)°$	210	240	270	300	330	360
$\tan \theta$	0.58	1.7	E	−1.7	−0.58	0

b)

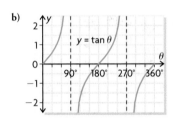

c) no; it's a periodic function
d) period: 180°; axis: $y = 0$; no amplitude because there are no maximum or minimum values; vertical asymptotes at 90° and 270°

Lesson 6.4, pp. 348–353

1. Ferris wheel C: max. height 15 m; amplitude/radius = 7 m; speed = 0.55 m/s; period 80 s longer ride than A and B; speed between A and B; higher than A and B

2.

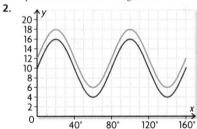

3. a)

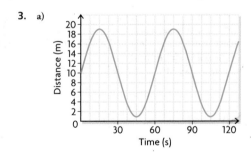

b)

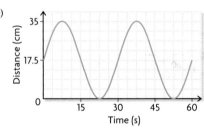

c)

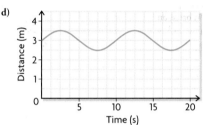

d)

4. a) $y = 0$; the brief instance in time when you are between inhaling and exhaling
b) 0.8 L/s
c) period: 6 s; amount of time to breathe in and out, one cycle
d) domain: $\{t \in \mathbf{R} \mid t \geq 0\}$; range: $\{V \in \mathbf{R} \mid -0.8 \leq V \leq 0.8\}$

5. a) deeper breaths; period is the same
b) amplitude
c) +0.16 L/s

6. A: $r = 2$; period = 6 s; speed = 6.3 m/s, height of axle = 1.5 m;
B: $r = 3$; period = 9 s; speed = 2.1 m/s, height of axle = 2.5 m

7. a) Experiment 1 and graph c); Experiment 2 and graph b); Experiment 3 and graph a); Experiment 4 and graph d)
b) graph a): $y = 20$; graph b): $y = 30$; graph c): $y = 30$; graph d: $y = 20$; for all graphs the equation of the axis represents the height of wheel axle above ground
c) graph a): 20 cm; graph b): 20 cm; graph c): 30 cm; graph d): 10 cm
d) graph a): 125.5 cm; graph b): 188.5 cm; graph c): 188.5 cm; graph d): 125.5 cm; circumference of the wheel
e) $r = 30$ cm
f) $r = 20$ cm
g) $h = 25$ cm
h) straight line

8. a)

b) 40° latitude (approximately): period: 365 days; axis: $h = 12$; amplitude: 3
60° latitude (approximately): period: 365 days; axis: $h = 12$: amplitude: 6.5

c) higher latitude results in larger deviations in the number of hours of sunlight throughout the year

9. periods are the same; period = 0.5 s, equations of the axes are the same; $d = 0$, amplitudes differ; one is 2 cm, the other is 1.5 cm; implications: greater wind speed makes the pole vibrate farther from left to right

10. a)

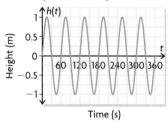

b) 60 s; based on period
c) radius = 1; based on amplitude
d) centre is at water level; based on the equation of the axis

11. a)

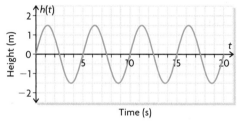

b) 5 s ; look at the time that passes between peaks
c) 12 waves; Since it takes 5 s for 1 cycle, 12 cycles would give you 60 s (1 min).
d) 3 m; distance between max and min values

12. a)

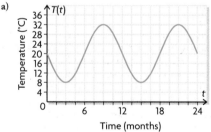

b) 12 months or 1 year
c) between 8° C and 32° C
d) 20 °C

13. Look at the periods, amplitudes, domains, ranges, maximum, minimum, etc.

14. a) one cycle, or $\pi d = 50\pi$
b)

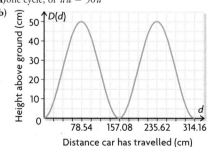

c) approximately 50 cm
d) between 170 and 180 cm

15. a) clockwise
b) 6 s
c) blue: smaller gear; red: larger gear

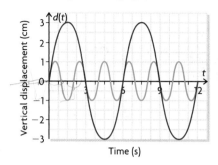

d) 3 cm
e) approximately 0.8 m
f) $d = 0$

Mid-Chapter Review, pp. 357–358

1. a) yes   **b)** no   **c)** yes   **d)** no   **e)** yes
2. Cycle–time it takes to complete one action or activity. E.g., dishwasher cycle, turn of Ferris wheel.
Period—change in maximum values—time it takes to go from one peak of the wave to the next peak of the wave.
Amplitude—vertical distance from its axis to its maximum (or minimum) value.
Equation of the axis—the value that is halfway between maximum and minimum value.
Maximum—highest value during the cycle.
Minimum—lowest value during the cycle.

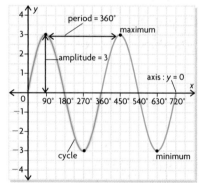

3. a) yes   **b)** 60 psi   **c)** 120 psi   **d)** 80 s   **e)** 20 s
4. a) 1.2 s   **b)** 50 times   **c)** $\{p \in \mathbf{R} | 80 \leq p \leq 120\}$
5. a) 2 s; the time it takes for the pendulum to swing one full cycle
b) $y = 0$, resting position of the pendulum
c) 0.25 m; how far the pendulum swings left or right from the resting position
d) 0.15 m

6.

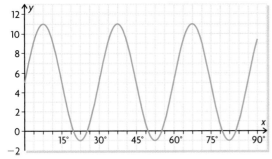

7. a) period: 4; amplitude: 1; axis: $y = 3$
   b) period: 180°; amplitude: 4; axis: $y = 8$
8. a) Monique: period: 3; amplitude: 2; axis: $d = 5$; Steve: period: 3.5, amplitude: 3; $d = 5$
   b) Monique is swinging faster.
   c) Monique: $\{d \in \mathbf{R} \mid 3 \le d \le 7\}$;
      Steve: $\{d \in \mathbf{R} \mid 2 \le d \le 8\}$
9. a)

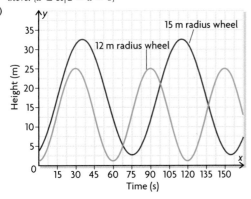

b) First Ferris wheel: perio: 60 s; amplitude: 12 m; axis: $y = 14$
   Second Ferris wheel: period: 75 s; amplitude: 15 m; axis: $y = 18$
c) The 2 Ferris wheels are traveling at the same speed. This can be calculated by finding the circumference of both Ferris wheels and dividing each by the time each Ferris wheel takes to travel its circumference. This gives the distance travelled per second for each Ferris wheel, which is the same.

Lesson 6.5, pp. 365–367

1. a) horizontal translation of −40; domain: $\{x \in \mathbf{R}\}$;
      range: $\{y \in \mathbf{R} \mid -1 \le y \le 1\}$
   b) vertical translation of 8; domain: $\{x \in \mathbf{R}\}$;
      range: $\{y \in \mathbf{R} \mid 7 \le y \le 9\}$
   c) horizontal translation of 60; domain: $\{x \in \mathbf{R}\}$;
      range: $\{y \in \mathbf{R} \mid -1 \le y \le 1\}$
   d) vertical translation of −5; domain: $\{x \in \mathbf{R}\}$;
      range: $\{y \in \mathbf{R} \mid -6 \le y \le -4\}$

2.

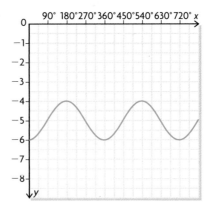

3. a) $f(x) = \sin(x + 70°) + 6$
   b) amplitude: 1; period = 360°; axis: $y = 6$
   c) domain: $\{x \in \mathbf{R}\}$; range: $\{y \in \mathbf{R} \mid 5 \le y \le 7\}$
4. a) horizontal translation of 20

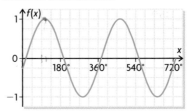

b) vertical translation of 5

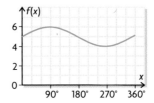

c) horizontal translation of 150 and a vertical translation of −6

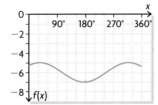

d) horizontal translation of −40 and a vertical translation of −7

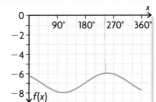

e) horizontal translation of -30 and a vertical translation of -8

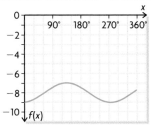

f) horizontal translation of -120 and a vertical translation of 3

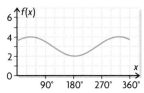

5. period, amplitude, and domain
6. **a)** $f(x) = \sin(x - 15°) + 4$
   b) $f(x) = \sin(x + 60°) - 7$
   c) $f(x) = \sin(x - 45°) + 3$
7. $f(x) = \sin(x + 30°) - 5$
8. **a)** shifted vertically 4 units; shifted horizontally 90°
   b) shifted vertically -3 units
9.

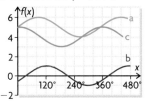

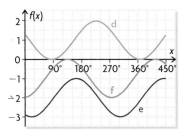

10. **a)** 360°
   b) axis: $y = 1$; tire axle; vertical translation
   c) 1; position of pebble above the ground
   d) negative horizontal shift
11. **a)** same vertical transformation; same amplitude
   b) The first is shifted horizontally $-45°$, and the other is horizontally shifted 90°.
   c) The tires are the same size. The initial positions of the pebbles are different.
12. Answers may vary. E.g., $f(x) = \sin x + 4$; $f(x) = \sin(x - 90°) + 4$; $f(x) = \sin(x + 145°) + 4$
13.

Function	Horizontal Shift	Vertical Shift
$f(x) = \sin(x + 80)° - 7$	$-80°$ or 80° to the left	-7 units or 7 units down
$f(x) = \sin(x - 10)° + 3$	10° or 10° to the right	3 units or 3 units up
$f(x) = \sin(x + 25)° + 3$	$-25°$ or 25° to the left	9 units or 9 units up

14. If a number has been added or subtracted to the x-value, then there is a horizontal shift. If there is a number added or subtracted to the end of the funtion, then there is a vertical shift.
15. **a)** $f(x) = \sin(x + 45°) + 3$
   b) $f(x) = \sin(x - 45°) - 2$
16. All three. $\sin(x - 90°) = \sin(x - 450°) = \sin(x - 810°)$

Lesson 6.6, pp. 373–376

1. **a)** amplitude 3 times larger
   b) reflection in the x-axis; amplitude 2 times larger
   c) amplitude 0.1
   d) reflection in the x-axis; amplitude $\frac{1}{3}$
2. $f(x) = -5 \sin x$
3. **a)**

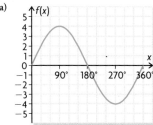

   b) period: 360°; amplitude: 4; axis: $y = 0$
   c) domain: $\{x \in \mathbf{R}\}$; range: $\{y \in \mathbf{R} | -4 \le y \le 4\}$
4. **a)** vertical stretch   **d)** vertical stretch
   b) vertical compression   **e)** vertical compression
   c) vertical compression   **f)** vertical stretch
5. **a)** horizontal translation of $-20°$; vertical stretch of 3
   b) reflection in the x-axis; vertical translation down 3
   c) horizontal translation of 50°; vertical stretch of 5; vertical translation of -7
   d) vertical stretch of 2; reflection in the x-axis; vertical translation of 6
   e) horizontal translation of $-10°$; vertical stretch of 7; reflection in the x-axis
   f) horizontal translation of 30°; vertical compression of 0.5; reflection in the x-axis; vertical translation of 1
6. $f(x) = 3 \sin(x + 20)$:
   a) amplitude $= 3$; period $= 360°$; axis: $y = 0$
   b) domain: $\{x \in \mathbf{R}\}$; range: $\{y \in \mathbf{R} | -3 \le y \le 3\}$
   $f(x) = -\sin x - 3$:
   a) amplitude $= 1$; period $= 360°$; axis: $y = -3$
   b) domain: $\{x \in \mathbf{R}\}$; range: $\{y \in \mathbf{R} | -4 \le y \le -2\}$
   $f(x) = 5 \sin(x - 50) - 7$:
   a) amplitude $= 5$; period $= 360°$; axis: $y = -7$
   b) domain: $\{x \in \mathbf{R}\}$; range: $\{y \in \mathbf{R} | -12 \le y \le -2\}$
   $f(x) = -2 \sin x + 6$:
   a) amplitude $= 2$; period $= 360°$; $y = 6$
   b) domain: $\{x \in \mathbf{R}\}$; range: $\{y \in \mathbf{R} | 4 \le y \le 8\}$
   $f(x) = -7 \sin(x + 10)$:
   a) amplitude $= 7$; period $= 360°$; $y = 0$
   b) domain: $\{x \in \mathbf{R}\}$; range: $\{y \in \mathbf{R} | -7 \le y \le 7\}$
   $f(x) = -0.5 \sin(x - 30) + 1$:
   a) amplitude $= 0.5$; period $= 360°$; axis: $y = 1$
   b) domain: $\{x \in \mathbf{R}\}$; range: $\{y \in \mathbf{R} | 0.5 \le y \le 1.5\}$
7. **a)** vertical stretch of 3; vertical translation of $+5$
   b) vertical stretch of 2; reflection in the x-axis; vertical translation of -1

8. **a)** reflection in the *x*-axis   **c)** horizontal translation
    b) vertical stretch   **d)** vertical translation

9. **a)** $f(x) = \sin(x - 135°) + 5$
    b) $f(x) = \sin(x + 210°) - 7$

10. **a)**

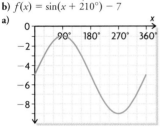

    b) $f(x) = 4 \sin x - 5$

11. **a)**

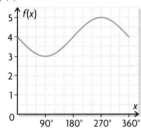

    b) $f(x) = -\sin x + 4$

12. **a)** max: 3; min: −3   **c)** max: 10; min: 2
    b) max: 2; min: −2   **d)** max: −2.5; min: −3.5

13.

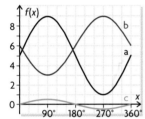

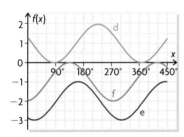

14. period
15. **a)** height of axle
    b) vertical translation
    c)

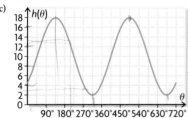

    d) $\{y \in \mathbf{R} \mid 2 \le y \le 18\}$
    e) 8; represents radius of Ferris wheel
    f) 13.4 m

16.

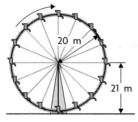

17. **a)** h: $f(x) = 4 \sin x + 6$
      k: $f(x) = -\sin x + 5$
      f: $f(x) = \sin x$
      g: $f(x) = -0.5 \sin x - 1$
    b) horizontal stretch of $\frac{1}{2}$

18. You can shift the graph vertically and then horizontally and/or compress/stretch.

19. **a)** 720°   **b)** 1440°   **c)** 180°   **d)** 36°

20. **a)** $y = 8$; height of axle
    b) 6 m; length of windmill blade
    c) 18 s; rotation speed of windmill
    d) horizontal stretch
    e) $y = 6 \sin(20(x - 4.5°)) + 8$
    f) The period would be longer.
    g) The graph would shift up 1 m.

Chapter Review, pp. 378–379

1. **a)** yes
    b) yes, once it reaches its constant speed
    c) yes, assuming an ideal wave
    d) yes

2. **a)** periodic   **d)** 1.5 s
    b) $\frac{1}{2}$ s   **e)** The flat line would be longer.
    c) 3 cm

3. **a)** −0.5 cm; makes sense because the function represents height relative to the ground, but it's digging and therefore below the ground
    b) $y = 30$; height of axle on the rotating drum relative to the ground
    c) 1 s; time for the drum to complete one full rotation
    d) 35 cm; distance from the axle to the tip of the digging teeth

4.

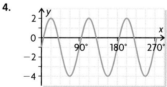

5.

6. **a)** $\{d \in \mathbf{R} \mid -1.5 \le d \le 1.5\}$; $\{d \in \mathbf{R} \mid -0.5 \le d \le 0.5\}$
    b) 0.02 s; 0.025 s; one is idling/vibrating faster than the other
    c) $y = 0$ for both; resting position
    d) 1.5 mm, 0.5 mm; how much they shake to the left and right of their resting positions

7. **a)** amplitude: 1; period: 360°; axis: $y = 3$; max: 4; min: 2

b) amplitude: 1; period: 360°; axis: $y = -2$; max: -1; min: -3

8. a)

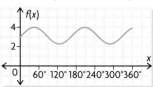

domain: $\{x \in \mathbf{R}\}$; range: $\{y \in \mathbf{R} \mid 2 \leq y \leq 4\}$

b)

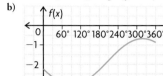

domain: $\{x \in \mathbf{R}\}$; range: $\{y \in \mathbf{R} \mid -3 \leq y \leq -1\}$

9. a) $f(x) = -0.5 \sin x + 4$

b)

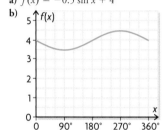

10. a) amplitude = 3; period = 360°; axis: $y = 0$; max: 3; min: -3
b) amplitude = 2; period = 360°; axis: $y = 0$; max: 2; min: -2
c) amplitude = 4; period = 360°; axis: $y = 6$; max: 10; min: 2
d) amplitude = 0.25; period = 360°; axis: $y = 0$; max: 0.25; min: -0.25
e) amplitude = 3; period = 360°; axis: $y = 0$; max: 3; min: -3

11. a) domain: $\{x \in \mathbf{R}\}$; range: $\{y \in \mathbf{R} \mid 2 \leq y \leq 8\}$
b) domain: $\{x \in \mathbf{R}\}$; range: $\{y \in \mathbf{R} \mid -3 \leq y \leq -1\}$
c) domain: $\{x \in \mathbf{R}\}$; range: $\{y \in \mathbf{R} \mid -1.5 \leq y \leq -0.5\}$
d) domain: $\{x \in \mathbf{R}\}$; range: $\{y \in \mathbf{R} \mid -2 \leq y \leq 2\}$
e) domain: $\{x \in \mathbf{R}\}$; range: $\{y \in \mathbf{R} \mid 2.5 \leq y \leq 3.5\}$

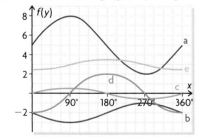

12. a)

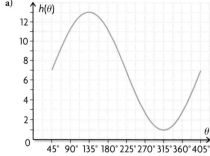

b) $\{h \in \mathbf{R} \mid 1 \leq h \leq 13\}$
c) 6; radius of Ferris wheel
d) 6.5 m

Chapter Self-Test, p. 380

1. a)

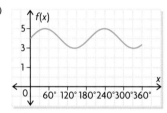

amplitude: 1; period: 360°; equation of the axis: $y = 4$; domain: $\{x \in \mathbf{R}\}$; range: $\{y \in \mathbf{R} \mid 3 \leq y \leq 5\}$

b)

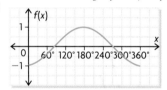

amplitude: 1; period: 360°; equation of the axis: $y = 0$; domain: $\{x \in \mathbf{R}\}$; range: $\{y \in \mathbf{R} \mid -1 \leq y \leq 1\}$

c)

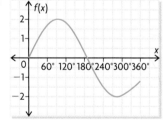

amplitude: 2; period: 360°; equation of the axis: $y = 0$; domain: $\{x \in \mathbf{R}\}$; range: $\{y \in \mathbf{R} \mid -2 \leq y \leq 2\}$

d)

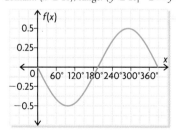

amplitude: 0.5; period: 360°; equation of the axis: $y = 0$; domain: $\{x \in \mathbf{R}\}$; range: $\{y \in \mathbf{R} \mid -0.5 \leq y \leq 0.5\}$

2.

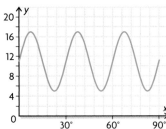

3. a) period: 60 min for both; the amount of time it takes for the minute hand on each clock to make one complete revolution;
axis: $y = 27$; $y = 18$; how far the centre of the clock face is away from the ceiling;
amplitude: 18 cm; 9 cm; length of the minute hands
   b) pointing up at the 12
   c) $\{d \in \mathbf{R} \mid 9 \le d \le 45\}$; $\{d \in \mathbf{R} \mid 9 \le d \le 27\}$
   d) 22 cm for the smaller clock; 36 cm for the larger clock
   e)

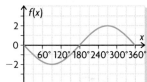

4. a) $f(x) = -2 \sin x$
   b)

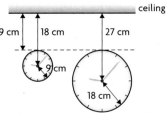

   c) amplitude: 2; axis: $y = 0$; period: $360°$; domain: $\{x \in \mathbf{R}\}$;
range: $\{y \in \mathbf{R} \mid -2 \le y \le 2\}$

5.

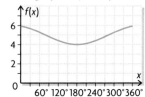

6. a) $f(x) = 4 \sin x$
   b) $f(x) = -\sin x + 2$
   c) $f(x) = -2 \sin x - 1$
   d) $f(x) = 3 \sin x + 4$

Cumulative Review Chapters 4-6, pp. 382–385

1. (b)	**6.** (d)	**11.** (a)	**16.** (d)	**21.** (b)	**26.** (c)
2. (d)	**7.** (a)	**12.** (a)	**17.** (b)	**22.** (c)	**27.** (b)
3. (b)	**8.** (c)	**13.** (b)	**18.** (a)	**23.** (d)	**28.** (b)
4. (d)	**9.** (c)	**14.** (d)	**19.** (a)	**24.** (b)	**29.** (c)
5. (d)	**10.** (d)	**15.** (c)	**20.** (b)	**25.** (a)	

30. a) $f(x) = -0.0732x^2 + 45.75$
   b) 6.24 m
31. a) 224.93 m   **b)** 428.98 m   **c)** $27.67°$

32. a)

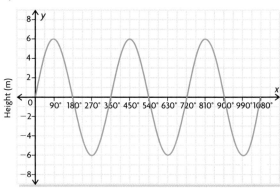

Angle of rotation (degrees)

   b) amplitude: 6 m; period: 1 revolution; axis $y = 0$;
range: $\{y \in \mathbf{R} \mid -6 \le y \le 6\}$
   c) The graph would be shifted to the right 0.25 revolutions. Also, from 0 revolutions to 0.25 revolutions, the height relative to the platform would go from -6 m to 0 m, and the shifted graph would end at 3 revolutions, not 3.25 revolutions.

Chapter 7

Getting Started, p. 388–390

1. a) (viii)   **c)** (iii)   **e)** (i)   **g)** (ii)
   b) (vi)   **d)** (v)   **f)** (iv)   **h)** (vii)
2. a) 5.9   **b)** 7.8   **c)** 4.8   **d)** 6.7
3. a) $5 \times 5 \times 5 \times 5 = 625$
   b) $(-4) \times (-4) \times (-4) = -64$
   c) $-2 \times 2 = -4$
   d) $(5 \times 2)(5 \times 2)(5 \times 2) = 1000$
   e) $\dfrac{3}{4} \times \dfrac{3}{4} = \dfrac{9}{16}$
4. a) 512   **c)** 15 625   **e)** 1024
   b) 14 641   **d)** 361   **f)** 1024
5. a) 25   **c)** -8 **e)** 10 000
   b) -25 **d)** -8 **f)** $-10\ 000$
6. a) $-19\ 683$ **c)** 6561   **e)** 729
   b) $-19\ 683$ **d)** 6561   **f)** -729
7. a) -7 **c)** 2500   **e)** -200
   b) 110   **d)** 400   **f)** -1
8. a) 9   **b)** 2   **c)** 3
9. a) $x = 4$ **c)** $y = 3$ **e)** $n = 3$
   b) $m = 1$ **d)** $x = 3$ **f)** $c = 3$
10. a) $-\dfrac{5}{28}$ **c)** $\dfrac{5}{6}$ **e)** $\dfrac{2}{3}$
   b) $\dfrac{35}{36}$ **d)** $1\dfrac{11}{12}$ **f)** $1\dfrac{4}{5}$
11. a) linear   **b)** quadratic   **c)** quadratic

Lesson 7.1, p. 394

1. The curve would eventually be the same as the room temperature.
2. a) It would take longer to cool down.
   b) It would cool down much faster.

3. a)

Year	Annual Payment ($1000)
1	5
2	10
3	20
4	40
5	80
6	160
7	320
8	640
9	1280
10	2560

b)

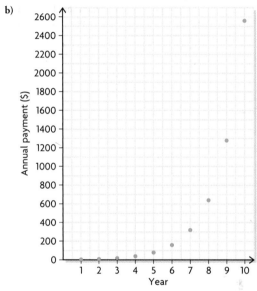

c) They are both exponential. The cooling curve is decay and the payment curve is growth.

Lesson 7.2, p. 399–401

1. a) 4^8 **b)** 13^{14} **c)** 7^3 **d)** 12^9

2. a) 7 **b)** 6^5 **c)** 9^2 **d)** 5^3

3. a) 2^{12} **b)** 12^9 **c)** 10^{28} **d)** 3^{24}

4. a) Answers may vary. E.g., $3^2 \times 3^3 = 3 \times 3 \times 3 \times 3 \times 3 = 3^5$
$$3^2 \times 3^3 = 3^{2+3} = 3^5$$
b) Answers may vary. E.g., $\dfrac{2^3}{2} = \dfrac{2 \times 2 \times 2}{2} = 2 \times 2 = 2^2$
$$2^3 \div 2 = 2^{3-1} = 2^2$$
c) Answers may vary. E.g., $(4^2)^3 = 4^2 \times 4^2 \times 4^2 = 4^{2+2+2} = 4^6$
$$(4^2)^3 = 4^{2 \times 3} = 4^6$$

5. a) 3^3 **c)** 7^5 **e)** 9
   b) 10^4 **d)** 6^5 **f)** 8^2

6. a) 2^{18} **c)** 7^{18} **e)** 10^4
   b) 5^3 **d)** 8^2 **f)** 4

7. a) 10^3 **b)** 8^3 **c)** 13^{12} **d)** 5^3

8. a) $\left(\dfrac{4}{3}\right)^3 = \dfrac{64}{27}$ **c)** $\left(\dfrac{2}{5}\right)^4 = \dfrac{16}{625}$

   b) $\left(\dfrac{1}{9}\right)^2 = \dfrac{1}{81}$ **d)** $\left(\dfrac{5}{4}\right)^2 = \dfrac{25}{16}$

9. a) x^8 **c)** y^{21} **e)** a^6
   b) m^2 **d)** a^8 **f)** b^5

10. a) 2^6 **c)** 3^{15} **e)** $\left(\dfrac{1}{2}\right)^6$

   b) 3^{10} **d)** $(-2)^{12}$ **f)** $\left(\dfrac{1}{5}\right)^6$

11. a) 4 **b)** 1 **c)** $10\,000$ **d)** 625

12. a) Clare multiplied the bases.
   b) $3^2(2^2)^2 = 3^2(2^4) = 9(16) = 144$

13. a) Multiply all the terms in the numerator by adding the exponents on the terms with the same base. Then divide the numerator by the denominator by subtracting the exponents on the terms with the same base.
   b) x^3y^2 **c)** -72

14. a) Since n will equal $a + b$ and n is even, $a + b$ must be even. Therefore, $a + b$ must be even.
   b) m must be even

15. a) No, they do not have the same value.
$$(-5)^2 = 25, \quad -5^2 = -25$$
   b) $(-5)^2 = 25, \quad (-5)^3 = -125$
   Conjecture: If the sign of the base is negative and the exponent is odd, the value of the power will be negative. If the sign of the base is negative and the exponent is even, the value of the power will be positive.

16. a) $n = 3$ **b)** $m = 14$

17. a) $16x^6$ **b)** $125x^9y^{12}$ **c)** $\dfrac{2}{3x^6y^2}$

18. a) Answers may vary. E.g., $x = \dfrac{1}{2}$

   b) $x = \dfrac{1}{\sqrt{2}}$
   c) Answers may vary. E.g., $x = 1$
   d) not possible

Lesson 7.3, p. 407–409

1. a) $\dfrac{1}{4^6}$ **b)** $\left(\dfrac{3}{7}\right)^5 = \dfrac{3^5}{7^5}$ **c)** 8^2 **d)** $\dfrac{1}{(-3)^2}$

2. a) 1 **b)** $\dfrac{1}{125}$ **c)** $\dfrac{8}{27}$ **d)** $\dfrac{1}{16}$

3. a) $0.015\,625$ **b)** $0.015\,625$ **c)** 0.064 **d)** -8

4. a) $\dfrac{1}{100}$ **c)** 32 **e)** $\dfrac{1}{81}$

   b) $\dfrac{1}{16}$ **d)** 343 **f)** 1

5. a) 9^4 **c)** 8^{11} **e)** $(-3)^1$
   b) 6^2 **d)** 17^2 **f)** $(-4)^0$

6. a) 2^{12} **d)** 3^2

   b) $(-5)^{-11} = \dfrac{1}{(-5)^{11}}$ **e)** $9^{-5} = \dfrac{1}{9^5}$

   c) $(-12)^{-10} = \dfrac{1}{(-12)^{10}}$ **f)** 7^{24}

7. a) 11^9 **c)** 4^3 **e)** $(-8)^0 = 1$
b) $\dfrac{1}{9^{12}}$ **d)** 10^6 **f)** 5

8. a) $13^{-1} = \dfrac{1}{13^1}$ **c)** $10^4 = 10\,000$ **e)** $2^{-6} = \dfrac{1}{2^6} = \dfrac{1}{64}$
b) $3^4 = 81$ **d)** $6^0 = 1$ **f)** $5^3 = 125$

9. a) 0 **c)** $\dfrac{5}{64}$ **e)** $\dfrac{32}{3}$
b) $\dfrac{15}{16}$ **d)** 5 **f)** $\dfrac{1}{2}$

10. Negative exponents indicate that you need to move the decimal place to the left and this will give you very small numbers.

11. a) $x^0 = 1$ **c)** w^6 **e)** a^5
b) m^{18} **d)** a^{12} **f)** $b^{-15} = \dfrac{1}{b^{15}}$

12. a) $x = 0$ **c)** $n = -4$ **e)** $b = -5$
b) $m = 5$ **d)** $k = -1$ **f)** $a = 3$

13. a) Sasha's solution is incorrect because he multiplied the bases to get 4. He added the exponents correctly. Then, he correctly evaluated 4^{-1}, except that he made the whole number negative. Negative exponents do not change the sign of the number. Vanessa multiplied the exponents instead of adding them. After this error, the expression was evaluated correctly.
b) $2^{-2} \times 2 = 2^{-1} = \dfrac{1}{2^1} = \dfrac{1}{2}$

14. a) $2^{12} = 4^6 = 8^4 = 16^3 = 4096$
b) Write each base using powers of two:
$16^3 = (2^4)^3 = 2^{12}$
$8^4 = (2^3)^4 = 2^{12}$
$4^6 = (2^2)^6 = 2^{12}$
c) Answers may vary. E.g., $3^6 = 9^3 = 27^2$

15. $\dfrac{y^{-4}(x^2)^{-3}y^{-3}}{x^{-5}(y^{-4})^2}$, $(y^{-5})(x^5)^{-2}(y^2)(x^{-3})^{-4}$, $\dfrac{x^{-3}(y^{-1})^{-2}}{(x^{-5})(y^4)}$

Lesson 7.4, p. 415–417

1. a) $\sqrt{49} = 7$ **c)** $\sqrt[4]{81} = 3$ **e)** $\sqrt[4]{16} = 2$
b) $\sqrt[3]{-125} = -5$ **d)** $\sqrt{100} = 10$ **f)** $-\sqrt{144} = -12$

2. a) $1024^{\frac{1}{2}} = 32$ **c)** $27^{\frac{4}{3}} = 81$ **e)** $16^{\frac{1}{4}} = 2$
b) $1024^{\frac{1}{5}} = 4$ **d)** $(-216)^{\frac{5}{3}} = 7776$ **f)** $25^{-\frac{1}{2}} = \dfrac{1}{5}$

3. a) 2.05 **c)** 1.75 **e)** 0.22
b) 0.5 **d)** 3.46 **f)** 0.12

4. a) -2 **c)** 2 **e)** $-\dfrac{1}{3}$
b) -2 **d)** -32 **f)** 3

5. length of the side of the cube, 0.3 cm

6. a) $7^{\frac{5}{2}}$ **c)** 16^4 **e)** $10^{-\frac{5}{4}}$
b) $3^{\frac{7}{2}}$ **d)** $12^{\frac{1}{2}}$ **f)** 2^5

7. a) $\dfrac{1}{3^2}$ **c)** 8^5 **e)** 3
b) $\dfrac{1}{5^4}$ **d)** $\dfrac{1}{4^{1.2}}$ **f)** $\dfrac{1}{16}$

8. $10^{1.5} \doteq 31.55$; $10^{-0.5} \doteq \dfrac{1}{3.16}$

9. a) 10 **c)** 7 **e)** 5
b) 6 **d)** $\dfrac{1}{2}$ **f)** 18

10. a) $16^{\frac{11}{2}}$ **c)** $12^{\frac{-1}{4}} = \dfrac{1}{12^{\frac{1}{4}}}$ **e)** 4^2
b) $8^{-2} = \dfrac{1}{8^2}$ **d)** $11^{\frac{-13}{8}} = \dfrac{1}{11^{\frac{13}{8}}}$ **f)** $16^{\frac{1}{4}}$

11. a) $4^{\frac{13}{8}} = \sqrt[8]{4^{13}}$ **c)** $10^{\frac{1}{4}} = \sqrt[4]{10}$ **e)** $\dfrac{1}{\sqrt[5]{5}}$
b) $9^{\frac{19}{30}} = \sqrt[30]{9^{19}}$ **d)** $8^{\frac{1}{15}} = \sqrt[15]{8}$ **f)** $\sqrt[4]{4^9}$

12. a) 4.932 **c)** 0.358 **e)** 0.028
b) 11.180 **d)** 0.158 **f)** 0.164

13. $(-8)^{\frac{1}{3}} = -2$, $-(4)^{\frac{1}{2}} = -\sqrt{4} = -2$
They give the same answer.

14. a) $m^{\frac{5}{3}}$ **c)** $c^{\frac{5}{2}}$ **e)** $\dfrac{1}{s^{3.25}}$
b) $x^{\frac{7}{3}}$ **d)** b^2 **f)** $m^0 = 1$

15. a) $\dfrac{1}{t^{\frac{6}{5}}}$ **c)** $\dfrac{1}{y^3}$ **e)** x^1
b) $x^{\frac{41}{24}}$ **d)** a^3 **f)** $\dfrac{1}{b^3}$

16. $64^{\frac{-5}{3}} = \dfrac{1}{64^{\frac{5}{3}}} = \dfrac{1}{(\sqrt[3]{64})} = \dfrac{1}{y^3} = \dfrac{1}{1024}$

17. a) $M = 1.05^3 = 1.02$ **d)** $x = \sqrt[3]{125} = 5$
b) $T = 2.5^4 = 39.06$ **e)** $x = 8$
c) $N = 3^5 = 243$ **f)** $y = 81$

18. a) $1000^{\frac{2}{3}} = 100$ **b)** $64^{\frac{16}{3}} = 12\,816$

19. $32^{\frac{4}{5}} = (32^{\frac{1}{5}})^4 = 2^4 = 16$ or $32^{\frac{4}{5}} = (32^4)^{\frac{1}{5}} = 1\,048\,576^{\frac{1}{5}} = 16$

20. a) $\dfrac{17}{15}$ **b)** $-\dfrac{2}{3}$ **c)** $\dfrac{1}{\sqrt[5]{4}}$

21. $\dfrac{1}{\sqrt{9261}}$

22. (a)

Mid-Chapter Review, p. 419

1. a) 5^5 **c)** 16^{10} **e)** $\left(\dfrac{1}{10}\right)^2$ or $\dfrac{1}{10^2}$
b) 8^2 **d)** $\dfrac{1}{(-4)^9}$ **f)** 7^{10}

2. a) $\dfrac{1}{x^2}$ **c)** $\dfrac{1}{b^5}$ **e)** n^{14}
b) $\dfrac{1}{m^8}$ **d)** y^2 **f)** $\dfrac{1}{y^4}$

3. a) $\dfrac{7}{50}$ **c)** 0 **e)** $-\dfrac{7}{64}$
b) $\dfrac{49}{64}$ **d)** 45 **f)** $\dfrac{1}{5}$

4. a) $\dfrac{3}{2}$ **c)** $\dfrac{9}{4}$ **e)** $\dfrac{2}{3}$
b) $-\dfrac{125}{8}$ **d)** 256 **f)** $\dfrac{1}{64}$

5. a) $a^{\frac{13}{15}}$ **c)** $\dfrac{1}{c^5}$ **e)** e^{11}

b) $b^{\frac{1}{2}}$ **d)** d^2 **f)** $\dfrac{1}{f^{\frac{5}{12}}}$

6.

	Exponential Form	Radical Form	Evaluation of Expression
a)	$100^{\frac{1}{2}}$	$\sqrt{100}$	10
b)	$16^{0.25}$	$\sqrt[4]{16}$	2
c)	$121^{0.5}$ or $121^{\frac{1}{2}}$	$\sqrt{121}$	11
d)	$(-27)^{\frac{5}{3}}$	$\sqrt[3]{-27^5}$ or $(\sqrt[3]{-27})^5$	-243
e)	$49^{2.5}$	$\sqrt{49^5}$ or $(\sqrt{49})^5$	16 807
f)	$1024^{0.1}$ or $1024^{\frac{1}{10}}$	$\sqrt[10]{1024}$	2
g)	$\left(\dfrac{1}{2}\right)^{\frac{9}{3}}$ or $\left(\dfrac{1}{2}\right)^3$	$\sqrt[3]{\left(\dfrac{1}{2}\right)^9}$	$\dfrac{1}{8}$ or 0.125

7. a) 3.659 **c)** 0.072 **e)** -2.160

b) 46.062 **d)** 1.414 **f)** 25

8. a) 1 953 125 **b)** 16 **c)** 3 **d)** 4

Lesson 7.5, p. 423–424

1. a) exponential **b)** linear **c)** quadratic

2. a) i) y-intercept is 1

ii)

iii) domain: $\{x$ is any real number$\}$
range: $\{y > 0\}$

iv) $y = 0$

b) i) y-intercept is 1

ii)
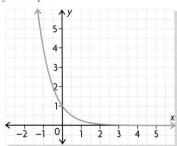

iii) domain: $\{x$ is any real number$\}$
range: $\{y > 0\}$

iv) $y = 0$

c) i) y-intercept is -1

ii)
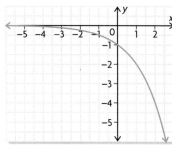

iii) domain: $\{x$ is any real number$\}$
range: $\{y < 0\}$

iv) $y = 0$

d) i) y-intercept is 2

ii)
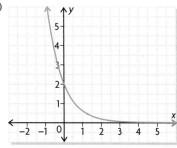

iii) domain: $\{x$ is any real number$\}$
range: $\{y > 0\}$

iv) $y = 0$

e) i) y-intercept is -2

ii)
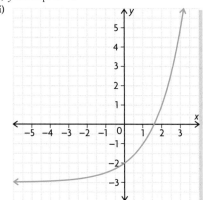

iii) domain: $\{x$ is any real number$\}$
range: $\{y > -2\}$

iv) $y = -3$

f) i) y-intercept is 9

ii)
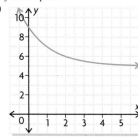

iii) domain: {x is any real number}
   range: {$y > 5$}
iv) $y = 5$

3. a) $h(x)$ is exponential. It has a variable as an exponent.
  b) The first differences are not constant but are related to each other by a factor of 4.
  c) $f(x)$ is linear because it has a degree of 1. $g(x)$ is quadratic because it has a degree of 2.
  d) The first differences for $f(x)$ are constant. The second differences for $g(x)$ are constant.

4. a) (ii) **c)** (i) **e)** (v)
  b) (vi) **d)** (iii) **f)** (iv)

Lesson 7.6, p. 429–432

1. a) approximately $6400
  b) around $12 500
  c) Yes, it is approx. 18 years.
2. a) 0.25 is the initial coverage of the pond.
  b) 1.1 is 1 + 10%. 10% is the rate of growth.
  c) 65%
3. a) 364 552 **c)** 19 **e)** 551 222
  b) 2.2% per year **d)** $P = P_0(1.022)^t$
4. a)

	Initial Value	Growth Rate	Number of Growth Periods
i)	18	5%	22
ii)	64 000	10%	12
iii)	750	100%	4

  b) i) 52.65 **ii)** 200 859.42 **iii)** 12 000
5. a) 20 cells **c)** 15 h **e)** The doubling times are the same for any amount chosen.
  b) 80 cells **d)** 15 h
6. a) For: $N = 12(1 + 0.04)^t$, N represents the total number of guppies after t weeks. There are 12 guppies to start with and they grow at a rate of 4% per week.
  b) 17 guppies
  c) No. We assume the growth rate will slow as the aquarium becomes more crowded.
7. a) $V = 5000(1.0325)^n$; V represents value and n represents the number of years
  b) $P = 2500(1.005)^t$; P represents population and t represents the number of years
  c) $P = (2)^d$; P represents population and d represents the number of days
8. a) $V = 2000(1.06)^n$ **b)** $2524.95
  c) $2676.45, $151.50 subtract $2524.95 from $2676.45
  d) $6414.27 and $6799.13; $384.86
  e) fourth year: $151.50, twentieth year: $384.86
     Money grows faster the longer that it has been invested for.

9. a)

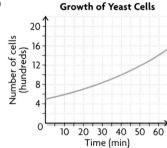

Growth of Yeast Cells

  b) It takes about 63 min for the number of cells to triple.
  c) Extend the horizontal axis to the left and extend the graph to the left.
10. a) 200%, initial population = 24 000 **c)** 648 thousand
  b) $N = 24\,000(3)^n$ **d)** 99 ants
11. a) $y = 1.33^x$ **b)** 67 641 stores
12. a)

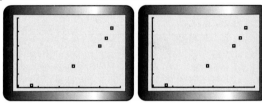

  b) $P(n) = 6(1 + 0.08)^n = 6(1.08)^n$
  c) It should be close to the data.
  d) The equation comes close to this population.
13. a) 4.8 m
  b) The height of the tree should get greater each year but slow down as the tree gets older.
14. approximately 3.9%
15. on the 29th day

Lesson 7.7, p. 437–440

1. a)

	Initial Value	Decay Rate	Number of Decay Periods
i)	100	25%	28
ii)	32 000	44%	12
iii)	500	2.5%	20

  b) i) 0.03 **ii)** 30.44 **iii)** 301.34
2. a) approximately 58% **b)** approximately 50 years
3. a) $15 102.24 lost **b)** $14 613.22
4. a) 19.62 **b)** 94.15 **c)** 1
5. a) $V = V_0(1 - 0.032)^d$; V represents the volume of air in the ball after d days. V_0 represents the original volume of air. The ball loses 3.2% of its volume each day.
  b) approximately 737.53 cm^3
  c) The ball will eventually have no volume of air. As a result, the equation will not fit once there is no air in the ball.
6. a) approximately 18 g **b)** 5 min
7. a) $I = (0.96)^n$; I represents the intensity of the light as a percent of the original intensity with n gels in front of the spotlight. With each gel, 96% of the light gets through.
  b) approximately 88.5%
  c) 7 gels

8. **a)** $25\ 000$ **c)** $15\ 353.13$ **e)** 1957.52
b) -15% **d)** 3750 **f)** approximately 4.25 years

9. **a)** The base of the power is $\frac{1}{2}$, which is less than 1.
b) 22
c) It is the time required for the temperature to reduce by $\frac{1}{2}$.
d) $68\ ^\circ\text{C}$
e) $37\ ^\circ\text{C}$

10. **a)** 48.4% **b)** 4800 years

11. **a)** $V = 25000(0.75)^n$. The car *depreciates* (loses value) at a rate of 25% per year.
b) $R = R_0\left(\dfrac{1}{2}\right)^{\frac{t}{4.5\times 10^9}}$. The amount of U_{238} is halved each time period.
c) $I = I_0(0.88)^n$. The light intensity is reduced by 12% with each metre of depth.

12. **a)** The population *declined* each year by an average *percent*.
b) $P = 13\ 700(0.95)^n$; 13 700 was the initial population, 0.95 represents a 5% decline each year, and there are n years.
c) just after 2019

13. **a)**

Bounce	Max. Height after Bounce (m)
Initial height	5
1	$5 \times 0.80 = 4$
2	$4 \times 0.80 = 3.2$
3	2.56
4	2.048

b) approximately 1.05 m

14. **a)** The shirt loses a percent of its colour with each wash. $C = 100(0.995)^n$; C is the percent of colour left. The shirt has 100% of its colour to begin with. The shirt has 99.5% of its colour left after each wash. There are n washes. Algebraically by using the equation above and guessing and checking until you get a value of 85%. Graphically, by graphing the function and determining when the line falls below the 85% level.
b) 32 washes

15. **a)** 20% lost **b)** $M = 50(0.80)^d$ **c)** approximately 30.3 kg

Chapter Review, p. 444–445

1. **a)** 3^{13} **c)** $\dfrac{1}{11^4}$ **e)** $4^0 = 1$
b) $(-5)^2$ **d)** $(-9)^{10}$ **f)** $\dfrac{1}{6^2}$

2. **a)** 12^2 **c)** $\dfrac{1}{20^{16}}$
b) $(-8)^0 = 1$ **d)** 10^4

3. **a)** a^2 **c)** $\dfrac{1}{c^6}$ **e)** $\dfrac{1}{e}$
b) b^7 **d)** d^{18} **f)** f^6

4.

a)
b)

7. **a)** quad...
b) linear

8. **a)** approximately 178 ...
b) 8 years

9. **a)**

Growth of Yeas...

(graph: *Number of cells (hundreds)* vs *Time (min)*, y-axis 0 to 200, x-axis 30, 60, 90, 120, 150, 180, 210)

b) 8000 cells **c)** Extend the graph to the left to -180.

10. **a)** $V(t) = 2500(1.05)^t$
b) $P(t) = 750(1.02)^t$
c) $V(10) = \$4\ 072.24$; value of coin in 2010; $P(10) = 914$; number of students enrolled in this school in 2013

11. **a)** $500 -$ card value in 2000
$0.07 = 7\%$ appreciation rate
b) 1934.84
c) about 10 years and 3 months

12. **a)** I is the intensity, in percent, at a depth of x metres. *100* is the percent of light at the surface. *0.94* is obtained by subtracting 6% from 1. 6% is the rate of decay of light intensity per metre.
b) approximately 37%

13.

Function	Exponential Growth or Decay	Initial Value (y-intercept)	Growth/ Decay Rate
$V = 125(0.78)^t$	decay	125	loss of 22%
$P = 0.12(1.05)^t$	growth	0.12	gain of 5%
$A = (2)^x$	growth	1	gain of 100%
$Q = 0.85\left(\dfrac{1}{3}\right)^x$	decay	0.85	loss of $\dfrac{2}{3}$

14. **a)** $A(n) = 500(1.05)^n$
b) 578.81
c) 78.81
d) No, in 6 years, he saves $67...

1. **a)** $\dfrac{1}{125}$
b) $\dfrac{16}{9}$

2. **a)** 6^3
b) 4^5

3.

4.

c) 2 e) −1

d) $\frac{1}{8}$ f) $\frac{1}{1000}$

c) 10^5 e) $\frac{1}{a^5}$

d) 7^2 f) $\frac{1}{b^3}$

2

$y = 2^x$ and $y = 0.5^x$

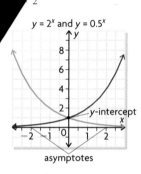

5. a) Car A starts at $20 000 while car B has an initial value of $25 000. Car B's value declines more quickly than car A's.
  b) Since car B's value falls faster, it has a higher depreciation rate.
6. 23%
7. a) $P = 1600(1.015)^n$. The population, P, is initially 1600 and grows at 1.5% for n years.
  b) $n = 2008 - 1980 = 28$, so $P \doteq 2428$

Chapter 8

Getting Started, pp. 450–452

1. a) (ii) **b)** (i) **c)** (iv) **d)** (iii)
2. a) 0.35 **b)** 0.67 **c)** 0.085 **d)** 0.0275
3. a) 11.25 **b)** 51 **c)** 2.1 **d)** 145
4. a) 12 288, 0.000 73 **c)** 9, 4
  b) 320, 0.078 125 **d)** 24 379.89, 29 718.95
5. a) $\frac{6}{73}$ **c)** $\frac{2}{3}$ **e)** $\frac{25}{13}$
  b) $\frac{1}{2}$ **d)** $\frac{80}{73}$ **f)** $\frac{3}{2}$
6. a) 42 **b)** 117 **c)** 1825 **d)** 3
7. a) 0.35 **c)** 0.146 **e)** 0.0275
  b) 0.05 **d)** 1.15 **f)** 0.1475
8. a) $4.88 **c)** $247.50 **e)** $3.00
  b) $3.92 **d)** $30.00 **f)** $2000.00
9. a) 3.14 **c)** 1464.10 **e)** 1754.87
  b) 0.49 **d)** 2220.06 **f)** −281 825.99

10. a)

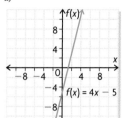

b)

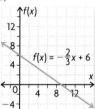

c)

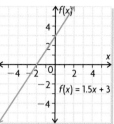

d)

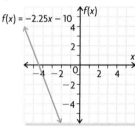

11. a)

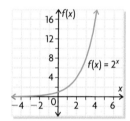

b)

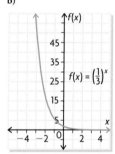

c)

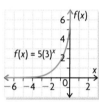

d)

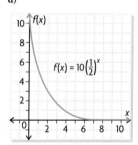

12. $13.83
13. $10 462.10

Lesson 8.1, pp. 459–461

1. a) $3.00 **b)** $4.50
2. a) Interest: $97.88, Amount: $772.88
  b) Interest: $1171.78, Amount: $5432.78
3. a) Interest: $94.58, Amount: $694.58
  b) Interest: $112.03, Amount: $862.03
4. a) Interest: $120.00, Amount: $620.00
  b) Interest: $480.00, Amount: $2480.00
  c) Interest: $56.25, Amount: $1306.25
  d) Interest: $23.08, Amount: $1023.08
  e) Interest: $30.14, Amount: $5030.14

5.

	Regular-Interest CSB, $11\frac{1}{4}$%, 5 years, $500			Compound-Interest CSB, $11\frac{1}{4}$%, 5 years, $500		
Year	Interest Earned ($)	Accumulated Interest ($)	Amount at End of Year ($)	Interest Earned ($)	Accumulated Interest ($)	Amount at End of Year ($)
1	56.25	56.25	556.25	56.25	56.25	556.25
2	56.25	112.50	612.50	62.58	118.83	618.83
3	56.25	168.75	668.75	69.62	188.45	688.45
4	56.25	225.00	725.00	77.45	265.90	765.90
5	56.25	281.25	781.25	86.16	352.06	852.06

6.

	Principal, P ($)	Interest Rate, r (%)	Time, t	Simple Interest, I ($)
a)	735.00	$5\frac{1}{2}$	27 days	2.99
b)	2548.55	8.25	240 days	138.25
c)	182.65	6.75	689 days	23.28
d)	260.00	38.08	2 months	16.50

7. 156 days

8. 19.7%

9. $3157.89

10. **a)** $1152.00   **b)** $1297.76

11. **a)** Compound interest earned is $740.17. Simple interest earned is $700. Therefore, the GIC that earns compound interest earns more.
      b) $40.17

12. Simple interest is calculated on the original principal. Compound interest is calculated on the principal plus interest earned so far.

13. **a)** $750.00   **b)** $795.80

14. $12 104.00

15. $5619.87

Lesson 8.2, pp. 468–470

1. **a)** $i = 0.09$, $n = 4$ **c)** $i = 0.0225$, $n = 8$
    b) $i = 0.045$, $n = 12$ **d)** $i = 0.0075$, $n = 36$

2.

	Principal ($)	Annual Interest Rate (%)	Time (years)	Compounding Frequency	Rate for the Compounding Period, i (%)	Number of Compounding Periods, n	Amount ($)	Interest Earned ($)
a)	400	5	15	annually	0.05	15	831.57	431.57
b)	750	13	5	semi-annually	0.065	10	1407.85	657.85
c)	350	2.45	8	monthly	0.002	96	424.00	74.00
d)	150	7.6	3	quarterly	0.019	12	188.01	38.01
e)	1000	4.75	4	daily	0.0001	1460	1157.19	157.19

3. **a)** $i = 0.0575$, $n = 3$ **c)** $i = 0.014\ 375$, $n = 12$
    b) $i = 0.02875$, $n = 10$ **d)** $i = 0.004\ 792$, $n = 24$

4.

	Principal ($)	Interest Rate (%)	Years	Compounding Frequency		i	n	Amount ($)	Interest Earned ($)
a)	800	8	10	annually		0.08	10	1727.14	927.14
b)	1500	9.6	3	semi-annually		0.048	6	1987.28	487.28
c)	700	$3\frac{1}{2}$	5	monthly		0.002 92	60	833.83	133.83
d)	300	7.25	2	quarterly		0.018 125	8	346.36	46.36
e)	2000	$4\frac{1}{4}$	$\frac{1}{2}$	daily		0.000 116 4	183	1157.19	157.19

5. $19 535.12

6. $1560.14

7. Years ago

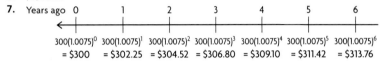

$300(1.0075)^0$ $300(1.0075)^1$ $300(1.0075)^2$ $300(1.0075)^3$ $300(1.0075)^4$ $300(1.0075)^5$ $300(1.0075)^6$
= $300   = $302.25   = $304.52   = $306.80   = $309.10   = $311.42   = $313.76

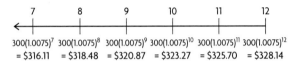

$300(1.0075)^7$ $300(1.0075)^8$ $300(1.0075)^9$ $300(1.0075)^{10}$ $300(1.0075)^{11}$ $300(1.0075)^{12}$
= $316.11   = $318.48   = $320.87   = $323.27   = $325.70   = $328.14

Interest earned = $328.14 − $300 = $28.14

8. 0.126 24 or 12.625%

9. 0.06 or 6%

10. $n = 6.637$ or about 7 years

11. $13 040.38

12. **a)** Bank: $A = \$5255.80$; Dealer: $A = \$5265.95$
    b) Take the dealer loan because the effective annual interest rate is lower.

13. $500 invested at 1%/a compounded annually for x years.

14. **a)**

Compounding Frequency	Number of Compounding Periods per Year	Formula	Amount ($)
annually	1	$1000(1 + 0.1)$	1100.00
semi-annually	2	$1000(1 + 0.05)^2$	1102.50
quarterly	4	$1000\left(1 + \frac{0.1}{4}\right)^4$	1103.81
monthly	12	$1000\left(1 + \frac{0.1}{12}\right)12$	1104.71
weekly	52	$1000\left(1 + \frac{0.1}{52}\right)^{52}$	1105.06
daily	365	$1000\left(1 + \frac{0.1}{365}\right)^{365}$	1105.16
hourly	8760	$1000\left(1 + \frac{0.1}{8760}\right)^{8760}$	1105.17

    b) The largest increase in amount occurs when the compounding frequency is changed from annual to semi-annual. After the frequency change from annual to semi-annual, the amount increases minimally.
    c) It appears that $1105.17 is a maximum amount. It would not be feasible to compound interest by the minute.
    d) Banks have many accounts and investment vehicles. Computers are used to process the financial information. Hourly compounding frequencies would require too much computer time. The amount of increase in amount from daily to hourly would not be considered a significant benefit by bank customers.

15.

	Formula	P ($)	Compounding Frequency	i (%)	n	Annual Interest Rate (%)	Number of Years	A ($)	I ($)
a)	$145(1 + 0.0475)^{12}$	145	annually	0.047 5	12	4.75	12	253.06	108.06
b)	$850(1 + 0.195)^5$	850	annually	0.195	5	19.5	5	2 071.37	1221.37
c)	$4500\left(1 + \frac{0.0525}{365}\right)^{1095}$	4500	daily	0.000 144	1095	5.25	3	5 267.55	767.55
d)	$4500\left(1 + \frac{0.15}{12}\right)^{78}$	4500	monthly	0.012 5	78	15	6.5	11 858.37	7358.37
e)	$4500\left(1 + \frac{0.03}{4}\right)^{20}$	4500	quarterly	0.007 5	20	3	5	5 225.33	725.33

16. Answers may vary. E.g., larger terms offer the best rate of interest, but customers may prefer smaller terms if they want access to their money earlier.

17. 10.4%

18. $7472.58

19. $n = 9.75$ or about 10 years

Lesson 8.3, pp. 476–479

1. a) $P = \$86.38$ **b)** $P = \$417.78$

2.

	Future Value ($)	Annual Interest Rate (%)	Time Invested (years)	Compounding Frequency	i (%)	n	Present Value ($)	Interest Earned I = A − P ($)
a)	4 000	5	15	annually	0.05	15	1924.07	2075.93
b)	3 500	2.45	8	monthly	$\frac{2.45}{1200}$	96	2877.62	622.38
c)	10 000	4.75	4	daily	$\frac{4.75}{36\,500}$	1460	8269.69	1730.31

3.

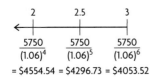

Years ago 0   1   2   3   4

$\frac{150}{(1.05)^0}$ $\frac{150}{(1.05)^1}$ $\frac{150}{(1.05)^2}$ $\frac{150}{(1.05)^3}$ $\frac{150}{(1.05)^4}$

$= \$150$ $= \$142.86$ $= \$136.05$ $= \$129.58$ $= \$123.41$

4.

	Future Value ($)	Annual Interest Rate (%)	Time Invested (years)	Compounding Frequency	i (%)	n	Present Value ($)	Interest Earned I = A − P ($)
a)	8000	10	7	annually	0.1	7	4105.26	3894.74
b)	7500	13	5	semi-annually	0.065	10	3995.45	3504.55
c)	1500	7.6	3	quarterly	0.019	12	1196.74	303.26

5.

Years ago 0   0.5   1   1.5

$\frac{5750}{(1.06)^0}$ $\frac{5750}{(1.06)^1}$ $\frac{5750}{(1.06)^2}$ $\frac{5750}{(1.06)^3}$

$= \$5750$ $= \$5424.53$ $= \$5117.48$ $= \$4827.81$

2   2.5   3

$\frac{5750}{(1.06)^4}$ $\frac{5750}{(1.06)^5}$ $\frac{5750}{(1.06)^6}$

$= \$4554.54$ $= \$4296.73$ $= \$4053.52$

6. $8609.76
7. $4444.74
8. a) $1717.85, $8282.15   b) $1655.57, $8344.43
9. $74.35
10. $2862.81
11. $2768.38
12. a) $98 737.24   b) $151 262.76   c) $129 795.92
13. a) i) $21 500 ii) $13 081.40
    b) i) $100 000 ii) $28 008.45

14.

	Future-Value Formula	A ($)	Compounding Frequency	i (%)	n	Annual Interest Rate (%)	Number of Years	Present Value ($)
a)	$280\,000 = P(1 + 0.0575)^{24}$	280 000	annually	0.057 5	24	5.75	24	73 186.17
b)	$16\,000 = P(1 + 0.20)^{5}$	16 000	annually	0.20	5	20	5	6 430.04
c)	$10\,000 = P\left(1 + \frac{0.0425}{365}\right)^{1460}$	10 000	daily	0.000 12	1460	4.25	4	8 436.73
d)	$9500 = P\left(1 + \frac{0.15}{12}\right)^{50}$	9500	monthly	0.012 5	50	15	4 years 2 months	5 104.72
e)	$1500 = P\left(1 + \frac{0.03}{4}\right)^{24}$	1500	quarterly	0.007 5	24	3	6	1 253.75

15. Savings account: $4347.68 must be invested now; GIC: $4369.20 must be invested now. Marshall should invest in the savings account since the present value required is less.
16. $2000
17. $911.39
18. $75 305.12

Lesson 8.4, pp. 486–488

1. Solutions are **in bold**.

	N	I%	PV	PMT	FV	P/Y	C/Y
a)	8	4.5	−600	0	**858.27**	1	4
b)	**11.6**	2.5	−6 000	0	8000	1	2
c)	5	**14.1**	−20 000	0	40 000	1	4
d)	1	0.6	**−847.62**	0	900	1	52

2. $54 059.13
3. Plan A
4. a) $16 288.95   b) $26 532.98   c) $43 219.42
5. a) present value: $4444.98, interest: $555.02
   b) present value: $10 625.83, interest: $2874.17
   c) present value: $8991.83, interest: $2208.17
   d) present value: $77 030.40, interest: $51 469.60
   e) present value: $797.31, interest: $52.69
   f) present value: $4849.55, interest: $1375.45
6. 5.95%
7. 11.032 years
8. 4.25%
9. 2 years
10. $4758.14

11. $5200.00
12. $4514.38
13. $3427.08
14.

Month	0	1	2	3
a) A ($)	465	471.98	479.05	486.24
b) I ($)	0	6.98	14.05	21.24

15. Answers will depend on online calculators chosen. Similarities should include basic variables such as present value, future value, payments, i, n; differences in format, and so on.
16. Between 321.7% and 357.3%
17. Dealer: $P = 32\,000$, $i = 0.002$, $n = 60$, $R = \$566.51$; Bank: $P = 29\,000$, $i = 0.0045$, $n = 60$, $R = \$552.60$; The monthly payments for the loan to the bank are less than those to the dealer.
18. $16 637.84

Mid-Chapter Review, pp. 491–492

1.

	Principal ($)	Annual Interest Rate (%)	Time	Simple Interest Paid ($)	Amount ($)
a)	250	2	3 years	**15.00**	**265.00**
b)	**399.98**	2.5	200 weeks	38.46	**438.44**
c)	1000	3.1	18 months	46.50	**1046.50**
d)	5000	5	30 weeks	**144.23**	**5144.23**
e)	**750**	4.2	5 years	157.50	**907.50**
f)	**1500**	3	54 months	202.50	**1702.50**

2. 3 years

3.

	Your Investment (10% Simple Interest)		Friend's Investment (10% Compound Interest)	
Year	Interest Earned ($)	Accumulated Interest ($)	Interest Earned ($)	Accumulated Interest ($)
1	50.00	50.00	50.00	50.00
2	50.00	100.00	55.00	105.00
3	50.00	150.00	60.50	165.50
4	50.00	200.00	66.55	232.05
5	50.00	250.00	73.21	305.26
6	50.00	300.00	80.52	385.78
7	50.00	350.00	88.58	474.36
8	50.00	400.00	97.43	571.79
9	50.00	450.00	107.18	678.97
10	50.00	500.00	117.90	796.87

4.

	Principal ($)	Annual Interest Rate (%)	Years Invested	Compounding Frequency	Amount ($)	Interest Earned ($)
a)	400.00	5	15	annually	**831.57**	**431.57**
b)	350.00	2.45	8	monthly	**425.70**	**75.70**
c)	**420.05**	3.5	5	quarterly	500.00	**79.95**
d)	120.00	**3.2**	7	semi-annually	150.00	**30.00**
e)	2500.00	7.6	**1 yr 9 mths**	monthly	**2 850.00**	350.00
f)	10 000.00	7.5	3	quarterly	**12 497.16**	**2497.16**

5. $270.58

6. $1225.35

7. Option 1: A = $60 110.84; Option 2: A = $60 047.92; Option 1 is the better choice because it earns $62.92 more interest.

8. $7413.72

9. $752.31

10. $3233.46

11. 2.2 %

12. 14 years

13. 33 years

14. Bank A: $563.58, Bank B: $ 573.76; She should go with Bank B as it will provide Sarah with the most interest.

Lesson 8.5, pp. 498–500

1.

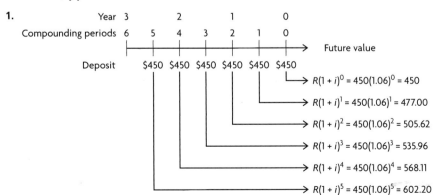

2. Geoff's investment earns $391.87 more than Marilyn's.

3. **a)** $4607.11   **b)** $29 236.22

4.

	Payment ($)	Interest Rate	Compounding Period	Term of Annuity	Amount ($)
a)	1000	8% per year	annually	3 years	**3 246.40**
b)	500	$7\frac{1}{2}$ % per year	quarterly	8.5 years	**23 482.98**
c)	200	3.25% per year	monthly	5 years	**13 010.95**

5. $348.92

6. $51 960.58; $1960.58

7. **a)** $451 222.88   **b)** $416 222.88

8. No; entertainment costs $2848.86, investment is worth $2743.74.

9. $7599.64

10. No. At the end of 8 months, he will have only $1601.57.

11. $3788.00

12. **a)**

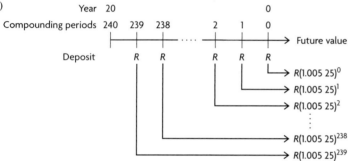

b) $167.08   **c)** $268.12

13. Answers will vary; may include 8% compounded annually, annual payments of $400; 5% compounded monthly, monthly payments of $40.

14. **a)** $5466.13

b) $5330.00; Annuity (a) will earn the greater amount.

15. In the formula $A = \dfrac{R[(1 + i)^n - 1]}{i}$, n represents the number of compounding periods. If there were more payments than compounding periods, the formula would assume n payments and therefore miss out the extra payments. Similarly, if there were fewer payments, the formula would include payments that were not made.

16. $433.28

17. $556.05

Lesson 8.6, pp. 506–508

1.

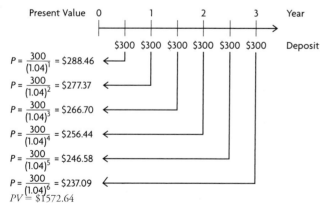

2.

	Withdrawal ($)	Annual Interest Rate (%)	Compounding Period	Term of Annuity	R ($)	i	n	PV ($)
a)	750	8	annual	3 years	750	0.08	3	1 932.82
b)	450	7.5	quarterly	8.5 years	450	0.018 75	34	11 238.20
c)	225	3.25	monthly	5 years	225	$\frac{3.25}{1200}$	60	12 444.69

3. a) $2850.55   **b)** $3464.76
4. $721.37
5. $2081.32
6. $92.48
7. a) $2537.48   **b)** $128 747.84
8. $706.82
9. $783.49
10. $7770.40
11. a) $389.47
    b) $23 368.36
    c) Cash offer; payments to bank are $386.66, which is less than dealer's finance payments.
12. a) $961.63   **b)** $108.37
13. Answers may vary. E.g., If an annual rate of 5% is used, R = $1060.66. For an annual rate of 7.5%, R = $1187.02. For an annual rate of 10%, R = $1321.51.
14. a) Answers may vary. E.g., Quarterly payments of $2500 for 9 years at 20%/a; annual payments of $2500 for 36 years at 5%/a
    b) $41 367.13 for all annuities matching the given information
15. Answers may vary. E.g.,
    a) Find quarterly payments at 8% for a 3-year annuity worth $2800 at maturity.
    b) Find amount of a 5-year loan annuity at 6% interest with monthly payments of $350.
16. a) Answers may vary. E.g., suppose that there is an annuity that has an annual interest rate of 12% compounded quarterly with quarterly payments of $400 paid out over 6 years.

$$A = \frac{400\left[\left(1 + \frac{0.12}{4}\right)^{24} - 1\right]}{\frac{0.12}{4}}$$

$$= \$13\ 770.59$$

Now suppose the duration of the annuity were doubled from 6 years to 12 years. The amount of the annuity would be

$$A = \frac{400\left[\left(1 + \frac{0.12}{4}\right)^{48} - 1\right]}{\frac{0.12}{4}}$$

$$= \$41\ 763.36$$

Therefore, doubling the duration of an annuity does not double the amount of the annuity at maturity. Rather, it more than doubles the amount of the annuity.

    b) Answers may vary. E.g., suppose that there is an annuity that has an annual interest rate of 12% compounded quarterly with quarterly payments of $400 paid out over 6 years.

$$A = \frac{400\left[\left(1 + \frac{0.12}{4}\right)^{24} - 1\right]}{\frac{0.12}{4}}$$

$$= \$13\ 770.59$$

Now suppose the payment amount were doubled from $400 to $800. The amount of the annuity would be

$$A = \frac{800\left[\left(1 + \frac{0.12}{4}\right)^{48} - 1\right]}{\frac{0.12}{4}}$$

$$= \$27\ 541.18$$

Therefore, doubling the payment amount does double the amount of the annuity at maturity.
17. a) $21 278.35   **b)** 8.17%
18. 5 years 3 months

Lesson 8.7, pp. 517–519

1. a) Interest = previous period's balance × interest rate per compounding period
    b) Annual contribution + interest
2. a) $997.48
    b) 19.41 periods or approximately 4 years 10 months; minimal impact on the duration of the loan
    c) a): $19 949.60; b): $19 361.09
3. Variables to be determined are **in bold**.

	N	I%	PV	PMT	FV	P/Y	C/Y
a)	**156**	0	10 000	−350	0	12	12
b)	0	**7.5**	15 000	−500	0	12	12

4. a) $216.25
    b) $653.23
    c) The payment in part (b) is about 3 times the payment in part (a) because payments are made $\frac{1}{3}$ as often. It is slightly more because payments earn interest on the interest, and with monthly payments there are more compounding periods for this to happen.
5. 4.96%
6. a) $597.58
    b) $24 834.63; $19 282.07; $13 313.30; $6897.12; $0
    c) $5854.85; $30 000
7. 10.33 years or 10 years 4 months
8. $18 649.01
9. $29 127.96
10. a) $307.75   **b)** $3464.88   **c)** $2286.34
11. Get low interest rate—reduces the amount of interest paid; Borrow as little as possible—just what you need—reduces the present value of the loan; Pay as much as possible each period—principal is paid down faster, interest amounts are reduced at a faster rate.
12. a) Answers will vary; for example, a car, a computer, a vacation
    b) Answers will depend on information available.
    c) Schedule should be calculated using similar formulas to those in Reflecting part F.
13. a) $459.35
    b) about 31 more payments
    c) $1603.20

Chapter Review, pp. 522–523

1.

	Principal ($)	Annual Interest Rate (%)	Years	Compounding Frequency	Amount ($)	Interest Earned ($)
a)	400.00	5	15	semi-annually	**839.03**	**439.03**
b)	450.00	4.5	10	monthly	**705.15**	**255.15**
c)	**622.87**	3.4	10	weekly	875.00	**252.13**
d)	508.75	3.7	3	semi-annually	568.24	**59.49**
e)	10 000.00	2.34	**4**	quarterly	**11 000.00**	1000.00

2. Answers may vary; for example:

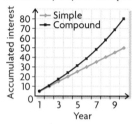

The relationship between accumulated interest and year is linear for simple interest and exponential for compound interest, because simple interest is earned on the principal only, while compound interest earns interest on the interest as well.

3. a) $6929.29 **b)** $9603.02

4. $7559.90

5. 8%/a

6. 17 years 6 months

7. a)

Months ago $\;0\qquad\qquad\qquad\qquad\qquad\qquad 120$

$750(1 + 0.02)^0$ $750(1 + 0.02)^{40}$

$= \$750$ $= \$1656.03$

b) $45 301.49, $15 301.49

8. $11 000.73

9. a) 16 years old **b)** A little over 1 year less

10. a) 5-year term: $622.75; 8-year term: $439.51 **b)** $4827.96

11. $60 744.31

12. $9324.48

13.

	a)	b)	c)	d)
N	5	180	36	32
I	4.5	3.75	0	9
PV	0	0	10 000	0
PMT	−1500	0	−334.54	−2000
FV	0	200 000	0	0
P/Y	1	12	12	4
C/Y	1	12	12	4

14. $8530.20

15. $358.22

16.

Loan	$10 000		Start Year	0
Annual Rate	0.08		End Year	5
Rate per Period	0.0067		Number of Payments	60
Compounding Periods per Year	12		Contribution	$202.76

Payment Number	Payment ($)	Interest	New Balance $
0			10 000.00
1	202.76	66.67	9 863.91
2	202.76	66.09	9 727.24
3	202.76	65.17	9 589.65
4	202.76	64.25	9 451.14
5	202.76	63.32	9 311.70
6	202.76	62.39	9 171.33
7	202.76	61.45	9 030.02
8	202.76	60.50	8 887.76
9	202.76	59.55	8 744.55
10	202.76	58.59	8 600.38
11	202.76	57.62	8 455.24
12	202.76	56.65	8 309.13
13	202.76	55.67	8 162.13
14	202.76	54.69	8 014.06
15	202.76	53.69	7 864.99

17. $1548.95

Chapter Self-Test, p. 524

1. In simple interest, only the principal amount earns interest, whereas in compound interest, the interest earns interest as well.

2. $12 717.67, $2217.67

3. $13 367.01, $8132.99

4. $16 783.48

5. $15 169.64

6. 1.96%

7. 390 weeks or 7 years 6 months

8.

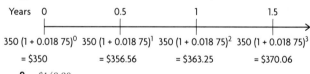

Years   0   0.5   1   1.5

$350(1 + 0.018\ 75)^0$ $350(1 + 0.018\ 75)^1$ $350(1 + 0.018\ 75)^2$ $350(1 + 0.018\ 75)^3$

$= \$350$ $= \$356.56$ $= \$363.25$ $= \$370.06$

9. $148.90

10. $60 880.81

11. An amortization table is used to show the amount of a regular payment, how much of the payment is interest, how much is used to reduce the principal, and the outstanding balance after each payment. A spreadsheet is useful in creating amortization tables and for analyzing the effects of changing the parameters of a loan problem.

Cumulative Review Chapters 7–8, pp. 526–527

1. (a) **3.** (b) **5.** (c) **7.** (d) **9.** (a) **11.** (c)
2. (c) **4.** (c) **6.** (a) **8.** (c) **10.** (c) **12.** (b)

13. **a)**

Bounce	Height (m)
0	6.00
1	3.60
2	2.16
3	1.30
4	0.78
5	0.47

b)

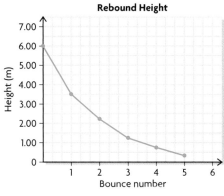

Rebound Height

c) $f(n) = 0.6 \times f(n-1)$ or $f(n) = 6(0.60)^n$
d) $f(12) \doteq 0.013$
e) 6th bounce
14. $1028.24
15. Monthly payments: $348.39; total interest: $2722.72; value of car after four years: $7912.13

Appendix A

A-1 Operations with Integers, p. 530

1. **a)** 3   **c)** -24 **e)** -6
   b) 25   **d)** -10 **f)** 6
2. **a)** $<$ **c)** $>$
   b) $>$ **d)** $=$
3. **a)** 55   **c)** -7 **e)** $\frac{15}{7}$
   b) 60   **d)** 8   **f)** $\frac{1}{49}$
4. **a)** 5   **c)** -9 **e)** -12
   b) 3   **d)** 76   **f)** -1
5. **a)** 3   **c)** -2
   b) -1 **d)** 1

A-2 Operations with Rational Numbers, pp. 531–532

1. **a)** $-\frac{1}{2}$ **c)** $-\frac{19}{12}$ **e)** $-\frac{41}{20}$
   b) $\frac{7}{6}$ **d)** $-8\frac{7}{12}$ **f)** 1

2. **a)** $-\frac{16}{25}$ **c)** $\frac{2}{15}$ **e)** $-3\frac{2}{5}$
   b) $-\frac{9}{5}$ **d)** $\frac{3}{2}$ **f)** $32\frac{7}{24}$

3. **a)** 2   **c)** $\frac{16}{9}$ **e)** $\frac{15}{2}$
   b) $-4\frac{3}{4}$ **d)** $-\frac{9}{2}$ **f)** $\frac{2}{3}$

4. **a)** $\frac{1}{5}$ **c)** $\frac{1}{15}$ **e)** $\frac{36}{5}$
   b) $\frac{3}{10}$ **d)** $-\frac{1}{18}$ **f)** $-\frac{3}{8}$

A-3 Exponent Laws, p. 533

1. **a)** 16   **b)** 1   **c)** 9   **d)** -9 **e)** -125 **f)** 0.125
2. **a)** 2   **b)** 31   **c)** 9   **d)** $\frac{1}{18}$ **e)** -16 **f)** $\frac{13}{36}$
3. **a)** 9   **b)** 50   **c)** 4 194 304   **d)** $\frac{1}{27}$
4. **a)** x^8 **b)** m^9 **c)** y^7 **d)** a^{bc} **e)** x^6 **f)** $\frac{x^{12}}{y^9}$
5. **a)** x^5y^6 **b)** $108m^{12}$ **c)** $25x^4$ **d)** $\frac{4u^2}{v^2}$

A-4 The Pythagorean Theorem, pp. 534–535

1. **a)** $x^2 = 6^2 + 8^2$ **c)** $9^2 = y^2 + 5^2$
   b) $c^2 = 13^2 + 6^2$ **d)** $8.5^2 = a^2 + 3.2^2$
2. **a)** 10 cm   **b)** 11.5 cm   **c)** 7.5 cm   **d)** 7.9 cm
3. **a)** 13.93   **b)** 6   **c)** 23.07   **d)** 5.23
4. **a)** 11.2 m   **b)** 6.7 cm   **c)** 7.4 cm   **d)** 4.9 m
5. 10.6 cm
6. 69.4 m

A-5 Evaluating Algebraic Expressions and Formulas, p. 536

1. **a)** 28   **b)** -17 **c)** 1   **d)** $\frac{9}{20}$
2. **a)** $\frac{1}{6}$ **b)** $\frac{5}{6}$ **c)** $\frac{-17}{6}$ **d)** $\frac{-7}{12}$
3. **a)** 82.35 cm^2 **b)** 58.09 m^2 **c)** 10 m   **d)** 4849.05 cm^3

A-6 Finding Intercepts of Linear Relations, p. 538

1. **a)** x-int: 3, y-int: 1   **d)** x-int: 3, y-int: 5
   b) x-int: -7, y-int: 14   **e)** x-int: -10, y-int: 10
   c) x-int: 6, y-int: -3 **f)** x-int: $-\frac{15}{2}$, y-int: 3
2. **a)** x-int: 7, y-int: -7 **d)** x-int: -10, y-int: 6
   b) x-int: -3, y-int: 2   **e)** x-int: -7, y-int: $\frac{7}{2}$
   c) x-int: -3, y-int: 12   **f)** x-int: 2, y-int: $-\frac{12}{5}$

3. a) x-int: 13

b) y-int: -6

c) x-int: 7

d) x-int: -10

e) y-int: $-\dfrac{3}{2}$

f) y-int: 8

4. a) x-int: 1.5, y-int: 6.75

b)

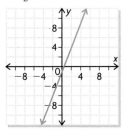

c) The foot of the ladder is on the ground 1.5 from the wall and the top of the ladder is 6.75 up the wall.

A-7 Graphing Linear Relationships, p. 540

1. a) $y = 2x + 3$

c) $y = -\dfrac{1}{2}x + 2$

b) $y = \dfrac{1}{2}x - 2$

d) $y = 5x + 9$

2. a)

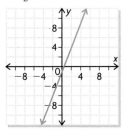

c)

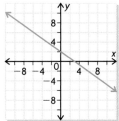

b)

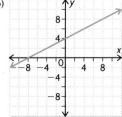

d)

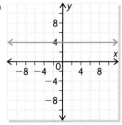

3. a) x-int: 10, y-int: 10

b) x-int: 8, y-int: 4

c) x-int: 5, y-int: 50

d) x-int: 2, y-int: 4

4. a) x-int: 4, y-int: 4

b) x-int: 3, y-int: -3

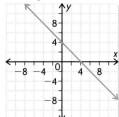

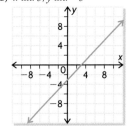

c) x-int: 5, y-int: 2

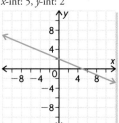

d) x-int: 4, y-int: -3

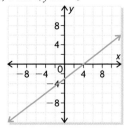

5. a)

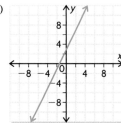

c)

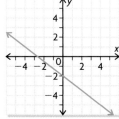

b)

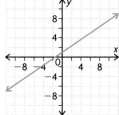

d)

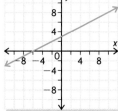

6. a)

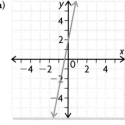

c)

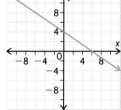

b)

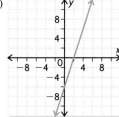

d)

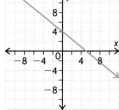

A-8 Graphing Quadratic Relations, p. 542

1. **a)**

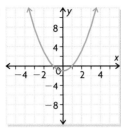

i) $(0, -1)$ **ii)** $x = 0$ **iii)** -1 **iv)** $-1, 1$

b)

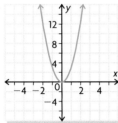

i) $(0, 0)$ **ii)** $x = 0$ **iii)** 0 **iv)** 0

c)

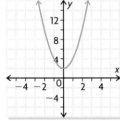

i) $(0, 2)$ **ii)** $x = 0$ **iii)** 2 **iv)** none

d)

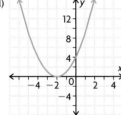

i) $(-2, 0)$ **ii)** $x = -2$ **iii)** 4 **iv)** -2

e)

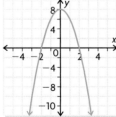

i) $(0, 8)$ **ii)** $x = 0$ **iii)** 8 **iv)** $-2, 2$

f)

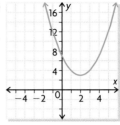

i) $(2, 3)$ **ii)** $x = 2$ **iii)** 7 **iv)** none

g)

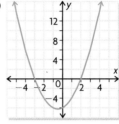

i) $(-0.5, -6.25)$ **ii)** $x = -0.5$ **iii)** -6 **iv)** $-3, 2$

h)

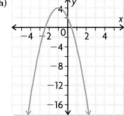

i) $(-1, 4)$ **iii)** 2
ii) $x = -1$ **iv)** $-1 + \sqrt{2}, -1 - \sqrt{2}$

2. **a)**

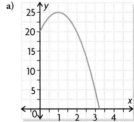

b) 25 m; 1 m
c) the height of the cliff
d) $1 + \sqrt{5}, 1 - \sqrt{5}$; The negative x-intercept represents a negative time, which is not possible. The positive x-intercept represents the time when then stone hits the ground.

A-9 Expanding and Simplifying Algebraic Expressions, p. 543

1. **a)** $-2x - 5y$ **c)** $-9x - 10y$
   b) $11x^2 - 4x^3$ **d)** $4m^2n - p$
2. **a)** $6x + 15y - 6$ **c)** $3m^4 - 2m^2n$
   b) $5x^3 - 5x^2 + 5xy$ **d)** $4x^7y^7 - 2x^6y^8$
3. **a)** $8x^2 - 4x$ **c)** $-13m^5n - 22m^2n^2$
   b) $-34h^2 - 23h$ **d)** $-x^2y^3 - 12xy^4 - 7xy^3$
4. **a)** $12x^2 + 7x - 10$ **c)** $20x^2 - 23xy - 7y^2$
   b) $14 + 22y - 12y^2$ **d)** $15x^6 - 14x^3y^2 - 8y^4$

A-10 Factoring Algebraic Expressions, p. 545

1. a) $a(b + 2)$ **c)** $3y(1 - 3x)$ **e)** $11b^2(7b + 5)$
   b) $2(2x + 3)$ **d)** $a(7 + a)$ **f)** $3a(7 + 2b - 5a)$
2. a) $(x - 1)^2$ **c)** $(x - 7)(x + 4)$ **e)** $3(a + 1)^2$
   b) $(a + 1)(a + 2)$ **d)** $(z - 4)(z + 2)$ **f)** $5(x - 3)(x + 1)$
3. (a), (c), (e)
4. a) $(x - 9)(x + 9)$ **c)** $(2a - 1)(2a + 1)$ **e)** $369 - 4x^2$
   b) $2(2 - 9z^2)$ **d)** $16(x - 1)(x + 1)$ **f)** $16(25 - xy)$

A-11 Solving Linear Equations Algebraically, p. 546

1. a) $x = 9$ **c)** $m = 3$ **e)** $y = 6$
   b) $x = 0.8$ **d)** $m = -4$ **f)** $r = \frac{23}{10}$
2. a) $x = 100$ **c)** $m = \frac{2}{3}$ **e)** $y = \frac{7}{18}$
   b) $x = 20$ **d)** $y = 21$ **f)** $m = -\frac{6}{5}$
3. a) 6 cm **b)** 16 m
4. 147 student and 62 adult

A-12 Pattern Recognition and Difference Tables, pp. 547–548

1. a) i)   **ii)** 1, 3, 5, 7 **iii)** 9
   b) i)

   ii) 1, 4, 9, 16 **iii)** 25

   c) i)   **ii)** 2, 4, 6, 8 **iii)** 10
   d) i)

   ii) 1, 4, 9, 16 **iii)** 25

   e) i)

   ii) 2, 8, 18, 32 **iii)** 50

   f) i)   **ii)** 8, 10, 12, 14 **iii)** 16

2. a) i) 4, 10, 16, 22 **ii)** 28
   b) i) 4, 12, 24, 40 **ii)** 60
   c) i) 7, 13, 19, 25 **ii)** 31
   d) i) 3, 9, 18, 30 **ii)** 45
   e) i) 7, 22, 45, 72 **ii)** 115
   f) i) 24, 30, 36, 42 **ii)** 48
3. a) i) linear **ii)** 17
   b) i) quadratic **ii)** -301
   c) i) quadratic **ii)** -0.4
   d) i) linear **ii)** 565
4. a) Constant difference is -1; linear
   b) Second constant difference of 2; quadratic
5. Constant difference of $6; linear
6. a) 13 **b)** -100 **c)** $\frac{43}{24}$ **d)** 4.8

A-13 Creating Scatter Plots and Lines or Curves of Good Fit, p. 550

1. a) i)

ii) The data display a strong positive correlation.

b) i)

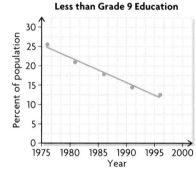

ii) The data display a strong negative correlation.

2. a)

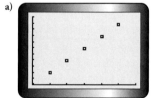

b) The motion sensor's measurements are consistent since the curve goes through several of the points.

A-14 Interpolating and Extrapolating, p. 553

1. a) 53 m, 54 m **b)** 73 m, 77 m
2. a)

b) 24.5 m/s ; 34.3 m/s; 46.55 m/s
c) 58.8 m/s; 88.2 m/s; 98 m/s
3. Extrapolation is a prediction based on the data, while interpolation is taken directly from the data.

4. a)

b) The average for 0 absences is 90%. For every time absent, the average mark drops by about 3%.

c) 72%

d) 13 days absent

5. 82 m; 24 m/s

A-15 Trigonometry of Right Triangles, p. 557

1. 27 m

2. a) $\sin A = \dfrac{12}{13}$, $\cos A = \dfrac{5}{13}$, $\tan A = \dfrac{12}{5}$

   b) $\sin A = \dfrac{4}{5}$, $\cos A = \dfrac{3}{5}$, $\tan A = \dfrac{4}{3}$

   c) $\sin A = \dfrac{8}{17}$, $\cos A = \dfrac{15}{17}$, $\tan A = \dfrac{8}{15}$

   d) $\sin A = \dfrac{3}{5}$, $\cos A = \dfrac{4}{5}$, $\tan A = \dfrac{3}{4}$

3. a) 4.4   **b)** 6.8   **c)** 5.9   **d)** 26.9

4. a) 39°   **b)** 54°   **c)** 41°   **d)** 65°

5. a) 12.5 cm   **b)** 20.3 cm   **c)** 19.7 cm   **d)** 24°

6. a) 12.4   **b)** 5.7   **c)** 27°   **d)** 46°

7. 8.7 m

8. 84.2 m

9. 195 m

Index

A

Acute angle, 260
Acute triangles. *See also* Trigonometry
    cosine law application in, 294–298
    problem-solving using, 302–308
    sine law application in, 283–287
Adjacent side, 260, 263, 270, 291
Algebra of quadratic expressions
    common factoring of polynomials, 88–92
    factoring of (special cases), 111–114
    factoring of $ax^2 + bx + c$, 104–108, 118–119
    and factoring quadratic expressions, 95–99
    quadratic expressions and, 78–84
Amortization table creation, 509–517, 521
Amount, 456
Amplitude, 336, 345, 355
Angle of depression, 267–269
Angle of elevation, 264–267
Annuities
    determination of future value of, 493–498, 521
    determination of present value of, 501–506, 521
Area
    parallelogram/trapezoid, 86
    triangle, 277–278, 305–306
Asymptote, 421
$ax^2 + bx + c$, factoring of, 104–108, 118–119

B

Base, 388
Bearing, 302
Binomials
    algebraic expressions and, 79–83
    common factoring of, 91
Braking distance, 17–20

C

Cash outflow/inflow, 485
Circumference, 388
Coefficient, 90, 102
Common factors, algebraic expressions and, 75, 88–92
Completing the square, 208–212
Compound interest, 454, 458
    determination of future value, 462–467
    determination of future value of compound interest, 489–490

determination of present value, 471–475
determination of present value of compound interest, 489–490
difference from simple, 489
problem-solving using, 479–485
Contained angle, 296
Cosine law, 294–298, 313
    and problem-solving using acute-triangle models, 302–303, 308
Cosine ratio, 260, 270
    and right-triangle models for problem-solving, 275–276
Cube root estimates, 388–389
Curve of good fit, 170, 243–244, 247–249, 253
Cycle on graphs, 324

D

Decay. *See* Exponential decay
Decimal rational exponent, 412–413
Decomposing a number, 106–107
Decreasing function, 388, 433–437
Degree of a polynomial, 18, 23
Dependent variables, 6, 12
Difference of squares, 113, 119
Discriminant of quadratic formula, 228–232
Distance, as function of time, 22
Distributive property, expansion of algebraic expressions and, 75, 83–84, 101
Domain
    exponential model and, 393
    quadratic functions and, 7, 12, 59–63, 67
    sinusoidal functions and, 377

E

Equation of the axis, 336, 345, 355
Exponent, 388, 403–405
Exponential decay, 393, 433–437, 443
Exponential functions
    and creation of amortization tables, 509–517
    data collection for, 392–393
    decay, 433–437, 443
    determination of future value of compound interest, 462–467
    determination of future value of regular annuities, 493–498, 521
    determination of present value of compound interest, 471–475
    determination of present value of regular annuities, 501–506, 521

integer exponents and, 402–407
interest/rates of change and, 454–459
and laws of exponents, 395–399
problem-solving using compound interest, 479–486
problem-solving using exponential growth, 425–429, 443
properties of, 420–422, 442
rational exponents and, 410–415
Exponential growth, 393
    problem-solving using, 393, 425–429, 443
Exponent laws, algebraic expressions for multiplication/division and, 74
Extrapolate, 320, 551

F

Factored form of quadratic functions
    and creation of quadratic model from data, 170–176, 181, 185
    definition, 134
    exploring situations using, 130–131
    of quadratic functions, 181
    related to standard forms of quadratic functions, 132–139, 181
    relating factored/standard forms of, 181
    solving by factoring, 156–161
    solving problems involving, 164–167, 181
    solving using graphs, 143–149
Factoring, 88–92
    to determine zeros from quadratic function in vertex form, 200
    quadratic expressions, 95–99
    quadratic expressions (special cases), 111–114
    quadratic functions and, 156–161
    trinomials, 104–108
Function. *See* Quadratic functions
Function notation, 27–32
Future value of compound interest, 462–467, 489–490
Future value of regular annuities, 493–498, 521
$f(x)$. *See* Function notation

G

Guaranteed Investment Certificate (GIC), 458

Credits

This page constitutes an extension of the copyright page. We have made every effort to trace the ownership of all copyrighted material and to secure permission from copyright holders. In the event of any question arising as to the use of any material, we will be pleased to make the necessary corrections in future printings. Thanks are due to the following authors, publishers, and agents for permission to use the material indicated.

Chapter 1 Opener, pages viii–1: clockwise from top, Stuart & Michele Westmorland/Getty Images, Adam Hart-Davis/Photo Researchers, Inc., Rob Casey/Getty Images, Waterlily Pond: Pink Harmony, 1900 (oil on canvas), Monet, Claude (1840-1926) / Musee d'Orsay, Paris, France, Lauros / Giraudon / The Bridgeman Art Library, David wall/Alamy, Andre Jenny/Alamy, MalibuBooks/Shutterstock, Jarno Gonzalez Zarraonandia/Shutterstock, Ambient Images Inc./Alamy; page 5: ML Harris/Getty Images; page 15: Chris Fourie/Shutterstock; page 16: Sudheer Sakthan/Shutterstock; page 17: Scott Pehrson/Shutterstock; page 22: Photos.com; page 25: Comstock Premium/Alamy; page 33: Michelle Milano/Shutterstock; page 35 (top) Graham Prentice/Shutterstock, (bottom) Bettman/CORBIS; page 38: Art Resources, NY; page 41: Steve P./Alamy; page 50: Andrea Danti/Shutterstock; page 51: Michael Newman/Photo Edit, Inc.; page 64: Danijel Micka/Shutterstock; page 65: James Steidl/Shutterstock; page 71: from left, salamanderman/Shutterstock, Sue Cunningham Photographic/Alamy, Manor Photography/Alamy, Doug Wilson/CORBIS

Chapter 2 Opener, pages 72–73: Laurence Gough/Shutterstock; page87: Brand X Pictures/Alamy; page 88: Westend61/Alamy; page 95: Ian Shaw/Alamy; page 123: Dana Heinemann/Shutterstock

Chapter 3 Opener, pages 124–125: GJON MILI/Getty Images; page 130: Talia Arkut Berkok; page 136: Duomo/CORBIS; page 142: (top) Franz Aberham/Getty Images, (bottom) Gilbert Lundt;Temp Sport/CORBIS; page 143: Alan Freed/Shutterstock; page 150: Joe Gough/Shutterstock; page 151: (top) Chris Laurens/Alamy, (bottom) max blain/Shutterstock; page 156: Andreas Pollock/Getty Images; page 162: Clive Rivers/Alamy; page 163: HO/Reuters/CORBIS; page 164: Brian Summers/First Light; page 169: Nigel Shuttleworth/Life File/Getty Images; page 178: World Pictures/Alamy; page 179: Alison Wright/CORBIS; page 185: (left) Photos.com, (right) Bill Brooks/Alamy; page 188: Photo Edit; page 189: Jack Sullivan/Alamy

Chapter 4 Opener, pages 190–191: © Masterfile Royalty Free; page 194: Joe McBride/CORBIS; page 204: Louis Gubb/CORBIS SABA; page 215: Ola Moen/Alamy; page 222: NASA,ESA,The Hubble Heritage Team(STScI/AURA), J.Bell(Cornell University) and M.Wolff(Space Science Institute); page 233: Ryan McVay/Getty Images; page 236: Ros Drinkwater/Alamy; page 237: Lester Lefkowitz/CORBIS; page 240: ©Randy Faris/Corbis; page 241: (top) Fukuoka Irina/Shutterstock, (bottom) Sergei A. Tkachenko; page 242: Licensed under Health Canada copyright; page 243 & 244: Image Courtesy of Colin Palmer- www.buyimage.co.uk; page 247: Duncan Soar/Alamy; page 251: David R. Frazier Photolibrary, Inc./Alamy; page 255: Kit Cooper-Smith/Alamy; page 256: Brand X Pictures/Alamy; page 257: Susan Quindland-Stringer/Shutterstock

Chapter 5 Opener, pages 258–259: top left clockwise, Bettman/CORBIS, Jose Luis Pelaez inc/Getty Images, Superstock/Alamy, Karen Moskowitz/Getty Images, Image Source Pink/Getty Images, paul ridsdale/Alamy, Nick Dolding/Getty Images; page 272: (top left) Dvoretskiy Igor Vladimirovich/Shutterstock, (top right) Christopher Konfortion/Shutterstock, (middle left) Gregory James Van Raalte/Shutterstock, (middle right) Kameel4u/Shutterstock, (bottom) Mark Richards/Photo Edit; page 273: Chris Thomaidis/Getty Images; page 274: Superstock/Alamy; page 276: Radius Images/Alamy; page 281: (top) jaxpix/Alamy, (bottom) Photos.com; page 292: john t. fowler/Alamy; page 293: Dennis Hallinan/Alamy; page 301: Niall Benvie/Alamy

Chapter 6 Opener, pages 318–319: photolibrary.com/First Light; page 323: Puzant Apkarian/First Light; page 326: Roger Ressmeyer/CORBIS; page 327: Photodisc/Getty Images; page 328: untitled/Alamy; page 332: (top) Lester Lefkowitz/CORBIS, (bottom) Bill Freman/Photo Edit; page 333: (left & right) Barrett & MacKay Photography; page 340: Kevin Fleming/CORBIS; page 342: Richard Hartmier/First Light; page 348: Carl Stone/Shutterstock; page 349: PhotoCreate/Shutterstock; page 376: Jarno Gonzalez Zarraonandia/Shutterstock; page: 381: Avind Balaraman/Shutterstock; page 385: Gary & Sandy Wales/Shutterstock

Chapter 7 Opener, pages 386–387: image100/First Light; page 391: From the archives of the National Soccer Hall of Fame, Oneonta, NY; page 392: Elemental Imaging/Shutterstock; page 394: Mike Danton/Alamy; page 395: Toronto Star/First Light; page 402: Stockbyte/Getty Images; page 408: NASA/CORBIS; page 410: Museum of History & Industry, Seattle; page 416: bobo/Alamy; page 420: Darren Matthews/Alamy; page 425: John Czenke/Shutterstock; page 430: Hisham Ibrahim/Getty Images; page 431: Nature's Images/Photo Researchers, Inc; page 433: Chris A Crumley/Alamy; page 436: Luca DiCecco/Alamy; page 438: Samuel Acosta/Shutterstock; page 439: Kmitu/Shutterstock; page 440: PHOTOTAKE Inc./Alamy; page 441: Johanna Goodyear/Shutterstock; page 447: Alex Segre/Alamy

Chapter 8 Opener, pages 448–449: Audrius Tomonis/banknotes.com; page 453: Blend Images.com/Alamy; page 454: (top) Clive Watkins/Shutterstock, (bottom) Courtesy of Bank of Canada; page 461: Wesley Hitt/Alamy; page 462: Gehry Partners, LLP/Art Gallery of Ontario/Sean Weaver; page 469: J A Giordano/CORBIS SABA; page 470: max blain/Shutterstock; page 471: Wendy Shiao/Shutterstock; page 474: (left) Ternovoy Dmitry/Shutterstock, (right) Wojcik Jaroslaw/Shutterstock; page 477: Andres Rodriguez/Shutterstock; page 478: Zastol'skiy Victor Leonidovich/Shutterstock; page 479: Dave Starrett; page 487: dean sanderson/Shutterstock; page 488: Silvia Bukovac/Shutterstock; page 493: Lisa F. Young/Shutterstock; page 499: Robert Manley/Shutterstock; page 500: Mark Lewis/Alamy; page 501: Comtock/Alamy; page 504: PhotoStockFile/Alamy; page 507: (top) Edyta Pawlowska/Shutterstock, (bottom) Aflo Foto Agency/Alamy; page 508: Semenov Gleb/Shutterstock; page 509: Tomasz Pietryszek/Shutterstock; page 518: Rudy Sulgan/CORBIS; page 519: Suprijono Suharjoto/Shutterstock; page 520: Chris Howes/Wild Places Photography/Alamy; page 524: blue jeans images/Getty Images; page 525: (top) Photos.com, (bottom) Stock Connection Distribution/Alamy; page 527: Carlos Caetano/Shutterstock

Appendix B pages 558–575: Reproduced by permission of Texas Instruments.